FUNDAMENTALS OF SURVEYING

William Horace Rayner, C.E., M.S.
Professor of Civil Engineering, Emeritus
University of Illinois

Milton O. Schmidt, Ph.D.
Professor of Civil Engineering
University of Illinois

VAN NOSTRAND · REINHOLD COMPANY
NEW YORK   TORONTO   LONDON   MELBOURNE

VAN NOSTRAND REGIONAL OFFICES: *New York, Chicago, San Francisco*

D. VAN NOSTRAND COMPANY, LTD., *London*

D. VAN NOSTRAND COMPANY (Canada), LTD., *Toronto*

D. VAN NOSTRAND AUSTRALIA PTY. LTD., *Melbourne*

Copyright © 1937, 1943, 1955, 1963, by
D. VAN NOSTRAND COMPANY, INC.
Copyright © 1969, by AMERICAN BOOK COMPANY

Published simultaneously in Canada by
D. VAN NOSTRAND COMPANY (Canada), LTD.

*No reproduction in any form of this book, in whole
or in part (except for brief quotation in critical articles or
reviews), may be made without written authorization
from the publishers.*

Library of Congress Catalog Card No. 69-11139

Fundamentals of Surveying is the expanded
fifth edition of the authors' *Elementary Surveying.*

Caption for photograph on title page:
TOPOGRAPHIC SURVEYING. Courtesy U.S. Geological Survey

PRINTED IN THE UNITED STATES OF AMERICA

Preface

General works on surveying can make little claim to originality except, perhaps, with respect to manner of treatment. Rather significant changes in sequence and emphasis have been made in *Fundamentals of Surveying* which merit fresh recognition. This book is by no means simply the fifth edition of the authors' *Elementary Surveying,* as was originally proposed. In the complete rewriting and reorganization to increase its instructional effectiveness, it has developed into a truly new book.

The first five chapters deal with the fundamental measurements of surveying and provide the logical prelude for advancing in subsequent chapters into the full range of operations categorized by the American Society of Civil Engineers as Land, Cartographic, and Engineering Surveying.

The treatment has been modernized, strengthened, and developed at a pace which leads the student through progressively more involved concepts and techniques.

All of the problems have been redesigned, and numerous sample calculations are presented to assist the student. An abundance of exciting illustrations dramatically portray the role of surveying in engineering technology. Thus, the student is constantly reminded of the practical applications of his newly acquired knowledge and motivated to study more diligently. Should he choose to expand his studies, a limited number of critically selected references are appended at the ends of most chapters.

The attention of the teacher is directed to the inclusion of various topics which are either new or are being given the special emphasis they deserve. They include (1) field safety, (2) field communications, (3) notekeeping, (4) monumentation and referencing, (5) procurement of horizontal and vertical control data, and other subjects. A brief but adequate introduction to electronic distance measurement is provided, an illuminating explanation is made of significant figures, and new equipment developments like the gyro-theodolite, the Transit-Lite (laser instrument), and the *electronic* desk calculator are capsuled. The use of the *electric* desk calculator is underscored.

This title, *Fundamentals of Surveying,* reflects the role of the new book in helping to prepare those who are planning to enter the broadening field of civil engineering technology. The work still maintains, however, its appropriateness for those pursuing pre-professional curricula in Forestry, Landscape Architecture, Agriculture, and related areas. Furthermore, the treatment is adequate in depth and more than ample in breadth to satisfy the needs of students enrolled in professional civil engineering schools where surveying course work has been generally abbreviated in recent years. The borrowing of some material from the authors' *Surveying: Elementary and Advanced* has broadened the scope and given the text qualified value as a reference work for the practitioner.

The authors extend their thanks for the use of numerous illustrations supplied by their friends in the agencies and organizations cited in the captions below the figures.

January 1969

W.H.R.
M.O.S.

Contents

	PAGE
Preface	v

1 Introduction   **1**
1-1. Surveying Defined 1-2. Classes of Surveys 1-3. Geodesy and the Earth's Shape 1-4. Geodetic Surveying 1-5. Plane Surveying 1-6. Kinds of Surveys 1-7. Historical Origins 1-8. Developing Competency 1-9. Surveying Measurements 1-10. Field Notes 1-11. Errors and Adjustments 1-12. Sources of Errors 1-13. Kinds of Errors 1-14. Checks 1-15. Precision and Accuracy 1-16. Law of Compensation 1-17. Professional Surveying Practice 1-18. Role of the Surveying Technician 1-19. Surveying Agencies 1-20. Surveying Societies

2 Direct Linear Measurements   **20**
2-1. Introduction 2-2. Units of Length 2-3. Standards of Length 2-4. Tapes 2-5. Taping Accessories 2-6. Care of Equipment 2-7. Standardization 2-8. Taping Procedures 2-9. Taping Over Level Ground 2-10. Taping Over Sloping or Uneven Ground 2-11. Taping Corrections 2-12. Mistakes 2-13. Checks 2-14. Accuracy 2-15. Specifications 2-16. Summary Problems

3 Angular Measurements   **46**
3-1. Introduction. GENERAL 3-2. Units of Angular Measurement 3-3. Horizontal Angles 3-4. Bearings and Azimuths 3-5. Vertical Angles 3-6. Zenith Distances 3-7. Tripods. ENGINEER'S TRANSIT 3-8. General Characteristics 3-9. Definition of Terms 3-10. Surveying Telescope 3-11. Bubble Tube 3-12. Verniers 3-13. Measurement of Horizontal Angles 3-14. Repeating an Angle 3-15. Measurement of Vertical Angles 3-16. Adjustments of Transit 3-17. Uses of Transit 3-18. Care of Transit 3-19. Errors in Angles 3-20. Mistakes. OPTICAL THEODOLITES 3-21. Definitions 3-22. General Characteristics 3-23. Types and Operation 3-24. Care and Adjustment Problems

v

CONTENTS

PAGE

4 Indirect Linear Measurements   **88**

4-1. Introduction. TACHEOMETRIC MEASUREMENTS 4-2. Tacheometry 4-3. Stadia Method 4-4. Stadia Theory 4-5. Reduction of Stadia Notes 4-6. Errors in Stadia 4-7. Subtense Method. ELECTROMAGNETIC MEASUREMENTS 4-8. General 4-9. Principles of Electromagnetic Measurements 4-10. Index of Refraction 4-11. Microwave Instruments 4-12. Light-Wave Instrument 4-13. Length Calculations Problems

5 Elevation Measurements   **109**

5-1. Introduction 5-2. Basic Concepts 5-3. Curvature and Refraction 5-4. Methods of Leveling 5-5. Leveling Instruments 5-6. Leveling Accessories 5-7. Adjustments of Level 5-8. Care of Level and Rod 5-9. Differential Leveling 5-10. Precautions 5-11. Field Notes 5-12. Grades of Leveling 5-13. Errors in Differential Leveling 5-14. Mistakes 5-15. Reciprocal Leveling 5-16. Precise Levels 5-17. Vertical Control Data 5-18. Surveying Monumentation 5-19. Trigonometric Leveling 5-20. Barometric Leveling 5-21. Profile Leveling 5-22. Cross-Section Leveling Problems

6 Transit Traverse   **162**

6-1. Introduction. TRAVERSE 6-2. Traverse Configuration 6-3. Route Selection 6-4. Stations 6-5. Party Organization 6-6. Connections to Existing Control 6-7. Linear Measurements 6-8. Angular Measurements 6-9. Angular and Linear Accuracies 6-10. Checking Traverses 6-11. Grades of Traverse 6-12. Traverse Control Data 6-13. Other Transit Surveys. FIELD NOTEKEEPING 6-14. Importance 6-15. Field Notebooks 6-16. Fundamental Requirements 6-17. Memorandum of Survey. FIELD SAFETY 6-18. General Considerations 6-19. Promoting Safety. FIELD COMMUNICATIONS 6-20. Hand Signals 6-21. Radio Problems

7 Computations   **189**

7-1. Introduction. GENERAL 7-2. Basic Considerations 7-3. Computational Aids 7-4. Desk Calculators 7-5. Significant Figures 7-6. Rounding Off 7-7. Checks. TRAVERSE COMPUTATIONS 7-8 Rectangular Coordinates 7-9. Forward and Inverse Problems 7-10. Latitudes and Departures 7-11. Traverse Computation 7-12. Traverse Adjustment. AREA COMPUTATIONS 7-13. Area by Coordinates 7-14. Area of Irregular Tracts 7-15. Area by Polar Planimeter. VOLUME COMPUTATIONS 7-16. Average-End-Area Method 7-17. Prismoidal Formula 7-18. Borrow Pits. OTHER TOPICS 7-19. Cutoff Line 7-20. Auxiliary Traverse 7-21. Electronic Calculator Problems

CONTENTS vii

PAGE

8 Errors and Adjustments 218
8-1. Introduction 8-2. Errors 8-3. Precision and Accuracy 8-4. Distribution of Accidental Errors 8-5. Measures of Precision 8-6. Propagation of Accidental Errors 8-7. Weighted Measurements 8-8. Adjustments Problems

9 Triangulation 234
9-1. Introduction 9-2. Triangulation Systems 9-3. Classification of Triangulation 9-4. Reconnaissance 9-5. Signals and Towers 9-6. Station Markers 9-7. Angle Measurement 9-8. Base Measurement 9-9. Baseline Reduction Calculations 9-10. Angle Adjustment 9-11. Length and Position Calculations 9-12. Network of Triangulation in the United States 9-13. Uses of Triangulation Data 9-14. Trilateration 9-15. Three-Point Problem Problems

10 Observations for Meridian 262
10-1. Introduction. MAGNETIC OBSERVATIONS 10-2. Magnetic Compass 10-3. Declination Changes. ASTRONOMICAL OBSERVATIONS 10-4. Celestial Sphere 10-5. Apparent Motion of Celestial Sphere 10-6. Time 10-7. Time Signal Service 10-8. Latitude and Longitude 10-9. Position of a Celestial Body 10-10. Local Hour Angle 10-11. The Astronomical Triangle 10-12. Polaris 10-13. The Azimuth Problem 10-14. Field Procedure 10-15. Ephemerides 10-16. Reduction Procedures 10-17. Azimuth from Polaris at Any Time 10-18. Quality of Azimuth Determination 10-19. Latitude from Polaris at Any Time Problems

11 State Coordinate Systems 285
11-1. Introduction 11-2. National Network of Horizontal Control 11-3. Beginnings of State Coordinate Systems 11-4. Advantages of State Coordinate Systems 11-5. Quality of Surveys 11-6. Local Plane Coordinates 11-7. Map Projections 11-8. Tangent Plane Coordinates 11-9. Lambert Projection 11-10. Mercator Projection 11-11. Transverse Mercator Projection 11-12. State Coordinate Systems 11-13. Computation of Plane Coordinates on Lambert Grid 11-14. Computation of Plane Coordinates on Transverse Mercator Grid 11-15. Grid Azimuths 11-16. Computation of Grid Azimuths 11-17. Determination of Geodetic Distance from Ground Distance 11-18. Determination of Grid Distance from Geodetic Distance 11-19. Grid Factor. TRAVERSE COMPUTATIONS 11-20. Preliminary Remarks 11-21. Control Data 11-22. Computation of Grid Coordinates 11-23. Some Uses of State Plane Coordinate Systems Problems

CONTENTS

12 Photogrammetry — **PAGE 315**

12-1. Introduction   12-2. Definitions   12-3. Perspective Principles of Vertical Photographs   12-4. Scale of a Photograph   12-5. Number of Photographs Required   12-6. Image Displacement Caused by Ground Relief   12-7. Computed Length of Line Between Points of Different Elevations   12-8. Control for Photogrammetric Maps   12-9. Radial-Line Method   12-10. Ground and Map Control   12-11. Base Map   12-12. Plotting the Line of Flight   12-13. Marking the Photographs   12-14. Locating Map-Control Points   12-15. Plain Templets   12-16. Transferring Photographic Detail   12-17. Stereoscopic Vision   12-18. Stereoscopic Fusion   12-19. Parallax in Aerial Stereoscopic Views   12-20. Space-Coordinate Equations for Aerial Stereoscopic Views   12-21. Difference in Elevation by Stereoscopic Parallaxes   12-22. Parallax Computations   12-23. Effects of Changes in Elevation and Parallax   12-24. Mosaics   12-25. Photographic Mission   12-26. Parallax Bar   12-27. Stereocomparagraph   12-28. Kelsh Plotter and Wild A-7 Stereoplotter   Problems

13 Land Surveying — **351**

13-1. Introduction. GENERAL   13-2. Boundaries   13-3. Land Transfers   13-4. Kinds of Land Surveys   13-5. Rural and Urban Land Surveys. U.S. PUBLIC LANDS SURVEYS   13-6. Historical   13-7. U.S. Public Lands Survey System   13-8. Extent of the Rectangular System   13-9. Benefits of the Rectangular System   13-10. Instruments and Methods   13-11. Conditions Affecting Early Surveys   13-12. Field Notes and Plats   13-13. Meander Lines   13-14. Subdivision of Sections   13-15. Kinds of Corners   13-16. Corner Materials   13-17. Corner Accessories   13-18. Perpetuation of Land Corners   13-19. What Present Surveys Reveal   13-20. General Rules Governing Surveys of Public Lands. RURAL AND URBAN LAND SURVEYS   13-21. Land Descriptions   13-22. Requirements of a Valid Description   13-23. Interpretation of Deed Descriptions   13-24. General Scope of a Land Survey   13-25. Search for Corners   13-26. Rural Surveys   13-27. Urban Surveys   13-28. Survey for a Deed   13-29. Excess and Deficiency   13-30. Adverse Possession   13-31. Highways and Streets   13-32. Riparian Rights   13-33. Riparian Boundaries   13-34. Alluvium, Reliction, and Accretion   13-35. Avulsion   13-36. Extension of Riparian Boundary Lines   13-37. Legal Authority of the Surveyor   13-38. Liability   13-39. Registration of Land Surveyors

CONTENTS

		PAGE
14	**Cartographic Surveying**	**398**

14-1. Introduction 14-2. Uses of Topographic Maps. REPRESENTATION OF RELIEF 14-3. Contours 14-4. Characteristics of Contours 14-5. Other Relief Symbols 14-6. Systems of Contour Points 14-7. Contour Interpolation. PLANE TABLE 14-8. Plane Table 14-9. Setting Up and Orienting 14-10. Traversing 14-11. Intersection 14-12. Resection 14-13. Locating Details. TOPOGRAPHIC SURVEYING 14-14. Topographic Survey 14-15. Horizontal and Vertical Control 14-16. Scale and Contour Internal 14-17. Selecting Contour Points 14-18. System A—Coordinates 14-19. System B—Tracing Out Contours 14-20. System C—Controlling Points 14-21. System D—Hand Level Method 14-22. Aerial Mapping. MAP DRAFTING 14-23. Plats and Maps 14-24. Plotting 14-25. Titles 14-26. Symbols 14-27. Topographic Expression. MAP USE AND PROCUREMENT 14-28. Map Studies 14-29. National Topographic Mapping 14-30. Procurement. HYDROGRAPHIC SURVEYING 14-31. General 14-32. The Hydrographic Survey 14-33. Sounding Datums 14-34. Topographic and Shoreline Surveys 14-35. Equipment for Hydrography 14-36. Sounding Operations 14-37. Reduction of Soundings 14-38. Plotting Soundings Problem

15	**Engineering Surveying**	**452**

15-1. Introduction. ROUTE SURVEYING 15-2. General 15-3. Horizontal Curves 15-4. Tables of T and E 15-5. Principle of Deflection Angles 15-6. Calculation and Layout 15-7. Intermediate Setup on Curve 15-8. Parabola as a Vertical Curve 15-9. Calculation of Vertical Curve 15-10. Other Operations. CONSTRUCTION SURVEYING 15-11. General 15-12. Construction Grid System 15-13. Grade Stakes 15-14. Building Layout 15-15. Bridges and Tunnels 15-16. Laser Instrument. INDUSTRIAL SURVEYING 15-17. Optical Tooling 15-18. Electron Accelerator Surveys Problems

Appendix—Tables I–XXII	**481**
Index	**527**

Tables

		PAGE
I	Greenwich Hour Angle of Polaris	481
II	Increase in GHA for Elapsed Time Since 0^h GTC	482
III	Bearing of Polaris at All Local Hour Angles	483
IV	Polar Distance of Polaris	485
V	Corrections to Preliminary Bearings of Polaris as Obtained from Table III	485
VI	Corrections to be Applied to Altitude of Polaris to Obtain Latitude	486
VII	Stadia Table	487
VIII	Radii for Circular Curves	491
IX	Tangents and Externals to a 1° Curve	493
X	Inclination Correction for 100-Ft Tape	495
XI	Elevation Factors	496
XII	Elevation Corrections, in Feet, per 1000 Ft	496
XIII	Natural Sines and Cosines	497
XIV	Natural Tangents and Cotangents	506
XV	Greek Letters	518
XVI	Value of R, y', and Scale Factors—Minnesota North Zone	519
XVII	Values of θ—Minnesota North Zone	520
XVIII	Values of H and V—Illinois East Zone	521
XIX	Values of b and c—Illinois Zones	522
XX	Values of g—Illinois Zones	523
XXI	Values of Scale Factors—Illinois East Zone	524
XXII	Trigonometric Formulas	525

1 Introduction

1-1. Surveying Defined Surveying is the science and art of making such measurements as are necessary to determine the relative position of points above, on, or beneath the surface of the earth, or to establish such points in a specified position. Surveying operations are not confined to land areas. They are conducted over vast reaches of water as well as in planetary space. The measurements of surveying are essentially those of distance, both horizontal and vertical, and direction.

The data procurement phase of surveying is commonly known as the *field work*. Because virtually all such data must be analyzed, reduced to useful form by mathematical calculation, adjusted, and frequently converted to graphical modes of expression, such as maps and plans, it is customary to speak of this companion activity as the *office work of surveying*. Both phases of activity properly comprise surveying operations.

1-2. Classes of Surveys The term, survey, frequently connotes a critical inspection, such as of a traffic situation, in order to obtain adequate information needed for a solution. Here it essentially means any orderly process of acquiring data relating to the physical characteristics of the earth with particular reference to the determination of the relative positions of points and sizes of areas.

Various divisions or groupings of such surveys can be made based on a variety of distinguishing elements. Probably the most simple classification is that which recognizes three major classes of surveys described as follows:

1. *Land surveys* are those surveys which are associated with the fixing of property lines, the calculation of land areas, and the transfer of real property from one owner to another. The earliest known surveys were land surveys.

2. *Engineering surveys* comprise the operations of acquiring the

data needed to plan and design an engineering project and providing the necessary position and dimensional control at the site so that the bridge, building, or highway is built in the proper place and as designed.

3. *Informational surveys* are exemplified by those governmental surveys that obtain data concerning the topography, drainage, and man-made features of a large area. These data are portrayed through the medium of maps and charts.

In a subsequent section the distinctions between typical surveys in these three broad classes will be clarified by describing the nature of the surveying operations associated with them. For the moment it is emphasized that this classification has more pedagogic than engineering significance and that any grouping of survey types will be unsatisfactory because the differences in objectives and dissimilarities in procedure which are employed to distinguish between them are frequently so vague and negligible as to permit a survey to be classified in several ways.

1-3. Geodesy and the Earth's Shape Surveying operations are carried out on or near the surface of a spheroidal body, the earth, which has a mean radius of 3960 statute miles and within the lower levels of a gaseous region, termed the atmosphere, that envelops this planet. Some of the most significant factors that affect such surveying measurements and their subsequent computations are the size, shape, and composition of the earth and the various physical characteristics of the atmosphere, such as refraction which will be treated in Chapter 5.

Since time immemorial, man has manifested a consuming curiosity in the nature of the earth upon which he lives. He has always had an exciting interest in its size and shape. The speculation of the early Greeks ranged from the concept advocated by Homer that the earth was a flat disk to the idea of a spherical body advanced by Pythagoras. Eratosthenes (ca. 276–194 B.C.) is generally credited with being the first to determine the size of the earth. His studies led to a rather crude value of the earth's circumference but they were significant in heralding a succession of investigations that added progressive refinement to man's knowledge of the earth's dimensions.

It has been established, since at least the beginning of the nineteenth century, that the earth (more specifically its average or mean sea-level surface) is closely approximated by a mathematical figure called an oblate spheroid of revolution with the polaris axis being slightly shorter than the equatorial axis. The lengths of these axes are 41,709,784 and 41,851,664 ft, respectively, for a particular figure called the Clarke Spheroid of 1866 which is used in the United States. The average radius of curvature in this country is 20,906,000 ft.

Geodesy is commonly described as the science that treats of the investigations of the form and dimensions of the earth. It also embraces mathematical principles and procedures utilized in the precise determination of the positions of points on the surface of the earth. These points may be separated by distances of intercontinental magnitude. Geodesy is the foundation on which geodetic surveying rests.

1-4. Geodetic Surveying Geodetic surveying is that branch of surveying which employs such field practices and office calculations as are indispensable in the execution of precise surveying operations over areas of such considerable extent that the curvature of the earth must be considered. Geodetic surveying involves the most accurate distance and angle measurements. It embraces a wide variety of activities, such as triangulation, trilateration, leveling, and astronomic direction fixing, all of which will be treated in subsequent chapters. Its fundamental mission is the determination by complex calculations based on the formulas of geodesy of the exact horizontal and vertical position of widely distributed points on the earth's surface. It is pertinent to add that although the measurements are made on the earth's surface, sometimes aptly called the topographic surface, the expressions of horizontal position refer to the point where the plumb line passing through the survey station pierces the mean sea-level surface.

The principles of geodetic surveying give adequate recognition to the true spheroidal shape of the earth. Its computations, for example, make proper provision for the convergency of the true meridians and for correcting the length of a baseline measured on the topographic or ground surface to its equivalent length at the average height of the sea.

1-5. Plane Surveying Plane surveying is that branch of surveying which considers the surface of the earth to be a plane. Curvature is thereby ignored and calculations are performed using the formulas of plane trigonometry. All meridians are parallel and the direction of the plumb line is assumed to be the same, i.e., perpendicular to the plane, at all points within the limits of the survey.

Plane surveying principles are applied to surveys of limited extent or to such situations where the required accuracy is so low that the corrections for curvature would be negligible as compared with the errors of measurement. For small areas plane surveying methods can be expected to yield accurate results, but the quality will decrease as the project area increases in size. It is impracticable to state with universal application the maximum distance a plane survey can be extended from an origin with satisfactory results.

There is one exceedingly important characteristic common to both geodetic and plane surveying. Both branches of surveying express the vertical position of points in terms of height above a curved reference surface which is usually that of mean sea level.

It is to be emphasized that the principles of plane surveying play a dominant role in the treatment of surveying in this book and that almost exclusively plane surveying field and office practices provide the substance of the surveying operations described in later chapters.

1-6. Kinds of Surveys Some of the principal kinds or branches of surveys embraced by the three major classes of Art. 1-2 and the operations associated with them will be briefly identified. The term, kinds of surveys, has been arbitrarily selected and signifies merely various subcategories of the major classes.

Land surveying entails many operations. Among other things it comprises the determination of the location of land boundaries and land boundary corners, the preparation of maps showing the shapes and areas of tracts of land, their division into smaller parcels, and the preparation of official plats of land subdivisions in compliance with the laws of the state in which the work is done.

Cadastral Surveys are those land surveys executed by the federal government in connection with the disposal of the vast land areas known as the public domain.

Route Surveys are made for the purpose of designing and constructing a wide variety of engineering projects associated with transportation and communication. These include highways, railroads, pipelines, canals, and transmission lines.

Industrial Surveying, or optical metrology, as it is more frequently called, refers to the use of surveying techniques in the aircraft and other industries where very accurate dimensional layouts are necessary. Modified versions of conventional surveying equipment, such as the jig transit, have been designed especially for optical metrology.

Topographic Surveys are executed in order to obtain field data needed in the preparation of topographic maps. A wide variety of field and office operations are involved which culminate in the editing and printing of multicolor contour maps portraying the terrain; lakes and rivers; and highways, railroads, bridges, and other works of man.

Hydrographic Surveys comprise the operations necessary to map the shorelines of bodies of water; to chart the bottom areas of streams, lakes, harbors, and coastal waters; to measure the flow of rivers; and to assess other factors affecting navigation and the water resources of the nation.

Mine Surveys are necessary to determine the position of the underground works and surface structures of mines, to fix the positions and directions of tunnels and shafts, and to define the surface boundaries of all properties.

Aerial Surveys make use of photographs taken with precise cameras mounted in specially designed airplanes. These photographs are invaluable in supplementing the information obtained by other surveying operations and serve a wide variety of purposes. The results of the aerial survey are usually mosaics of matched vertical photographs, oblique views of the landscape, and topographic maps drawn from the photographs. Aerial surveying has several important advantages as compared with ground surveying and it is widely and extensively employed.

1-7. Historical Origins The early history of surveying merges with that of astronomy, astrology, and mathematics. The first theories of mathematics developed from the practical use of numbers required in the life of the ancient communities. The Egyptians, Greeks, and Romans employed surveying and mathematical principles in the establishment of land boundaries, the staking out of public buildings, and the measurement and calculation of land areas. The intimate relationship between mathematics and surveying is indicated by the name given to one of the earliest branches of mathematics, geometry, which is derived from Greek words meaning "earth measurements."

The remainder of this article highlights only a few of the most notable developments in surveying. Appropriate mention will be made of other historical events in subsequent chapters.

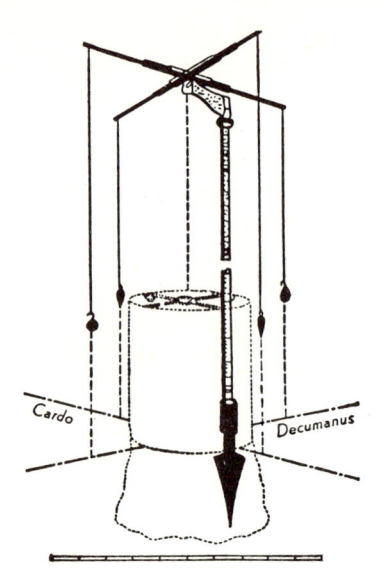

Fig. 1-1. Roman Groma.

Roman Instruments. The Roman surveyors were known as *gromatici* from their use of the *groma,* Fig. 1-1. It consisted merely of crossarms fixed at right angles and pivoted eccentrically upon a vertical staff. From the ends of the arms plumb lines were suspended. The whole purpose and use of the instrument was to establish upon the ground two lines at right angles to each other.

The *chorobates* was an instrument for leveling. It was a wooden bar about 20 ft long, in which a groove was made at its middle about 1 in. deep and 5 ft long. When the bar was leveled so that water stood evenly in the groove, a horizontal line was established.

The Telescope. The discovery of the telescope is generally credited to Lippershey, about 1607. This invention when applied to surveying instruments is of the greatest importance in increasing the precision and speed of surveying measurements.

The Vernier. In 1631 Pierre Vernier, a Frenchman, published in Brussels the description of his device which, bearing the inventor's name, is in general use today as a means of accurately determining subdivisions of a graduated scale.

The Semicircumferentor. Before the telescope was applied to angle measuring instruments, the peep sight served to fix the line of sight on most instruments used in mine and land surveying. An example is shown in Fig. 1-2.

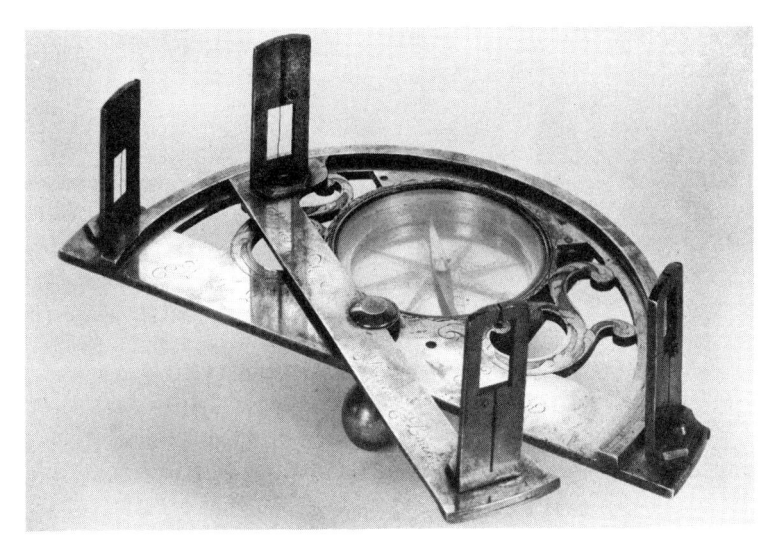

Fig. 1-2. Semicircumferentor.

Instruments of this type in the hands of George Washington, Thomas Jefferson, Mason and Dixon, and others of less renown, served to lay down the boundaries of the colonial possessions and boundaries of the thirteen original states, and to locate our canals and railroads in the early period of their development.

The Transit. Two men, Draper and Young, working independently, in the early years of the nineteenth century (about 1830) brought together in one instrument, Fig. 1-3, the essential parts of what has long been known as a "transit." This, with many later improvements, has been recognized as the surveyor's most important instrument. A modern transit is depicted in Fig. 1-4.

Units of Length. The commentaries of Caesar have made us familiar with the *mille passum* of the Roman armies, from which our terms "mile" and "pace" are clearly derived; also the *ager* or area of land that could be plowed in one day by a yoke of oxen has become our "acre." Again the Roman *pertica,* meaning a measuring stick or pole, became in English usage the "perch," "pole," "rood" or "rod" still used in our term "level rod"; also it is used as a unit of length (16½ ft) in land measure.

The necessity of standardizing the length of the rood was soon recognized, and in his book on surveying (1570) Koebel gives the

Fig. 1-3. First Transit Made in the United States. *Keuffel & Esser Co.*

Fig. 1-4. Modern Engineer's Transit. *Dravo Corporation*

Fig. 1-5. Determining the Length of a Rood.

following method: "A rood should by the right and lawful way, and in accordance with scientific usage, be made thus: sixteen men, short and tall, one after the other, as they come out of church (Fig. 1-5) should place each a shoe in one line; and if you take a length of exactly 16 of these shoes, that length shall be a true rood."

The area of an acre of land was established in England by an act of the Star Chamber in the 12th year of the reign of Henry VII, which "setteth down that an acre should be XL pole in length and IV pole in breadth." This was followed by an Act of Queen Elizabeth: "Commons or waste grounds lying within 3 miles of London shall not be enclosed. A mile shall contain 8 furlongs, every furlong 40 poles, and every pole shall contain 16 foot and an half."

To render land measure into a decimal system, Gunter devised a chain 66 ft long having 100 links. The link chain is now obsolete, but steel tapes, etched with these units, are still in use.

The Westward Movement in the United States, involving the construction of canals, roads, and railways and the settlement of new lands, created an enormous demand for surveying services in the nineteenth century. The military requirements of two world wars and other conflicts since that time have provided much of the motivation for vast improvements in the design of surveying equipment and in the execution of all kinds of surveying operations.

1-8. Developing Competency Plane surveying makes extensive use of geometry, algebra, and trigonometry. A working knowledge of these subjects, especially trigonometry, is expected of the student as he begins his study of surveying. The office work of geodetic surveying requires special training in computing.

Although the theory of plane surveying is quite simple, its practice requires resourcefulness, good judgment, and industry if the beginner desires to become proficient. The development of personal reliability in executing field and office assignments will be a direct consequence of giving close attention to the methodical performance of the numerous details associated with the subject. As experience is acquired, the young engineer or technician will become able to handle the larger problems connected with the selection of method and plan of organization. Since the practice of surveying is necessarily conducted under a wide variety of field conditions, it may be a challenging task to learn how to combat physical fatigue and bad weather.

1-9. Surveying Measurements All surveying operations are subject to the imperfections of the instruments and to the errors inherent in their manipulation. Therefore, *no surveying measurement is exact.* Accordingly, the nature and magnitude of the errors in the surveyor's work must be well understood if good results are to be ensured.

Obviously, there are many degrees of precision possible in any measurement. Thus, the distance between two established fence corners may be measured by estimation, by pacing, by stadia, or by taping. Each of these methods may be the best one to use for a given purpose because, ordinarily, it is a waste of time and money to attain unnecessarily high accuracy. On the other hand, a faulty survey results if the measurements are not sufficiently precise.

The best surveyor is not the one who makes the most precise measurements, but the one who is able to choose and apply just that degree of precision requisite to his purpose.

1-10. Field Notes No phase of the surveyor's work is of greater importance, or requires more careful attention, than keeping the field notes. The student should realize from the beginning that the quality of his work is reflected directly from his field record.

The field notes constitute the only reliable record that exists of the measurements and other items of pertinent information that are obtained in the field. Consequently, they must be recorded in the field, and they should be permanent, legible, complete, and capable of but one, and that the correct, interpretation.

The notes should be kept in a bound or loose-leaf book of good quality paper and recorded with a fairly hard (4H) drawing pencil. The pencil point should be sharp and sufficient pressure used to indent the paper slightly.

Field notes are recorded in the field. Records made on scratch paper and copied later, or other data recorded from memory, may be useful, but they are not field notes. The authority and reliability of field records are always under suspicion unless they have been entered in the book at the time and place the data were obtained.

To be complete, the notes should show all data, together with a sufficient interpretation to answer all questions that may properly be raised with respect to any given survey. Such completeness cannot be achieved unless the surveyor keeps clearly in mind not only the immediate uses of the data, but those which may reasonably be expected to arise at some future time. It often becomes desirable or necessary to retrace or extend surveys after many years, and after many physical changes have been made in the vicinity. The original field notes afford the only means of accomplishing the desired ends, and unless they are sufficiently complete, they may be worthless.

If field notes are to be useful, they must be legible. To this end they should be lettered instead of written.

A more complete treatment of notekeeping will be deferred until Chapter 6. By the time the student reaches that point in his instructional program he will have had some field practice and be better able to comprehend the need for adequate documentation of field operations.

1-11. Errors and Adjustments It has been stated previously that no surveying measurement is exact. The surveyor, therefore, is necessarily and continuously dealing with errors. If his work is to be well done, he must understand thoroughly the nature of the sources and behavior of the errors that affect his results. He must likewise recognize the need for making suitable adjustments of the raw data so that inconsistencies are removed and the most likely values of the desired quantities are obtained.

This chapter presents briefly some of the general principles of this important subject, and, although the treatment will be quite elementary throughout, the discussion will be continued in connection with nearly every subject in this book. It is hoped that this continued repetition will not only emphasize the importance of the subject, but will materially aid the beginner in the development of good judgment in his work.

1-12. Sources of Errors Errors in surveying measurements are sometimes classified as to their source, i.e., instrumental, personal, or natural.

Instrumental Errors are those due to imperfections in the instruments used, either from faults in their manufacture, or from improper adjustments between the different parts. The incorrect length of a steel tape, or the improper adjustment of the plate bubbles of a transit are examples. It is understood that the instruments are never perfect; therefore, proper corrections and field methods are applied to bring the measurements within the desired limits of precision. It may be added that some of the principal advances in the art of surveying within recent years have been effected by improvements in the design and manufacture of the instruments.

Personal Errors arise from the fallibility of the senses of sight and touch on the part of the observer. He must frequently estimate fractional parts of scale divisions, and he must manipulate the instruments with dexterity. These operations are never done exactly, and the magnitude of the resultant errors depends largely on the coordination of the senses of sight and touch, and on the skill of the observer. Reading the divisions on a graduated circle or fixing the line of sight of a transit on a given object are examples.

Natural Errors have their sources in the phenomena of nature, such as changes in temperature, differential refraction of the atmosphere, wind, curvature of the earth, etc. Such sources of error are quite beyond the control of the observer, but he can take proper precautions and adapt his methods to conditions so as to keep the resulting errors within proper limits.

1-13. Kinds of Errors In surveying measurements two kinds of errors are effective, namely, *systematic* and *accidental* errors. These are to be defined in terms of their behavior as to sign and magnitude both when the field conditions are constant and when the field conditions are changing.

Systematic Error. A systematic error is one which (a) for constant field conditions is constant both in sign and magnitude; and which (b) for changing field conditions is usually constant in sign but variable in magnitude, proportionate with the changing field conditions. Hence, the total error resulting from a series of systematic errors is the sum of the separate errors.

An example is the error resulting from the thermal expansion of a steel tape. It is assumed that the tape has a known length at a given (standard) temperature. If a distance is measured at a different, but constant, temperature the resulting error will be constant in magnitude and cumulative.

Accidental Error. An accidental error is one (a) whose sign is just as likely to be plus as minus under both constant and changing field conditions; (b) the magnitude of the separate accidental errors under all conditions vary according to the laws of chance; and (c) because of the changing signs and magnitudes of the errors their total effect on a measurement will be compensative, as stated in the *law of compensation* in Art. 1-16.

An example is that of setting pins at the end of the tape when measuring distance. The attempt is made to set the pin exactly opposite the end mark on the tape, but for a given measurement there is an equal probability that the pin will be set either beyond, or short of, the end mark, thus causing either a plus or a minus error, and the magnitude of the error has no proportional relation to the field conditions. Accordingly, the signs of the separate errors are either plus or minus, the magnitudes vary according to the laws of chance, and their total effect is compensative.

Discrepancy. A discrepancy is the difference between two measurements of a given quantity. Usually, although not always, it indicates the precision with which the measurements have been made.

Mistakes. Mistakes are gross differences from true values due to carelessness or inattention on the part of the observer. They are detected by checking the results.

An example is that of misreading the numbers on a tape. Thus, a measured distance of 76 ft may be misread and recorded as 79 ft. Such a mistake may be detected by a duplicate measurement in which, presumably, the mistake will not be repeated; or other known conditions may serve to reveal the presence of the mistake. The distinction between errors and mistakes, or blunders as they are frequently termed, is most important. In dealing with errors and making adjustment calculations, it is always presumed that all mistakes have been detected and their influence on the raw data has been eliminated.

1-14. Checks Because of the inherent errors and mistakes in surveying measurements, it is necessary to apply checks to the results (1) to detect mistakes, and (2) to determine the degree of precision attained. Since there is no assurance that any original measurement is free from mistakes, and since any important mistake is intolerable in final results, it is necessary that all work shall be checked; and, no work, either field or office, may be considered complete until it has been checked. It is difficult to convey the importance of this matter to the novice. Invariably, the engineer in charge of surveys of any magnitude is more concerned about the mischief caused by undetected mistakes than he is by the inherent errors of the measurements.

1-15. Precision and Accuracy It is necessary to distinguish between the meanings of the terms *precision* and *accuracy* as they are used in describing physical measurements and the subsequent computations.

Precision refers to the care and refinement with which any physical measurement is made. It relates to the expertness of manipulation on the part of the observer or to the capabilities of the instrument used.

Accuracy refers to the difference between the final measured value of a quantity and its absolute, or true, value.

The distinction can be illustrated by an example. Consider two 12-in., boxwood, engineers' scales of similar construction, one of which (scale A) has been kept in a damp place so that it has expanded 0.1 in. over its full length, and the other one (scale B) has very nearly its true length, say within 0.001 in. It is evident that if the same care and refinement are used in scaling a given dimension with each of these scales, the result obtained with scale B will be nearer the true value; in other words, of the two dimensions scaled with the same *precision,* one is more *accurate* than the other. Also, it may easily be imagined that a measurement with scale A may be more *precise,* but less *accurate,* than the measurement with scale B.

1-16. Law of Compensation As regards accidental errors, it has been found by experience that their behavior conforms to a principle which may be called the *law of compensation*. This may be stated as follows: *in a series of observations, the uncompensated or residual error is proportional to the square root of the number of opportunities for its occurrence.*

An example will illustrate further the distinction between systematic and accidental errors. Consider again the two sources of error mentioned previously—thermal expansion of a steel tape and the setting of pins in taping. Suppose that the distance between two section corners one mile apart has been measured and that the magnitude of each separate tape error was the same; i.e., the temperature was such that the 100-ft tape was 0.01 ft longer than its standard length, and that the average error in setting pins was also 0.01 ft. In a distance of one mile there will be 53 opportunities for each of these errors to occur, and the resultant errors from these two sources are calculated as follows:

The error due to temperature = $0.01 \times 53 = +0.53$ ft

The error due to setting pins = $0.01 \times \sqrt{53} = \pm 0.07$ ft

Thus it appears that for these two sources of error, each of which was of equal magnitude in each separate measurement, after 53 opportunities for the errors to occur, the magnitude of the systematic error was seven times as great as the total estimated value of the accidental error.

1-17. Professional Surveying Practice It would be unfortunate and incorrect for the student to conclude that the study, no matter how diligent, of any surveying textbook would be in itself sufficient to qualify him to engage in the professional practice of surveying. The design, execution, and administration of extensive surveying and mapping programs and the location of important land boundary lines requires both a specialized technical education of college level and extensive experience involving considerable responsibility.

As a consequence of thorough studies conducted by a Task Committee on the Status of Surveying and Mapping, the American Society of Civil Engineers in 1959 declared that the following four major categories in the field of activity commonly designated as surveying and mapping are a part of the Civil Engineering profession:

I. Land Surveying
II. Engineering Surveying
III. Geodetic Surveying
IV. Cartographic Surveying

In addition to this landmark policy pronouncement, the Society delineated the general characteristics of professional level positions and technician or pre-professional occupations within each category. Such job descriptions are recognized by civil service commissions in the classification of surveying personnel and are used as guidelines in the evaluation of the experience of those seeking membership in professional engineering societies.

The young man who desires a professional career in surveying would be well advised to obtain a sound, college-level, technical education in some branch of engineering, such as civil engineering, and to pursue a program of graduate study in surveying and the associated disciplines. Subsequent employment with a governmental surveying agency, a consulting engineering firm, or a registered land surveyor would provide the necessary background of experience needed to qualify for state registration as a professional engineer or land surveyor (Fig. 1-6). All fifty of the United States register engineers. Land surveyors are registered (1968) in 41 states.

Fig. 1-6. Surveying and Civil Engineering. *Dravo Corporation*

1-18. Role of the Surveying Technician The American Society for Engineering Education defines an *engineering technician* as one whose education and experience qualify him to work in the area of engineering technology, which is that part of the engineering field lying in the occupational area between the craftsman and the engineer. The engineering technician differs from the craftsman in his knowledge of engineering theory and methods; he differs from the engineer in that he has a more specialized background and uses technical skills in support of engineering activities. This general definition is illustrated with respect to surveying by emphasizing that, although there are many kinds of surveyors, there is a fairly clear distinction between those who are professional engineer-surveyors and those who are surveying technicians. This observation is by no means derogatory to the latter personnel. They play an extremely important supporting role in performing many surveying missions. They are truly members of the engineering team.

The formal training of the surveying technician is best provided by the appropriate curriculum, usually two years in length, of a technical institute.

1-19. Surveying Agencies Surveying and Mapping are indispensable first steps in the orderly inventorying of the natural resources of any nation. Such initial surveys are largely informational in character but serve a wide variety of subsequent needs. Since the surveying operations are of broad extent and benefit the general public, they are usually executed by governmental organizations.

The most important federal surveying and mapping agencies are as follows:

(*a*) *U.S. Geological Survey.* This organization, which was established in 1879, is charged with the responsibility for preparing the National Topographic Map Series covering the United States and its outlying areas.

(*b*) *U.S. Coast and Geodetic Survey.* This Bureau of the Environmental Science Services Administration, ESSA, celebrated the completion of 160 years of service in 1967. It publishes nautical charts of the coastal waters of the United States and its territorial possessions, executes the principal geodetic surveys of the country, and prepares and distributes the aeronautical charts needed by American civil aviation. President Thomas Jefferson and Ferdinand Hassler, noted Swiss scientist, were the founders of the Coast Survey.

(c) *U.S. Naval Oceanographic Office.* This agency performs essentially the same hydrographic charting functions as the Coast and Geodetic Survey but with respect to waters not contiguous to the United States and its possessions.

(d) *U.S. Lake Survey.* This is the nautical charting agency of the Corps of Engineers. It is concerned primarily with the publication of navigation charts for the Great Lakes.

(e) *U.S. Bureau of Land Management.* This organization is responsible for surveys of the public domain. Rectangular public surveys are still being executed in some of the western states and in Alaska.

(f) *U.S. Corps of Engineers (U.S. Army).* Each army engineer district office has a survey section that performs many kinds of surveying tasks associated with the control of navigable waters over which the Corps has jurisdiction.

(g) *U.S. Bureau of Reclamation.* The extensive construction program of this organization involves many surveying tasks ranging from the preliminary mapping of a proposed reservoir to the layout of a large dam.

1-20. Surveying Societies A number of technical and professional societies have been organized to advance the science and art of surveying and mapping in their various branches. They include the American Congress on Surveying and Mapping, American Society of Photogrammetry, Surveying and Mapping Division of the American Society of Civil Engineers, Canadian Institute of Surveying, and numerous state land surveyor organizations such as the Illinois Registered Land Surveyors Association. The high quality literature of these societies and their continued efforts to improve the public image of both the professional engineer-surveyor and the licensed land surveyor have helped to elevate the status of the professional in this area of American technology.

REFERENCES

1. Baldwin, J. T., Jr., Donald B. Clement, Herman R. Friis, and Robert H. Randall, "The History of Surveying in the United States," *Surveying and Mapping,* Vol. 18, No. 2, pp. 179–219, American Congress on Surveying and Mapping, Washington, D.C., 1958.
2. Burnside, R. S., Jr., "The Evolution of Surveying Instruments," *Sur-*

veying and Mapping, Vol. 18, No. 1, pp. 59–63, American Congress on Surveying and Mapping, Washington, D.C., 1958.
3. Clark, David, and James Clendinning, *Plane and Geodetic Surveying,* Vol. I, *Plane Surveying,* 5th ed., 1957; Vol. II, *Higher Surveying,* 5th ed., 1962, Frederick Ungar Publishing Co., New York.
4. Davis, Raymond E., Francis S. Foote, and Joe W. Kelly, *Surveying—Theory and Practice,* 5th ed., McGraw-Hill Book Company, New York, 1966.
5. Department of the Army, Technical Manual 5-232, *Elements of Surveying,* U.S. Government Printing Office, Washington, 1964.
6. Kissam, Philip, *Surveying for Civil Engineers,* McGraw-Hill Book Co., New York, 1956.
7. Merdinger, Charles J., "Surveying through the Ages"—Part I, "Early Developments"; Part II, "Development of Instruments," *The Military Engineer,* Vol. XLVI, Nos. 310 and 311, resp., Washington, D.C., 1954.
8. Multhauf, Robert P., "Early Instruments in the History of Surveying: Their Use and Their Invention," *Surveying and Mapping,* Vol. 18, No. 4, pp. 399–416, American Congress on Surveying and Mapping, Washington, D.C., 1958.
9. Surveying and Mapping Division, *Definitions of Surveying, Mapping and Related Terms, Manual No. 34,* American Society of Civil Engineers, New York, 1954.

2 Direct Linear Measurements

2-1. Introduction In order to trace the origin of presently accepted units of linear measurement such as the foot and the mile, one is led inevitably to the study of *metrology,* which is usually defined as the science of weights and measures. Research into the evolution of various linear measures begins with the written records of the earliest metrologists and with the examination and study of the remains of various ancient edifices such as the pyramids of Egypt, the Parthenon of Athens, and the Stonehenge of Britain. One of the outstanding measuring devices used by early civilization was the Nilometer, which was utilized to determine flood heights along the Nile.

The most primitive linear units were derived from the length of certain parts of the human body. The digit was the width of the first joint of the forefinger, the span was the stretch from the thumb to the little finger, the foot was the length of the pedal extremity, and the cubit was the distance along the forearm from the elbow joint to the tip of the middle finger.

Modern linear units had their genesis in the British Foot and Yard of 1855 and the French Toise of 1766, which had a length of about 6.4 English feet. The most significant unit of length, the meter, is associated with the development of a comprehensive decimal system of length now known as the metric system. The meter was originally defined as being 1/10,000,000 part of a meridional quadrant of the earth. Following the execution of surveys of geodetic accuracy and the deliberations of outstanding geodesists, an international treaty in 1875 provided for the creation of an International Bureau of Weights and Measures. At the first conference in 1889 new standards for the metric system were adopted. The meter was redefined in terms of the distance between certain markings on a platinum-iridium bar at 0°C. This is known as the International Prototype Meter.

Direct linear measurement, which is treated in this chapter, refers to the determination of length by the simple operation of ascertaining how many times a specific unit, such as a yard, is physically contained in the length of a given line. *Indirect linear measurement,* which is the subject of Chapter 4, is illustrated by the determination of a distance by timing the interval required for an acoustical wave to travel from one point to another and calculating the distance by multiplying the elapsed time by the known velocity of sound. Regardless of the type of field technique and the reduction calculation procedure, the ultimate product of length measurement in surveying is a horizontal distance. Occasionally, as in geodetic surveying, the projected length along the curved mean sea level surface may be desired.

Distance measurement by taping constitutes the entire substance of this chapter.

2-2. Units of Length The basic units of length used within the United States are the foot and the meter. The foot is of Anglo Saxon origin and is quite universally used in English-speaking countries. The meter is of French origin and has become the adopted unit for international and scientific usage. Its use in the United States in surveying work is limited practically to the precise measurements of geodetic surveys.

The decimal system of linear measurements, based on the foot unit, has been adopted for practically all surveying work. In engineering practice, the decimal system is extended to measurements of less than one foot, the foot unit being subdivided into tenths and hundredths. The building trades still use the older English units in which the foot is divided into inches, quarter-inches, etc.; hence, architects and engineers frequently have to convert units from one system to the other.

The rod (sometimes called a perch, or a pole) is a unit of 16½ ft, which has considerable use in land measure.

The unit of land measure is the acre, which has been standardized at 1/8 mile in length and 1/80 mile in width. It is, therefore, 660 ft long and 66 ft wide. So it was, by reason of these dimensions, that the Englishman, Gunther, made use of a 66-ft chain. This made the acre 10 chains long and 1 chain wide, or 10 sq chains in area, and so reduced land measure to a decimal system.

The equivalents of the different units are as follows:

1 mile = 5280 ft = 1760 yards = 320 rods = 80 chains.
1 chain = 66 ft = 4 rods.
1 meter = 39.37 in. = 3.2808 ft = 1.0936 yd.
1 vara = 33 in. (California) = 33⅓ in. (Texas).
1 acre = 43,560 sq ft = 10 sq chains.

In the metric system of weights and measures certain prefixes are so widely used as to deserve special mention here. *Kilo* means 1000 and *milli* indicates one-thousandth, respectively. To provide for multiples larger than 1000 and for subdivisions smaller than one-thousandth, the prefixes *mega* meaning 1,000,000 and *micro* meaning one-millionth have been generally recognized. A special term is *micron*, which means one-millionth of a meter or one-thousandth of a millimeter and is indicated by the Greek letter, μ. Hence, a *millimicron* (abbreviated as mμ) is one-thousandth of a micron or one-millionth of a millimeter. Table XV contains the Greek alphabet whose letters are widely used in engineering and surveying.

2-3. Standards of Length It may be desirable to clarify the distinction between a unit of length and a standard of length.

A *unit*, whether of length or otherwise, is a value or quantity in terms of which other values or quantities are expressed. In general, it is fixed by definition and is independent of such physical conditions as temperature.

A *standard* can be defined as the physical embodiment of a unit. Generally, it is not independent of physical conditions and it is the true embodiment of the unit only under certain specific conditions.

The primary standard of length in the United States is the distance between two engraved lines on national prototype meter bars Nos. 21 and 27 which are housed at the National Bureau of Standards at Gaithersburg, Maryland. These bars, which are of ×-shaped cross section, are duplicates of the International Prototype Meter kept at the International Bureau of Weights and Measures at Sèvres, France.

The United States Congress legalized the use of the metric system on July 28, 1866. This legislation was, however, only permissive. It did not make mandatory the use of the meter. Effective on April 5, 1893, all legal units of measure used in the United States were defined as exact numerical multiples of metric units. This action, known as the "Mendenhall Order," established the relationship between the foot and meter as follows:

$$1 \text{ U.S. yard} = \frac{3600}{3937} \text{ meter} = 0.91440183 \text{ meter}$$

or

$$39.37 \text{ in.} = 1 \text{ meter}$$

In 1960 the International Meter was officially defined in a supplementary manner as 1,650,763.73 wave lengths of the orange-red light of krypton 86, a rare gas extracted from the atmosphere. The U.S. inch thus becomes equal to 41,929.399 wave lengths of the krypton light. The new definition of the meter relates it to a constant of nature, the wave length of a specified kind of light, which is believed to be immutable and can be reproduced with great accuracy in any well-equipped laboratory.

The precision requirements in length measurements increased greatly following World War II and the difference between the U.S. inch and the British inch became especially important in gage-block standardization. As a result of several years of discussion the directors of the national standards laboratories of the United Kingdom and the United States entered into an agreement effective July 1, 1959, whereby uniformity was established for use in the scientific and technical fields. The new relationship between the yard and meter became

$$1 \text{ yard} = 0.9144 \text{ meter}$$

or

$$1 \text{ inch} = 25.4 \text{ millimeters}$$

Thus, the new value for the yard is smaller by 2 parts in one million than the 1893 yard. However, it is emphasized that any data expressed in feet derived from and published as a result of geodetic surveys within the United States will continue to bear the original relationship as defined in 1893, viz.

$$1 \text{ foot} = \frac{1200}{3937} \text{ meter}$$

The foot unit defined by this equation is referred to as the *U.S. Survey Foot*. In all survey work except that of geodetic accuracy, the distinction between the two definitions of the foot can be disregarded.

At various times bills have been introduced in the Congress of the United States to effect a mandatory conversion to national use of the metric system of weights and measures, but there has been little support for such legislation.

2-4. Tapes Distance measurement is an important element in the conduct of most surveys. Distance can be determined by pacing, pedometer, odometer, stadia, subtense, triangulation or trilateration, and electronic distance measuring devices, but taping is still the predominant means for making length measurements. Occasionally, the term *chaining* is employed to express a taping operation even though the 66-ft linked chain has been obsolete for many years.

The most common surveying tapes are made of a steel ribbon of constant cross section bearing graduations at regular intervals. Others are made of a steel alloy or of a metallic or nonmetallic cloth. There is great diversity of tapes with respect to lengths and widths and manner of graduation.

The 100-ft steel tape is the surveyor's favorite device for measuring distance. Its width varies from ¼ to ½ in. and its thickness from 0.020 to 0.025 in. The graduations and the identifying foot numbers are either stamped on soft (babbit) metal previously embossed on the tape at the foot divisions or etched into the metal of the tape. Loops are riveted to each end of the tape and usually rawhide thongs are attached to the loops. When not in use, the tape should be stored on its reel. If not kept on a reel, it should be gathered up in 5-ft lengths so as to form a figure eight and then thrown into a single small circle.

Nickel-steel alloy tapes, known as *Invar* or *Lovar*, have a coefficient of thermal expansion about one-thirtieth that of steel and are used when the taping specifications prescribe a high order of accuracy. The relative insensitivity of these tapes to temperature changes and the fact that they do not readily tarnish or rust on exposure to the elements make them particularly suitable for important survey tasks. However, the alloy metal is relatively soft and special care must be taken to prevent these tapes from becoming kinked or broken.

Woven tapes are ⅝ to ⅞ in. wide and most commonly 50 or 100 ft long. The nonmetallic type is made of synthetic yarns without metallic threads. It offers excellent wearing properties, high tensile strength, and dimensional stability. The metallic type contains very fine, noncorrosive metallic strands woven in with the yarn. The woven tape is impregnated with a paint-like material for protection and graduations are applied to the surface. A leather-covered metal case with a built-in reel is used to wind up the tape when not in use. Because the materials in this kind of tape are susceptible to temperature and moisture changes, woven tapes should be used only when relatively low accuracies can be tolerated.

Steel tapes are graduated in various ways and care must be taken to read the graduations properly in order to prevent making a blunder. Of special significance is the location of the end marks, particularly the zero mark. It may be located in different places on the different tapes. On some tapes, zero is at the end of the tape ribbon and the attached loop is not included in the graduated portion. On others, the zero is at the very end of the loop. On still others, the zero is marked at some distance from the end of the ribbon and there is a blank piece of tape between the zero and the loop. Some tapes have the first foot subdivided into tenths or hundredths and others have an extra foot graduated from the zero toward the loop. Various kinds of end arrangements are shown in Fig. 2-1. Both tapemen must make a careful examination of all tapes before using them and assure themselves of the position of the end marks, especially the zero mark, so that mistakes are prevented.

2-5. Taping Accessories Various accessory equipment is usually used with tapes in order to accomplish the distance measuring mission. Some of the more common items will be briefly described and a few are depicted in Fig. 2-1.

Steel pins with a ring at one end and pointed on the other are called *chaining pins* or *taping arrows*. They are used for marking tape ends on the ground and for tallying the number of tape lengths in a given line.

The *tension handle* is used to apply the appropriate tension to the tape when fairly careful measurements are to be made.

The *clamping handle* is employed to grip the flat ribbon of steel tape without kinking it when less than a full tape length is being measured.

In addition to such special equipment as the transit and the hand level which will be described in subsequent chapters and which are used to convert taped slope distances to their horizontal equivalents, other miscellaneous accessories include *range poles* for alining the taping, plumb bobs, thermometers, and cutting tools like hatchets or machetes for clearing the line of vegetation.

One exceedingly handy device is the 6-ft folding wood rule which is available with subdivisions in feet and inches on one side and in feet and hundredths of a foot on the other.

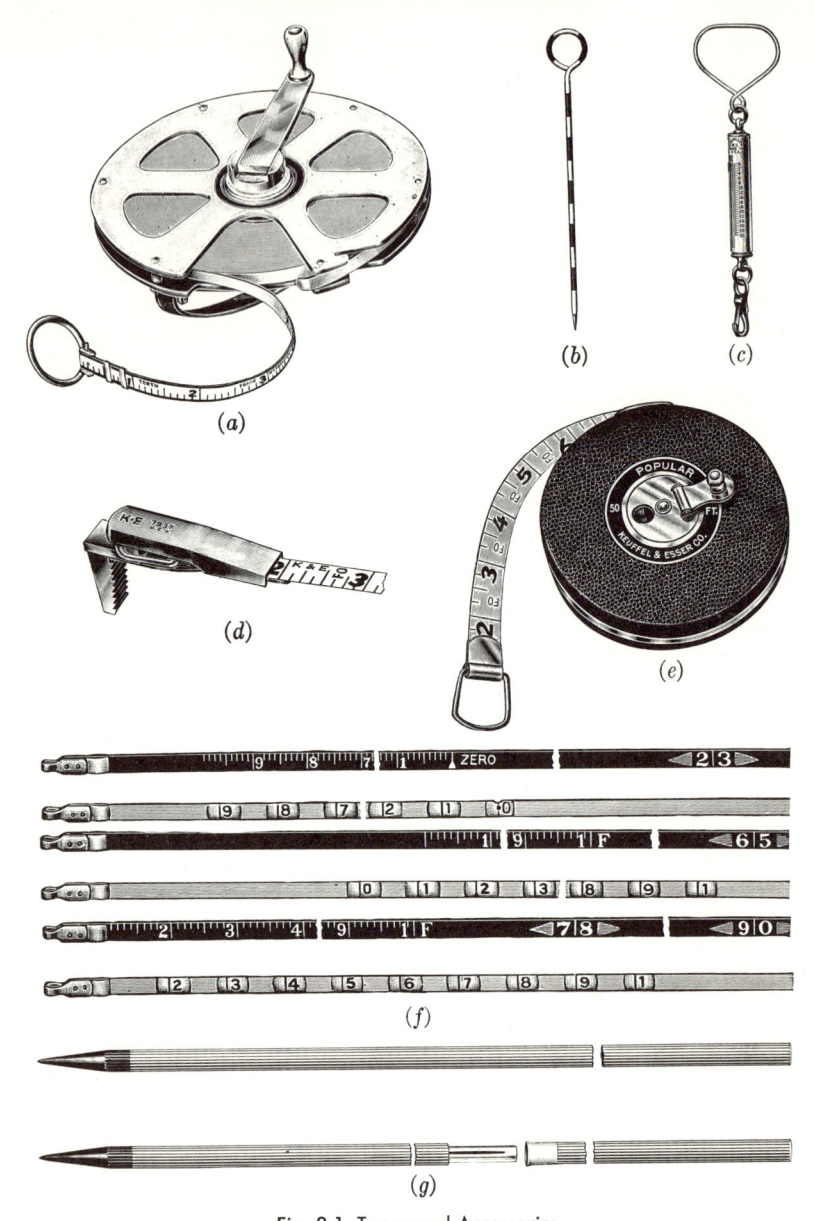

Fig. 2-1. Tapes and Accessories
(a) Steel Tape (100 ft) on Reel; (b) Steel Marking Pin; (c) Tension Handle; (d) Clamping Handle; (e) Woven Tape; (f) End Arrangements for Steel Tapes; (g) Steel Range Poles. *Keuffel & Esser Co.*

2-6. Care of Equipment Although the steel tape is relatively tough, it will readily break if, when it is kinked, it is subjected to a strong pull. Therefore, the tape should be kept straight when in use. It should never be jerked, pulled around corners, or bent in order to obtain a better hand grip on it. Vehicles of any kind should never be permitted to run over it even though the tapemen feel the tape will remain flat on the pavement surface. When the inexperienced or careless tapeman unwinds the steel tape hastily, it may uncoil itself suddenly and twists will develop. These must be carefully removed before any tension is applied. Steel tapes rust readily and should be wiped dry after use.

Range poles should not be used to loosen stakes or stones or as a javelin. Such practice will blunt the steel point, bend the shaft, and render the pole useless for accurate use as a target for transit observations. To avoid losing pins, a piece of brightly colored cloth can be tied to the ring of each.

The zero end of a woven tape receives the roughest treatment and is most subject to wear. In order to decrease the chances for breakage here, most woven tapes are specially reinforced with a plastic-rubber which is laminated to the first 6 in. However, tapemen should be especially alert to the susceptibility of the woven tape to break near the end ring by avoiding violent or hard pulls.

2-7. Standardization Standardization refers to the comparison of an instrument or device with a standard to determine the value of the instrument or device in terms of an adopted unit. A tape is considered to be standardized when the distance between its end marks is determined by comparing it, under prescribed conditions, with a standard which represents that unit.

All steel survey tapes are carefully graduated by the manufacturer under controlled conditions of temperature, tension, and mode of support. When they are used in the field, the conditions are different. For example, a so-called 100-ft tape will hardly ever be exactly 100.00 ft long except under the most fortuitous circumstances. For low accuracy surveys the amount of error in the length of the tape under average field conditions may be disregarded but for higher quality taping it may be essential to know the exact length of the tape. The standard for comparison purposes may be a master tape that is not used for making field measurements in order to protect it against injury or a local baseline, one tape length long, whose ends are solidly monumented and whose length was carefully determined to the nearest 0.001 ft with a master tape. The standardization of master tapes and those to be used on high accuracy surveys is best accomplished by the National Bureau of Standards which issues a certificate of length for specific conditions of tension, temperature, and support. A typical NBS Report of Calibration is shown in Fig. 2-2.

2-8. Taping Procedures Taping procedures are variable because of differences in project requirements, terrain, kind of tape, and such other factors as the personal preferences of party chiefs and the established practices of surveying organizations.

In general, there are two basic methods for measuring distances with a tape. These methods are termed *horizontal taping* and *slope taping*. When the horizontal taping method is employed, the tape is held horizontally and the positions of the end or intermediate marks are transferred to the ground. In the slope taping method, the tape's full or partial length is transferred to an inclined surface that supports the tape, the slope of the tape is determined, and the corresponding horizontal distance is calculated. Accurate taping by either method requires proper support of the tape, careful alinement, the application of correct tension, skillful use of plumb bobs and placement of arrows, and a knowledge of other factors, such as temperature, that affect the quality of taping.

Detailed descriptions of taping operations follow in the next two sections.

2-9. Taping Over Level Ground The minimum taping party consists of two technicians, a head tapeman and a rear tapeman. The front tapeman usually is the senior member of the party and also acts as recorder, and the rear tapeman is responsible for keeping the tape properly alined. The easiest taping operation is measuring over level ground. The equipment includes a 100-ft steel tape, a set of 11 taping pins or arrows, and one or two range poles. It is assumed that the tape has the first foot subdivided.

Having arrived at the initial point, the head tapeman goes forward, if necessary, and sets the range pole for lining-in purposes. Meanwhile, the rear tapeman unwinds the tape, laying it out in the general direction of the line to be measured, being careful that it is not looped or unduly twisted. The head tapeman then takes the *zero* end of the tape, hands one pin to the rear tapeman and moves for-

DIRECT LINEAR MEASUREMENTS

U.S. DEPARTMENT OF COMMERCE
NATIONAL BUREAU OF STANDARDS
WASHINGTON, D.C. 20234

NATIONAL BUREAU OF STANDARDS
REPORT OF CALIBRATION
100-Foot
STEEL TAPE

Maker: Keuffel & Esser Co.   NBS No. 13710
No.   Submitted by
University of Illinois
Department of Civil Engineering
Urbana, Illinois 61801

This tape has been compared with the standards of the United States, and the horizontal straight-line distance between the terminal points of the indicated intervals have the following lengths at 68° Fahrenheit (20 °Celsius) when subjected to horizontally applied tensions and conditions of support as indicated below:

Supported on a horizontal flat surface:

Tension	Interval	Length
10 pounds	0 to 100 feet	100.004 feet
20 pounds	0 to 50 feet	50.006 feet
20 pounds	0 to 100 feet	100.011 feet

Supported as indicated below with all points of support in the same horizontal plane:

Tension	Points of Support	Interval	Length
20 pounds	0, 50, and 100 feet	0 to 50 feet	50.002 feet
20 pounds	0, 50, and 100 feet	0 to 100 feet	100.002 feet
20 pounds	0 and 100 feet	0 to 100 feet	99.977 feet

The values given for the lengths of the indicated intervals are not in error by more than 0.001 foot.

See paragraphs 3(a), 3(d), 4, 5(b), and 6 on the reverse side of this report of calibration.

For the Director,

J. A. Beers
for
T. R. Young
Chief, Length Section
Metrology Division, IBS

Test No. 212.21/G-37913
Date January 23, 1967

Fig. 2-2. Tape Calibration Report.

---

ward along the line. When the 100-ft end of the tape comes up even with the point of beginning, the rear tapeman calls out "halt." At this signal the head tapeman halts and quickly places himself on line with the aid of right or left signals from the rear tapeman.

As soon as the tape has been placed on line, the rear tapeman holds his end of the tape exactly even with the initial point. The head tapeman takes his position just to the left of the line (not on the line), kneels, and applies tension (about 10 lb) to the tape with his left arm bearing against his leg. His right hand is then free to place the pin on line and at the zero mark of the tape. The pin may be set vertically, but more often it is given a slant at right angles to the tape, by which it can be placed more conveniently and accurately in position.

When the head tapeman sets his pin, he should be assured that the rear tapeman is holding his end of the tape precisely on the mark, and the rear tapeman must not pull his pin until the head tapeman has finished setting his pin. Hence, before setting his pin, the head tapeman waits for the signal "right here" from the rear tapeman. As soon as he has set the pin he also calls out "right here," which is the signal for the rear tapeman to pull his pin. It is important that the signals be carefully observed.

At the initial point, marked, let us say, by a transit stake, the rear tapeman holds one pin and the head tapeman begins with ten pins on his ring. As soon as the head tapeman sets his first pin, the pin which the rear tapeman holds indicates the fact that one tape length has been measured. When the next pin is set, the rear tapeman pulls his pin and he now has two pins, indicating that two tape lengths have been measured. Accordingly, the number of pins that the rear tapeman holds in his hand, not counting the pin set in the ground, indicates the number of full tape lengths that have been measured. When the head tapeman sets his tenth or last pin, he calls out "tally." The rear tapeman now has ten pins which he brings forward, and the taping proceeds. Thus, the number of tallies indicates the number of thousands of feet that have been measured.

If the terminus of the line being measured is a previously fixed point, the last measurement will be a fractional tape length. It is here that mistakes in taping most frequently occur, and care must be exercised that the procedure is orderly and always the same, to avoid confusion.

When the end of the line is reached, the head tapeman halts and the rear tapeman comes up to the last pin set. The tape is quickly adjusted so that a full foot mark is opposite the pin, and the terminus falls within the end foot length which is subdivided into tenths. Tension is applied, the head tapeman observes the number of tenths, estimating hundredths if necessary, that extend beyond the terminus, and the rear tapeman observes the number of the foot mark he is holding at the pin. The number that the head tapeman observes is subtracted from the number the rear tapeman reads to obtain the measured fractional distance. For example, the head tapeman observes 0.28 ft as that part of the tape which extends beyond the terminus, and the rear tapeman observes his foot mark to be 35 ft. The head tapeman then calls out "Cut twenty-eight hundredths," the rear tapeman calls out "Thirty-five," and they both make the subtraction mentally, and check each other on the result, 34.72 ft. In case the tape had an extra foot subdivided from the zero mark toward the end of the tape, this fractional distance would be read directly.

If the rear tapeman holds seven pins in his hand, not counting the one in the ground, the total distance is 734.72 ft.

If the taping is done on a hard surface, such as a sidewalk, steel rail, or pavement, the position of the end of the tape is marked with a colored lumber crayon, called *keel*. In this case the number of the tape length is recorded beside the mark as a means of keeping the count of tape lengths measured. To avoid mistakes, the rear tapeman calls out the number of his mark just before the head tapeman records the next number.

2-10. Taping Over Sloping or Uneven Ground When taping over rough and undulating terrain a combination of horizontal and slope taping may be executed. If the terrain is uneven but the slopes are moderate, the tape may be used in a horizontal position. If the ground is quite smooth and slopes uniformly, it will be advantageous to conduct the measurements with the tape lying on the ground. Additional items of equipment are plumb bobs and a hand level.

In Fig. 2-3 consider the two points A and B located on a slope,

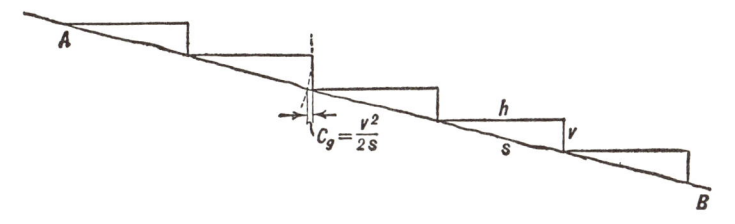

Fig. 2-3. Measuring on Slope.

and between which the horizontal distance is 500 ft. Let $h =$ a horizontal distance of 100 ft; $v =$ the vertical fall of the slope for each tape length; $s =$ the slope distance corresponding to the horizontal 100-ft distance; and $C_g =$ the correction, or difference between the horizontal and the slope distance.

Evidently, the correct horizontal distance between A and B will be measured if the tape is held horizontally for each tape length and if the position of the end of the tape is projected vertically to the ground by means of a plumb line; also, if the measurement should be taken on the slope and the proper corrections applied. Each method will now be described and the merits of the two methods will be discussed presently.

First Method; the Tape Level. When, on sloping ground, the tape is held horizontally, each tapeman should carry a plumb bob, and possibly a hand level, for leveling the tape.

If the slope is downhill, the head tapeman estimates the vertical distance v, holds his end of the tape at that height above the ground, applies the tension, and by means of the plumb bob transfers the position of the end of the tape to the ground, where a pin is set or, on a hard surface, a keel mark is made. If accurate work is being done, the tape is stretched a second time and the mean of the two measurements taken.

It is not convenient, or good practice, to hold the tape more than 5 ft above the ground, and hence, if the slope is more than 5 ft per tape length it is necessary to "break" tape, as the process is called. In this case the head tapeman pulls the tape forward until the rear end comes up to the rear tapeman. He then goes back until he reaches a point where the vertical distance v is not more than 5 ft. Here, at some full foot mark, he plumbs down to the ground. The rear tapeman then comes up, holds the foot mark at the ground point, and the head tapeman goes forward until another point is found for which v is approximately 5 ft, where he again plumbs down and fixes a new ground point. This process is repeated until the full tape length has been measured. It may be noted that it is immaterial what foot marks on the tape are used, and no record of them is kept.

The head tapeman may find that it is less work to use a different method, by which he goes forward only until he reaches the first point where the tape is broken. The rear tapeman reads the foot mark and they proceed again. The full tape length is then found by adding together the separate fractional lengths. This practice is strongly to be condemned, because it too often results in mistakes being made in reading the tape and in adding the lengths together.

On less accurate surveys, a range pole may be used to plumb from the end of the tape to the ground. In going uphill, of course, the rear tapeman must hold his end of the tape above the ground and its position, likewise, is found by plumbing from the ground point.

In taping over rough ground where there is dense vegetation such as cornstalks, weeds or underbrush, it may be difficult or impracticable to hold the tape on the ground even though the slope is negligible. In this case, a plumb line must be used at each end of the tape.

Second Method; Tape on the Slope. Wherever the tape can conveniently be held on the ground, no matter how steep the slope, it should be done; because, as will be found when the sources of error are discussed, this method is more accurate and rapid than attempting to hold the tape horizontal and plumbing to the ground. The only difference beween this method and that of taping on level ground is that a correction must be applied, the magnitude of which will now be considered.

In Fig. 2-3 it is evident that the value of the correction C_g is the difference between s and h, the hypotenuse and vertical leg of the right triangle whose sides are s, h, and v.

The ratio of the sides v/h is called the *gradient* of the slope and is usually expressed in per cent; i.e., the rise or fall in a distance of 100 ft. Thus, a 1% grade means one for which the vertical rise v is 1 ft for a horizontal distance of 100 ft. The gradient is sometimes expressed in degrees of arc, indicating the vertical angle between the horizontal and the slope, but this practice is not common in taping.

Evidently, the correction C_g is equal to the difference $s - h$, which can be found exactly from the right triangle as follows: $s^2 = h^2 + v^2$, or $s^2 - h^2 = v^2$, from which

$$(s - h)(s + h) = v^2$$

or

$$(s - h) = \frac{v^2}{s + h} \qquad (2\text{-}1)$$

The usual condition for which the value of C_g is desired is that in which the value of v is known (measured in the field) and the slope distance is 100 ft; hence, h is unknown. In the right-hand member the ratio $v^2/(s + h)$ is usually a small number, and since s and h are nearly equal in magnitude, the error introduced will be small if s and h are assumed to be equal. With this assumption, the equation becomes

$$C_g = \frac{v^2}{2s}. \qquad (2\text{-}2)$$

This correction to reduce a slope measurement to the horizontal is always negative. Under certain circumstances, however, when it may be necessary to set points one full "station" or 100 ft apart (horizontally), the following different manner of applying the correction should be noted. As the tapemen proceed in the field, the head tapeman estimates the slope, either up or downhill, makes a mental calculation of the correction, and sets the pin, or makes his mark, at the calculated distance beyond the end of the tape. This establishes a distance whose horizontal projection is 100 ft and, therefore, the same horizontal distance that would have been measured if the ground were level. By this procedure no tabulation in the notebook or subsequent corrections are necessary. Table X will prove helpful for calculating grade corrections.

It may be added that a third method of measuring on slopes is that by which the distance is measured on the slope, the inclination of which is found on the vertical arc of a transit, or a clinometer (see Fig. 2-4). Obviously then, $h = s \cos\alpha$, where α is the measured

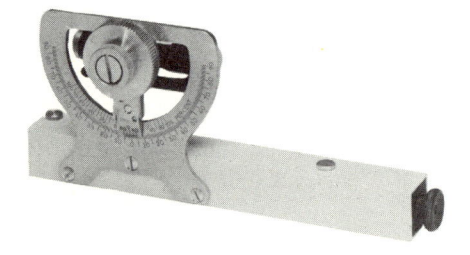

Fig. 2-4. Clinometer. *Keuffel & Esser* Co.

angle of inclination. This relation yields exact results and is easy to apply where the vertical angle can be measured conveniently.

2-11. Taping Corrections The prescribed relative accuracy of a taping measurement will dictate the degree of the refinement of the corrections that are applied to the raw field data. In general, every measured distance must be corrected in order to obtain the true length.

The major sources of error in taping work can be identified in terms of the following corrections:

(*a*) *Length of Tape.* The length of a tape varies with temperature, tension, and manner of support. The difference between the nominal length of a tape and its effective length under the conditions of standardization is known as the length correction. The calculations of taping corrections always begin with the standard length. The conditions of use in the field will then determine the magnitude and sign of other corrections that are to be applied to the observed values.

A steel tape purchased from reputable makers usually will have a length not different from the standard by more than 0.01 ft, but a tape that has been in use may have become kinked or patched so that its length has been appreciably altered. Such tapes are frequently in error as much as 0.02 ft and sometimes 0.1 ft. Hence, no tape that has not been standardized should be used on important work. If no standard is available in the vicinity, the Bureau of Standards will make the comparison for a small charge.

The effect of this source of error is greatly reduced by applying the proper corrections, but it is not entirely eliminated, because the comparison with the standard is not exact.

The matter of applying corrections due to an incorrect length of tape should receive careful attention. While measuring with a 100-ft tape, its length is assumed to be 100 ft exactly. Hence, the measured length of a line, i.e., the value observed and recorded in the notebook, is that for a tape exactly 100 ft long. Then, if the actual length of the tape, when compared with a standard, is found to be 100.02 ft, the true length is 100.02, although the "recorded" length is 100.00 ft. Hence, if the tape is too long, the correction must be added to the recorded length.

For example, if a distance is measured with the tape just mentioned and found to be 705.76 ft, the resultant error would be 7.05 × 0.02 = 0.14 ft, and the corrected length, therefore, would be 705.76 + 0.14 = 705.90 ft.

Likewise, if the tape is too short, the correction must be subtracted from the recorded length.

It should be noted that these corrections are made when a distance is being measured between existing end marks. If two end marks are to be established on the ground at a previously determined distance, then the signs of the corrections will be reversed. For example, if, in staking out a city subdivision, it is necessary to set two iron pins exactly 600 ft apart, and if the true length of the tape is 100.02 ft, the measured distance with this tape will be $600.00 - 0.12$ ft $= 599.88$ ft.

(*b*) *Temperature.* The standard length of a tape is its length at a temperature of 68°F (20°C). When the temperature of a steel tape is less than 68°F, the length of the tape will be less than its standard length; and conversely, when the temperature exceeds 68°F, the length of the tape will be greater than its standard length. The correction to be applied to the observed length of a survey line because of the effect of temperature on the steel tape can be evaluated from

$$C_t = 0.0000065(T_1 - T_0)L \qquad (2\text{-}3)$$

where 0.0000065 is the coefficient of thermal expansion of steel per 1°F, T_1 is the field temperature, T_0 is the standardization temperature, and L is the length of the line. For example, if $T_0 = 68°$ and $T_1 = 83°$, the temperature correction for a 100-ft steel tape would be

$$C_t = 0.0000065(83 - 68)100 = 0.01 \text{ ft}$$

Thus it is noted that a 100-ft tape will have its length changed 0.01 ft for each change of 15°F in the temperature. For small ranges of temperature and on ordinary work, this error may not be important, but the inexperienced engineer is apt to underestimate the importance of this source of error when even the ordinary temperatures of winter and summer measurements are encountered. For example, a summer temperature of the tape of 100°F and a winter temperature of 25°F are not uncommon. This difference in temperature of 75°F causes a change in the length of the tape of 0.05 ft, which makes a discrepancy of 2.6 ft in a mile. This error is greater than that permitted from all sources combined, on many surveys, and yet it is frequently disregarded entirely.

(*c*) *Slope.* When a measurement is made with the tape in an inclined position, the slope distance is always greater than the horizontal distance. Failure to hold the tape horizontal or to determine the slope correctly will produce errors whose magnitude can be calculated as explained in Art. 2-10.

(d) *Alinement.* The effect of inaccuracy in keeping the tape on line is the same in nature and magnitude as that due to slope. However, it is much more easily controlled than is the effect of slope, and the resulting errors are usually much smaller. Of course, the tape should be kept on line, within proper limits, but when it is remembered that a 1% slope causes the same error as a 1-ft error in alinement, it shows considerable ignorance on the part of the tapemen if they use an undue amount of time to "line in" the tape to the nearest 0.1 ft and disregard entirely a slope of perhaps 3 or 4%.

(e) *Mode of Support.* A tape supported only at the ends will sag in the center by an amount that is related to its weight per unit length and the pull. The shortening effect of the sag is essentially the difference between the axial length of the tape and the chord distance between the ends. Sag causes the recorded distance to be greater than the actual length being measured. When the tape is supported at its mid-point, the effect of sag in the two spans is considerably less than when it is supported at the ends only. Furthermore, as the number of equally spaced intermediate supports is increased, the distance between the end marks will closely approach the length of the tape when supported through its length.

The sag correction, C_s, can be calculated by the equation

$$C_s = \frac{W^2 L}{24 P^2} \tag{2-4}$$

in which W = the weight (in pounds) of tape between supports,
L = the interval between supports,
P = the tension (in pounds) on the tape.

EXAMPLE 2-1: A 100-ft steel tape weighs 2 lb and is held supported at the ends only with a pull of 12 lb. Find the sag correction.

$$C_s = \frac{2^2 \times 100}{24 \times 12^2} = -0.11 \text{ ft}$$

If the tension were increased to 20 lb, the shortening is reduced from 0.11 ft to 0.04 ft, which shows the desirability of using a higher tension on the tape when unsupported, and also the fact that the error in any case is considerable. The error is, of course, reduced by applying corrections, but they are not readily determined; therefore it is better practice, when conditions permit, to avoid the effects of sag by taping on the ground.

The most practicable way of dealing with sag is to standardize the tape with a specific pull, say 20 lb, when supported at the ends and use the tape in the field in this manner whenever it has to be elevated above the ground. Very rarely will it be necessary to evaluate the sag effect through the use of Eq. 2-4.

(*f*) *Setting Pins.* The errors due to setting the forward pin or to positioning the rear mark of the tape by the last pin are accidental in character and corrections cannot be computed to eliminate their effect.

(*g*) *Tension.* Since a steel tape is elastic to a small extent, its length is changed by variations in the tension applied. This change in length is not to be associated with the effect on the sag of the tape due to variations in tension but rather with the elastic deformation of the tape.

It can be calculated from the expression

$$C_p = \frac{(P_1 - P_0)L}{AE} \tag{2-5}$$

in which C_p = the elongation of the tape of length, L, in feet,
P_1 = the applied tension, in pounds,
P_0 = the standard tension, in pounds,
A = the cross-sectional area of the tape, in square inches,
E = the modulus of elasticity of the tape material (for steel 29,000,000) in pounds per square inch.

It will be sufficient here to state that an ordinary 100-ft steel tape will stretch about 0.01 ft for a change of 15 lb in tension. Since there is little occasion for any but slight changes in tension, and since the tension may be assumed to vary either above or below the standard tension, it is an accidental error and may be disregarded in all but precise measurements. Ordinarily, tensions of 10 lb and 20 lb are used for conditions of full support and end support, respectively.

(*h*) *Wind.* If a tape is stretched unsupported and a strong wind is blowing, the center of the tape will be carried to one side of the line joining the two ends. This condition causes an effect similar to, but usually much less than, sag.

A review of the previous discussion shows that (1) most of the errors in taping are systematic, and accordingly they vary nearly with the distance measured; (2) the magnitudes of the errors can be greatly reduced by applying simple corrections; and (3) the effects of temperature and slope are likely to receive too little, and alinement too much, consideration. Furthermore, it is to be noted that the sign of any error is always opposite to the sign of the corresponding correction.

2-12. Mistakes Some of the common mistakes made in taping and recording are listed below:

1. *Omitted Tape Length.* The serious mistake of omitting or adding a tape length is to be prevented by orderly procedure and careful attention to the work. The manner of checking is stated in the following article.

2. *Misreading the Tape.* A frequent mistake is that of misreading the tape, as when 6 is read for an inverted 9. Thus 86 is read for 89, or vice versa. Also, when the numbers on a tape become worn, an 8 may be read as a 3, etc. Mistakes of this kind are prevented if the tapemen will form the habit of looking at the number of the next adjacent mark before and after the reading has been made.

3. *Calling and Recording Numbers.* Numbers are easily reversed or misunderstood when they are called out to be recorded. The zero digit and the decimal point are most likely to cause mistakes. Thus the number 40.4 should be called as "forty, point four." Otherwise, it may be misunderstood as "forty four" and recorded as 44.0. The recorder should always repeat such numbers as are called to him, before recording them.

4. *One-Foot Mistakes.* It is easy to make a mistake of one foot when measuring a fractional tape length by subtracting incorrectly.

5. *Mistaking the End Mark.* The end marks are differently arranged on the different tapes. Hence, the tapemen should assure

themselves of the position of the end marks of each tape before it is used.

2-13. Checks In general, the field check that can be applied to the measurement of distance consists of a duplicate measurement. However, the engineer must be careful to remember that any discrepancies found between measurements made under similar conditions do not reveal any systematic errors. Thus duplicate measurements of a distance of one mile might show a discrepancy of 1 ft, but if the tape were 0.1 ft too long there would be an error of 5 ft from that source alone, which would not be indicated by the discrepancy. Since most errors in taping are systematic, under ordinary conditions, it is important that the engineer be not deceived with regard to the apparent precision indicated by small discrepancies between duplicate measurements that, in reality, are seriously in error. Careful attention must be given to the various sources of error such that repeated checks under various conditions will show results within the desired accuracy.

2-14. Accuracy The variety of conditions the engineer meets in the field prevents the making of any definite statement as to the accuracy that may be expected by the use of the different methods discussed here, yet it is desirable that anyone engaged in a survey should have a knowledge of the approximate degree of accuracy that the different methods should yield.

Because, for ordinary work, the principal errors in measuring distances are systematic, the resultant error is nearly proportional to the distance measured, and the accuracy of results is expressed by the ratio of the error to the distance. Thus, the ratio 1/3000 expresses that accuracy in which the error is 1 part to 3000 parts of distance. For comparative purposes, such ratios are always expressed with the numerator as unity, and with the denominator in round numbers only. Thus, an error of 3.4 ft in a distance of 4346.8 ft would be expressed as 1/1300.

For average conditions in open country, good taping is represented by the ratio of 1/5000. For fair results the accuracy may be taken as about half, or 1/2500. Rough taping may be taken as 1/1000.

2-15. Specifications To indicate more definitely what is meant by each of the three grades of taping mentioned above, specifications are given below which may be expected to yield the desired accuracy for the assumed conditions.

Conditions. It is assumed that the average conditions as to weather, equipment, and personnel obtain; that the line is measured across country where the ground is rolling or hilly, partly open and partly wooded, so that some of the taping is done with the tape on the ground and some with it unsupported; that the tape has its standard length at 68°F supported throughout and under a tension of 10 lb.

Three ratios of accuracy are specified—1/5000, 1/2500, and 1/1000. An accuracy of 1/5000 signifies that the total linear error due to taping a course may be expected to be not greater than 1 ft in 5000 ft.

1/5000. (a) *Length of Tape.* The length of the tape should be determined within ±0.01 ft and the proper correction applied.
 (b) *Temperature.* The temperature of the tape should be determined within ±10°F and the proper corrections applied.
 (c) *Slope.* All slopes should be estimated within 2% and the proper corrections applied.
 (d) *Alinement.* The alinement should be correct within 1 ft.
 (e) *Sag.* When unsupported, the tension on the tape should be 20 lb within ±3 lb.
 (f) *Tension.* Disregard variation, but use a tension of 10 lb when taping on the ground.
 (g) *Setting Pins.* End of tape to be marked within ±0.01 ft.

* * *

1/2500. (a) *Length of Tape.* Length of tape to be determined within ±0.01 ft, and if necessary, the proper corrections to be applied.
 (b) *Temperature.* Disregard ordinary temperatures but apply corrections for those above 90° or below 30°F.
 (c) *Slope.* Disregard slopes of 1% or 2%, but apply corrections or break tape for others, the errors of estimation not to exceed 2%.
 (d) *Alinement.* Alinement to be correct within 1 ft.
 (e) *Sag.* Disregard variations, but use a tension of 20 lb.
 (f) *Tension.* Use a tension of 10 lb when supported throughout.
 (g) *Setting Pins.* End of tape to be marked within ±0.02 ft.

* * *

1/1000. (a) *Length of Tape.* Tape to be of average quality in good condition but not standardized.
 (b) *Temperature.* Disregard.
 (c) *Slope.* Disregard slopes up to 5% and break tape on others.
 (d) *Alinement.* Ordinary care to be exercised.
 (e) *Sag.* Disregard variations but use 20 lb tension.
 (f) *Tension.* Use a tension of 10 lb when supported throughout.
 (g) *Setting Pins.* Use ordinary care.

2-16. Summary The field notes must contain all the information needed to convert the field data into horizontal distances. The accuracy of the survey will establish the kind of taping to be performed, the precision of the measurements, and the scope of supplementary data that must be obtained. A relatively simple form of taping notes is shown in Fig. 2-5.

Course	TAPING OVER LEVEL GROUND					Locker 32	May 19, 1968 31 R. Hansmeir, Tape J.P. Joyce, "
	Distance			Correct Dist.			
	For'd	Back	Mean.	Cor.			
A-B	178.20	178.16	178.18	+0.04	178.22		Temp. 80°F Cloudy
B-C	289.81	289.87	289.84	+0.07	289.91		
C-D	362.12	362.22	362.17	+0.08	362.25		Tape compared with standard
D-E	311.03	311.05	311.04	+0.07	311.11		Tape U.S.B.S. No. 3248 and found
E-A	222.16	222.12	222.14	+0.05	222.19		to be 100.01 ft at 60°F supported
	1363.32	1363.42					throughout. Tension = 10 lb.

Discrepancy between forward and back measurement = 0.1 ft

$$\text{Precision} = \frac{1}{13000}$$

Fig. 2-5. Taping Notes.

In spite of the fact that the techniques of taping are not difficult and standard operating procedures can be easily evolved and should be carefully observed, attention is directed to some closing comments with respect to direct linear measurements.

The instruments used in taping are of simple construction; and to the layman who observes it, the process of measuring appears elementary. Consequently, the notion is generally held that anyone, regardless of his ignorance or inexperience in the work, is qualified to serve as a tapeman on a survey. It is true that under many conditions a person of average intelligence can be given a few simple instructions and serve acceptably on the survey. Yet it must be said that in most work the proper execution of the distance measurements is accompanied with more and greater difficulties than are angle measurements. The engineer's transit, quite universally used for the measurement of angles, is an instrument of precision such that, with a little experience, the novice can measure the angles of a survey with greater relative accuracy and ease than the tapeman can measure the distances. Furthermore, mistakes are usually more frequent and more difficult to detect in the distance than in the angle measurements. Therefore, faulty results of surveying work are more commonly the result of the taping than the measurement of angles.

The transitman is usually given the responsibility of directing the work of the survey party in the field, and he has the important duty of keeping the notes; hence, he usually is a man of broader knowledge and experience than the tapeman. Insofar as the execution of manual duties is concerned, however, the experienced engineer in charge of a survey is more concerned with the qualifications of his tapemen (especially the head tapeman) than he is of his transitman.

The beginner is likely to be so impressed with the more complicated transit and level instruments that he takes a superficial interest in the taping work. It is hoped that these remarks will dispel this attitude and that he will give his most serious and careful attention to the measurement of distance.

PROBLEMS

Suggestion: It is important that the subjects of *significant figures* and *rounding off* in computations (see Arts. 7-5 and 7-6) be well understood. Much unnecessary labor and resulting mistakes will be avoided in computing the corrections to taped distances if it is noted that (a) most of the corrections are small quantities which may be computed mentally or by slide rule, (b) the total corrections are more easily found if they are computed for one tape length and then multiplied by the number of tape lengths, and (c) the different corrections may be computed independently of each other.

2-1. The recorded length of a line was 2217.31 ft. Later the tape was standardized and found to be 99.96 ft long. Find the correct length of the line.

2-2. A distance of 821.15 ft was measured on an average slope of 4.5% with the tape lying on the ground. Find the horizontal distance.

2-3. A steel tape was standardized and found to be 100.02 ft long at 68°F. The recorded length of a line measured with this tape at an average temperature of 22°F was 2814.28 ft. Find the correct length of the line.

2-4. In staking out a city subdivision it is necessary to establish a distance of 450.00 ft. The 100-ft steel tape is known to be 0.03 ft short at 68°F and the field temperature is 90°F. What nominal length will establish the desired distance?

2-5. A standardized steel tape has the following characteristics: length—100.01 ft at 68°F, supported throughout under a tension of 10 lb; weight—2 lb. A recorded distance of 3285.78 ft was measured with this tape on a 3% slope with the tape on the ground and at a temperature of 16°F. Compute the separate corrections for length of tape, slope, and temperature, and find the corrected length of the line.

2-6. While engaged in a forest survey, you measured the distance between two section corners and found it to be 81.76 chains. The steel chain is correct length at 70°F but the survey was conducted at 95°F. Calculate the corrected distance to the nearest 0.1 ft.

2-7. What error (nearest 0.01 ft) results from having one end of a 100-ft tape
  (a) Too high by 4.2 ft?
  (b) Off line by 1.0 ft?

2-8. A steel tape weighing 1.75 lb has a length of 100.00 ft under a pull of 10 lb while supported throughout its length. Calculate the effect of sag (nearest 0.01 ft) for a tension of 20 lb when supported at the end marks.

2-9. How much error (nearest 0.01 ft) is introduced in effecting the layout of the 1000-ft side of a large mill building if the nominal length of the 100-ft steel tape is considered as its true length? The field temperature is $-15°F$ and the tape was actually 99.95 ft long at 68°F.

2-10. A distance was measured along a uniform slope with the tape supported throughout its length. If the recorded distance is 814.23 ft and the slope angle is $5°20'$, what is the horizontal distance?

2-11. The distance between two established points in a desert was measured at a temperature of 105°F with a 100-ft steel tape that was true length at 68°F and was recorded as 1920.10 ft.
  (a) What is the corrected distance?
  (b) What would probably have been the observed distance if measured at 25°F?

2-12. A 100-ft steel tape weighs 1.10 lb and has a cross-sectional area of 0.0030 sq in. At 68°F when supported at the ends only under a pull of 15 lb, it is 100.002 ft long. Calculate the distance between the end marks for a pull of 20 lb at a temperature of 80°F with the manner of support remaining the same.

2-13. What error (nearest 0.01 ft) results from having one end of a 300-ft tape
  (a) Too low by 5.1 ft?
  (b) Off line by 1.5 ft?

2-14. A 50-meter steel tape was known to be 49.987 meters long at 68°F. If the recorded length of a line measured with this tape at $-10°F$ was 987.436 meters, find the corrected length.

2-15. What nominal distance (nearest 0.01 ft) should be measured to establish a 100-yd straightaway if the 50-ft steel tape is 50.05 ft long at 68°F and it is used at a field temperature of 88°F.

REFERENCES

1. Judson, Lewis V., *Weights and Measures Standards of the United States,* NBS Miscellaneous Publication 247, U.S. Government Printing Office, Washington, D.C., 1963.
2. Thomas, Paul D., "Linear Measures in the Evolution of the Mile," *Journal of the Coast and Geodetic Survey,* No. 4, pp. 12–22, Department of Commerce, Washington, D.C., December 1951.

3 Angular Measurements

3-1. Introduction Fundamentally, the purpose of surveying is to determine the relative locations of points on or near the surface of the earth. In order to fix the position of a point, both distance and angular measurements are usually required. Such angular measurements are either horizontal or vertical, and they are most commonly accomplished with instruments of similar design called *transits* or *theodolites*.

Although the ancients developed devices for angular measurements, it seems probable that it was not until 1571 that the concept of the modern transit was evolved. In that year Thomas Digges, an English mathematician and surveyor, published one of the earliest treatises on surveying in which he described his "Topographical Instrument," which was a forerunner of today's engineers transit.

Prior to 1800 practically all surveying instruments were brought to America from France and England. As the frontier moved westward, the demand for surveying services greatly increased and equipment soon was in short supply. The high prices of European equipment and the long delay in delivery were prime factors stimulating American manufacture of such instruments. The semicircumferentor, depicted in Fig. 1-2, became the principal surveying instrument in the early years of the new nation. David Rittenhouse, who was born in Germantown, Pennsylvania, in 1732 and became an accomplished astronomer and surveyor at an early age, is reputed to have made the first surveying telescope in this country and, independent of European practice, was the first to have used lines of spider web in the focal plane of the telescope. A compass bearing his name and made for George Washington can be found in the Smithsonian Institution.

Various claims have been made as to the origin of the first American transit but it seems quite likely that both William J. Young and

Edmund Draper should be credited, since available records indicate each produced independently a transit instrument in approximately 1830.

The scope of this chapter includes an introduction to the basic concepts of angles and directions and a detailed treatment of the transit and theodolite which are the principal instruments for measuring angles. Other devices occasionally used under rather special circumstances include the sextant, clinometer, and plane table. Small hand-held angle mirrors and rectangular prisms are sometimes used for the approximate establishment of 90° angles in the field.

GENERAL

3-2. Units of Angular Measurement An angle between two lines at a point is given by the difference in the directions of the lines. Only plane angles are considered here. The magnitude of an angle can be expressed in different units, all of which are basically derived from the division of the circumference of a circle in various ways. The principal systems of units are as follows:

1. *Sexagesimal System.* The circumference is divided into 360 parts. The basic unit is the degree (°), which is further subdivided into 60 minutes (60'), and the minute is subdivided into 60 seconds (60''). This system is used exclusively in surveying practice in the United States.

2. *Centesimal System.* The circumference is divided into 400 parts called *grads*. Hence $100^g = 90°$. The grad is divided into 100 centesimal minutes (100^c) and a centesimal minute is divided into 100 centesimal seconds (100^{cc}). The centesimal system has wide usage in Europe.

3. The *mil* is 1/6400 part of a circumference of a circle and will subtend very nearly one linear unit in a distance of 1000 such units. It is used in military operations.

4. The *radian* is the angle at the center of a circle subtended by an arc having exactly the same length as the radius. One radian equals $360°/2\pi$ or approximately 57.30°. Radians can be employed for certain calculations, such as determining the length of circular

arcs. The radian, in contrast with the other units of circular measure previously mentioned, is sometimes referred to as the natural unit of angle because there is no arbitrary number, like 360, in its definition.

3-3. Horizontal Angles Various kinds of horizontal angles can be used to express difference of direction. In the case of a closed figure survey along the perimeter of a tract of land, *interior angles* are frequently measured. When an open traverse is executed, two types of angles can be employed. Figure 3-1 depicts the *deflection*

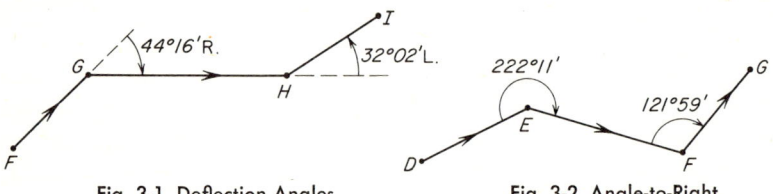

Fig. 3-1. Deflection Angles.   Fig. 3-2. Angle-to-Right.

angle. This is the angle that any line makes with the prolongation of the preceding line. Such angles must be identified as right or left to express whether the angle is turned to the right (clockwise) or to the left (counterclockwise) from the preceding line produced.

Figure 3-2 shows the *angle-to-right*. This is the angle turned clockwise from the back line to the forward line.

3-4. Bearings and Azimuths. It is frequently convenient to choose or fix a reference line to which the directions of all the lines of a survey are referred. Such a reference line is called a *meridian*, of which there are four kinds: *magnetic, true, grid,* and *assumed*.

A *magnetic* meridian is the direction in the horizontal plane taken by a magnetized needle when it comes to rest in the earth's magnetic field.

A *true* meridian is that meridian through a given point joining the north and south poles of the earth's axis.

A *grid* meridian is a line parallel to the central meridian or "Y" axis of a system of plane-rectangular coordinates.

An *assumed* meridian is a direction chosen by considerations of convenience for any particular survey or locality.

The acute angle which a line makes with a reference meridian is called its *bearing*, and since for any given line there is more than one such angle, this term must be further defined.

It is customary to refer directions with respect to the magnetic meridian to both the north and the south ends of the meridian and also both to the west and the east. Thus, in Fig. 3-3 the magnetic

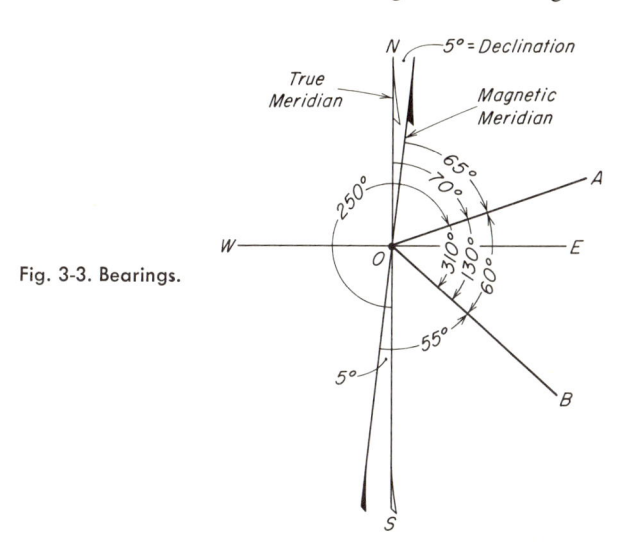

Fig. 3-3. Bearings.

bearing of *OA* is given as N. 65° E., and of *OB* as S. 55° E., and accordingly the magnitude of a bearing is never greater than 90°.

In the figure it is assumed that the magnetic meridian makes an angle of 5° with the true meridian, and this angle is called the *declination* of the needle. It is evident, then, that the true bearing of line *OA* is N. 70° E., but a more common way of designating direction with respect to the true meridian is simply to give the value of the clockwise angle which the line makes with the north end of the meridian, and this angle is called the line's *azimuth*. Thus, the true azimuth of the line *OA* is 70°, and there is no need for the use of letters referring to points of the compass. The magnitudes of azimuths vary from 0° to 360°; hence the azimuth of *OB* is 130°. Sometimes azimuths are referred to the south end of the meridian, in which case the azimuths of *OA* and *OB* are 250° and 310°, respectively. But for any given survey, azimuths are referred to but one end of the meridian only. *In this book, unless otherwise stated, the term azimuth will refer to the north end of the meridian.*

From the above discussion we have the following definitions:

The *declination* of the needle is the angle that it makes with the true meridian. The declination is either east or west, depending upon whether the north end of the needle points to the east or west of the true meridian.

The *magnetic bearing* of a line is the acute angle that a line makes with the magnetic meridian.

The *grid bearing* of a line is the acute angle that the line makes with the grid meridian.

The *true bearing* of a line is the acute angle that the line makes with the true meridian.

The *azimuth* of a line is the clockwise angle that a line makes with the north end of the selected meridian.

Every line has two directions, differing from each other by 180°, and depending on the point of view of the observer, i.e., at which end of the line he is stationed. Thus the magnetic bearing of OA, with the observer at O, is N. 65° E.; but at A, the bearing is S. 65° W. This value is termed the back bearing of OA. Likewise, the true azimuth of OB is 130° and its back azimuth is $130° + 180° = 310°$.

3-5. Vertical Angles A vertical angle is the angle between two intersecting lines in a vertical plane. In surveying practice it is ordinarily implied that one of these lines is horizontal, and a vertical angle to a point is the angle in a vertical plane between a line to the point and the horizontal. When the point sighted is above the horizontal plane, the vertical angle is called an *angle of elevation* and is considered a positive angle. When the point sighted is below the horizontal plane, the angle is termed an *angle of depression* and is considered a negative angle.

3-6. Zenith Distances With increasing frequency in surveying operations involving the use of the theodolite the reference line flanking the vertical angle is the vertical defined by the plumb line extending upward to the zenith. This vertical angle is termed the *zenith distance*. An angle of elevation of 20° would be equivalent to a zenith distance of 70°. A vertical angle of $-20°$ would equal a zenith distance of 110°.

3-7. Tripods A very common feature of several kinds of surveying instruments is the tripod. This is essentially a three-legged stand for supporting the instrument and maintaining its stability during observations. It consists of an upper element or head to which the instrument it attached, three metal or wooden legs hinged at the head, and pointed metal shoes on the legs which can be pressed into the ground by pushing on spurs situated at the lower ends of the legs. The instrument may be screwed to the tripod head or it may be secured to it by a special threaded bolt.

Tripods are of two general types—the *extension leg tripod* and the *fixed leg tripod*. Both types are available in a special design called *wide frame* (see Fig. 1-6), which has the advantages of minimizing wind vibration and of promoting torsional rigidity.

Great care should always be taken to secure a stable setup of the tripod in order to assure the safety of the instrument mounted on it. On level terrain this is accomplished by having each leg form an angle of approximately 60° with the ground. When pressing upon the tripod shoes, especially when the setup is over hard soil, caution must be exercised to apply the pressure parallel to the legs and not vertically in order to avoid breakage of the legs. If the setup is over a hard smooth surface, the tripod leg hinges should be more snugly tightened and the tripod shoes should be placed, if at all possible, in cracks to prevent the tripod from collapsing. When setting up on sloping ground, stability will be increased by having one leg pointed uphill. If the tripod is being transported or is in storage, its head should always be covered with the leather or metal cap to protect the threads.

The foregoing comments, although aimed immediately to the use of the tripod for supporting the transit and theodolite, apply equally well to the tripod when bearing a level, a subtense bar, a target, or any other surveying device.

ENGINEER'S TRANSIT

3-8. General Characteristics Angle measuring instruments employed in surveying are classified into two broad categories—*transits* and *theodolites*. Although it is true that certain advantages of the theodolite have caused it partially to replace the vernier-reading transit, the latter is still very widely and effectively used in many phases of American surveying operations. Because of its versatility and capabilities, the engineer's transit has been frequently called the universal surveying instrument. Although it is used principally to measure horizontal angles and prolong straight lines, it may be employed to measure vertical angles and stadia distances, observe celestial bodies for direction determination, and ascertain differences of elevation.

A typical transit is shown in Fig. 3-4. It consists of three major parts, viz., the *leveling head,* the *lower plate,* and the *upper plate.*

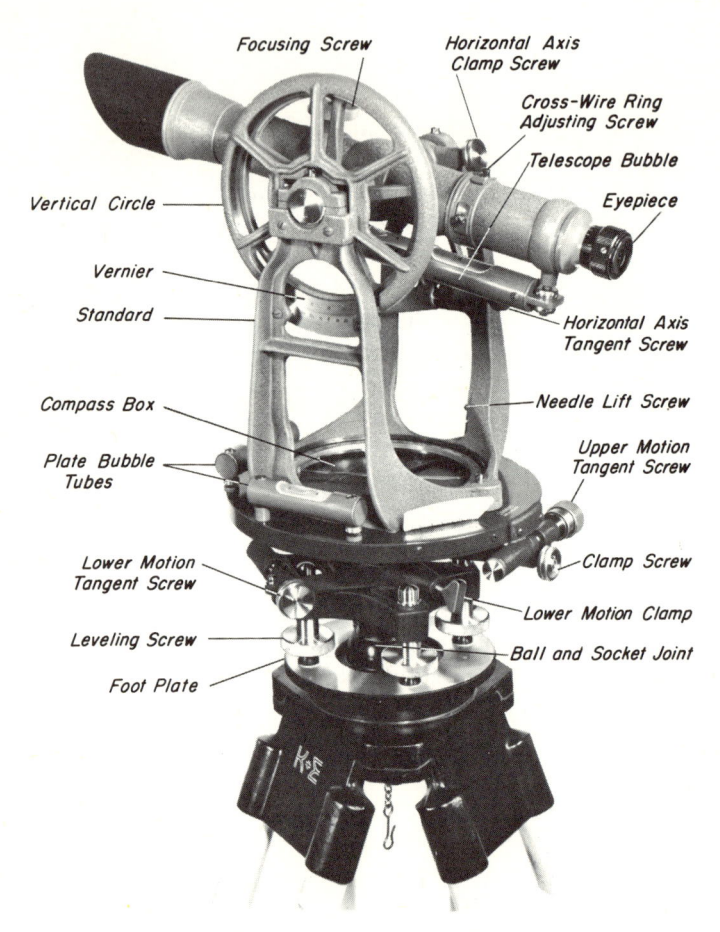

Fig. 3-4. Engineer's Transit. *Keuffel & Esser Co.*

(*a*) *Leveling Head.* The leveling head is the assembly that supports the instrument on the tripod and provides a means for leveling the instrument. The number of leveling screws, either three or four, used to level the instrument provides the basis for denoting the head as the three-screw or four-screw type. The four-screw head is usually used with the engineer's transit and is so constructed as to permit the instrument to be shifted laterally on the foot plate in order to accomplish centering over a specific point on the ground.

(*b*) *Lower Plate.* The lower plate consists of a hollow spindle that is accurately fitted to a socket in the leveling head and carries the graduated horizontal circle. The rotation of the lower plate is controlled by a *clamp screw,* which provides a means for locking it in place. A slow motion *tangent screw* permits the lower plate to be rotated a small amount relative to the leveling head.

(*c*) *Upper Plate.* The upper plate consists of a spindle that is attached to a circular plate bearing the verniers, the standards that support the telescope, the plate bubble tubes, and usually a magnetic compass. This spindle coincides with the socket in the lower plate spindle and the two are held together by a clamp screw that may be loosened to permit movement of the upper plate relative to the lower. A small rotation of the upper plate is effected by the use of a tangent screw.

The engineer's transit has two verniers, A and B, situated 180° apart, which are used to measure horizontal angles. The A vernier is located beneath the eyepiece of the telescope when the latter is in the normal position.

3-9. Definition of Terms A few terms commonly used in connection with transit work may be defined as follows:

Orientation. As applied to transit work, this term refers to the fixed position of the horizontal plate with respect to an established line through the instrument station. Thus, the transit is said to be oriented with respect to a given line when, with the line of sight directed along it, the vernier reads a given angle.

Backsight. A backsight in transit work refers to a sight taken on a point, usually the last preceding, so as properly to orient the transit.

Foresight. A foresight with the transit is a sight taken to fix the direction of a line.

Normal or Direct Position. The normal or direct position of the transit is that in which the eyepiece is above the *A* vernier and the attached bubble tube is below the telescope.

Inverted Position. The inverted position of a transit is that in which the attached bubble tube is above the telescope and the eyepiece is over the *B* vernier.

To reverse the Instrument. To reverse the instrument means to rotate the upper plate, including the standards and telescope, approximately 180° about the vertical axis.

Horizontal Axis. The horizontal axis, accurately speaking, is the imaginary axis of the trunnion that supports the telescope. However, the trunnion itself is commonly spoken of as the horizontal axis.

Vertical Axis. The vertical axis is the center line of the inner spindle.

Upper Motion. The part of the instrument that rotates on the inner spindle and includes the upper plate, the standards, and the telescope is commonly called the *upper motion.* It is controlled by the upper-motion clamp and tangent screw.

Lower Motion. The part of the instrument that rotates on the outer spindle and includes the graduated circle is commonly called the *lower motion.* It is controlled by the lower-motion clamp and tangent screw.

3-10. Surveying Telescope The surveying telescope serves the two purposes of fixing the direction of the line of sight and of magnifying the apparent size of objects in the field of view, and its proper use requires a brief description of its essential parts.

The *line of sight* is the line fixed by the intersection of the cross-wires and the optical center of the objective lens.

The principal features of the surveying telescope are the objective lens, the cross-wires, and the eyepiece. These parts and their relations to each other are illustrated in Fig. 3-5.

[3-10] ANGULAR MEASUREMENTS 55

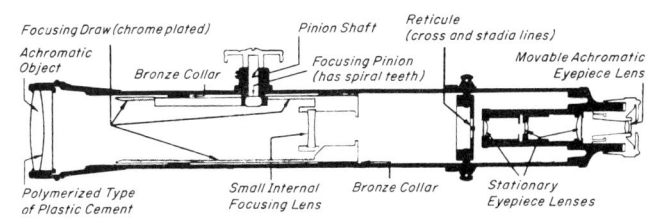

Fig. 3-5. Surveying Telescope. Keuffel & Esser Co.

The Objective Lens. The objective lens forms an image of any object within its field of view. This image is a real image and lies in a plane, within the telescope, at a distance from the lens which depends on the curvature of the lens surfaces and the distance to the object. The distance from the lens to the image is called the focal distance f, for that particular object. If the object is at a great distance, the image will be formed at a distance from the lens called the *principal focal length F*. These focal lengths will, of course, be different for different telescopes. Thus, the objective lens may be compared with the lens of a camera which forms an image at a definite distance from the lens, i.e., upon the photographic plate or film. Since this distance varies for different objects, a means of changing the focal distance, i.e., of *focusing* the lens, is provided on the telescope by the focusing screw shown in the illustration.

The use of an achromatic objective and eyepiece will correct lens aberrations that cause imperfect images with edges fringed with rainbow colors.

The Eyepiece. The image that is formed by the objective lens is inverted and small in size. A system of eyepiece lenses is used, therefore, to magnify the image and, in many telescopes, to re-invert or to yield an erect image of the object. The eyepiece, then, may be thought of as a microscope with which to view the image formed by the objective lens. An image may be viewed through the eyepiece only when it lies in the focal plane of the eyepiece; and, accordingly, a small focusing adjustment is also provided for the eyepiece. An image which has thus been brought into the common focal plane of both the objective and eyepiece lenses will appear to be magnified and distinct.

When any distant object is viewed through the telescope, the crosswires will, at the same time, appear sharply only when they lie in the common focal plane of the objective and eyepiece lenses. If the plane of the cross-wires is very close to, but not coincident with, the common focal plane of the lenses, the cross-wires may appear to be quite distinct, but they will seem to move about on the object with the slightest movement of the eye of the observer. A similar phenomenon is the apparent movement of the window sash with respect to any out-of-door object if the observer moves his head slightly. This condition in the telescope is called *parallax* and is to be prevented by careful focusing of the eyepiece on the cross-wires before viewing any distant objects.

An eyepiece yields a direct or an inverted image, depending on the arrangement of the lenses. An inverting telescope is superior in its optical properties and is preferred by many engineers. However, the inverted image causes some confusion at first and may easily cause mistakes until the engineer has become thoroughly familiar with its use. Accordingly, an instrument to be used by one person only may have either type of eyepiece, but where two or more instruments are to be interchangeable, they should all be equipped with the same kind, preferably the erecting type.

Figure 3-5 is a typical surveying telescope of the internal focusing type with erecting eyepiece. The objective is fixed at the far end of the telescope and focusing is accomplished through the use of the small internal focusing lens, which moves in response to rotation of the knob on the pinion shaft. Two major advantages of the internal-focusing telescope are (1) the interior of the telescope is virtually free from moisture and dust because both ends of the telescope are closed, and (2) an instrumental constant, which is relevant to stadia observations (Chap. 4) is practically eliminated.

The Cross-Wires. From what has been said, it is evident that, when the image of any object is seen plainly, it lies in the common focal plane of both the objective and the eyepiece lenses; also, that the position of this common plane can be altered by moving (focusing) the eyepiece. Now, if cross-wires are placed in this common focal plane, they will appear to be projected upon the object viewed and will serve to fix the line of sight upon any point of the object. This condition will be effected if the cross-wires are fixed in a stationary position and if the eyepiece is then focused upon them, and, finally, if the image of the object to be viewed is brought into the common focal plane by focusing the objective lens.

For proper use of the telescope, the eyepiece must first be focused upon the cross-wires and then the objective lens be focused upon the object to be observed. Since the position of the cross-wires is fixed, it will be necessary to focus the eyepiece only once for the day's work, unless the position of the eyepiece somehow becomes altered.

The manner of mounting the cross-wires is illustrated in Fig. 3-6.

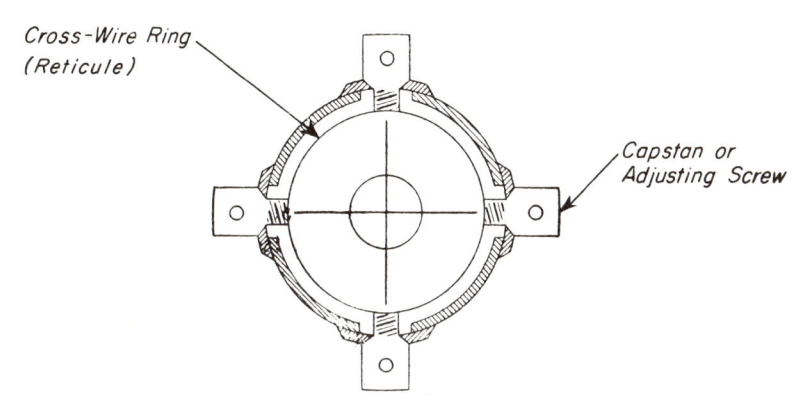

Fig. 3-6. Cross-Wire Ring.

The cross-wires consist of finely drawn platinum wire, glued in position upon a heavy brass ring. Also, lines may be etched on a glass diaphragm. Four threaded holes are drilled into the edge of the ring to receive capstan headed screws that are inserted through slots in the barrel of the telescope tube. The heads of these screws bear against curved washers. By this arrangement the cross-hair ring is held suspended by the capstan screws and within the telescope tube. It is thus held securely and firmly in place, but is subject, as occasion arises, to a small amount of lateral movement by turning the capstan screws. This movement is used for purposes of adjustment, to be described later, and for this reason they are commonly called adjusting screws. Some reticule designs incorporate two additional horizontal cross-wires that are situated equal distances above and below the main horizontal cross-wire. They are called *stadia lines*.

Magnification. The magnification of a telescope is fixed by the ratio of the focal lengths of the objective and the eyepiece lenses. It can be determined closely for any telescope by viewing, at a close range, a graduated rod with one eye looking through the telescope and with the other naked eye. Thus two images are seen, one being the magnified and the other the natural size of the rod. By a suitable adjustment of these images the observer may count the number of divisions on the unmagnified rod which is covered by one of the magnified divisions. This number is the measure of the magnification of the telescope, expressed as diameters. The magnification for telescopes for transits varies from 15 to 30 diameters, and for levels from 25 to 40 diameters.

High magnification, beyond proper limits, is a disadvantage, because it limits the field of view and reduces the illumination or brightness of the objects viewed. Accordingly, the size of the aperture of the objective lens and the qualities of the lens system are as important considerations in a good telescope as is the magnification.

The *field of view* of a telescope refers to the angle at the eye of the observer subtended by the arc whose magnitude is the width of the field viewed through the telescope. The field of view varies from approximately 1°30′ for a magnifying power of 20 to 45′ for a magnification of 40.

The *illumination* of the image depends upon the magnifying power, the quality and number of the lenses, and the effective size of the objective. The use of coated lenses substantially increases the amount of light transmitted through the telescope. The advantages are (1) a sharper and brighter image, and (2) the elimination of stray light produced by internal reflections. Coated lenses are easily recognized by their purple color.

Care of Telescope. The optical system of a telescope is particularly vulnerable to severe shocks that may disturb the lens alinement and to dust and dirt that may not only accumulate on the exposed metal surfaces but also penetrate into the interior of the tube. Ordinarily, the only parts of the telescope visible to the instrumentman will be the faces of the objective and eyepiece lenses, the focusing devices such as knobs and knurled rings, and the capstan heads of the reticule adjusting screws, because the telescope is never dismantled in the field.

The proper care of the lenses is especially important. Optical glass is not particularly hard and may easily be scratched. Dust should be carefully brushed away with a clean camel's-hair brush, and it is preferable to hold the lens downward when brushing so that any loosened dirt will fall away. Fingerprints on lens surfaces should be avoided because the oil transmitted from the human skin to the glass will tend to retain dust. If further cleaning is necessary, use special lens-cleaning tissue which can be lightly wadded and gently passed over the lens surface. Rubbing must be avoided. Lens-cleaning fluid, obtainable from an optician, can be used with lens tissue for removal of dust not yielding to the first treatment. Usually, breathing on the lens will provide sufficient moisture for ordinary cleaning. Acids, alcohols, or other solvents must not be used. When dust has worked its way into the interior of the telescope, it is necessary to have the instrument serviced at the factory.

3-11. Bubble Tube The engineer's transit and various other surveying instruments are equipped with one or more level or bubble tubes to define the horizontal plane.

The bubble tube is a glass vial, the inside surface of which is ground accurately to a curved surface so that a longitudinal section shows, on the upper half, a circular arc of long radius. The inside of the tube is nearly filled with ether or some nonfreezing liquid, the remaining volume being a vapor space called the bubble. The buoyancy of the liquid lifts the bubble to a position symmetrical with the highest point in the tube, and since this point is on the arc of a vertical circle, the tangent at that point (whether it be at the mid-point of the tube or not) will be truly horizontal and perpendicular to the direction of gravity. Now, if divisions are marked on the tube symmetrical with its mid-point, then when the bubble is centered, the tangent at the mid-point will always be a horizontal line and parallel with the geometrical axis of the tube. Hence, this tangent at the mid-point of the tube is called the *axis of the bubble tube*. See Fig. 3-7.

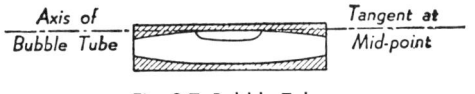

Fig. 3-7. Bubble Tube.

The *sensitivity* of the bubble tube is determined by the radius of the circular arc and is expressed by the seconds of arc subtended by one division of the bubble tube.

The sensitivity of bubble tubes varies over a considerable range and must be stated in terms of the length of a bubble tube division. Formerly, this was usually 0.1 in. Modern instrument design favors bubble division lengths of 2 mm. Ambiguity is eliminated by giving the radius of curvature of the bubble tube. For a transit of the general type depicted in Fig. 3-4, the sensitivity of both plate bubble tubes is commonly 70″ per 2 mm division and that of the telescope vial is 30″.

If the radius of the bubble tube is large, a small vertical movement of one end of the tube will be accompanied by a large displacement of the bubble; if the radius is small, the displacement will be small. For a given length of bubble division, the radius of the level vial will vary inversely with the sensitivity expressed in seconds. Furthermore, a 10″ bubble tube is twice as sensitive as a 20″ bubble tube. A simple field test to determine the sensitivity of the bubble tube of an engineer's level will be treated in Chapter 5.

Circular bubble vials are to be found on various kinds of surveying instruments. They are less sensitive than bubble tubes and are used for approximate leveling only.

3-12. Verniers The vernier is a device for reading subdivisions of a graduated scale. Such a scale may be rectilinear, as that of a level rod, or it may be circular, as that of a transit. The principle is the same in either case.

The Linear Scale. Figure 3-8a shows a decimal scale with a vernier alongside that is movable with respect to the scale. The vernier is divided into 10 parts, which cover 9 parts on the scale. Hence, the value of one vernier division is equal to 9/10 of one scale division; therefore, in reading upward from 6.00 on the scale, the first division on the vernier falls 1/10 of a division short of the first mark on the scale. Likewise, the second division mark on the vernier falls 2/10 of a division short of the second scale mark, etc.

It will be seen that, if the vernier is moved upward until the first mark on the vernier coincides with a scale mark, the zero, or index, of the vernier will have moved upward 1/10 of a scale division, in which position the vernier reading would be 6.01. Similarly, if the vernier is moved upward until the second mark on the vernier coincides with a scale, the index will have moved 2/10 of a division and the reading would be 6.02.

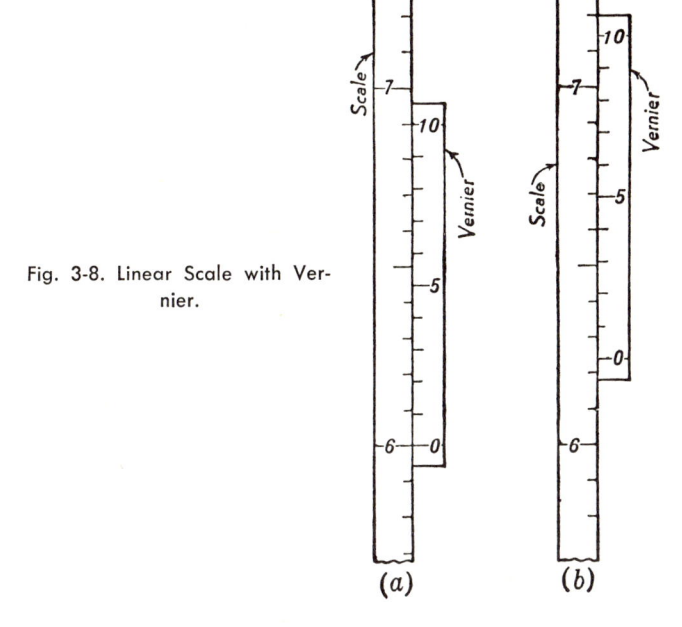

Fig. 3-8. Linear Scale with Vernier.

In Fig. 3-8b it is seen that the index of the vernier has been moved upward (above 6.00) past two scale divisions and that the fourth mark of the vernier coincides with a scale division. Accordingly, the reading of the vernier is 6.24.

Evidently, the smallest subdivision that can be read with this vernier is 1/10 of a scale division and is determined by the number of parts on the vernier. For example, if a vernier were arranged with 15 parts to cover 14 parts on the scale, the smallest subdivision that could be read would be 1/15 of a scale division. Hence, the general principle of all verniers may be stated thus: if n equals the number of vernier divisions, and if these n *vernier* divisions cover $(n - 1)$ scale divisions, the smallest subdivision of the scale that can be read with the vernier is equal to the value of a scale division divided by n. Or,

$$D_s = \frac{s}{n}. \tag{3-1}$$

in which D_s is the smallest subdivision of the scale that can be read, s is the value of a scale division, and n is the number of divisions on the vernier.

The application of the vernier to a circular scale is shown in Fig. 3-9. Two arrangements are shown, the vernier at (*a*) reading to

(a) GRADUATED 30 MINUTES READING TO ONE MINUTE DOUBLE DIRECT VERNIER

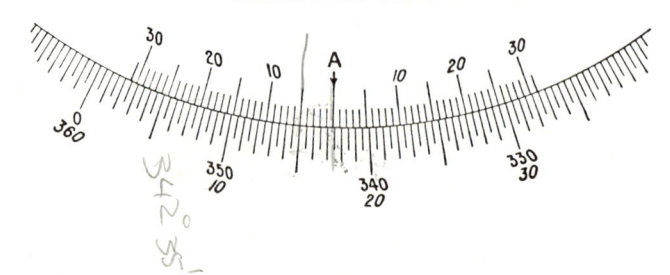

(b) GRADUATED 20 MINUTES READING TO 30 SECONDS DOUBLE DIRECT VERNIER

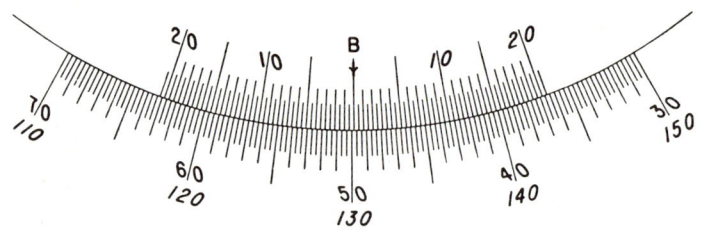

Fig. 3-9. Transit Verniers. Keuffel & Esser Co.

minutes and the one at (*b*) reading to half minutes, or thirty seconds. Each one is a double vernier, i.e., it is arranged to read in either direction, to the right or to the left, depending upon the direction of rotation in measuring the angle.

The fact that vernier (*a*) reads to minutes may be determined by Eq. 3-1:

$$D_s = \frac{s}{n} = \frac{30'}{30} = 1'$$

The value of an angle is found by first reading the scale, then the vernier, and then adding the two readings together.

Thus, in Fig. 3-9*a* the clockwise angle reading (inner row) is $342°30' + 05' = 342°35'$. The counterclockwise angle reading (outer row) is $17° + 25' = 17°25'$. Similarly, in Fig. 3-9*b* the clockwise angle reading (inner row) is $49°40' + 10'30'' = 49°50'30''$.

The counterclockwise angle (outer row) is 130°00′ + 9′30″ = 130°09′30″. It is important to note that the vernier is always read in the same direction as the scale. This relation is indicated by the slope of the numbers, both on the vernier and on the scale. Thus, the numbers that slope to the left on the scale and on the vernier are to be taken together.

Various arrangements of verniers will be found on the different instruments and the transitman should be careful to determine correctly the characteristics of each vernier before it is used.

In obtaining a vernier reading it is essential that the instrumentman have his eye directly over the line of coincidence. It will be helpful in identifying this line to note that the flanking lines will differ from coincidence by equal amounts and in opposite directions. Blunders can be avoided by making a rapid preliminary reading of the angle by estimating the fractional part of the circle division.

3-13. Measurement of Horizontal Angles Before a horizontal angle can be measured, it is necessary to make a stable setup of the tripod, center the instrument directly over the fixed ground point, and effect its leveling through the use of the plate bubbles. The wing nuts on the top of the tripod legs should be turned up snugly at the beginning of transit operations and left so. They should not be alternately tightened and loosened. It is assumed in this explanation that the transit is like that depicted in Fig. 3-4, i.e., it has a four-screw leveling head and two plate bubble tubes.

First of all, with the instrument "head" approximately centered on the foot plate and the leveling or foot screws equalized, the tripod legs are so positioned relative to the contour of the ground surface that the foot plate is nearly horizontal and the plumb bob is close to the survey point. Then, the departure from the point is noted and the transit is bodily lifted and shifted laterally until it is closer to the point, which is commonly marked by a tack in a stake. Pressing on the spurs of each tripod leg will serve to make the setup stable as well as aid in bringing the plumb bob closer to the tack. To effect the final centering, two adjacent foot screws should be loosened, the head shifted, and then the same two foot screws again turned, but in the opposite direction, until they bear snugly on the foot plate.

Leveling is now accomplished by rotating the instrument about its vertical axis until each plate bubble tube is approximately parallel to the line joining a pair of diagonally opposite leveling screws. Each plate bubble is separately centered by turning uniformly in opposite directions the foot screws that control it. It is to be noted that the bubble will move in the same direction as the left thumb. If, at any time, all the foot screws become so tight that an opposite pair of them cannot be turned simultaneously, a remedy usually can be effected by loosening one screw only. This will loosen the entire assembly and permit normal leveling procedures to be performed.

Centering the transit over the survey station is an important step preliminary to the measurement of the angle. The *plumb bob,* already mentioned in connection with taping, is commonly employed to accomplish centering. It usually has a body and removable cap made of brass weighing from 10–18 oz with a replaceable point made of hardened steel and a means for attaching a plumb line in a central position.

In the case of some transits and virtually all theodolites, an optical plumbing assembly, sometimes called an *optical plummet,* is an integral part of the instrument. It consists of a small prismatic telescope, with a small circle reticule, having its line of sight coincident with the vertical axis. After the transit is leveled, a sight through the plummet will enable the instrumentman to complete the centering procedure. The plummet is especially helpful in windy weather.

Another centering device consists of a telescoping metal rod with its upper end attached to the tripod head and with the lower pointed end placed precisely on the station mark. When the bubble of the attached circular level is centered, the rod is plumbed.

With the preliminaries of centering and leveling the instrument completed and with parallax removed by careful focusing of the eyepiece on the cross-wires, the measurement of the horizontal angle between two points, C and D, is performed by executing in a systematic manner the following steps:

1. Loosen the lower clamp. Using the upper clamp and tangent screw, set the A vernier at zero.
2. Using the lower clamp and tangent screw, point the vertical cross-wire precisely on point C.
3. Loosen the upper clamp and sight approximately at point D. Using the upper clamp and tangent screw, point the vertical wire precisely on point D.
4. Read the angle with the A vernier.

The mechanics of the foregoing procedure indicate that the clamps are used for approximate settings of the vernier and for rough pointings and the tangent screws fulfill the function of perfecting such settings and pointings. It is obvious that a tangent screw belonging to a particular clamp must be used with that clamp only.

The B vernier is used when greater angular accuracy is desired as explained later.

The pointings of the transit are made to various objects or targets which are placed concentrically over the distant points C and D. Ordinarily, sights are taken to range poles which are held plumb over the points. If sight distances are only a few hundred feet, a pencil may serve as a sharper target. A plumb line will prove to be a very accurately placed target but may be difficult to see unless a suitable background such as a white card is provided.

Upon the completion of transit operations at a given survey station, the head should be centered, the foot screws equalized, the upper or lower clamp loosened, and the telescope turned either up or down.

3-14. Repeating an Angle In order to obtain a more refined value of the angle than that imposed by the limitations of the vernier, two or more measures of the angle are accumulated on the horizontal circle and the final value is equal to the total angle turned through divided by the number of measures. For example, an angle whose true value is 30°21′26″ would be read on a single-minute transit as 30°21′. If the angle is doubled, the second reading (total angle turned) to the nearest minute, would be 60°43′. One-half the double angle or 30°21½′ agrees quite closely with the true value.

In order to repeat a horizontal angle, leave the upper and lower plates clamped together after measuring the first angle. Then, loosen the lower clamp, turn back to the initial point, and sight on it using lower clamp and tangent screw. Loosen the upper clamp, sight at the second target, and precisely place the vertical wire on it with the upper clamp and tangent screw. This operation can be repeated up to, say, five times, but practice with the engineer's vernier transit favors only one repetition or a doubling of the angle. Furthermore, the second backsight upon the initial target should be made with the telescope inverted in order to nullify the effects of nonadjustment of the instrument.

Doubling of the angle, including reading of the B vernier also, not only serves to increase the accuracy of angular measurement but provides an important safeguard against the introduction of blunders.

3-15. Measurement of Vertical Angles When the vertical angle to any point is desired, it is necessary to define the direction of a horizontal line and then measure the vertical angle between this line and that to the object. The direction of the horizontal line is fixed by the axis of the bubble tube attached to the telescope when that bubble is centered, and the vertical circle serves to measure angles in the vertical plane.

In order to measure a vertical angle, whether of elevation or depression, the transit is carefully set up and leveled. The telescope is pointed to the object and the middle horizontal cross-wire set exactly on the point using the vertical clamp and tangent screw. The vertical angle, with the proper prefix, is read and recorded. Without rotating the telescope in azimuth, the telescope is brought to the horizontal, the telescope bubble carefully centered, and the reading of the vertical vernier obtained. This angle, termed the *index reading*, will ordinarily be zero in the case of a properly adjusted and operated transit. The index reading should be recorded with the proper prefix.

To illustrate the application of the index reading to an observed vertical angle, assume that an angle of $+4°21'$ was recorded and that the index reading was found to be $+0°01'$. The corrected vertical angle is $+4°20'$. Also, suppose that a depression angle of $3°49'$ was recorded and the index reading was $+0°02'$. The corrected vertical angle is $-3°51'$.

If the vertical arc of the transit is a full circle and if the object is observed first with the telescope normal and then with the telescope inverted, the mean of the two readings will be the correct value of the vertical angle, the index error having thus been eliminated.

Vertical angles obviously cannot be repeated.

3-16. Adjustments of Transit Although the transit is an instrument of precision and is manufactured with great care, ordinary field usage will periodically require its testing and adjustment. Rough treatment, whether accidental or deliberate, will accelerate the need for such adjustment. Shop adjustments which can be performed only in the factory are not treated here. Only those adjustments that can be checked in the field are described. The instrumentman should be familiar with

1. The relationships that should exist between various geometric lines in a properly adjusted transit.
2. The procedure employed to test whether these relationships exist.
3. The manner of adjustment if the test reveals the relationship does not exist.
4. The effect of nonadjustment of the transit on field results.
5. The field procedures that will most nearly nullify the effects of nonadjustment.

The following testing and adjustment program should be undertaken under favorable atmospheric conditions, over terrain permitting solid setups, and preferably with the instrument in the shade. Adjusting pins that properly fit the capstan screws should be used.

1. Adjustment of the Plate-Bubble Tubes

Relation. The axis of each plate-bubble tube should lie in a plane perpendicular to the vertical axis.

Test. The transit is set up and carefully leveled with each bubble tube parallel with a diagonal pair of leveling screws. The plate is then rotated on its vertical axis until each tube is turned end for end over its pair of leveling screws. If the correct relations exist, each bubble will remain centered; but, if not, the bubbles will be displaced and the amount of the displacement will be double the error of adjustment because of the reversal of conditions which has been made.

Adjustment. If, upon reversal, a bubble is displaced, say four divisions, the adjustment is made by bringing it back two divisions by means of the capstan adjusting screw at the end of the bubble tube. The bubble is then centered with the foot screws and the test repeated for verification.

2. Adjustment of the Cross-Wire Ring

Relation. The vertical cross-wire should lie in a plane perpendicular to the horizontal axis.

Test. The transit is set up and the vertical cross-wire sighted on a definite point in the field of view. The telescope is then rotated slightly about its horizontal axis. If the correct relation exists, the cross-wire will apparently remain on the point. If not, the point will appear to move off the cross-wire as the telescope is rotated from top to bottom of the field of view.

Adjustment. To adjust the cross-wire ring, both pairs of capstan screws that hold the ring in position are loosened slightly so that it may be rotated by means of pressure of the fingers on the screws, or by tapping with a pencil, until the correct position has been obtained.

3. *Adjustment of the Line of Sight*

Relation. The line of sight should be perpendicular to the horizontal axis.

Test. Set the instrument up and level it carefully. Take a backsight on a point as A, Fig 3-10a, with the telescope in the normal position. Invert the telescope and set a point at D. Reverse the horizontal axis end for end, turning the plate about the vertical axis, and take a second backsight on A with the telescope in the inverted position, Fig. 3-10b. Reinvert the telescope and set a point at E.

The lack of perpendicularity between the line of sight and the horizontal axis is represented by angle α in the illustration; and it is evident that, upon inverting, the line of sight is deflected from the true prolongation of line AB by an angle equal to 2α. Accordingly, after reversal of the horizontal axis and inversion of the telescope the second time, the angle between the two foresights D and E is equal to 4α.

Adjustment. The adjustment is made by sighting point E, Fig. 3-10c, and then, by loosening one capstan screw of the horizontal pair and tightening the other, the vertical cross-wire is fixed on a point F which is set ¼ of the distance from E toward D.

After this adjustment the test is repeated; if the adjustment is perfect, the line of sight will fall on point C both before and after reversal of the horizontal axis.

ANGULAR MEASUREMENTS

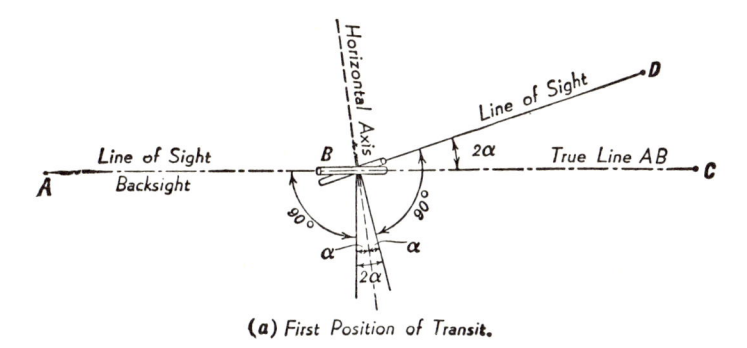

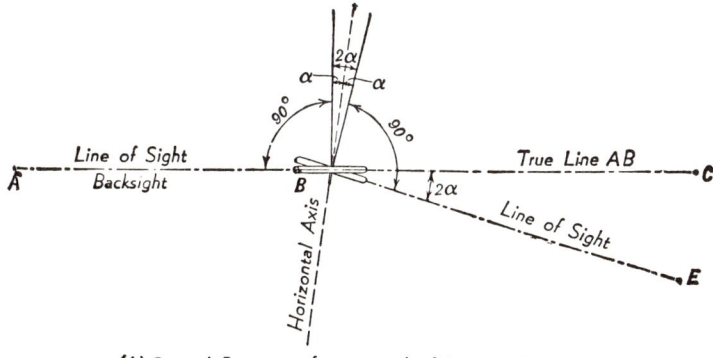

(a) First Position of Transit.

(b) Second Position, after reversal of Horizontal Axis.

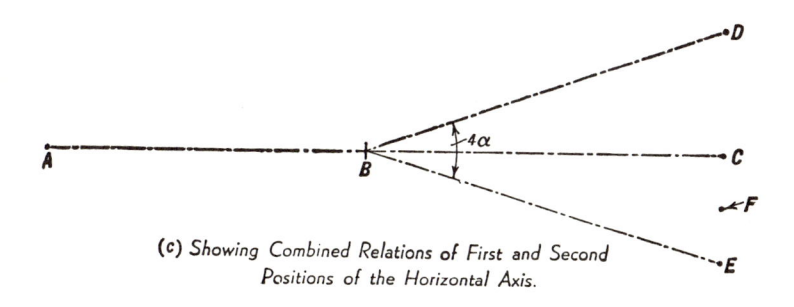

(c) Showing Combined Relations of First and Second Positions of the Horizontal Axis.

Fig. 3-10. Adjustment of Line of Sight.

4. *Adjustment of the Standards*

Relation. The horizontal axis should be perpendicular to the vertical axis.

Test. Set the transit near a building where a definite point can be sighted that requires the telescope to be elevated through a large vertical angle. Level the plate carefully and sight the elevated point. Depress the telescope and set point A near the ground. Reverse the horizontal axis end for end by turning the plate about the vertical axis, invert the telescope, and sight the elevated point again. Depress the telescope a second time; if the adjustment is perfect, the line of sight will fall on point A, previously set. If not, a second point B is set near the ground.

Adjustment. The adjustment is made by raising or lowering one end of the horizontal axis until, after repeated reversals, the line of sight falls on the same point near the ground. The horizontal axis rests in journals, and provision is made for raising or lowering one of them, usually by first loosening setscrews on top of the standards and then by turning a capstan screw under the journal and between the two legs of the standard.

There is no way of telling exactly how much the standard is to be adjusted, and it becomes a trial-and-error method until the proper adjustment has been made.

5. *Adjustment of the Telescope Bubble Tube*

Relation. The axis of the bubble tube attached to the telescope should be parallel to the line of sight.

Test. This is accomplished essentially in the same manner as the "peg test" for the engineer's level explained in Art. 5-7. An alternate procedure would begin with setting up the transit, leveling it with the plate bubbles, and driving a stake at a taped distance of 150 ft. With the bubble of the level tube attached to the telescope carefully centered, a reading to the nearest 0.01 ft would be obtained on a level rod held on the stake. Another stake would be set in the opposite direction and at the same distance from the instrument and be progressivly hammered into the ground until the reading on the rod now placed on it is exactly the same. Here again the bubble of the attached level tube must be carefully centered when the reading is obtained. It is now obvious that the tops of both stakes are at the same elevation even though the axis of the telescope bubble tube was inclined with the horizontal.

The transit is now set up close to either stake (minimum focusing distance is usually 5–10 ft) and a reading is obtained on the rod held on that stake. The rod is removed to the distant stake where the same reading must be obtained in order to define a horizontal line.

Adjustment. The adjustment is made by placing the middle horizontal wire at the correct reading on the distant rod by using the vertical clamp and tangent screw and then centering the bubble of the attached level tube by means of its capstan adjusting screws.

6. *Adjustment of the Vertical Arc*

When the plate bubbles and the bubble on the telescope are centered, the vertical arc should read zero. If adjustment is necessary, provision is made for moving the vernier until the index reads zero on the vertical circle.

3-17. Uses of Transit The engineer's transit has many uses but only a few of the basic operations performed with it will be described here. Its fundamental role in the execution of transit surveys will be treated in Chapter 6 and numerous applications of its capabilities to a wide variety of surveys will be discussed in later chapters.

Prolonging a Straight Line. One of the important uses of a transit is that of prolonging a straight line. Three or four methods are available but, for ordinary purposes, the best one is that called the *method of double sights* or *double centering,* which will now be explained.

It may be supposed that the two points A and B fix the direction of the line that is to be prolonged. The instrument is set up over B and a backsight is taken on A with the telescope in the normal position. The telescope is inverted and a temporary point C is marked on a stake set on the line of sight and at a distance as far as can conveniently be seen. The instrument is then reversed, and a second sight is taken on A with the telescope remaining in the inverted position. The telescope is then made direct and a second temporary point D is marked on the foresight stake. A point midway between C and D is now found and fixed as a permanent point on the true prolongation of line AB.

If the points C and D should happen to coincide, that is the permanent point sought.

Possibly the two temporary points will not fall on one stake. If not, two or more stakes may be required to complete the process, care being taken not to disturb a point that has once been set, and that the final point is properly fixed.

Wiggling-In. Occasionally it may be necessary to establish a point on a line that is defined by termini which cannot be occupied by a transit and are not intervisible because of some obstruction like a hill. The general procedure for placing the intermediate point B on line is indicated in Fig. 3-11. Initially, an estimate is made of the position

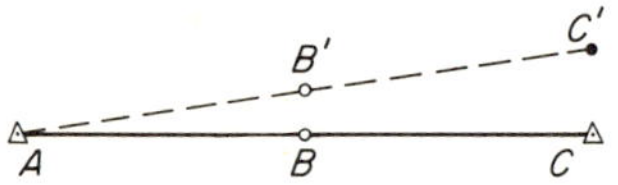

Fig. 3-11. Wiggling-In.

of the line and the transit is set up on the hill at point B' and a backsight taken to point A. Then, the telescope is inverted, a sight taken toward point C, the distance CC' measured, and the instrumentman estimates the approximate distance the transit must be moved laterally in order to place it on line. This trial-and-error procedure continues until the final movement of the instrument is effected by shifting the head on the foot plate. Furthermore, the final location of B should be checked by the method of double centering.

Sometimes the term "plunging the telescope" is employed in the literature to express the rotation of the transit telescope from the direct to the inverted position.

Random Line. When it is desired to connect with a straight line two distant points that are not intervisible, it is frequently accomplished by projecting from one point a straight line that is estimated to fall nearly upon the other point. When the line thus projected has been run out, its position relative to the distant point is measured, usually by a swing offset, and from these data the direction of a true line between the two points is calculated and established on the ground. The projected or trial line is called a *random* line. This procedure is much used in land surveying to establish intermediate points on a straight line between two corners.

Intersection of Two Lines. A problem frequently encountered, especially on route surveys, is that of finding the point of intersection of two lines. Figure 3-12 depicts the situation. The intersection of

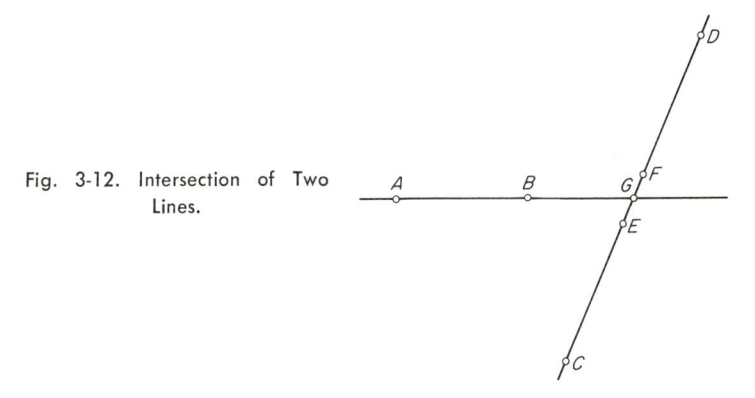

Fig. 3-12. Intersection of Two Lines.

lines *AB* and *CD* is to be found. The transit is set up on point *C*, a foresight taken on *D*, and stakes *E* and *F* are located in positions estimated to flank within a few feet the location of the point of intersection, *G*. A cord is stretched tightly between points *E* and *F*. With the transit now at point *A*, the foresight through *B* will intersect the stringline, *EF*, at point *G*. An obvious alternate to this procedure, provided two transits are available, is to occupy points *A* and *C*, take foresights to *B* and *D*, respectively, and progressively direct a tapeman, holding a range pole, to each line until the point *G* is reached.

Laying Out an Angle. When an angle of a given design magnitude is to be established in the field and the limitations of the vernier of the only available transit make it unlikely that the angle can be laid out with sufficient accuracy by turning a single angle, the following procedure can be employed. It may be supposed that angle *COD* is to be established in the field, that line *OC* is fixed, and that the transit is in position at *O*. The *A* vernier is set at 0°, the instrument is sighted on *C*, the upper motion is released, and angle *COD* is set with the *A* vernier. A temporary point D', which is approximately correct, is set on this line (Fig. 3-13).

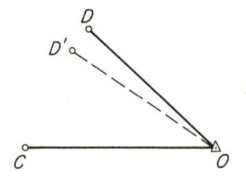

Fig. 3-13. Laying Out an Angle.

Now a more accurate value of the angle COD' is determined by the method of repetition involving the accumulation of six measures of the angle on the plates. One-sixth of the total angle through which the telescope was turned will yield this value.

The difference between COD' and COD is then a small angle to be established by a linear measurement from D', perpendicular to OD' and equal to the tangent of the angle DOD' times the distance OD.

3-18. Care of Transit Transits and similar surveying instruments are precision made and must be handled with care if quality results are to be obtained and costly repairs avoided. The following list of precautions and maintenance suggestions includes only the most important topics with which the instrumentman should be familiar.

1. When removing an instrument from its carrying case, observe how it was secured in the case so that it can be correctly returned.

2. Make certain the transit is properly fastened to the tripod but avoid overtightening.

3. When carrying the instrument in a building, through doorways, and beneath low-hanging trees, hold the transit waist high in a horizontal position with the tripod trailing the head. The clamp screws should be loosened so that shock due to impact with any solid object is lessened.

4. Never leave the instrument unattended when it is set up.

5. Turn snugly all clamp and leveling screws to a firm bearing. Do not "check" them for tightness by giving them an extra twist. Capstan and other adjusting screws are particularly susceptible to damage due to overstress.

6. Have a waterproof hood available in case of sudden rain.

7. Always use the sunshade. If it is to be removed or replaced, do so with a clockwise rotary action so that the objective lens is not loosened.

8. Between regular maintenance performed at the factory, the threads of leveling, clamp, and tangent screws should be cleaned and lightly lubricated.

9. The vertical circle and vernier should be cleaned with a chamois or very soft cloth. Do not touch the graduations with the fingers.

10. When returning an instrument to its carrying case, place the dust cap over the objective, center the head on the foot plate, and equalize the foot screws. If the carrying case does not close readily, find out what is wrong. Never use force to close it.

3-19. Errors in Angles The errors of angles measured with a transit can be classified into three major categories as already mentioned in Art. 1-12. Following the listing of the major sources of error in these groups, appropriate comments will be made as to the relative importance of the errors and the means that can be utilized to minimize them.

(*a*) *Instrumental Errors.* The most significant instrumental errors that may affect the quality of horizontal angles are those associated with nonadjustment of the instrument. Such errors are eliminated or reduced to negligible magnitudes by proper adjustment and by correct observational procedures. However, the effect of parallax and failure of the horizontal circle to be truly horizontal cannot be so eliminated. It is essential to focus carefully the eyepiece on the cross-wires and to keep the plate bubbles centered.

As regards the use of the transit, the measurement of vertical angles is of minor importance (except in topographic surveying), the principal uses being to measure horizontal angles and to prolong straight lines. Accordingly, it will be desirable to observe the effect of each of the adjustments on these principal uses.

It may be noted that the imperfect adjustment of the plate-bubble tubes and of the standards have the same effect, i.e., to incline the horizontal axis with respect to a truly horizontal position. In this condition, when the telescope is rotated about the horizontal axis, the line of sight traces an inclined instead of a vertical line. Hence, in measuring the horizontal angle between two points of some difference in elevation, when the telescope is raised or lowered, the line of sight will follow an inclined, instead of a vertical, line, and the amount of this inclination will be introduced as an error in the magnitude of the horizontal angle. On level ground these imperfect adjustments will have no appreciable effects.

In inverting the telescope when prolonging a straight line, it is evident that, as regards the adjustments under consideration, the same conditions obtain as when measuring horizontal angles. Therefore, the effects of the imperfect adjustments of the plate bubbles and the standards may be stated as follows: (1) they are not important either in measuring horizontal angles or in prolonging lines on level ground, but (2) they do have important effects on both uses of the transit when sighting between points of considerable difference in elevation.

If the line of sight is not perpendicular to the horizontal axis, the measurement of horizontal angles on level ground will not be affected, and the effect will be only slight when the points sighted are of different elevations. But the full effect is present whenever the telescope is inverted between a backsight and a foresight, as when prolonging a straight line.

It will be remembered that the process given for prolonging a straight line was that of double sights, with the instrument reversed between backsights. It is now evident that this procedure is used to eliminate any effect of the line of sight not being perpendicular to the horizontal axis, for, if the line of sight is deflected to the right when first inverted, it will be deflected to the left when inverted the second time, and the mean of the two sights will be free from error.

It may be added that although the errors resulting from imperfect adjustments are systematic, they are, in general, rendered accidental in the process of taking backsights and foresights; further, most of the instrumental errors will be eliminated by the use of double sights with the instrument reversed between them.

(*b*) *Personal Errors.* These errors are minimized by proper training and careful adherence to specifically prescribed field instructions. They are

1. Eccentricity in setting up instrument over occupied station.
2. Eccentricity in erecting target over distant station.
3. Failure to plumb target.
4. Faulty pointings to targets.
5. Faulty reading of circle and verniers.

6. Failure to achieve or recognize coincidence when reading circle and vernier.

Brief comments will be made on the foregoing. The error introduced by not plumbing exactly over a point is similar in character and magnitude to that associated with failure to point on the signal or target. A rule of thumb worth remembering is that a chord 1 in. long will subtend an angle of very nearly one minute of arc at a distance of 300 ft. In windy weather an optical plummet is a very desirable feature of a transit or theodolite, but it is essential that the instrumentman realize that the circular or bull's-eye bubble must be centered if the plummet is to produce correct results.

The signal, range pole or otherwise, must be just as carefully centered over the distant survey station as the transit is positioned over the occupied point. Furthermore, the signal must be plumb. Unless these two conditions are satisfied, bisection of the target, no matter how carefully performed, will not give a correct direction. Furthermore, the quality of the measured angle, even though the vernier can be read to a fraction of a minute, will be adversely affected.

Errors of achieving coincidence and reading verniers will be minimized by using both A and B verniers, employing a magnifying glass, and doubling the angle.

Occasionally, the transitman may inherently make pointings consistently a trifle to the left or to the right. It is apparent that the effect of such erroneous pointings (and also of eccentric setups of the instrument) will be most significant for short sight distances. Accordingly, special attention should be given to centering the instrument, the signal, and checking the verticality of the latter when short sights are involved.

(c) *Natural Errors.* These errors are associated with environmental conditions. They include the following:

1. Instability of instrumental support due to soft or yielding ground.
2. Atmospheric conditions giving rise to temperature effects on the instrument and to horizontal refraction.

Swampy or thawing ground will make it impossible to obtain good results with a transit unless special instrumental supports are provided. Wood posts or heavy stakes (2 in. × 4 in.) driven to a sound bearing will prove very useful. Differential temperatures in the instrument giving rise to unequal expansions may be controlled by shading the instrument if very intense sunlight prevails.

Horizontal refraction refers to the bending in a horizontal plane of the sight line between the transit and target. This effect can be most pronounced if the line of sight passes close to buildings that are radiating heat.

3-20. Mistakes The mistakes commonly made in the use of the transit are as follows:

1. *Misreading the Vernier,* which refers to forgetting to add the complete scale reading to the vernier reading; e.g., if the index reads 13°30′ and the vernier reads 06′, the total reading is 13°36′, but is sometimes mistakenly read as 13°06′.

2. *Reading the Wrong Vernier.* This refers to the mistake of reading the right-hand vernier when the left-hand one should be used.

3. *Reading the Wrong Circle.* When the magnitude of an angle is near 180° it is a frequent mistake to read the wrong circle, i.e., the inner instead of the outer circle, or vice versa.

4. *Reading the Circle Incorrectly.* The circle may be read incorrectly, especially for values near the 10° divisions, e.g., 59° is misread as 61°, etc.

5. *Using the Wrong Tangent Screw.* A frequent mistake with beginners is to use the wrong tangent motion, i.e., the upper tangent screw with the lower clamp, or vice versa.

6. *Recording.* This refers to the usual mistakes of recording, such as transposing numbers, etc.

OPTICAL THEODOLITES

3-21. Definitions The term, theodolite, originally referred to any precision angle-measuring surveying instrument consisting of a high-quality telescope and an accurately graduated circle, and equipped with the necessary levels and reading devices. There are two general classes of theodolites—*direction theodolites* and *repeating theodolites*. Frequently, they are called *direction instruments* and *repeating instruments,* respectively.

The horizontal circle of a direction theodolite remains fixed during a series of observations. The telescope is pointed on a number of signals in succession, the direction of each object is read on the circle, and the horizontal angles are calculated as the differences of successive directions. The initial direction is arbitrary. The best grades of direction instruments are used almost exclusively on higher-order, horizontal control surveys as described in Chapter 9. They are not suitable for engineering layout work.

A repeating theodolite is designed so that successive measures of a horizontal angle may be accumulated on the graduated circle. The final value of the angle is obtained by dividing the total arc passed through by the number of times the angle was measured. Although the repeating instrument is an instrument of precision, its use is secondary to that of the direction instrument if the highest accuracy is desired.

Former designs of both types of instruments required the use of verniers or micrometer microscopes for reading the circle graduations. Modern theodolites of all qualities are most notably characterized by an optical-reading system that permits the observer to obtain readings of both horizontal and vertical circles through an eyepiece situated beside the telescope. Such optical reading theodolites are of many types and designs. They are manufactured by several European firms with which American instrument companies cannot economically compete because of wide differences in labor costs. The use of optical theodolites, also called *optical transits,* is increasing greatly in the United States where they are employed, particularly the repeating type, not only on precise surveys but also on construction and other surveys of ordinary quality.

3-22. General Characteristics Although European optical theodolites differ somewhat in certain features of design, all of them possess the following general characteristics:
 1. Compared with the American vernier transit the head or alidade is compact, lightweight, and of reduced height. The weight (without tripod) of the American transit is approximately 16–18 lb. Its European counterpart will weigh about 10 lb.

2. The telescope is short, has a large objective, is usually inverting, has sight lines etched on glass, and frequently has rifle sights mounted on top of the telescope for preliminary pointing. It is of the internal-focusing type so that in stadia work (Chapter 4) the quantity $(F + c)$ is zero and the multiplying factor is 100. For the American market a telescope of the erecting type is available.

3. Three leveling screws support the instrument.

4. It is equipped with an optical plummet.

5. The horizontal and vertical circles are made of glass with etched graduations and numerals.

6. The circles and all optical and mechanical systems are completely enclosed so that the instrument is moistureproof and dustproof.

7. The instrument has a circular level, a single tubular level in the plane of the horizontal circle, and an index level attached to the vertical circle.

8. All readings, both of horizontal and vertical circles, and observations of the bubbles can be made from the eye end of the telescope. There is no need for the instrumentman to walk about the theodolite in order to obtain the readings.

9. Usually the upper part of the head can be separated and lifted out from the centering and leveling base which is called the *tribrach*. This permits easy interchange of head with targets and subtense bar (see Chapter 6) on certain operations.

10. A wide range of accessories can be used with the theodolite and thus increase its usefulness and versatility.

11. Most instruments are housed in light, streamlined metal cases into which the theodolite can be fastened very securely and adequately protected when not in use.

12. The reticle and reading systems are provided with internal illumination for night operations.

3-23. Types and Operation Optical theodolites of various types and capabilities are available, but primary attention is directed here to those that are most nearly equivalent to the single minute and thirty-second vernier transits of American manufacture. Accordingly, a brief treatment will be provided of two representative designs of optical theodolites which are suitable for engineering surveys of ordinary accuracy. Theodolites of higher precision are described in Chapter 9.

(*a*) The one-minute theodolite depicted in Fig. 3-14 may be

[3-23] ANGULAR MEASUREMENTS 81

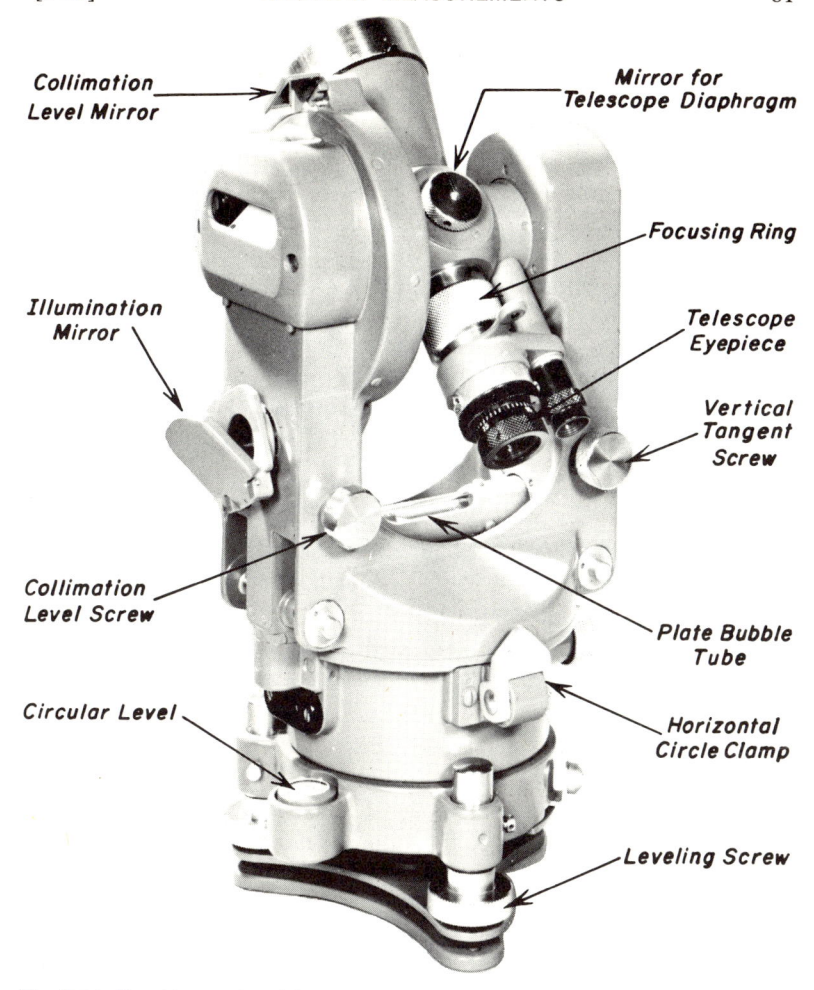

Fig. 3-14. One-Minute Theodolite (Wild T-16ED). *Wild Heerbrugg Instruments, Inc.*

used either as a direction or a repeating instrument. This versatility is provided by the appropriate manipulation of the horizontal circle clamp or repeating lever. The horizontal circle is graduated clockwise and counterclockwise and both circles are read directly (see Fig. 3-15) with an optical microscope to the nearest minute and by interpolation to 15″ or 20″.

The telescope is reversible and produces an erect image. Its magnification is 28 power. The reticule (Fig. 3-16) has a set of stadia

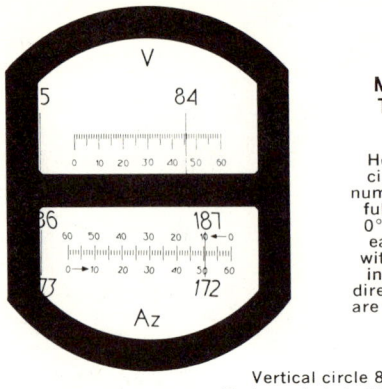

MODEL T16ED

Horizontal circle (Az) numbered two full rows—
0° to 360° each way, with figures inclined in direction they are to be read.

Vertical circle 84° 45′ 45″
Horizontal circle 172° 50′ 30″
or 187° 9′ 30″

Fig. 3-15. Horizontal and Vertical Scales. *Wild Heerbrugg Instruments, Inc.*

ticks on both the horizontal and vertical crosslines and two parallel vertical lines in the lower half of the field. These are used for precise pointings on distant range poles and plumb lines which would be otherwise obscured by the use of a single vertical sight line. For daytime use tilting mirrors provide illumination for both circles.

The tribrach or base (see Fig. 3-17) is a detachable part of the theodolite and contains the three leveling screws and the circular level. These screws are completely enclosed and dustproof. A locking device holds the instrument head and tribrach together.

The vertical circle, or collimation, level is built in and adjacent

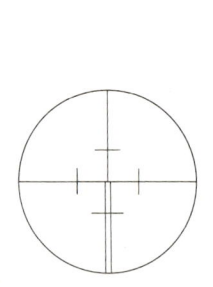

Fig. 3-16. Theodolite Reticule. *Wild Heerbrugg Instruments, Inc.*

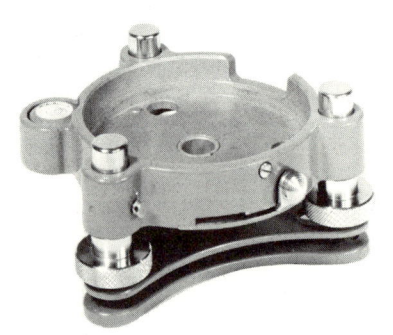

Fig. 3-17. Tribrach. *Wild Heerbrugg Instruments, Inc.*

to the vertical circle. Its bubble is known as a coincidence type or split bubble and is viewed from the eyepiece end of the telescope. Centering of the bubble is accomplished (see Fig. 3-18) by bringing

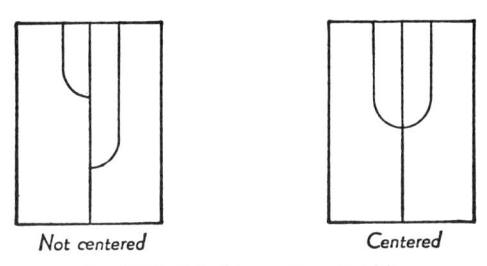

Not centered *Centered*

Fig. 3-18. Coincidence-Type Bubble.

the two halves together until the ends coincide. The vertical circle reads 0° when the telescope is pointed at the zenith or directly overhead and 90° when the telescope is horizontal and in the direct position. These angular values are zenith distances and must be converted into vertical angles.

The manufacturer's service manual should be thoroughly studied by the instrumentman before using the theodolite.

(*b*) A somewhat similar type of optical theodolite is shown in Fig. 3-19. It is a repeating instrument which employs a micrometer to permit direct readings of both circles to 20″ of arc and by interpolation to 5″. The horizontal circle is graduated clockwise from 0°–360°. This instrument does not have a vertical circle control bubble. The vertical circle is automatically controlled by an adjusting device that effectively eliminates the index error. The horizontal and vertical scales for this theodolite are shown in Fig. 3-20. Again, careful study of the service manual is recommended before operating the instrument.

The brief foregoing descriptions indicate that, although the two theodolites are similar in many respects, each has certain advantages for specific tasks. With respect to the important and commonplace function of measuring horizontal angles, it is to be emphasized that for best results the angles should be turned with the telescope in both positions, direct (D) and reverse (R), and repeated if desired.

3-24. Care and Adjustment Theodolites are delicate instruments and care should be exercised that they are not subjected to impact

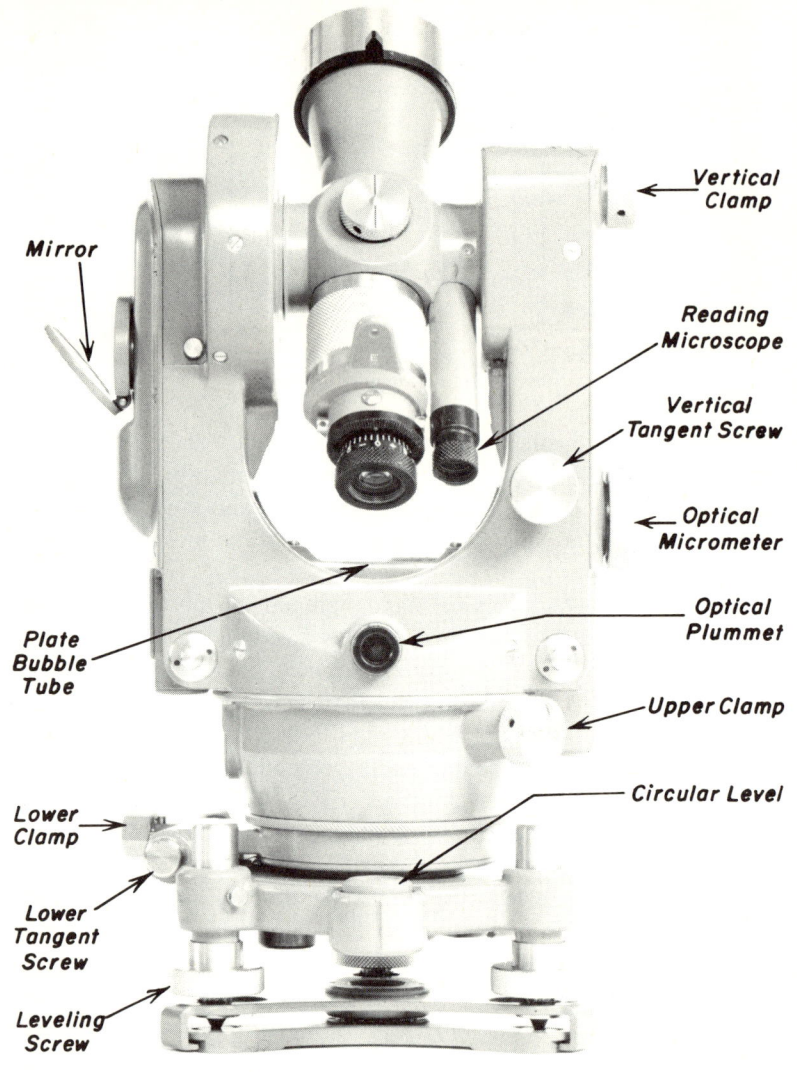

Fig. 3-19. Twenty-second Theodolite (Wild T-1AE). *Wild Heerbrugg Instruments, Inc.*

shocks. These generally occur when the instrument is dropped or bumped against any firm object. They should be kept clean and dry. Accumulations of dirt and foreign material will eventually penetrate into the motions and cause sticking. Painted surfaces should be wiped

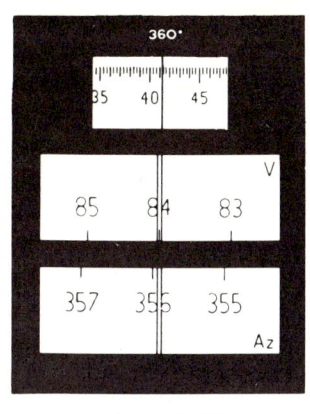

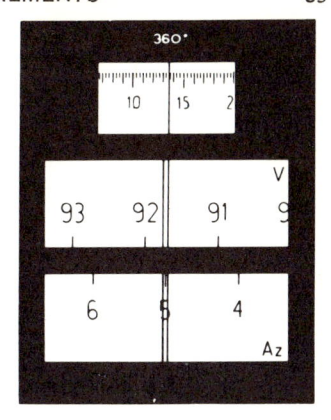

[3-24] ANGULAR MEASUREMENTS 85

84° 41'15"   5° 13'35"

Fig. 3-20. Reading the Wild T-1AE. *Wild Heerbrugg Instruments, Inc.*

with a clean cloth. Special care should be taken in removing dust from the lenses as previously explained in Art. 3-10.

Accidents leading to damage of the theodolite are particularly likely to occur when the instrument is being removed from or returned to its carrying case, fastened to or unloosened from the top of the tripod, and during transport. When the theodolite is carried over rough terrain, it should be in its carrying case. When transported in a vehicle the carrying case should be held on somebody's lap.

The theodolite, like a transit, must be in correct adjustment if the best results are to be obtained. The only field adjustments are those associated with the various level vials. They are, therefore, similar to the adjustments for the conventional American transit. For the details of making the tests and effecting the adjustments, the reader is referred to the manufacturer's manual.

In general, an adjusted instrument will retain the desired relationships if handled properly. If subjected to hard usage and abuse, adjustment will obviously be required more frequently. Such excessive manipulation of the adjusting screws will cause premature wear and result in the theodolite going more easily out of adjustment.

PROBLEMS

3-1. The bearing of line, AB, is N. 16°50′ E. At points B and C, the deflection angles are 69°07′ R and 20°56′ L, respectively. Find the bearing of lines BC and CD.

3-2. The measured interior angles in a closed figure are as follows:

Point	Angle
A	50°10′
B	164°07′
C	122°13′
D	48°50′
E	223°09′
F	111°31′

If the bearing of *AB* is N. 42°16′ E., find the bearings of the remaining lines. As an aid in interpreting the data and drawing a sketch, it is to be noted that *BC* has a northeasterly bearing.

3-3. Calculate the deflection angle and the angle-to-right at points *B* and *E* of Problem 3-2.

3-4. If the magnetic bearing of a line is S. 37½° E. and the declination is 2° W., find the true bearing.

3-5. The magnetic bearing of a line was recorded as N. 72°20′ E. in 1870 at a place where the prevailing declination was 10°30′ W. What is the present magnetic bearing if the declination is now 2°50′ E.?

3-6. The azimuth of a line is 297°31′. What is the back bearing?

3-7. The true bearing of a line was determined astronomically to be S. 89°22′ W. The magnetic bearing was measured as N. 89°42′ W. Find the magnitude and direction of the declination.

3-8. Calculate the azimuths of all the survey lines in Problem 3-2.

3-9. The horizontal circle of a transit is divided into 20′ intervals and 60 divisions on the vernier are equal to 59 on the circle. What is the least count of the vernier?

3-10. Because of brush obstructing the sight line, transit pointings were made on the top of a range pole which was subsequently found to be 0.1 ft out of plumb perpendicular to the line of sight. What angular error does this displacement represent for distances of 900, 500, and 100 ft?

3-11. High winds made it difficult to center a transit with the plumb bob over a survey station. If the plumb bob was eccentric by ¼ in. (0.02 ft), what is the maximum directional error represented by this displacement for sight distances of 1000, 200, and 20 ft?

3-12. In making a layout of a large mill building, a 90° angle was turned as carefully as possible with a 30″ transit. The angle was then measured by repetition and found to be 90°00′15″. What offset (nearest 0.01 ft) should be made at a distance of 600 ft from the transit in order to establish the true line?

3-13. Observations were made on a tall radio mast to check its verticality. With the transit telescope direct, the top of the mast was sighted, the telescope rotated about the horizontal axis, and a point was set on the ground 0.37 to the left of the base of the mast. The same procedure was followed with the telescope in the inverted position and a point was set .49 ft to the right. How much is the top of the mast out of plumb and in which direction, left or right?

REFERENCES

1. Rayner, W. H., and M. O. Schmidt, *Surveying—Elementary and Advanced,* Chapter 29, D. Van Nostrand Co., Inc., Princeton, N.J., 1957.
2. Higbee, Lester C., *Notes on Surveying Instruments.* Unpublished. Troy, N.Y., 1952.

4 Indirect Linear Measurements

4-1. Introduction Traditionally distances have been measured by direct comparison with some established unit of length as exemplified by the operation of chaining or taping. Other procedures involving the measurement of quantities from which the distance is indirectly obtained by calculation can be employed. Some of these methods, such as the theory of the stadia method, have been well established for many years. Others, such as the *electro-optical* method of determining distance, are of fairly recent origin.

The scope of this chapter is restricted to the basic treatment of certain kinds of tacheometric and electronic measurements of distance. The methods will be described and their capabilities compared with measurement by taping. The details of application to specific surveying operations will be deferred to later chapters.

TACHEOMETRIC MEASUREMENTS

4-2. Tacheometry The word *tacheometry* is derived from the Greek and has the meaning of "rapid measurement." The term is generally applied to those operations in which the distance is obtained from one position of the instrument, usually a transit or theodolite, through the establishment or the measurement of a small angle lying opposite a known base. The principles of tacheometry were first recognized in 1639 by the English astronomer, William Gascoigne.

Tacheometric instruments may either incorporate a base within themselves or make use of an external base. A leading example of the internal base type is the military range finder. Instruments utilizing an external base may be classified as (1) those that measure an intercept on a distant graduated rod by means of a fixed angle, and (2) those that measure the angle subtended by a fixed base. The field procedures usually associated with the foregoing two categories are commonly identified as the stadia method and the subtense method, respectively.

4-3. Stadia Method The stadia method of measuring distance is rapid and its results are sufficiently dependable for certain surveying missions. Under favorable conditions the error will not exceed 1/500. On surveys of ordinary accuracy conducted with the steel tape, the stadia method can be employed to provide a ready check on distances so that blunders cannot remain undetected. In combination with the measurement of vertical angles, height differences can be calculated. The stadia method is utilized on topographic and hydrographic surveys although, in general, its usefulness can be said to be declining because of notable advances in certain areas of surveying activity such as aerial mapping.

The instrumental equipment for stadia measurements includes a graduated rod and an engineer's transit the telescope of which is provided with two additional stadia cross-wires. These are mounted on the cross-wire ring, one above and the other below the middle horizontal cross-wire. The rod divisions are usually in units of feet and tenths of feet and arranged in various patterns, one of which is shown in Fig. 4-1. Of the many patterns of rods that have been designed, probably none is better for general use than that shown here. For short sights, up to about 500 ft, the ordinary leveling rod may be used; beyond that distance the fine markings of a level rod become indistinct, and a pattern of larger divisions is necessary.

A reading is taken by setting the lower cross-wire on a full foot mark on the rod and noting where the upper cross-wire cuts the rod. The difference between the two readings is called the *interval,* or *intercept,* and is a measure of the distance from the instrument to the rod.

4-4. Stadia Theory The theory of the stadia method will be considered for the case of a horizontal sight with an *external-focusing* telescope. Such a telescope is easily identified because the objective lens will move in response to rotation of its focusing screw. If a treatment of the optics of the telescope is omitted, the theory of the

Fig. 4-1. Stadia Rod.

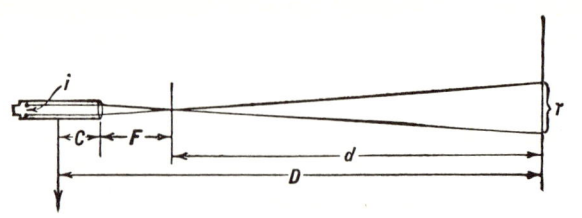

Fig. 4-2. Stadia Principle for Horizontal Sight.

stadia is simple. In Fig. 4-2 is illustrated a telescope, the plumb line, and a rod intercept, r. Two rays of vision are drawn, namely those emanating from the cross-wires and parallel with the optical axis. These rays are refracted by the objective lens and pass through a focal point at a distance F in front of the lens, and intersect the rod as shown.

If the distance between the cross-wires is represented by i, then in the two similar triangles we have the relation

$$\frac{d}{r} = \frac{F}{i}$$

from which $d = (F/i)r$, and from the figure, the total distance from the rod to the plumb line is

$$D = \left(\frac{F}{i}\right)r + (F + c) \tag{4-1}$$

The Constant $(F + c)$. The distance F is the focal length of the lens and is a constant. Also, the quantity $(F + c)$ is practically a constant, since the distance c varies but a small and negligible amount when the telescope is focused on different objects. The value of this constant varies somewhat for different instruments, but the range is between about 0.8 ft and 1.2 ft, or practically 1.0 ft. Since the uncertainties in determining distance by stadia, even for sights of moderate length, are at least 1 ft, inclusion of an allowance for $(F + c)$ is not critical. For most surveys this quantity can be safely neglected.

It may be added that most instruments of recent design have *interior focusing* telescopes, and for such telescopes the quantity $(F + c)$ can be disregarded under all conditions.

The Constant (F/i) *or* k. Since the principal focal length F and the distance i between cross-wires are constants for any telescope, the ratio (F/i) also is a constant and is represented by k. Then, neglecting $(F + c)$, Eq. 4-1 becomes $D = kr$. The quantity k is the factor by which each rod intercept r is multiplied to determine the distance D, and, for convenience, it is made to equal 100 as nearly as can be. It may be noted that it is of no importance to know the value of either F or i separately. The constant k is sometimes called the *stadia coefficient* or the *stadia multiplying factor* of the instrument. The term, cross-wire, should be interpreted to mean any kind of crosslines in the telescope.

Since all rod intervals are to be multiplied by this factor k, it is important that its value be determined with some precision. This may be accomplished conveniently as follows: set a series of pins by pacing at distances of about 100, 300, 500, and 700 ft from the instrument. Read the intercept carefully at each pin, estimating all readings to 0.01 ft. Finally, measure the distance to each pin with the tape. The ground should be nearly level, but the telescope bubble need not be centered. The bottom cross-wire should be set on a full foot mark to facilitate the intercept reading. Since considerable precision is necessary for this determination, $(F + c)$ may be included for an external focusing telescope and taken as 1 ft, and Eq. 4-1 applied. From these data, four values of k will be found, the mean of which may be accepted for use.

It should be added that, since the intercept r depends on the rod graduations, it is desirable that the accuracy of the divisions on the stadia rod be checked by a steel tape. This is especially important if the rods in use are "hand made," i.e., constructed by the organization using them. Such rods are likely to have their lengths appreciably affected by moisture and temperature changes.

Horizontal stadia sights in surveying practice are of infrequent occurrence. Hence, it is appropriate to extend considerations of stadia theory to the general case of inclined sights. On sloping ground the difference of elevation and the horizontal distance between two points can be found by the stadia method if, in addition to the rod interval, the angle of inclination of the line of sight is read on the vertical circle.

In Fig. 4-3, let it be supposed that the transit is in position at station A and that the rod is held vertically at station B, that the rod intercept is $mn = r$, and that the inclination of the line of sight as measured on the vertical arc is α; also that $m'n' = r'$ is the intercept which would be read on the rod if it were held perpendicular to the line of sight.

Since the imaginary line $m'n'$ is perpendicular to line of sight IO,

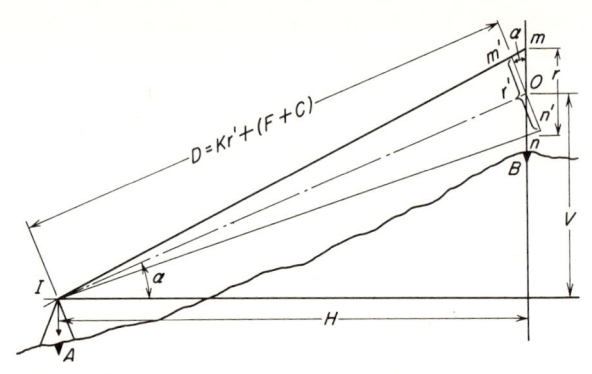

Fig. 4-3. Stadia Principle for Inclined Sights.

angle $Om'm = On'n$ is not exactly a right angle. However, in the following demonstration it will be so considered, and the error thus introduced is negligible. With this approximation understood, the following equations may be written:

$$m'n' = mn \cos \alpha, \text{ or } r' = r \cos \alpha \tag{4-2}$$

$$D = kr' + (F + c) \tag{4-3}$$

Substituting 4-2 in 4-3,

$$D = kr \cos \alpha + (F + c) \tag{4-4}$$

The difference in elevation between I and O is given by the vertical projection V of the slope distance D.

Hence,

$$V = D \sin \alpha \text{ or } V = kr \sin \alpha \cos \alpha + (F + c) \sin \alpha$$

from which

$$V = kr \tfrac{1}{2} \sin 2\alpha + (F + c) \sin \alpha \tag{4-5}$$

Also, the horizontal distance H is equal to the horizontal projection of the slope distance D, and

$$H = D \cos \alpha$$
$$= kr \cos^2 \alpha + (F + c) \cos \alpha \tag{4-6}$$

From the figure it is evident that the difference in elevation between I and O, which equals V, is the same as the difference in elevation between ground points A and B, if distance OB is taken equal to IA; also, horizontal distance H is equal to the horizontal distance between points A and B. Thus, Eqs. 4-5 and 4-6 permit the determination of the difference in elevation and of the horizontal distance between any two given stations.

For reasons stated above and for usual conditions, sufficiently satisfactory results will be secured if the constant $(F + c)$ is disregarded. Hence, Eqs. 4-5 and 4-6 become

$$V = kr\tfrac{1}{2} \sin 2\alpha \tag{4-7}$$

and

$$H = kr \cos^2 \alpha \tag{4-8}$$

Table VII has been computed by the use of these formulas.

4-5. Reduction of Stadia Notes *Tables.* To simplify the use of Eqs. 4-7 and 4-8 of the preceding article, Table VII has been computed. This table gives the values of H and V for various vertical angles α, and for the value $kr = 100$. For other values of kr, a direct proportion is to be used.

EXAMPLE 4-1: Given $r = 3.17$, $k = 100$, and $\alpha = 4°20'$. Find H and V.

In Table VII, the values of H and V for $\alpha = 4°20'$ and $kr = 100$ are 99.43 and 7.53, respectively. Hence, for $kr = 317$,

$$H = 3.17 \times 99.43 = 315$$
$$V = 3.17 \times 7.53 = 23.9$$

EXAMPLE 4-2: Given $r = 6.37$, $k = 101.2$ and $\alpha = 3°40'$. Find H and V.

$$kr = 6.37 \times 101.2 = 645$$

Then from the table, $H = 99.59$ and $V = 6.38$ for $kr = 100$. Accordingly,

$$H = 6.45 \times 99.59 = 642$$
$$V = 6.45 \times 6.38 = 41.2$$

For the unusual cases of high accuracy, when Eqs. 4-5 and 4-6 should apply, they can be transformed into Eqs. 4-7 and 4-8 by the relations

$$\left.\begin{array}{l}(F + c) \sin \alpha = 1 \text{ ft } \tfrac{1}{2} \sin 2\alpha \\ (F + c) \cos \alpha = 1 \text{ ft } \cos^2 \alpha\end{array}\right\} \text{very nearly}$$

and

Accordingly, Eqs. 4-5 and 4-6 become

$$V = (kr + 1)\tfrac{1}{2} \sin 2\alpha \qquad (4\text{-}9)$$

and

$$H = (kr + 1) \cos^2 \alpha \qquad (4\text{-}10)$$

from which it is evident that Eqs. 4-7 and 4-8 and Table VII may be used for this case if 1 ft is added to the value of kr before making the computations indicated in the examples given above.

It is strongly emphasized that the uncertainties in stadia work make it both illogical and misleading to express horizontal distances and vertical heights to closer than the nearest full foot and one-tenth foot, respectively.

Stadia Slide Rules. To avoid the use of tables and further to simplify the reductions of stadia notes, special slide rules have been designed. These need not be described in detail here, but they are most convenient and should be available if many transit-stadia notes are to be reduced. This remark applies in particular to plane table surveys where the reductions of notes are made in the field.

4-6. Errors in Stadia Some of the errors associated with stadia surveying are common to various basic operations with the transit and theodolite, such as measuring vertical angles. The determination of horizontal distances and vertical heights is, however, affected by the following sources of error.

Reading the Rod. The inability of the instrumentman to estimate exactly the rod intercept is a source of an accidental error. Its magnitude varies with the weather conditions and the length of sight, but for average conditions the average error will be, perhaps, for distances up to 300 ft, ± 0.01 ft, and ± 0.01 to ± 0.03 ft for distances between 300 and 800 ft. Beyond that distance the error is subject to large variations.

This error is minimized by limiting the length of sight, by repeating the reading, and by the use of targets on the rod.

Rod Not Plumb. If the rod is not plumb, the intercept is too large and the corresponding distance is too great. This is a systematic error.

The Constant k. Since all rod intercepts are multiplied by the constant $F/i = k$, any error in the determination of this constant is a systematic error. The resulting error in measured distances is directly proportional to the error in the value of k. Thus, an error of 1% in k will introduce an error of 1% in all measured distances. Accordingly, in comparison with the other errors, it is desirable that this error shall not exceed more than ±0.1 of 1%.

Length of Rod. It is obvious that any error in the length of a stadia rod is multiplied by k, or about 100, in the horizontal distance. Thus, an error of 0.1 ft in a 12-ft rod will introduce an error of $0.1 \times 100 = 10$ ft in a distance of 1200 ft. It is therefore important that the lengths of all stadia rods in use should be checked occasionally with a steel tape.

Parallax. Obviously, parallax, if present, is an important source of error in stadia work. It is to be prevented by keeping the eyepiece accurately focused on the cross-wires at all times.

Natural Errors. Such natural errors as wind, differential refraction, moisture, and temperature changes affect stadia measurements more or less depending on weather conditions. The two latter effects are usually not important. The wind is frequently troublesome both to the instrumentman and rodman, and accurate results cannot be expected when a high wind is blowing.

Differential refraction is that effect whereby the line of sight as fixed by one cross-wire is refracted more than the line of sight fixed by the other. This condition is caused by the difference in temperature and density of the air strata just above the ground. It is evidenced occasionally by so-called heat waves, but is present in lesser amount at all times. It may be minimized by keeping the line of sight fixed by the lower cross-wire, well above the ground; by reducing the lengths of sights when the heat waves are apparent; or, if practicable, by avoiding stadia work when this condition is serious.

4-7. Subtense Method This method essentially consists of measuring the angle lying opposite a fixed base and calculating the distance. It acquired its name from the fact that the angle is said to be subtended by the base. The equipment required for a subtense measurement consists of a transit or a theodolite and a *subtense bar*. Since angular errors of not more than a few seconds of arc must not be exceeded if suitable accuracy in the distance is to be obtained, it is virtually necessary that a high-quality optical theodolite, preferably of the direction type, be available.

The standard subtense bar is 2 meters long and consists of a metal tube having a target at each end. The distance between the targets is controlled by an invar wire in such a manner that the base length is unaffected by temperature variations.

The bar is mounted on a tripod as shown in Fig. 4-4 and is made

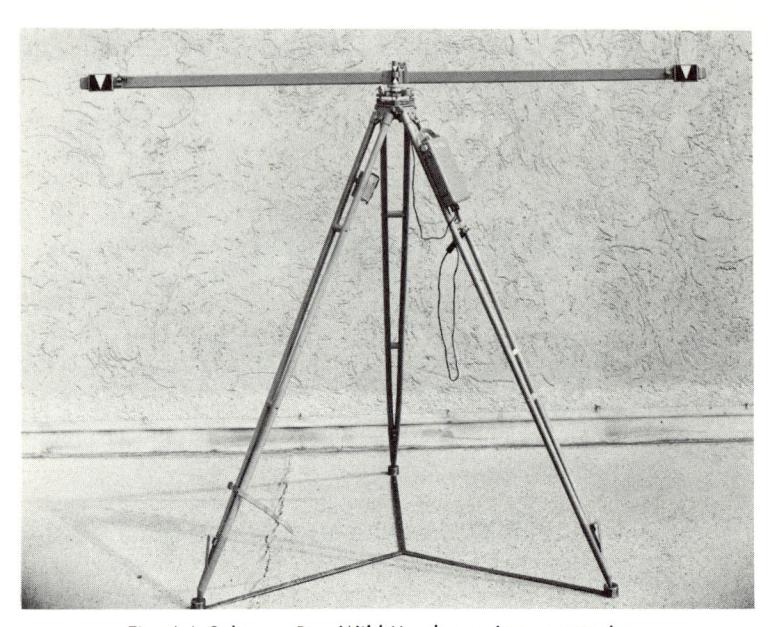

Fig. 4-4. Subtense Bar. *Wild Heerbrugg Instruments, Inc.*

horizontal with a circular level. A small telescope attached to the bar is used to sight at the theodolite station, thus making the bar perpendicular to the line of sight. The targets consist of small triangles or diamonds that can be internally illuminated for night work. For ease in transporting, the bars can be folded at a joint in the middle.

Figure 4-5 shows the principle of the subtense method. It is evident that the horizontal distance is given by the expression:

$$D = \tfrac{1}{2} S \cdot \cot \frac{\alpha}{2} \qquad (4\text{-}11)$$

Tables furnished by the maker of subtense equipment simplify the determination of the distance.

[4-7] INDIRECT LINEAR MEASUREMENTS

Fig. 4-5. Principle of Subtense Method.

In general, a single-second theodolite (see Chapter 9) should be used if reliable results are to be obtained. The uncertainty in the distance is given by

$$e_D = \frac{D^2}{S} e_\alpha \qquad (4\text{-}12)$$

where D is the distance, S is the length of the bar, and e_α is the angular uncertainty that must be expressed in radians. To convert seconds of arc to radians, multiply by 0.00000485.

EXAMPLE 4-3: A distance of approximately 400 ft is to be measured with a 2-meter subtense bar and a direction theodolite. Find the uncertainty in the length associated with an error of $1''$ in the angle.

$$e_D = \frac{(400)^2}{(2)(3.2808)} (.00000485)1 = \pm 0.118 \text{ ft}$$

The corresponding proportional error is .118/400 or 1/3400. When longer distances are to be measured, it is apparent that the accuracy attained will deteriorate very rapidly. However, if the larger lengths are subdivided into sections, more dependable results can be obtained.

Despite its limitations, the subtense method has several advantages over taping if only moderate accuracies are required. It can be used to measure distances over rough terrain, and across gullies and wide streams. Furthermore, since the horizontal angle subtended by the bar is independent of the inclination of the line of sight, the horizontal distance is obtained directly and no slope correction is necessary.

For an extended treatment of the errors in subtense work, including the derivation of Eq. 4-12, study of Ref. 7 at the end of this chapter is recommended.

With the introduction of electronic distance-measuring procedures, with which the subtense method can hardly compete, the applications of the subtense principle to the indirect determination of length have been greatly restricted.

ELECTROMAGNETIC MEASUREMENTS

4-8. General Until the middle of the twentieth century one of the most challenging and basic problems of surveying was the accurate determination of the lengths of long lines. There was no alternative to the laborious and costly procedure of methodically transferring successive lengths of a carefully calibrated invar tape to the ground, applying the appropriate corrections, and calculating the final value of the distance. When properly conducted, such measurements were accurate enough to satisfy the most exacting standards of geodetic surveying. Nevertheless, the quest for a less expensive and more rapid method for the accurate measurement of long lines stimulated research in the area of electronics as applied to length metrology. The development of an electro-optical distance-measuring device in 1949 can be considered one of the epic advances in the science of surveying and geodesy. The invention of the Geodimeter (see Fig. 4-6), the forerunner of a variety of electronic length-

Fig. 4-6. Geodimeter at Work. *Parsons, Brinckerhoff, Quade & Douglas*

measuring instruments, revolutionized the solution of the task of distance determination, both for ordinary engineering surveys and surveys of unusual accuracy.

The fundamental principle associated with the operation of all electromagnetic length-measuring devices is that distance is the product of velocity and time. Hence, if the velocity of a radio or light wave is known and if the time required for the wave to travel from one point to another is likewise known, the intervening distance can be calculated. This statement, even though a highly oversimplified explanation, affords an immediate and convenient approach to the basic concepts of electromagnetic measurements.

Electronic surveying is a term of wide connotation and embraces the treatment of such subject areas as *Loran* and *Shoran* and could include the use of many devices including the artificial geodetic satellite. However, the following brief treatment considers only the three major instruments used (1969) in the United States for the accurate determination of relatively short distances by electronic means, viz. *Tellurometer, Electrotape,* and *Geodimeter.* Prior to describing them and indicating their respective capabilities, a capsule account will be provided of the principles of microwave and optical measuring systems.

4-9. Principles of Electromagnetic Measurements Associated with the oscillation of energy of an alternating current flowing in an open circuit, such as an antenna, will be the formation of electric fields which radiate electromagnetic energy into space as electromagnetic waves. Both radio waves and light waves are electromagnetic waves having identical velocities in a vacuum.

In 1957 the International Union of Geodesy and Geophysics (IUGG) adopted as a standard the value of 299,792.5 ± 0.4 kilometers per second for the propagation velocity in *vacuo* of visible light and radio microwaves.

An important relationship in wave motion is that expressed by the equation

$$\lambda = \frac{V}{f} \tag{4-13}$$

where λ is the *wave length* or the distance traveled during the period of one cycle, V is the velocity of propagation, and f is the frequency or number of cycles per unit of time. The unit of frequency is the

Hertz (Hz) which is one cycle per second. A *megacycle* (Mc/s) is one million cycles per second. A frequency of 10 Mc/s indicates that the wave completes 10 million cycles per second. The wave form is sinusoidal.

Phase refers to a portion of a complete cycle of a wave. Since one complete cycle is represented by 360°, one-quarter cycle would be 90°. *Phase difference* is the time in electrical degrees by which one wave leads or lags another. Phase comparison procedures measure the phase of one wave and that of another at the same moment of time. For example, if the phase of wave *A* is 70° at the instant that the phase of wave *B* is 130°, a phase comparison will disclose a phase difference of 60°. If the complete wave length is 30 centimeters, the phase difference is equivalent to a distance of 60°/360° times 30 or 5 cm.

In order to measure electronically the distance between two points, an alternating signal travels from one point to the other where it is reflected and returned in some manner and compared at the transmitting station with the phase of the original signal to determine the travel time for the round trip. If the distance is to be directly calculated from the transit time with an accuracy of 0.5 ft, the interval of time would have to be correct to within one billionth of a second or one *millimicrosecond* (mμs). So small an interval would be extremely difficult to measure directly and the problem would become still more acute if the tolerance in distance was realistically smaller. The difficulty can be resolved by the physical possibility of making a very accurate phase measurement of the signal.

Electronic distance-measuring devices utilize a modulated signal carried by a radio microwave or a visible light beam. The basic function of such instruments is the determination of the number of wave lengths, at the modulation frequency, that are needed to bridge the distance between the termini of the line being measured. If the velocity of propagation is known and the frequency is carefully controlled, Eq. 4-13 permits the accurate calculation of the wave length. It is apparent that the wave length is the unit of length in terms of which the total distance is computed. Some instruments provide a direct read-out of distance.

4-10. Index of Refraction The quality of electronic distance measurements is directly dependent upon the accuracy with which the velocity of propagation of the signal is known. Although light-wave and radio-wave velocities are equal in a vacuum, they are markedly different under other conditions. Because the measurement of a line is conducted with the signal passing through the atmosphere, the velocity value used in the computations must be the actual velocity in the air. The relationship between the velocity in *vacuo* and that in air is given by

$$n = \frac{V_0}{V} \qquad (4\text{-}14)$$

where n is the *index of refraction*, V_0 is the velocity in *vacuo*, and V is the velocity in air. It is seen that the velocity in air varies inversely as the index of refraction. Different formulas are available (see Ref. 4) for computing the index of refraction for light waves and for radio microwaves. These formulas require the use of observed meteorological data, such as barometric pressure, temperature, and humidity. Since only the terminal points of the line being measured are occupied by the observers, these are the only locations at which meteorological conditions are usually assessed. The surface weather data at these points are averaged and assumed to represent conditions along the wave path.

The approximate value, 1.0003 for "average" conditions will serve to convey a general idea as to the magnitude of the index of refraction.

4-11. Microwave Instruments The microwave distance-measuring device is an electronic instrument whose basic operation is associated with the time required for a radio microwave signal to travel from the radiating station to a distant point and back after re-radiation from a similar instrument at that point. The time is measured in the form of a phase difference between the going and returning signal. Two types of microwave instruments will be described. It is underscored that a knowledge of electronic theory is not essential in the operation of these instruments. Even if the technician has no surveying experience, he can quickly learn to make accurate distance measurements.

(*a*) *Tellurometer.* The Tellurometer was invented by Dr. T. L. Wadley of the National Telecommunications Research Laboratory, South Africa, and appeared on the American market in 1957. The model, MRA 101, shown in Fig. 4-7, consists of two identical elec-

Fig. 4-7. Tellurometer. *Tellurometer, Inc.*

tronic units known as the Master and the Remote instruments. The Master transmits a modulated radio wave which is a combination of a carrier wave, usually of high frequency, and one or more, usually lower, frequency waves superimposed on it electronically. The modulated wave is received at the Remote instrument and retransmitted to the Master. There the phase of the received modulation is compared with the transmitted modulation. The resulting phase difference is a measure of the transit time of the radio wave over the path. This measure of time is then electronically converted into a metric expression of length which is read directly to the nearest centimeter by the operator. This value indicates the slope distance between the electrical centers of the two instruments.

The weight of each unit which is mounted on a tripod is 16 lb. The working range is 350 ft to 30 miles. The accuracy of the Tellurometer system depends upon meteorological and topographical factors and the calibration of the instruments and is said to be 1.5 cm ± 3 ppm which is equivalent to 0.05 ft $\pm 1/333{,}000$ of the distance. The major sources of error in Tellurometer measurements are

1. Multipath reflections
2. Incorrect index of refraction
3. Eccentricity between electrical center of instrument and point of suspension of the plumb bob

Multipath reflections are due to the fact that the emitted beam is not a pencil of radiation but a cone. Any surfrace struck by part of the beam will cause reflection. The reflections that reach the instrument will affect the readings, since the microwave receiver cannot distinguish between a direct and a reflected wave. The term applied to this condition which manifests itself in the form of a scattering of the readings is *ground swing*.

Faulty or nonrepresentative meteorological readings will render the calculated index of refraction defective.

Eccentricity can be evaluated by periodically testing the instruments over a baseline whose length has been accurately determined by taping. Such a baseline should be at least 500 ft long.

(*b*) *Electrotape.* The first electronic distance-measuring instrument to be produced in the United States for geodetic or general surveying purposes was one developed by the Cubic Corporation of San Diego in 1958. After a few years it assumed its present name, Electrotape. The 1967 model, designated as DM-20, is shown in the frontispiece.

The Electrotape system consists of two identical units, the Interrogator and the Responder. Each unit weighs 28 lb and during operation is supported on a surveying tripod. A phase-comparison technique is used to measure the slope distance which is read out directly to centimeters.

The accuracy of the Electrotape is said to be .03 ft $\pm 1/300{,}000$ and depends upon the same factors as those already mentioned for the Tellurometer. The minimum operating range is that at which the user can tolerate an error of .03 ft, and the maximum is 30 miles.

An index of refraction of 1.000320 has been applied to the internal circuitry of this instrument. If the existing atmospheric conditions result in an index that differs considerably from this value, a correction must be applied.

In the case of both the Tellurometer and Electrotape systems, operations are possible in all kinds of weather. Although the ends of the line being measured must have a clear line of sight between them,

Fig. 4-8. Geodimeter Reflectors. *Tennessee Valley Authority*

actual intervisibility at the time of observation is not necessary and measurements can be made at day or night. Inasmuch as both systems have beam widths of 6°, careful pointing at the distant station is not essential. A built-in telephone system affords the necessary communication between the operators at both units in both systems.

4-12. Light-Wave Instrument The Swedish geodesist, Dr. Erik Bergstrand, developed the first light-wave distance-measuring instrument which was named Geodimeter from the words *GEOdetic DIstance METER*. The first model became commercially available in 1952. The Geodimeter is a device that measures distance by the electronic timing of modulated light waves after they travel to a reflector at the opposite end of the line to be measured and return. The principle is basically similar to that of the Tellurometer and Electrotape except that, instead of microwaves, light is used as the carrier.

The Geodimeter system consists of a single measuring unit and a reflector unit. Only one operator is required. The reflector is an assembly of several retrodirective corner prisms mounted on a tripod. The number of prisms used depends on the distance and the visibility conditions. Figure 4-8 depicts a cluster of prisms set in a housing surmounted by a 6-in. range pole which serves as a theodolite target. The trihedral prisms will reflect light back to its source even though the housing is misoriented by as much as 20°. At comparatively short distances, mirrors, truck reflectors, and even reflectorized tape will return the light adequately.

Although three models of the Geodimeter differing primarily in accuracy are available, attention here is directed only to the latest version of a general-purpose instrument, the Model 6, which is shown in Fig. 4-9. This model has a digital read-out feature. Depending

Fig. 4-9. Geodimeter. *Harry R. Feldman, Inc.*

upon the conditions of visibility, the range of this instrument is said to be 1 to 2 miles in daylight and up to 10 miles in darkness. The average error is stated to be .03 ft $\pm 1/500{,}000$ of the distance.

Multipath reflections ordinarily cause no difficulties with an electro-optical system like the Geodimeter. The main source of error can be attributed to inaccuracies in the meteorological data as they affect the determination of the index of refraction. There is no assurance that conditions along the line of sight are the same as at the ends in view of the fact that the line of sight may be a considerable distance above the ground along most of its length. Measurements made at night are usually more satisfactory because the atmosphere is more stable at night than during the day. Furthermore, there will be less extraneous light. During daylight hours bright ambient light has a tendency to obscure the light beam and shorten the measuring range.

The Geodimeter requires an optically clear path between the line termini, and it must be aimed through the use of a telescopic sight toward the reflex station.

4-13. Length Calculations In the case of all electronic distance-measuring instruments, the measured length is the slope distance between the two stations. After the appropriate meteorological and instrumental corrections have been applied to the observed or preliminary slope distance, it is necessary to reduce this value to its horizontal equivalent. For lines of moderate length this calculation is easily accomplished through the use of Eq. 2-2. This presumes that the elevations of the ends of the line are known. Vertical angles can be measured and likewise utilized to obtain the horizontal distance. For an extended treatment of a refined calculation involving a long line and its reduction to the equivalent length at the elevation of mean sea level, the student can consult Ref. 2.

PROBLEMS

4-1. A field determination was made of the stadia multiplying factor, k, of the internal-focusing telescope of an engineer's transit. The stadia intercepts and the corresponding taped distances from the transit station were as follows:

Taped distance	116.3	305.1	528.0	714.8	834.1
Intercept	1.16	3.04	5.27	7.13	8.33

Calculate the mean value of k to the nearest 0.1.

4-2. Given the following data for two stadia observations with a transit having an external-focusing telescope: ($k = 100$, neglect $F + c$)

(a) $r = 6.12$, $\alpha = 3°10'$
(b) $r = 8.35$ $\alpha = 8°13'$

Find H and V by Eqs. 4-7 and 4-8 and compare with values obtained from Table VII.

4-3. Calculate the distance from a transit having an internal-focusing telescope with a stadia multiplying factor of 100.4 to the rod. The sight is horizontal and the half-stadia interval is 8.20.

4-4. The angle subtended by a 2-meter subtense bar was measured by an optical theodolite and found to be $2°17'25''$. Calculate the horizontal distance from the theodolite to the bar.

4-5. (a) What is the error in the calculated length of Problem 4-4 if the angle is uncertain by $\pm 2'''$?
(b) Express the error in the distance as a proportionate part of the distance such as, for example, 1/2100.

4-6. If the round-trip transit time for a radio microwave moving in a vacuum is 4021.60 millimicroseconds (mμs), what is the slope distance in feet?

4-7. If the round-trip transit time for a radio microwave traveling in air having an index of refraction of 1.00033 is 3287.40 mμs, what is the slope distance?

4-8. The slope distance from point A to point B was measured with a Geodimeter and found to be 1237.85 ft. If A is higher than B by 31.26 ft, find the horizontal distance.

4-9. The slope distance between two points was measured by an Electrotape and found to be 878.42 ft.
If the vertical angle between the two points is $10°15'$, find the horizontal distance.

REFERENCES

1. Berry, Ralph M., "Radio Microwave Length Measurements and Their Correction for Ambient Meteorological Conditions," *Surveying and Mapping,* Vol. 24, No. 1, pp. 55–68, American Congress on Surveying and Mapping, Washington, D.C., 1964.
2. Bouchard, Harry and Francis H. Moffitt, *Surveying,* pp. 331–335, 5th ed., International Textbook Co., Scranton, Pa., 1965.
3. Harrison, A. E., "Electronic Surveying: Electronic Distance Measurements," *Journal of the Surveying and Mapping Division,* American Society of Civil Engineers, Vol. 89, No. SU3, pp. 97–116, October, 1963.
4. International Association of Geodesy, *Bulletin Geodesique,* No. 58, p. 413, I.U.G.G., Paris, France, 1960.
5. Laurila, Simo, *Electronic Surveying and Mapping,* The Ohio State University, Columbus, Ohio, 1960.

6. Mussetter, William, "Stadia Characteristics of the Internal Focusing Telescope," *Surveying and Mapping*, Vol. 13, No. 1, pp. 15–19, American Congress on Surveying and Mapping, Washington, D. C., 1953.
7. ———, "Tacheometric Surveying—Methods and Instruments," *Surveying and Mapping*, Vol. 16, Nos. 2 and 4, pp. 137–156 and 473–487, American Congress on Surveying and Mapping, Washington, D.C., 1956.

5 Elevation Measurements

5-1. Introduction Measurements of height or elevation are associated with most surveys. In a sense, height measurements are linear measurements conducted along a vertical line, although only rarely, as in a mine shaft or tall building, would a tape be used routinely. In the design of most engineering works the calculation of the relative vertical positions of the component parts of a proposed structure or project is fundamental. During subsequent construction, surveying layout operations will fix or establish key points at the predetermined design elevations. The word *height* more frequently connotes altitude than elevation, and correct usage usually makes a distinction between the latter two terms. In surveying literature the word elevation is preferred and it will be used almost exclusively in this text.

The *elevation* of a point is its vertical distance above or below a datum or surface of reference. Hence, elevation of a point can be considered as its vertical coordinate. It is plus or minus, depending on whether the point is above or below the datum. The term *altitude* most properly applies to the vertical position of points in space, such as the altitude of a plane or a rocket. *Grade* is an expression of elevation that prevails in construction activities. Thus, the grade of the top of the curb is stated to be 92.25 ft.

The operation of determining difference of elevation is called *leveling*. This can be performed directly or indirectly. The theory and practice of *direct leveling* constitutes the chief substance of this chapter. Although specific applications of leveling concepts and techniques will be presented in subsequent portions of this book, especially Chapter 14, it is appropriate to mention here a few of the many engineering situations in which elevation data play important roles.

Engineers map or delineate the features of the earth's surface and must obtain and portray elevation information. They design and construct many projects that are built on calculated slopes or gradients.

Working under the supervision of the engineer, the technician will effect the field establishment of the physical position of points so that, for example, flow is promoted in the proper direction in a sanitary sewer or a roadside ditch. Prior to the placement of pavement concrete, he will utilize leveling procedures to set a manhole frame in the roadway area at the correct elevation so that its top will be flush with the surface of the concrete. Excavation for a building must be carried out to a prescribed depth or elevation and masonry footings to serve as foundations for the structure must be in the correct vertical as well as horizontal position. Timber piling must be cut off at the proper elevation, bridge piers must be constructed to a design elevation, and railroad bridges over highways must have the requisite minimum vertical clearance. A record of the settlement of existing buildings adjacent to a large excavation during construction operations may be extremely important data if a lawsuit takes place.

The scope of this chapter will not only include the various methods for determining difference of elevation but indicate also their relative merits. A brief account will be presented of the *vertical control system* or benchmark network of the nation and the importance of preventing disturbance of such survey marks will be stressed.

5-2. Basic Concepts A few general principles and definitions which apply to leveling are illustrated in Fig. 5-1. Here are repre-

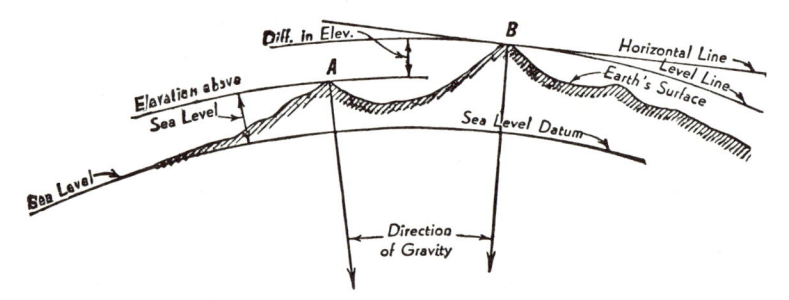

Fig. 5-1. Earth's Curvature and Elevation.

sented mean sea level and the adjacent land surface rising to two summits A and B. Mean sea level is universally regarded as the reference surface to which all elevations are referred. This surface conforms to the spheroidal shape of the earth's surface and is perpendicular to the direction of gravity at every point. It serves as a surface of reference for elevation and is called the *sea-level datum*. A cross section of this surface as represented in the figure is therefore a curved line. Mean sea level is a tidal datum which has been defined as the mean height of the sea. It is obtained by averaging hourly observations of the tide for a period extending over several years. The standard datum or reference surface for elevations in the United States is mean sea level as defined by the 1929 Fifth General Adjustment of the U.S. Coast and Geodetic Survey.

The elevation of summit A is measured by the vertical distance along the direction of gravity between the sea-level datum and point A, or between the sea-level datum and a parallel curved surface through A. Likewise, the elevation of summit B is equal to its vertical distance above the sea-level datum. Also, the difference in elevation between summits A and B is equal to elevation B minus elevation A, or it is equal to the vertical distance between the imaginary curved surfaces, parallel to sea level, one passing through A and the other through B.

From the above considerations we have the following definitions:

A *datum* is a surface of reference, coincident or parallel with mean sea level to which all elevations of a given region are referred.

A *level surface* is a surface which at every point is perpendicular to the plumb line or the direction in which gravity acts. A *level line* is a line in a level surface.

A *horizontal line* is a straight line tangent to a level line at any given point. It is perpendicular to the direction of gravity at the point of tangency only.

Leveling is the process of determining the difference in elevation between two points by measuring the vertical distance between the level surfaces passing through the points.

5-3. Curvature and Refraction It is essential to understand the nature of earth's curvature and atmospheric refraction as they affect leveling operations.

The definition of a level surface indicates that this surface parallels the curvature of the earth. A line of constant elevation, termed a level line, is likewise a curved line and is everywhere normal to the plumb line. However, a horizontal line of sight through a surveyor's telescope is perpendicular to the plumb line only at the point of observation. Hence, it should be carefully distinguished from a level line.

Because of atmospheric refraction, rays of light transmitted along the earth's surface are refracted or bent downward so that the actual line of sight is along a curve, which is concave downward. Unlike earth's curvature which is directly calculable from geometrical principles, refraction depends upon the state of the atmosphere and is variable. Ordinarily, it is considered as constant.

The effects of the earth's curvature and atmospheric refraction are illustrated in Fig. 5-2. For the distance AD, a level line parallel

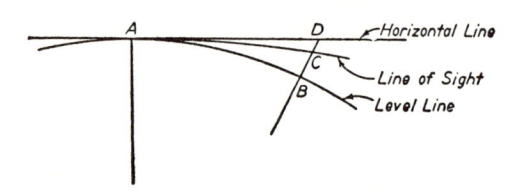

Fig. 5-2. Earth's Curvature and Atmospheric Refraction.

with the earth's curvature is deflected from a horizontal line the distance DB. The amount of this deflection is proportional to the square of the distance AD; and, for a distance of one mile, $DB = 0.667$ ft. Evidently this effect is to make a rod reading too great.

The line of sight is deflected from a horizontal line by atmospheric refraction by an amount indicated by the distance DC. The amount of this deflection is also proportional to the square of the distance sighted; and, although it varies slightly with atmospheric conditions, it may be taken, with negligible error, to be one-seventh of the magnitude of the earth's curvature and will, as shown, reduce the effect of the earth's curvature by this amount.

The combined effect of earth's curvature and atmospheric refraction can be closely approximated by the expression

$$C \& R = 0.021 S^2 \tag{5-1}$$

where S is the length of the sight in thousands of feet. For sight distances of 100, 200, 300, and 500 ft, C & R becomes 0.0002, 0.0008, 0.0019, and 0.0052 ft, respectively.

5-4. Methods of Leveling Leveling methods are commonly classified as direct or indirect.

Direct leveling is the operation of determining differences of elevation by measuring vertical distances directly on a graduated rod with the use of a leveling instrument. Formerly this method was termed *spirit leveling* because a level tube filled with spirits or alcohol provided the essential means for making the line of sight horizontal. With the advent in recent years of self-leveling instruments which automatically make the line of sight horizontal with a pendulum device, the term direct leveling has assumed, accordingly, a broader meaning.

The employment of direct leveling procedures to determine the elevations of points which are a substantial distance apart is frequently designated as *differential leveling*. In general, the terms direct and differential leveling are practically synonymous. Certain variants of the differential leveling method, such as profile leveling, will be described later.

Indirect leveling is exemplified by barometric and trigonometric leveling. *Barometric* leveling makes use of the phenomenon that differences in elevation are proportional to the differences in the atmospheric pressure. Accordingly, the readings of a barometer observed at various points on the earth's surface yield a measure of the relative elevations of these points. The method is explained in Art. 5-20.

The vertical distance between two points in a vertical plane, i.e., difference in elevation, can be determined by the measurement of a vertical angle and a horizontal distance, just as the length of any side in any triangle can be computed from the proper trigonometric relations. This procedure is called *trigonometric* leveling, and a modified form, called *stadia* leveling, is commonly used in mapping. Trigonometric leveling is treated in Art. 5-19.

5-5. Leveling Instruments The basic instrument used to measure differences of elevation is the engineer's level. Although of many types and designs, it consists essentially of a telescope for sighting and a leveling device for maintaining the line of sight in a horizontal position. This device may be a spirit level vial, whose bubble must be centered, or a pendulum. Differential leveling is accomplished by projecting optically a line, which is initially horizontal, from the instrument to a graduated rod held in a vertical position. When carefully leveled and rotated about its vertical axis, the line of sight of a level ostensibly generates a horizontal plane. The elevation of any nearby point, within the length of the rod below that plane, may be ascertained relative to the elevation of the line of sight of the instrument. The difference in elevation is merely the reading on the rod at the graduation where the line of sight cuts or intersects it.

The two most representative types of American-made leveling instruments are the *dumpy level* and the *wye level*. In both instruments, the spindle revolves in the socket of a leveling head that is controlled in position by four leveling screws. The leveling head is screwed to a tripod. At the lower end of the spindle is a ball-and-socket joint which permits a flexible connection between the instrument proper and the foot plate. Hence, when the foot screws are manipulated, the level is moved about this point as a center. The only significant difference between these two instruments is in the manner of supporting the telescope.

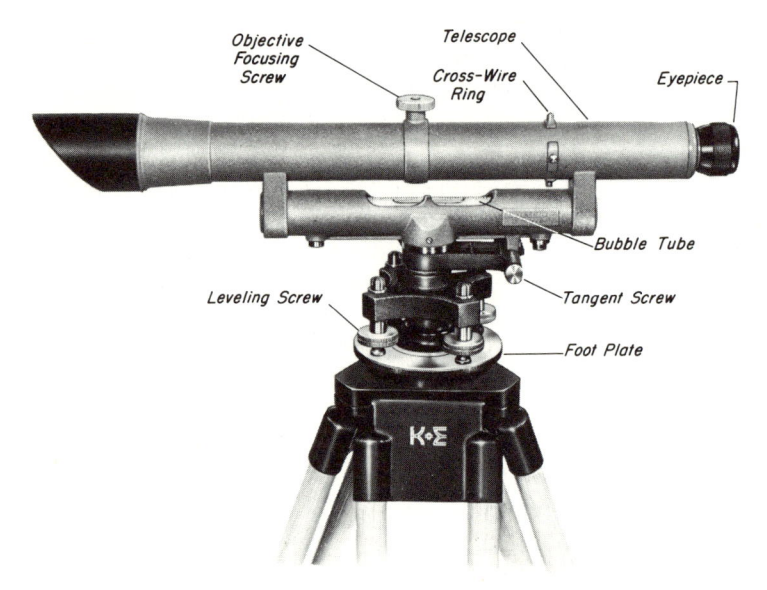

Fig. 5-3. Dumpy Level. Keuffel & Esser Co.

Dumpy Level. Figure 5-3 depicts a dumpy level with an internal focusing telescope providing an erect image. The telescope is rigidly attached to a horizontal bar, which houses the level tube. The level tube is adjustable in vertical position by means of a capstan screw at the eye end. The sensitiveness of the level tube is 20″ of arc per 2 mm graduation and the magnifying power of the telescope is 32 diameters. The use of a two-toned green surface finish facilitates the reflection of heat and minimizes glare. The meaning of the term "dumpy level" is associated with the dumpy appearance of older models of this type, which had inverting eyepieces and consequently shorter telescopes than wye levels of the same magnification.

Wye Level. The wye type of engineer's level received its name from the fact that the telescope rests in Y-shaped supports. The telescope, with the attached bubble, is independent of the other parts of the instrument and may be lifted from its supports and turned end for end; also it may be rotated about its axis in the wyes. The instrument depicted in Fig. 5-4 produces an erect image and the

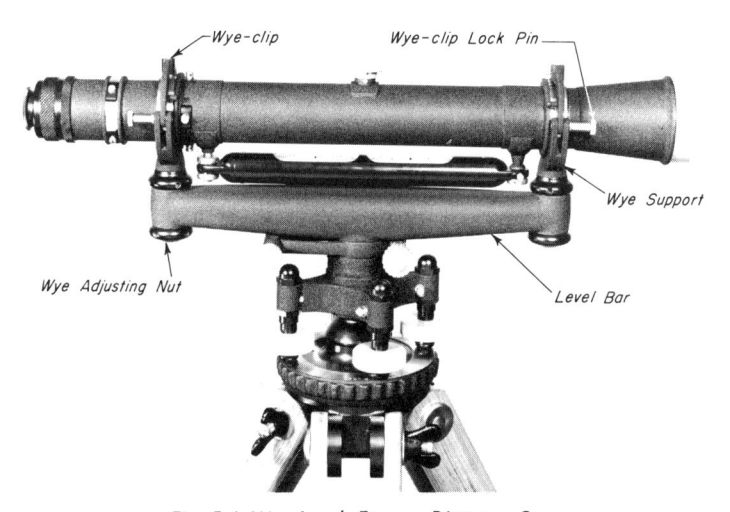

Fig. 5-4. Wye Level. *Eugene Dietzgen Co.*

magnifying power of the telescope and sensitiveness of the bubble tube are substantially the same as those of the previously described dumpy level. Although comparatively few wye levels are still manufactured, many of them remain in service in the United States.

Tilting Level. A *tilting level* is one whose telescope can be tilted or rotated about a horizontal axis through the use of a hinged joint. This design enables the operator quickly to center the bubble and thus bring the line of sight into the horizontal plane.

Figure 5-5 shows a level instrument of this type. A tilting screw is provided to raise or lower the eyepiece end of the telescope until

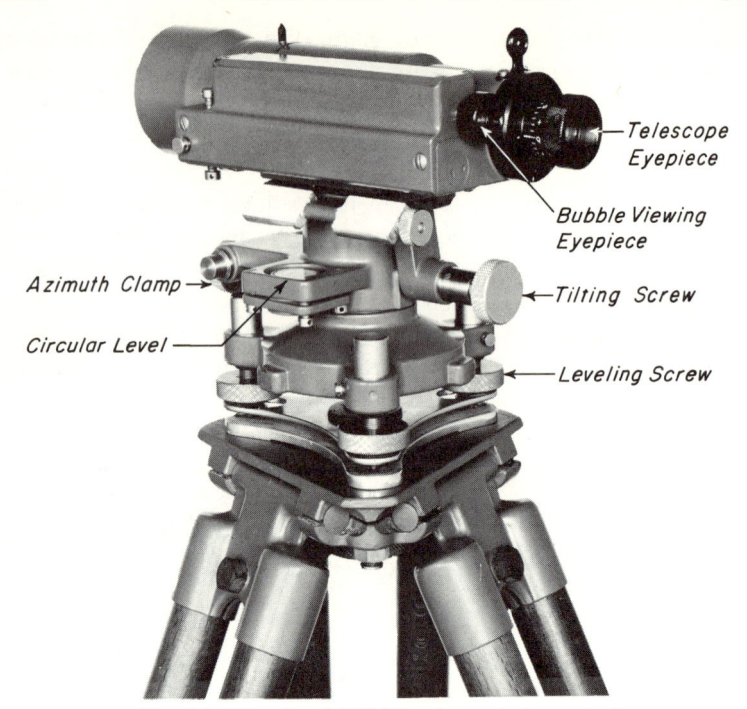

Fig. 5-5. Tilting Level. Wild Heerbrugg Instruments, Inc.

the bubble as viewed through the aperture on the left side of the telescope is centered. For preliminary leveling of the instrument, a circular level situated on the left side of the telescope is used. In order to minimize the effects of changes in temperature, the bubble tube is protected by a metal cover. This housing also covers two prisms that are located over the ends of the bubble tube. Hence, the observer sees an image of each end of the bubble. These images are split and separated by a thin line. As the bubble moves in its tube, the images move in opposite directions. The centering of the bubble is effected by causing the two images to become coincident as shown in Fig. 3-18. This device permits the accurate centering of the bubble from the normal observing position of the levelman. There is no need to walk about the level in order to view the bubble.

A distinguishing feature of the level shown in Fig 5-5 is the fact that both the inside upper and lower surfaces of the bubble tube are ground to the arc of a circle of the same radius and that the telescope can be rotated 180° about its optical axis together with the bubble tube. This permits checking the relationship between the line of sight and the axis of the bubble tube from a single setup of the level. The subject of adjustment is treated in Art. 5-7.

This instrument is supported by three foot screws that are manipulated to effect a preliminary leveling with the use of the spherical level. The magnification is 24 or 28 diameters.

The stadia constant is 100 and the image of the object sighted is inverted. Levels of this make are transported in handy, metal cases, which protect the instrument from dust, humidity, and damage.

Another tilting level is shown in Fig. 5-6. The micrometer screw

Fig. 5-6. Tilting Level with Micrometer Screw. *Keuffel & Esser Co.*

that accomplishes the rotation of the telescope in a vertical plane is graduated so that a particular reading of it can be associated with the centered position of the main bubble. The telescope has a magnifying power of 30 diameters and an erecting eyepiece. The stadia multiplying constant is 100 and the sensitiveness of the bubble tube is 20" per 2 mm movement. Note that this instrument has a conventional four-screw leveling head which for many years was favored by American engineers. However, most of the latest design of levels employ the three-screw type. The three-screw leveling head can be leveled rapidly and is relatively more stable than the four-screw type. Usually the instrument is first leveled along one pair of foot screws in the ordinary manner by turning both screws simultaneously in opposite directions. The telescope is then rotated 90° in azimuth and the centering of the circular bubble can be completed by using the third screw only.

Automatic Level. One of the most significant improvements in leveling instrumentation has been the *automatic* or *self-leveling level*. The distinctive feature of this type of level is an internal compensator that automatically makes horizontal the line of sight and maintains it in that position through the application of the force of gravity. As soon as the instrument is approximately leveled by means of a circular bubble, the movable component of the compensator swings free to a position that makes the line of sight horizontal. In addition to accelerating leveling operations, the automatic or pendulum level is very useful under conditions of unstable ground and high wind when the bubble of a spirit level must be continuously under observation to make certain it is centered.

The telescope of the Zeiss Ni2 depicted in Fig. 5-7 is of 32

Fig. 5-7. Automatic Level. *Keuffel & Esser Co.*

diameters magnification and, like most modern surveying instruments, has a glass reticule with the main crosslines and shortened stadia lines etched on it. The instrument produces an erect image, is supported on three foot screws, and is secured to the top of the tripod with a single holding screw. Other noteworthy features are an endless tangent screw for rotating the telescope in azimuth and a two-speed focusing mechanism for the objective lens.

Hand Level. For rough leveling and for sight distances of not more than 50 ft, a hand level has much utility. Most levels (see Fig. 5-8) of

Fig. 5-8. Hand Level. *Keuffel & Esser Co.*

this type consist of a metal sighting tube about 6 in. long equipped with a level vial whose bubble can be viewed and maintained in a central position when reading a rod. Hand levels are extremely useful in setting grade stakes to rough tolerances and aiding tapemen to determine whether the suspended tape is in a horizontal position.

5-6. Leveling Accessories Leveling operations require the employment of various accessories which will be briefly described.

Level Rods. Level rods are constructed in a variety of forms and patterns, but the one that has gained the widest use is the so-called Philadelphia rod, illustrated in two forms in Fig. 5-9. It is graduated in units of feet, tenths, and hundredths, the divisions being indicated by alternate black and white spaces of 0.01-ft dimension. A little study of the markings makes evident the exact subdivisions intended, and the pattern is excellently devised to permit readings to be taken quickly, easily, and with small chance for mistakes.

Form (*a*) is supplied with a target on which may be noted an auxiliary scale called a *vernier* by which subdivisions to thousandths of a foot may be read. (This device was described in Art. 3-12.) The pattern, however, is sufficiently legible to be read directly from the instrument. Furthermore, since rod readings are not taken to thousandths except in precise work, the target is seldom used, except for sights of unusual length or when vegetation or other obstructions to vision make the reading difficult. Accordingly, many levelmen remove the target, in which case all readings are taken from the instrument, and the rod, capable of such use, is called a *self-reading* rod. Obviously both forms (*a*) and (*b*) are of this type.

Form (*a*) is in two sections which, when folded together, form a continuously subdivided length of 7 ft. For readings requiring a

(a) *(b)*

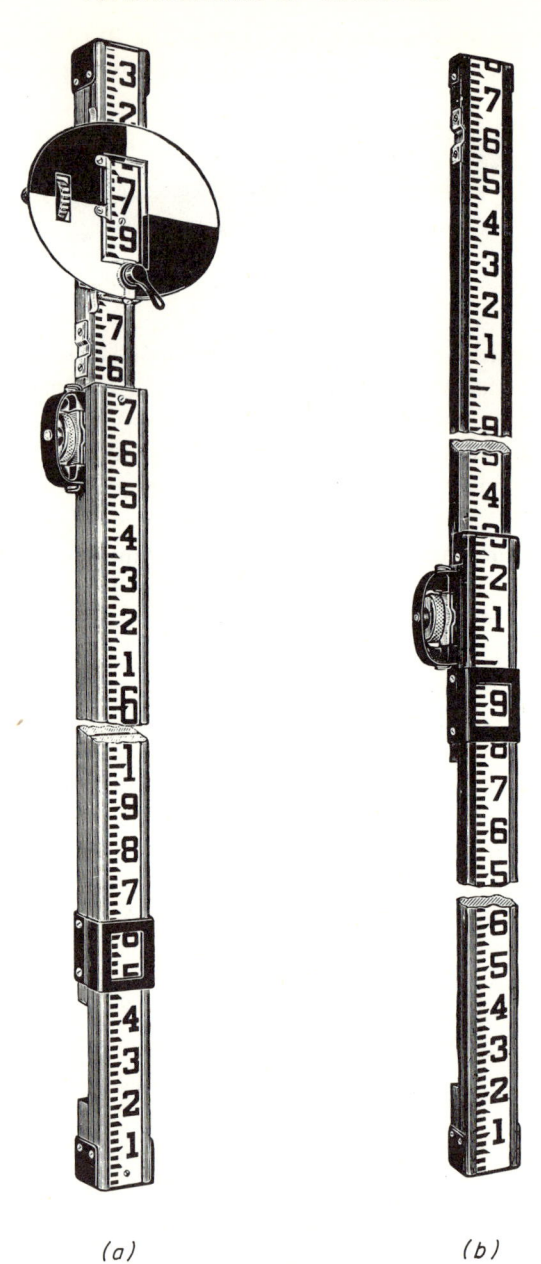

Fig. 5-9. Level Rods. *Keuffel & Esser* Co.

greater length, the rod is extended to its full length of 13 ft and clamped in this position. Form (*b*) is in three sections which permit it to be folded within a length of about 5 ft, and make it a more convenient rod to handle and to transport. It may be extended by successive lengths to its full length of 12 ft.

Rod Levels. It is imperative that the level rod be held vertically when a reading is obtained. Usually the rodman can plumb the rod accurately by balancing it upon the point on which it is held. In windy weather it is much more difficult to do this and a rod level will prove helpful. This device, a form of which is shown in Fig. 5-10, is

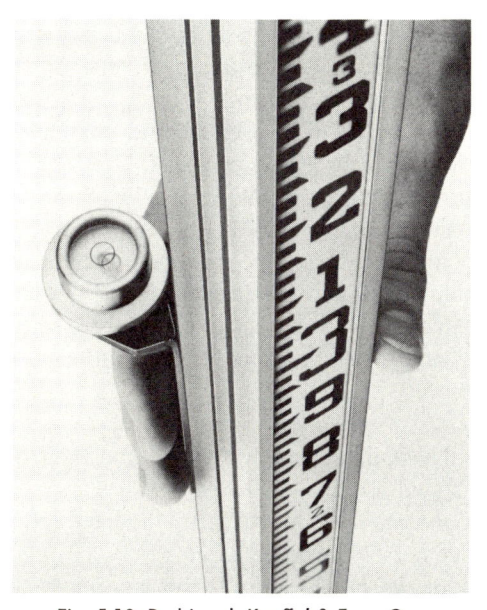

Fig. 5-10. Rod Level. *Keuffel & Esser Co.*

either attached to the side of the rod or held against the rod by the rodman.

Turning Points. The point on which a level rod is held in the course of direct leveling operations must be stable enough to hold an elevation temporarily and preferably should have a hard, rounded top. When such features as the top of a hydrant, a point on a curb, etc., are not available, a wooden stake or a steel pin must be used. A *turning pin* is a tapered steel spike with a round top and a ring through the shaft for ease in pulling the pin after it has served its purpose.

5-7. Adjustments of Level The adjustment of a level, like that of a transit, is the process of mechanical manipulation of the component parts of the instrument in order to obtain the desired relations between them. The opening statements of Art. 3-16 are just as valid for a level as a transit and should be reviewed.

The treatment of adjustment of a level will be restricted to that of the dumpy instrument only. The test and adjustment of other types are but logical extensions of the general principles and techniques described below. The levelman will find much useful information, however, regarding use and adjustment of other instruments in the literature furnished by the makers.

The student will be aided to a better understanding of the adjustments of the level if he will, so to speak, overlook the mechanical parts of the instrument and think first only of the simple essential relations that should exist between the elementary lines. Thus, he should understand that these line relations are not affected by any particular position in which the instrument, as a whole, happens to be. They remain fixed with respect to each other, whether the instrument is being carried on the shoulder or being carefully set up and leveled.

Definitions. These lines may be defined and their proper relations to each other stated as follows:

1. *The line of sight* is the line fixed by the intersection of the crosswires and the center of the objective lens.
2. *The axis of the bubble tube* is the tangent to the bubble tube at its mid-point.
3. *The vertical axis* is the axis of the spindle about which the telescope and bubble tube rotate.

Relations. The relations which should exist between these lines are as follows:

1. Insofar as the accuracy of results is concerned, there is but one essential relation, namely, that *the line of sight shall be parallel with the axis of the bubble tube;* but as a matter of convenience two other relations are desirable:

2. *The horizontal cross-wire should lie in a plane perpendicular to the vertical axis;* and
3. *The axis of the bubble tube should be perpendicular to the vertical axis.*

ADJUSTMENT OF THE DUMPY LEVEL

1. Adjustment of the Cross-Wire Ring

Relation. To make the horizontal cross-wire lie in a plane perpendicular to the vertical axis.

Test. Set up the level and sight some definite fixed point in the field of view. The bubble does not need to be exactly centered for this test and so, if necessary, the telescope may be moved up or down with the foot screws, and to the right or left on its vertical axis, until one end of the horizontal cross-wire is fixed on the point (see Fig. 5-11).

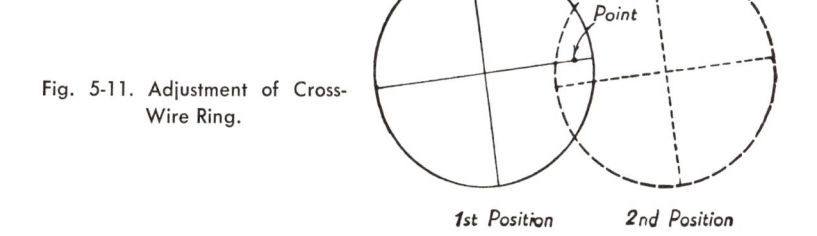

Fig. 5-11. Adjustment of Cross-Wire Ring.

1st Position   2nd Position

Then turn the telescope about the vertical axis until the other end of the cross-wire has reached the second position shown by dashed lines in the illustration. If the relation being tested exists, the point will appear to move along and will remain on the horizontal cross-wire. If not, as shown in the figure, the point will appear to move off the cross-wire.

Adjustment. The adjustment is made by slightly loosening both pairs of capstan screws which hold the cross-wire ring in position, and by turning the ring with pressure of the fingers or by tapping lightly with a pencil. When the adjustment is complete, the point will remain on the cross-wire as the telescope is moved slowly from side to side.

2. Adjustment of the Bubble Tube

Relation. To make the axis of the bubble tube perpendicular to the vertical axis.

Test. Center the bubble carefully over both pairs of leveling screws and bring it exactly to center over one pair. Then turn the telescope about, end for end, over the same pair of screws. If the correct relation exists, the bubble will remain centered. If not, it will move away from the center and the amount of the movement indicates double the error of adjustment. This relation between the axis of the bubble tube and the vertical axis both before and after reversal is shown in Fig. 5-12, (*a*) and (*b*), from which it is evident that

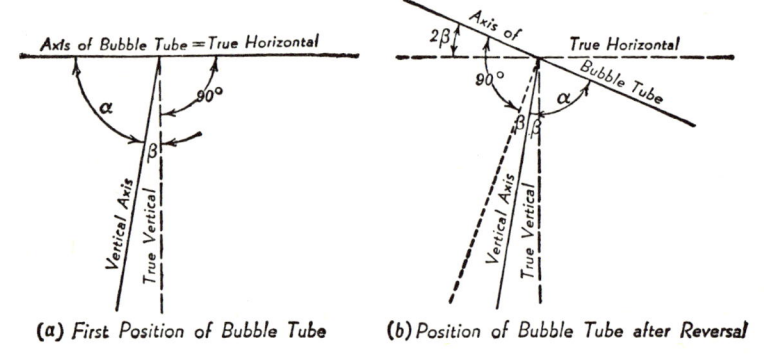

(a) *First Position of Bubble Tube*   **(b)** *Position of Bubble Tube after Reversal*

Fig. 5-12. Effect of Reversal of Bubble Tube.

the assumed error in the relationship is represented by the angle β, and that after reversal the inclination of the axis of the bubble tube is equal to 2β.

Adjustment. The adjustment is effected by bringing the bubble back halfway with the adjusting screw at one end of the bubble tube. The bubble is then brought to center with the foot screws. In other words, the bubble is adjusted by the capstan screw, the amount represented by the angle β, and then the vertical axis is moved through the same angle into its vertical position by means of the foot screws. Accordingly, if the above adjustment has been done correctly, the bubble will now remain centered both before and after reversal, thus proving that its axis is perpendicular to the vertical axis. The test should be repeated to verify the condition.

3. Adjustment of Line of Sight

Relation. To make the line of sight parallel with the axis of the bubble tube.

Test. Set the level up and drive one stake, *A*, at a distance of about 150 ft, and another stake, *B*, in the opposite direction at the same distance (see Fig. 5-13). Take a rod reading *a* on the first stake

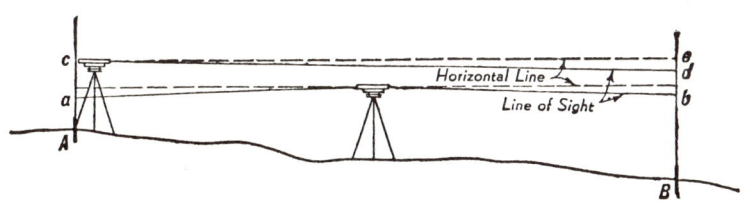

Fig. 5-13. Adjustment of Line of Sight.

and a reading *b* on the other stake. Obviously the difference in the readings, $b - a$, is the true difference in elevation between the two stakes, regardless of any error in the instrument.

Now set the instrument up near stake *A* so that the eyepiece will swing very close, i.e., within ½ in. of the rod, and read the rod by observing through the objective lens. The levelman will not be able to see the cross-wires, but a pencil held by the rodman will indicate the reading at *c*. This reading will be without error. Hence, if the true difference in elevation $(b - a)$ is added to the reading *c*, the sum will be the reading *e*, which may be called the *computed correct reading* on stake *B* that would be obtained if the line of sight were truly horizontal and therefore parallel with the axis of the bubble tube, i.e., in perfect adjustment. If the reading on *b* is less than on *a*, then the stake at *B* is higher than *A*, and, hence, the true difference in elevation $(a - b)$ will be subtracted from *c* to find the *computed correct reading* on stake *B*. Now when the levelman reads the rod on stake *B*, he has the direct evidence as to both the amount and the direction of the inclination of the line of sight for the distance between the two stakes.

Thus, assume readings as follows: $a = 2.04$, $b = 5.16$, and $c = 4.73$. Then $b - a = 3.12$ ft, the true difference in elevation between the stakes A and B. Since b is larger than a, the stake B is lower than A, and therefore $c + (b - a) = e$, or $4.73 + 3.12 = 7.85$ ft, the computed correct reading on stake B. Then the reading $d = 7.81$ is taken, and it is evident that the line of sight is inclined downward $7.85 - 7.81 = 0.04$ ft for the distance between the two stakes. This test is familiarly known as the *peg test*.

Adjustment. The adjustment is made with the instrument still in position at A, by keeping the bubble centered and by moving the horizontal cross-wire until it apparently cuts the rod at the correct reading. In the case assumed, the cross-wire would be moved until the rod reading is 7.85 ft.

This is accomplished by movements of the vertical pair of capstan screws which hold the cross-wire ring. One screw is loosened slightly and the opposite one is tightened a like amount, thus moving, apparently, the position of the horizontal cross-wire on the level rod. This is continued until the desired reading has been obtained. The horizontal pair of screws is left untouched to avoid disturbing the first adjustment.

The capstan screws should be turned with the greatest of care. The application of excessive pressure will easily strip the threads.

It is appropriate to introduce here the following miscellaneous tests which are related to leveling instruments:

(a) The Circular Level. The function of the circular level is to save time and effort in leveling the instrument. For the Zeiss level, it is particularly important that the circular bubble be in adjustment.

The test and ajustment for it are as follows: center the bubble, reverse the telescope, and if the bubble moves from the center, bring it halfway back by means of the adjusting screws in the rim of its housing. Recenter the bubble by means of the foot screws and repeat the test and adjustment, if necessary.

(b) The Reversing Point. On regular leveling operations the micrometer screw of a tilting level should not have to be rotated more than a small fraction of a turn after reversing the telescope through

180°, between a backsight and a foresight, in order to effect the final centering of the bubble. To make this possible and thus accelerate the progress of leveling, the *reversing point* on the micrometer head should be determined.

The reversing point is that reading on the micrometer screw at which, when the vertical axis of the level is truly vertical, the bubble will remain centered after reversal. This point is found by first centering the circular level bubble; then the main bubble is centered and the reading of the micrometer screw is observed. The telescope is then reversed, the bubble recentered, and the micrometer scale is again observed. The reversing point is then halfway between the two micrometer readings, and the micrometer screw should be set at that point whenever the instrument is being leveled.

This adjustment is merely an aid in quickly centering the bubble at all times. It has no effect on the accuracy of the rod readings.

(c) *Sensitivity of Bubble Tube.* In Art. 3-11 the sensitivity (also termed sensitiveness) of a bubble tube was defined as the number of seconds of central angle subtended by one division, usually 2 mm, of the bubble tube.

The field test to determine the sensitivity of the level vial of a dumpy level is conducted as follows: set up the instrument, sight the rod, and with the bubble in any noted position in the tube, carefully read the cross-wire on the rod which is at a taped distance, D, from the instrument. Then, by manipulating the foot screws, move the bubble an exact number, say five, of divisions and read the rod again. The difference between the rod readings, divided by 5, will be the intercept, i, in feet.

EXAMPLE 5-1: The intercept found by reading a level rod twice, at a distance of 300 ft, the bubble being moved five divisions between readings, is 0.15 ft. The length of a bubble division is 2 mm. Find the sensitivity, v. Here $D = 300$ ft and $i = 0.03$ ft. Hence,

$$\tan v = \frac{.03}{300} = 0.0001$$

and

$$v = 20'' \text{ (very nearly)}$$

The sensitivity can be interpreted to be the angle through which the line of sight will rotate in a vertical plane when the bubble is moved one division.

EXAMPLE 5-2: Determine the error in the reading on a level rod if the sight distance is 300 ft, the bubble is one-half division off center, and the sensitivity is 20″

$$\text{Error} = 300 \times \tan 10'' = 300 \times 0.00005 = 0.015 \text{ ft}$$

5-8. Care of Level and Rod Most of the remarks contained in Art. 3-18 refer just as aptly to the level as to the transit. A few additional comments will be made and certain important precautions will be restated.

Care of the level, as of any other instrument, begins with a general inspection of the instrument when it is removed from its carrying case at the start of a workday. Try all clamps, motions, and screws to make certain there is no binding or malfunctioning. Attach the head to the tripod securely but not so tightly that difficulty will be experienced in removing it. Always use the carrying case to transport the level over considerable distances or rough terrain during operations. Carry the level and tripod under the arm, not over the shoulder, when walking beneath trees, over rough ground, and in confined areas. The instrumentman must make certain that the tripod is set up in a stable manner and that there is no possibility of the tripod shoes slipping. *The instrument must never be left unattended.*

The quality of the leveling rods and the care given them contribute materially to the accuracy of leveling operations.

Care of the rod consists of keeping it clean, unwarped, and readable. Dragging it through the brush or sounding ditch depths with it will soon foul the face of the rod and render the graduations illegible. The Philadelphia rod should not be carried over the shoulder while fully extended because of the extreme flexibility of the rod in this condition. The rod, particularly in the extended condition, should not be left leaning against a wall or a tree, because of the possible damage that may ensue if it falls. It should be placed flat on the ground with the graduations facing upward. From time to time the continuity of the graduations from the top of the lower section to the bottom of the top section should be checked for accuracy and the sections clamped together correctly when "high rod" is used.

5-9. Differential Leveling Differential leveling is the most widely employed method for determining difference of elevation. Regardless of whether a spirit level or an automatic level is used, the theory is essentially the same. Consider a level set up on a horizontal surface, as illustrated in Fig. 5-14, with a rod held at an

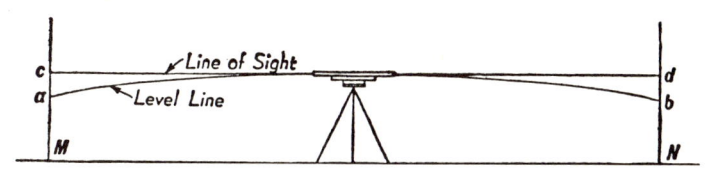

Fig. 5-14. Theory of Leveling.

equal distance on either side of the instrument at points M and N, respectively. Evidently the difference in elevation between these points is zero. Consider also that the level line passing through the telescope cuts the rod at points a and b, and that the line of sight cuts the rod at points c and d. It is evident that $a - b = c - d = 0$. Hence, we see that, for any two points which are equidistant from the instrument, the line-of-sight readings give the true difference in elevation regardless of the curvature of the earth or atmospheric refraction. The deflection of the level line from the line of sight is quite small for the ordinary lengths of sight used in leveling, being about 0.002 ft in 300 ft. For these two reasons the effect of the curvature of the earth and refraction will be very small for the usual conditions of ordinary leveling.

When the curvature of the earth and refraction are neglected, the theory of leveling is simple. With the level set up at any place, the difference in elevation between any two points within proper lengths of sight is given by the difference between rod readings taken on these points. By a succession of instrument stations and related readings, the difference in elevation between widely separated points is thus obtained.

By referring to Fig. 5-15, the common terms used in leveling may be defined.

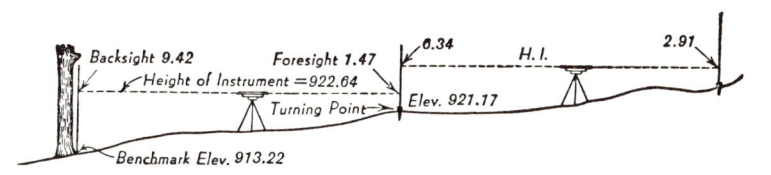

Fig. 5-15. Differential Leveling.

A *benchmark* is a permanent object of known elevation. It should be definite and located where it will have the smallest likelihood of being disturbed. Examples are a metal or concrete post planted in the ground, a notch cut in the root of a tree, a spike driven in a tree or pole, a definite corner of the masonry of a bridge or building, and a water hydrant.

A *turning point* is a temporary benchmark, i.e., a definite, firm object whose elevation is determined in the process of leveling, but which has served its purpose as soon as the necessary readings have been taken upon it.

Frequently, a point marked with lumber crayon (keel) on a sidewalk or street curb will serve as a turning point. The rod must never be held on the turf or any yielding or mobile object if it is to be used as a turning point.

A *backsight* is a rod reading taken on a benchmark or turning point of known elevation. It is the vertical distance from the benchmark or turning point to the line of sight.

The *height of instrument* is the elevation of the line of sight. It is found by adding the backsight to the elevation of the point on which the reading is taken.

A *foresight* is a rod reading taken on a turning point or other object whose elevation is to be determined. It is the vertical distance from the line of sight to the point observed.

The field procedure for executing differential levels is easily learned. Before making any readings it is essential to point the telescope toward the sky or any light background and turn the eyepiece until it is properly focused on the cross-wires and the latter appear clear, black, and distinct. This action followed by proper focusing of the objective lens will prevent parallax as explained in Art. 3-10.

In setting up the level, the tripod legs are adjusted so that the leveling head base is approximately level, and then the legs are firmly pushed into the ground. On a steep side hill, two legs are placed on the downhill side and one uphill. The foot screws are operated in diagonal pairs, and in each pair one screw is turned inward and the other outward, the proper direction being determined by the fact that the bubble moves in the same direction as the left thumb. If one screw is turned more rapidly than the other, either the pair will bind, or the leveling head will become unstable and may wobble from side to side. Either condition is remedied by turning one screw only until the proper bearing of both has been re-established.

The bubble is centered first over one pair of screws and then over the other pair, after which the telescope is sighted on the rod and the bubble is given its final centering. The beginner is likely to waste much time in the first two steps, for which the bubble need not be centered more nearly than within two or three bubble divisions. It is only when the rod reading is taken that the centering of the bubble becomes a source of error. The beginner is also likely to spend much time carefully centering the bubble only to find, when he sights the rod, that his line of sight is either below the bottom or above the top. The time thus spent is wasted and could largely have been prevented by quickly assuring himself, by roughly sighting the rod when the instrument is first set up, that it is not too high or too low.

The field procedure of running a line of levels is illustrated in Fig. 5-15, and the corresponding field notes are shown in Fig. 5-16.

\multicolumn{4}{c}{DIFFERENTIAL LEVELING}			
\multicolumn{4}{c}{FROM TURKEY RUN CREEK TO ROUTE 24}		Locker 47	J. E. Slattery π 33 W.D. Fooks Rod June 1, 1968
Sta.	B.S.	H.I.	F.S.	Elev.		
B.M.$_1$	9.42	922.64		913.22	Spike on Root of 10 in. Oak	
T.P.	6.34	927.51	1.47	921.17	200 ft east of Highway Bridge over Turkey Run Creek	
T.P.	10.15	934.75	2.91	924.60		
T.P.	9.78	943.39	1.14	933.61		
	5.38	948.04	0.73	942.66		
B.M.$_2$			3.07	944.97		
	41.07		9.32	913.22		
	9.32	check		31.75	Boulder at S.E. Fence Corner at Cross-roads, on Route 24	
	31.75					

Fig. 5-16. Field Notes for Differential Leveling.

The levelman sets up the instrument at a proper distance from the benchmark. He prepares his notebook with the headings, as shown, and enters the usual data regarding the party organization, date, weather, etc., also the elevation and careful description of the benchmark from which the line of levels begins. The rodman holds his rod on the initial benchmark, whose elevation is 913.22 ft above sea level, and the levelman finds the backsight reading to be 9.42 ft. This reading is recorded in the backsight (B.S.) column and on the line which pertains to the object observed. The rodman proceeds to a point properly located with respect to the instrument and the line of levels, selects or fixes a good turning point and holds his rod on it. While the rodman is going forward, the levelman adds the backsight reading to the elevation of the benchmark to determine the height of instrument (H.I.), which is 922.64. He now points the telescope at the rod which is held on the first turning point and centers the bubble. Generally the bubble will move when the telescope is rotated in azimuth from a backsight to a foresight. It is, however, absolutely essential that it be accurately centered whenever the rod is read.

The levelman then observes the rod and finds this foresight reading to be 1.47 ft. This is recorded in the foresight (F.S.) column and on the line pertaining to this turning point (T.P.). This reading subtracted from the H.I. gives the elevation of the T.P., 921.17. Thus the elevation of the first T.P. has been determined with respect to $B.M._1$.

The levelman now goes forward, selects a proper position and sets up his level. The rodman remains at the turning point until the levelman is ready, when he again holds his rod to be read by the levelman. This reading, 6.34, is a backsight and is entered in that column and on the line which pertains to the first T.P. It is added to the elevation of the T.P. to determine the new H.I., 927.51. The rodman now proceeds to establish another turning point. The levelman finds the new foresight to be 2.91, and so the work proceeds until a second benchmark, $B.M._2$ is reached. Here the foresight is 3.07 and the elevation is 944.97.

Thus the elevation of $B.M._2$ has been determined with respect to $B.M._1$ and obviously the process could, if necessary, be continued to obtain the elevations of other more widely separated points.

The operation which has been described requires the teamwork of both levelman and rodman in order to achieve satisfactory results.

Before placing the rod on any point, the rodman should clean the top of the point and the rod shoe to obtain proper contact. He should stand slightly behind the rod, face the instrument, and hold the rod with both hands in a balanced position. When windy weather prevails, the use of a rod level is required particularly in the case of an extended rod. If a rod level is not available, the rodman should *wave* the rod. The rod is waved by slowly rotating it about its point of support toward and away from the instrument. The smallest reading of the rod is that when it passes through the vertical position. This value is recorded. Waving the rod is not essential nor should it be necessary when the reading is on the lower section of the rod, there is no wind, and the rodman plumbs the rod by balancing it.

The proper length of sight will depend upon the distance at which the rod appears steady and the graduations legible to the levelman and also upon the accuracy desired. Under ordinary conditions when elevations to the nearest 0.01 ft are desired, the maximum length of sight is about 300 ft. Heat waves may severely restrict the lengths of the sights.

The engineer's transit may serve quite adequately for occasional leveling operations. The telescope bubble must be centered when the reading is obtained. Care must be taken not to confuse one of the stadia lines with the middle horizontal crossline.

5-10. Precautions The subjects of errors and mistakes are treated later, but three remarks should be made here.

1. *Centering the Bubble.* It has been emphasized that failure of the bubble to be centered is a source of error only when the rod is sighted, and time need not be wasted centering it at other times. At that instant, however, it is a principal source of error, and great care should be taken that the bubble is centered when the rod is read. The experienced levelman habitually looks at the bubble both before and after reading the rod to assure himself that the bubble was centered at that moment.

2. *Keeping the Rod Plumb.* The rodman must be careful to keep the rod plumb whenever a reading is taken.

3. *Equal Sights.* The backsight and foresight distances should be equalized insofar as field conditions permit. This is largely the responsibility of the rodman and is accomplished by estimation or by pacing. Sight distances should be limited to 300 ft.

4. *Turning Points.* Only durable points whose elevation will not change during the leveling operation should be used. Turning points on an expanse of masonry surface should be plainly marked, and the rod is held in the same position for backsight and foresight.

5-11. Field Notes Since the field notebook is the permanent record of the survey, the notes should be clear and legible. The main body of the field notes in differential leveling consists of a tabulation of the observed readings and of the computed elevations. Various engineering organizations have developed different notekeeping designs, but the format shown in Fig. 5-16 being neither unique nor foolproof, is quite representative of prevailing practice. The record of a line of differential levels must begin with the identification and elevation of a benchmark and with a suitable description of it.

Checking the Level Notes. The possibilities for mistakes in the additions and subtractions in the notes make it necessary to apply a check to these computations before any reliance can be placed on the final values. It may be noted that backsights are always added and foresights are always subtracted, and on this account the columns are commonly called (+) and (−), respectively. However, it is sometimes necessary to take rod readings on points that are higher than the line of sight. This will occur when leveling in a tunnel (see Fig. 5-17) where turning points are in the roof. Under such circumstances the rod is held in an inverted position and backsights are minus quantities and foresights are plus quantities. Obviously, the difference in elevation between the initial and the final benchmarks is equal to the difference in the sums of the backsight and foresight readings. Accordingly, no level notes may be regarded as complete until this check has been applied.

This arithmetical check is called a *page check*. It merely confirms the correctness of calculating heights of instrument and the elevations of turning points. Errors or mistakes in reading the rod would not be disclosed by a page check.

It is especially important that each benchmark be so clearly and completely described that anyone not familiar with the vicinity can find and use it at any subsequent time, possibly years later.

Checks on level work are frequently made in the field by closing circuits; consequently, the elevations of all benchmarks should be computed in the field as the work proceeds.

Fig. 5-17. Leveling Underground. *Chicago Metropolitan Sanitary District*

All records should be made in their proper place in the regular field notebook. The student is sometimes inclined to make temporary entries in the back of the book or on a loose-leaf sheet of paper, expecting to rectify and improve the record when it is copied into the notebook. He should avoid such bad practice, knowing that in actual work it would not be permitted.

Many times the elevation of a benchmark on one page in the notebook is copied from, or is related to, the same data on another page. In such a case, a cross-reference is necessary to facilitate comparisons at any time.

5-12. Grades of Leveling Leveling is classified in four orders of decreasing accuracy, according to the limits, stated below, for line closures and for the agreement between forward and backward runs, and also generally according to the methods and instruments used. A *line closure* is the discrepancy between the measured difference of elevation between two points of fixed elevation and the difference between the known elevations of those points. Another interpretation is that it is the amount by which the observed elevation of a terminal benchmark fails to equal the known or published value. Frequently, lines of levels are double run in order to make the resulting elevations more dependable. The accuracy specification with respect to such closures refers to the discrepancy between the results of the two runnings. Since practically all of the errors in leveling are accidental in their effect, the final error of closure in a level circuit is proportional to the square root of the number of rod readings. Accordingly, assuming that the number of readings per mile will generally be about the same, the accuracy or the size of the maximum permissible error of level work is expressed as some coefficient times the square root of the distance in miles.

The standards for quality of leveling work are stated in terms of the maximum allowable error of closure as follows:

$$0.017 \text{ ft } \sqrt{\text{miles}} = \text{First-order accuracy}$$
$$0.035 \text{ ft } \sqrt{\text{miles}} = \text{Second-order accuracy}$$
$$0.05 \text{ ft } \sqrt{\text{miles}} = \text{Third-order accuracy}$$
$$0.10 \text{ ft } \sqrt{\text{miles}} = \text{Fourth-order accuracy}$$

First- and second-order leveling is sometimes called geodetic leveling and its treatment is outside the scope of this text. Leveling of third- and fourth-order accuracy is most commonly associated with engineer-

Fig. 5-18. Leveling During Construction. *Dravo Corp.*

ing construction surveys (see Fig. 5-18). Sometimes, such leveling is called *ordinary leveling*. Because of its vague meaning the use of this term is not recommended.

5-13. Errors in Differential Leveling The principal sources of error that affect the results of leveling are listed below. Each is discussed briefly as to the nature of the source, the magnitude of the error, and the means by which it may be minimized or eliminated. Some remarks obviously do not refer to an automatic leveling instrument.

1. *Nonadjustment of the Instrument.* The purpose of the first adjustment is to cause the horizontal cross-wire to lie in a horizontal plane so that a rod reading taken anywhere along its length will be the same. If readings were always taken at the intersection of the cross-wires, this adjustment would be immaterial.

The second adjustment is merely for the convenience of the levelman and in no way affects the precision of results, for when the telescope is turned the bubble is always recentered.

The third adjustment is the one important adjustment that affects the precision of results, because, when the bubble is centered, if the line of sight is not parallel to the axis of the bubble tube, it will not give a correct reading of the rod. This error is systematic, but becomes accidental in the process of leveling.

The magnitude of the error depends on the inclination of the line of sight with respect to the axis of the bubble tube and is eliminated completely to the extent that backsight distances are made equal to foresight distances. Accordingly, this source of error is minimized by keeping the instrument in good adjustment and by equalizing backsight and foresight distances. This latter end may be attained within proper limits for most work by estimation. For more careful work the distances may be paced or obtained by stadia readings.

If the peg test of a spirit or an automatic level indicates that the line of sight is inclined upward or downward from the horizontal in an amount of, say, 0.005 ft or more, per 100 ft, an adjustment of the instrument should be made.

EXAMPLE 5-3: Calculate the error in the preliminary elevation of a benchmark resulting from leveling a distance of 24,000 ft up a uniform slope with backsights of 300 ft and foresights of 100 ft if it be assumed that when the bubble is centered the line of sight is inclined upward 0.004 ft per 100 ft. Disregard all other sources of error.

The solution indicates that the total error in the backsight readings is (60)(3)(.004) or +0.72 ft.

The total error in the foresight readings is (60)(1)(.004) or +0.24 ft.

Hence, the elevation of the terminal benchmark would be too high by 0.48 ft.

2. *The Bubble Not Centered.* If the bubble is not centered when the rod is read, an accidental error in the reading results. The magnitude of the error depends on the sensitiveness of the bubble tube. Thus, for the usual bubble tube of 15″ sensitiveness, the line of sight is displaced 0.02 ft at a distance of 300 ft when the bubble is one division off center. This source of error is minimized by carefully watching the bubble at the time the rod reading is taken. The experienced levelman looks at the bubble both before and after reading the rod to make sure the bubble was centered when the reading was taken.

3. *Incorrect Reading of the Rod.* This source of error is due to the fact that the eye is not able to judge exactly where the horizontal cross-wire apparently cuts the rod or the target. It is an accidental error and its magnitude depends on the distance to the rod, the qualities of the telescope, parallax, the weather conditions, etc. The error should not exceed 0.005 ft at a distance of 300 ft, and this may be taken as about the maximum distance for good results of ordinary precision. If weather conditions are adverse, the length of sight is shortened correspondingly. When long sights are necessary, as when crossing a wide river, the mean of a series of readings is taken.

4. *The Rod Not Plumb.* If the rod is not plumb when a reading is taken, a positive systematic error results. The magnitude depends on the size of the rod reading, i.e., it is greater near the top than near the bottom of the rod. Its value may be estimated by use of Eq. 2-2.

EXAMPLE 5-4: Calculate the error in an observed rod reading of 10.00 ft if the rod is out of plumb (in the plane joining rod and instrument) by 6 in. at that height.

$$\text{Error} = \frac{v^2}{2S} = \frac{(.5)^2}{2 \times 10} = +0.01 \text{ ft}$$

This source of error is minimized by carefully plumbing the rod for all readings. This is done by the rodman standing squarely behind the rod and balancing it between the finger tips of both hands. When the wind is blowing, this is more difficult to do. For this condition, and in fact at all times when readings are taken near the top of the rod, it is a good practice for the rodman to "wave the rod" slowly toward and away from the level, and for the levelman to take the minimum reading. The rod is plumbed in the direction transverse to the line of sight by means of the vertical cross-wire.

For more careful work a rod level is used to plumb the rod.

5. *Parallax.* Parallax is a source of accidental error. It is eliminated by focusing the eyepiece on the cross-wires.

The test for detecting parallax is very easy to perform and should be made before beginning any work with the level or transit.

6. *Curvature of the Earth and Atmospheric Refraction.* The nature of the effect of the earth's curvature and atmospheric refraction has been discussed in Art. 5-3. At 300 ft this source of error is 0.002 ft, which is quite negligible so far as a single reading is affected. In the long run, however, if a considerable number of unbalanced sights are taken, i.e., foresights longer than backsights, or vice versa, an appreciable error may result.

Curvature and refraction always has the effect of making a rod reading in differential leveling too great. Consider curvature and refraction effects in Example 5-3 as follows:

EXAMPLE 5-5: The constant error due to curvature and refraction alone per setup would be (see Art. 5-3) 0.0019–0.0002 or 0.0017 ft. For 60 setups the accumulative effect would be 0.10 ft. The combined error from these two sources is +0.58 ft.

This error is eliminated to the extent that foresight and backsight distances are made equal.

If a wide river or ravine necessitates a long sight, the error may be eliminated by the method of reciprocal leveling (see Art. 5-15).

7. *Incorrect Length of Rod.* Obviously a rod of incorrect length will cause a systematic error, but this again is rendered accidental

in the process of leveling, provided the differences in elevation are not great. However, if a line of levels is carried from the bottom to the top of a hill (or vice versa), a serious error will result if the rod has even a small error in its length. This error is similar to an error in the length of a tape in taping.

The error is to be minimized by testing the length of the rod from time to time by comparing its length with a steel tape. If necessary, computed corrections can then be made to benchmark elevations.

EXAMPLE 5-6: The preliminary difference of elevation between two points was found by differential leveling to be +404.68 ft. It was then discovered that the 12-ft rod was actually 12.025 ft long with the error distributed uniformly over the length of the rod. What is the corrected difference of elevation?

$$\text{Correction} = \frac{404.68}{12}(.025) = +0.84 \text{ ft}$$

The revised difference of elevation = +405.52 ft.

Sometimes an error is caused by improper joining of the sections of the rod. If this condition is detected, the rod should be discarded. This condition will also result if, in extending the rod, the rodman is not careful to extend the rod its full length.

8. *Settling of the Instrument.* In swampy ground, mud, or melting snow the level may settle in the interval of time between a backsight and a foresight reading. If so, a systematic error results which, in the ordinary process of leveling, is not rendered accidental, because the foresight reading will always be too small. For this condition extra precautions are taken to ensure the stability of the level, or the error can be rendered compensative if, on alternate setups, the foresight is taken before the backsight.

9. *Poor Turning Points.* Either because of soft ground or otherwise, a turning point may be faulty such that the bottom of the rod will not have the same elevation for the different readings taken upon it. For example, a turning point in the middle of a sidewalk would be a poor turning point unless it was definitely marked, because the rodman might be called away from the spot and upon returning he would not be able to recover the exact point. Also, even for levels of ordinary precision, a turning point is never taken on the bare ground because of the indefiniteness and instability of such

a point. Accordingly, this source of error is minimized by taking turning points on such definite and stable objects as the top of a stake, a solid stone, or the head of a steel pin driven in the ground.

10. *Heat Waves.* Under certain conditions there are noticeable "heat waves" which affect the precision of the rod readings. This is an accidental error to be minimized by limiting the lengths of sights. Under extreme conditions, leveling work is abandoned until the heat waves have subsided.

11. *Wind.* A high wind shakes the instrument, making it difficult to keep the bubble centered and to read the rod correctly. This is also an accidental error to be minimized by shortening the lengths of sights.

Because of the great variety of conditions encountered, it is obviously impossible to say which of the above sources are of greatest importance for all conditions. However, it may be said that for the ordinary conditions and precision of results, the first four listed are more important than those that follow. Also, from the above discussion it may be noticed that the most important precautions to observe are these: (1) keep the instrument in good adjustment; (2) keep the bubble centered when reading the rod; (3) keep the rod plumb; and (4) keep the foresight and backsight distances equal.

5-14. Mistakes Mistakes which are commonly made in leveling work are the following:

1. *Misreading the Rod.* This refers to reading as one number what in reality is another; e.g., reading the rod as 5.32, when the correct reading is 5.37. Or, what is most dangerous on long lines of levels, reading the wrong foot mark. This happens most often when the marks on the rod are obscured by leaves on trees, grass at the bottom of the rod, etc. This kind of mistake is to be prevented by habitually looking at the markings and numbers on both sides of the cross-wire reading.

2. *Not setting the rod on the same point for a foresight and the subsequent backsight.*

3. *Recording and Computing.* As in other kinds of work, mistakes may be made in calling numbers and in recording them. Such num-

bers should always be called back, to avoid mistakes. Also, since level notes require additions and subtractions for all results, there is a constant danger of mistakes in these computations. For this reason a page check should be made. However, the recording of a rod reading in the wrong column will cause a blunder that cannot be detected by a page check.

In many level parties, the rodman, as well as the levelman, keeps and computes the notes for all turning points and benchmarks, thus providing a further check against mistakes. The checks to be applied in leveling work are of two kinds; first, those that apply to the computations in the field notes, and second, those that apply to the field work itself. The principal check to the field notes is that indicated above.

The field work is checked by closing all level circuits either upon the initial benchmark or upon another benchmark whose elevation is known. All benchmarks along a line of levels should be used as turning points.

5-15. Reciprocal Leveling Where a line of levels crosses a wide and deep ravine, or a river, it is sometimes necessary to take sights much longer than are ordinarily permissible. For such sights the errors of reading the rod, the curvature of the earth, and the nonadjustment of the instrument become important, and special methods are employed to minimize their effects.

The error in reading the rod is reduced by using a target and taking the mean of a number of readings. The errors due to the nonadjustment of the instrument and the curvature of the earth are eliminated by a special method called "reciprocal leveling," which will now be described.

The procedure is as follows (Fig. 5-19): the instrument is set up a

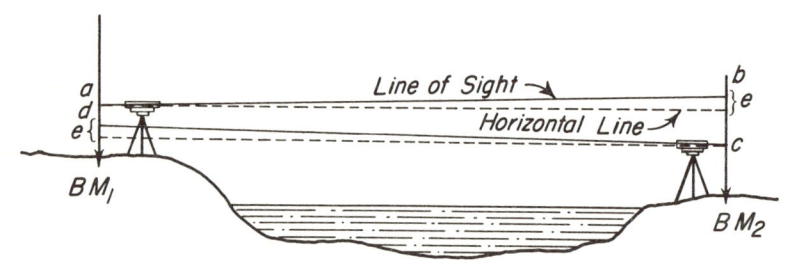

Fig. 5-19. Reciprocal Leveling.

short distance from $B.M._1$ and readings a and b are taken. Obviously the near reading a is without error and the far reading b is subject to the unknown error e due, it may be assumed, to the nonadjustment of the level. The instrument is then set up near $B.M._2$ and readings c and d are taken. Here, the reading c is without error and d is subject to the error e. For the readings taken at $B.M._1$ the true difference in elevation, D, between $B.M._1$ and $B.M._2$ is given by the equation $D = (b - e) - a$; also for the readings taken at $B.M._2$, $D = c - (d - e)$. Adding these equations, we have

$$D = \frac{(b - a) + (c - d)}{2} \tag{5-2}$$

from which the unknown error e has been eliminated. Since this error may be due to either or both sources mentioned above, the result will be free from errors of both kinds.

If it is plainly evident which benchmark is the higher, the matter of the signs of the rod readings should cause no difficulty. But, if it is doubtful which benchmark is the higher, it will be necessary to pay due regard to the signs of the readings. Since the elevation of $B.M._1$ is known (or may be assumed), the readings on it may be called backsights; and the readings on $B.M._2$ will then be foresights. Evidently, the above equation may be written

$$D = \frac{(a + d) - (b + c)}{2} \tag{5-3}$$

and if each backsight is given a plus sign and each foresight a minus sign, then the true difference in elevation will be

$$D = \frac{\text{(sum of backsights)} - \text{(sum of foresights)}}{2} \tag{5-4}$$

If the sign of the right-hand member of this equation is plus, then $B.M._2$ is higher than $B.M._1$, and vice versa.

It may be noted that for the conditions shown in Fig. 5-19 the sign of the equation is minus and, therefore, $B.M._2$ is lower than $B.M._1$.

5-16. Precise Levels There is no difference in principle between ordinary and precise leveling. In the former the distances between check points are relatively short, and the derived elevations are satisfactory for all routine purposes. However, in the case of precise leveling, the circuits may be of substantial length, and every effort should be made to control and limit, carefully, all possible sources of error. This requirement makes necessary the use of high-grade instruments and the employment of field procedures that will minimize instrumental and observational errors. Although precise leveling embraces the first three orders of leveling, the following comments refer to third-order leveling only. Ordinarily a good quality tilting level or an automatic level is used with a precise rod.

Third order levels may be run between two benchmarks of higher-order accuracy or between two benchmarks of adjusted third-order accuracy. Also, they may form a closed loop which begins and ends at the same point.

In starting or closing a new line of levels on a single benchmark, it is extremely important that the mark be positively identified from its description and that there be no evidence of its having been disturbed. If there is any doubt, a connection should be made to the nearest existing benchmarks at both ends of the line and the observed differences of elevation should agree with those calculated from the published values within the limits specified for this grade of work.

The accuracy of leveling depends as much on the quality of the rods as on the instrument. Differential leveling is really the measurement of vertical distances with the rod being used as a tape. The purpose, in part, of the instrument is to define a reference line, the line of sight, from which such measurements are made to the turning points or benchmarks. All too frequently experienced engineers will give little attention to the accuracy and condition of the rod but will insist upon a good instrument.

There are various types of *precise level rods*. They can be obtained with graduations in meters, yards, and feet and their respective subdivisions. All of them are characterized by a graduated metallic strip that is rigidly fastened to a metal foot piece, with the latter being secured to a wooden staff that provides backing for the strip. A spring at the top of the wooden staff serves to keep the metal strip under constant tension. The strips are made of invar or lovar. A circular level enables the rodman to hold the rod vertically, and some rods have a recess for mounting a thermometer.

Figure 5-20 shows two precise rods that are commercially available. Both are 12 ft long and of one-piece construction. These rods are called self-reading because they are read directly by the levelman and have no targets to be set by the rodman.

[5-16] ELEVATION MEASUREMENTS 145

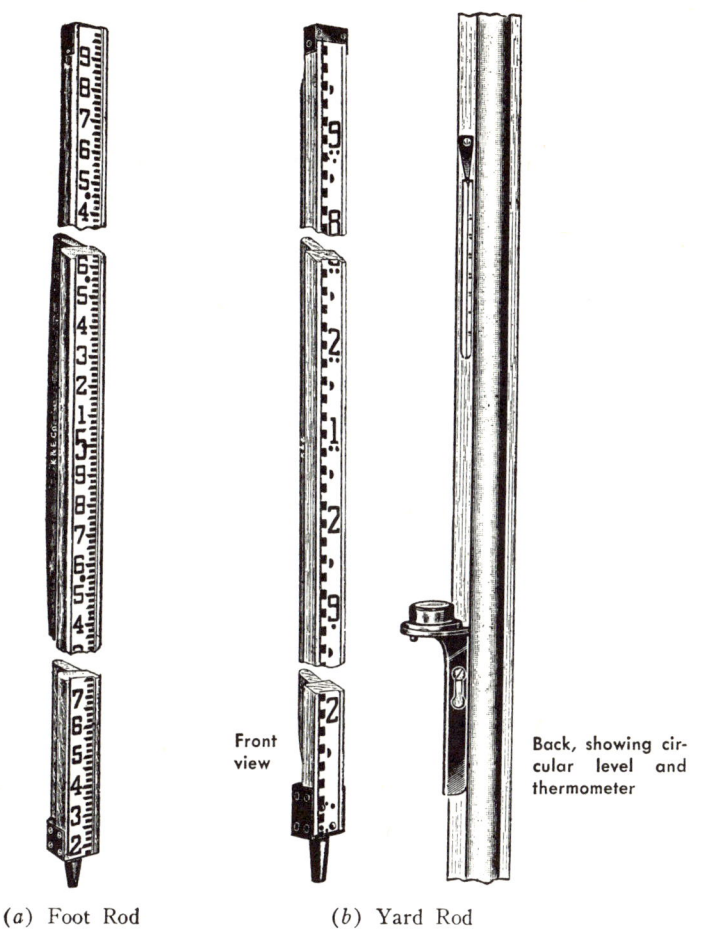

(a) Foot Rod   (b) Yard Rod

Front view

Back, showing circular level and thermometer

Fig. 5-20. Precise Leveling Rods. *Keuffel & Esser Co.*

Ordinary extensible level rods are usually unsuitable for precise leveling.

The maximum allowable length of sight depends upon the atmospheric conditions as well as the sensitiveness of the bubble and the quality of the telescope. Some surveying agencies limit the length of sight to 350 ft. The engineer who executes only occasionally precise levels will find it best to reduce drastically the sight distance to something like 150 ft.

Balancing of backsight and foresight distances is extremely important. These distances are readily obtained by stadia methods as part of a three-wire observing program.

The field record for a line of precise levels is somewhat different from the notes given in Art. 5-11. One important distinction is that there is no need to compute each H.I. and turning-point elevation. The desired difference in elevation between benchmarks is merely the difference between total backsight and foresight readings. Another significant difference is due to a change in the observing procedure. All three wires are read for each sight. The time required to obtain these extra readings is very nominal. The additional accuracy that is obtained easily justifies this procedure.

Figure 5-21 shows the manner of arranging the notes for a line of

THIRD-ORDER LEVELS From USGS "G-1" to USGS "6 LES 1948"						K&E No. 104326 Rods 10 and 11 Temp. 72°F Cloudy	Aug. 27, 1968 Weers-Levelman McCarthy-Recorder
B.S.			F.S.					
Thr'ds	Mean	Stadia	Thr'ds	Mean	Stadia		
1.984 1.338 0.688 4.010	1.337	0.646 0.650 1.296	4.719 4.027 3.335 12.081	4.027	0.692 0.692 1.384		
5.647 4.933 4.218 14.798	4.933	0.714 0.715 2.725	4.410 3.664 2.916 10.990	3.663	0.746 0.748 2.878	USGS "G-1"-on south side of East entrance to Civil Eng. Hall, University of Illinois; 18" above ground in wall; aluminum tablet stamped "G-1"	
5.802 5.021 4.242 15.065	5.022	0.781 0.779 4.285	5.822 5.083 4.348 15.253	5.084	0.739 0.735 4.352		
6.007 5.239 4.472 15.718	5.239	0.768 0.767 5.820	5.018 4.290 3.565 12.873	4.291	0.728 0.725 5.805		
	16.531			17.065			

Fig. 5-21. Record of Third-Order Levels (Foot Rod).

levels conducted with a level, such as that shown in Fig. 5-6, and a rod like that of Fig. 5-20a graduated in feet. The intervals between the middle and the top and bottom stadia wires are computed and recorded for each rod reading. These intervals should agree within 0.003 ft for all but the longest sights. The accumulative sum of the backsight and foresight intervals provides a convenient means for ascertaining quickly at any time how well the sight distances are balanced.

5-17. Vertical Control Data The basic vertical control system for the nation consists of an extensive network of lines of first- and second-order differential leveling executed by the U.S. Coast and Geodetic Survey. This network contains approximately 300,000 monumented benchmarks. In addition, many thousands of miles of third-order levels have been run by the U.S. Geological Survey between the first- and second-order lines. A portion of this network is shown in Fig. 5-22.

Vertical control data are available upon request from numerous federal agencies, such as the U.S. Coast and Geodetic Survey, U.S. Geological Survey, U.S. Corps of Engineers, and the Tennessee Valley Authority. In addition, there are such state agencies as the Department of Public Works of the Commonwealth of Massachusetts, and the Maryland Bureau of Control Surveys and Maps from which benchmark elevations can be obtained.

Occasionally the engineer will encounter, quite by accident, a standard benchmark and desire its elevation to control or check leveling operations. It cannot be too strongly recommended that the name of the organization on the tablet be carefully noted, as well as all the stamped numerals and identifying letters on the disc. A direct inquiry to this organization will result in the receipt of the latest, adjusted elevation on a specific datum, usually the 1929 Mean Sea Level Datum.

A typical benchmark tablet is shown in Fig. 5-23.

Some of the older government benchmark tablets were stamped with the elevation. This practice generally has been discontinued.

5-18. Surveying Monumentation Monumentation refers to the process of marking in an enduring manner the physical position of survey points. Proper monumentation, together with supporting records and reports, is the sole means for documenting surveys and providing the necessary relationship between field surveys and maps and charts. These comments refer not only to the preservation of benchmarks but also of points expressing horizontal position.

The value of a system of vertical control depends upon the distribution and permanency of the benchmarks that serve to monument the results of the leveling operations. As used herein, a benchmark is a permanent point whose elevation was established with third- or higher-order levels. Since benchmarks are the only visible evidence of the leveling operations, it is obvious that they should be of a substantial character, situated in locations favoring their subsequent

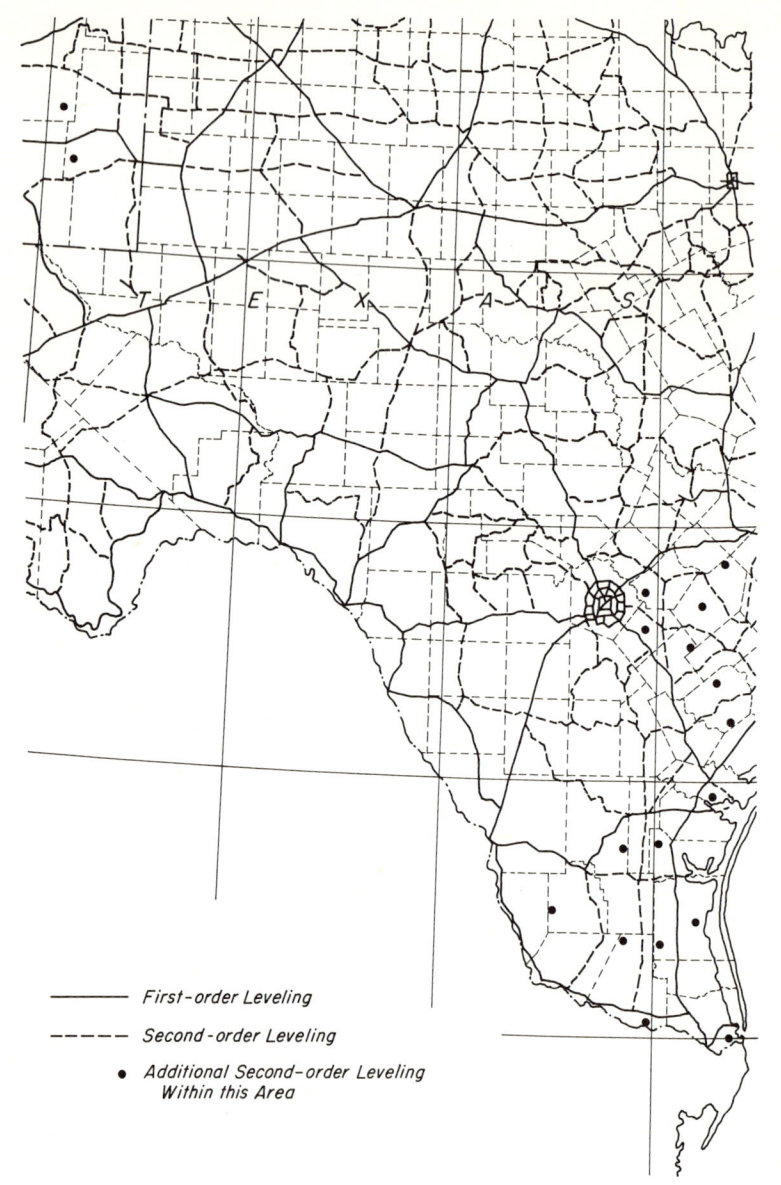

Fig. 5-22. Level Network. U.S. Coast and Geodetic Survey

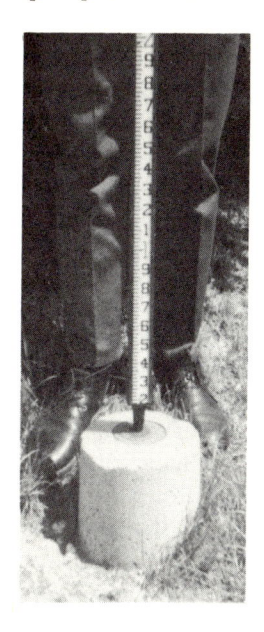

Fig. 5-23. Benchmark Tablet.

use, and so well described that they can be easily recovered and used for many years.

The civil engineer originating lines of differential levels at government benchmarks will find them to be of several types, set in a variety of locations, and at different spacings. In almost all cases the distinguishing feature of these benchmarks is the use of a standard metal tablet bearing the inscribed name of the organization that set the mark. An example of such a standard disc is the marker shown in Fig. 5-23. It is of bronze and about 4 in. in diameter. The split shank serves to fasten it to the object in which it is set. Such discs have been set in masonry structures, ledge rock, and buildings, but most frequently in concrete posts. Usually benchmark discs are set horizontally (shank vertical), but on occasion the discs have been set vertically (shank horizontal) in the face of a wall. In the latter case, special care must be taken in order to obtain accurate rod readings on the marker.

The selection of the site for a benchmark is an exceedingly important factor affecting the useful life of the mark. Accordingly specific instructions have been issued by the governmental surveying bureaus to the chiefs of level parties with respect to the most satisfactory locations for benchmarks. One of the most important considerations in the choice of a location for a benchmark is permanence and freedom from disturbance. Benchmarks should not be set where it is probable the reconstruction of bridges and buildings, the widening of highways, and the improvement of alinement at road or railroad intersections will disturb them.

The benchmarks set by the federal agencies are at various spacings ranging from one to three miles. In running supplementary levels between such monumented points the local engineer will frequently set temporary or intermediate benchmarks. These will be chiseled squares in masonry, bolts in steel structures, spikes in trees, and similar points. No benchmarks should be established in positions where over-hanging structures, tree limbs, etc., prevent the rod from being held in a vertical position on the mark.

The description of a benchmark is of the greatest importance for it provides the only means of recovering the mark. Much care has been taken in preparing adequate descriptions of the thousands of monumented marks in the national level net. Only by means of such descriptions can the civil engineer find a desired benchmark and be positive of its identity. Should he set monumented benchmarks for an engineering organization, it is considered essential to incorporate the following information in the descriptions of such points:

1. Distance and direction from a well known place.
2. Distance and direction from nearby objects.
3. Nature of the object in or on which the mark is placed.
4. Nature of the mark and how stamped.

The following is a typical U.S. Coast and Geodetic Survey benchmark description:

L 9—About 3.8 miles southwest along the Northern Pacific Railroad from the station at Blackduck, Beltrami County, *at Hines;* 84 feet northwest of the northeast corner of the station, 33.8 feet west of the west rail, and 39.9 feet east of the center line of U.S. Highway 71. A standard disk, stamped "L 9 1931" and set in the top of a concrete post. Elevation 1,404.174 ft.

An excellent type of benchmark is that shown in Fig. 5-24. It consists of a brass rod, one of whose ends is firmly embedded in the outside wall of a building and the other end having a bulb-like shape upon which the rod is easily held.

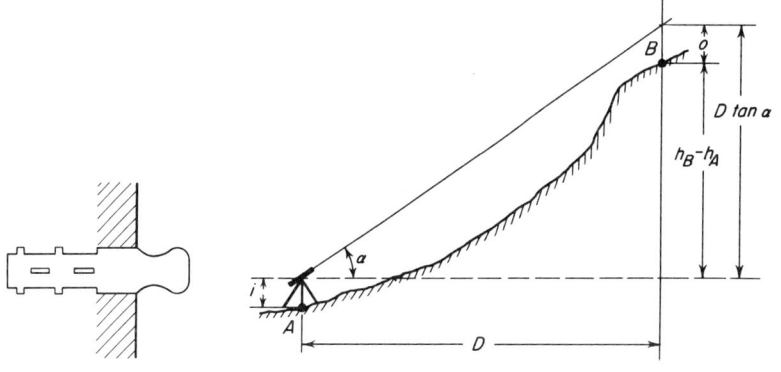

Fig. 5-24. Benchmark in Wall.

Fig. 5-25. Trigonometric Leveling—Short Sight.

5-19. Trigonometric Leveling The determination of differences of elevation from observed vertical angles and either horizontal or slope distances is termed *trigonometric leveling*.

Figure 5-25 depicts the usual situation: the sight distance is nominal, the horizontal distance, D, is known, and the vertical angle, α, has been measured with an engineer's transit as described in Art. 3-15. The height of the telescope above point A is denoted by i, and the vertical angle is read to a point that is situated the vertical distance o above station B.

The difference of elevation is given by the expression:

$$h_B - h_A = D \tan \alpha + i - o \qquad (5\text{-}5)$$

When sights longer than, say, 1500 ft are taken, it is advisable to incorporate the effects of curvature and refraction in the calculation of vertical height.

Figure 5-26 portrays the essential conditions. The separate effects of refraction and curvature are denoted by h_r and h_c. The net effect of both phenomena can be calculated by the usual expression for C & R (Eq. 5-1) and included in Eq. 5-5 to produce

$$h_B - h_A = D \tan \alpha + 0.021 S^2 + i - o \qquad (5\text{-}6)$$

Note that whereas for situations involving an angle of elevation (upward sight) the sign of the term for C & R is positive, the opposite would be true for an angle of depression (sight downward from the horizontal).

Fig. 5-26. Trigonometric Leveling —Long Sight.

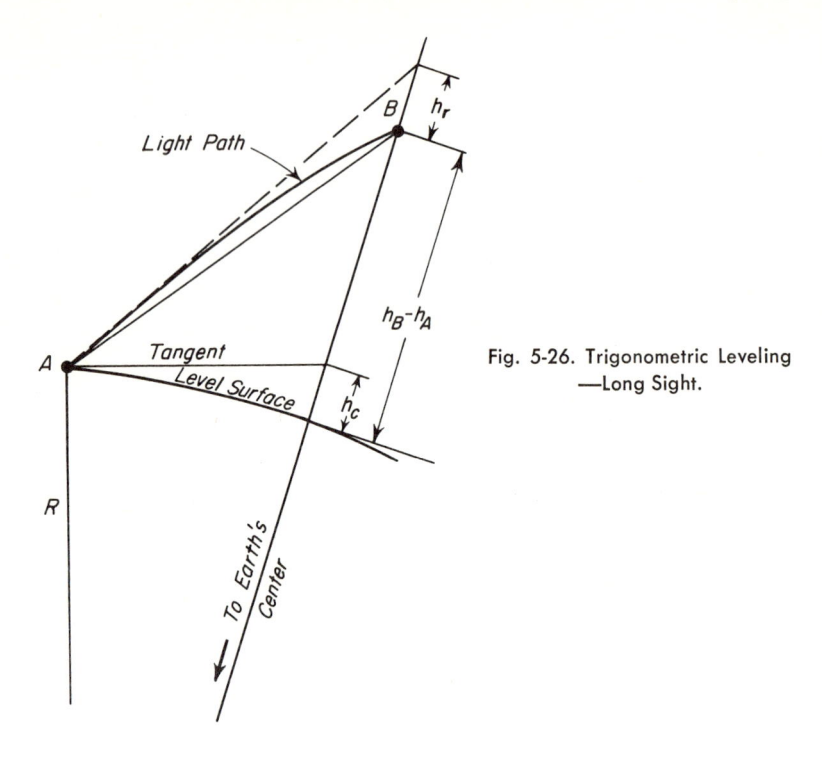

A form of trigonometric leveling, called stadia leveling, is frequently employed in certain mapping work. Inclined distances rather than horizontal distances are utilized together with vertical angles to obtain the difference of elevation. Stadia leveling is explained in Art. 14-15.

5-20. Barometric Leveling The practice of *barometric leveling,* or the determination of elevation by observations of atmospheric pressure, depends upon the basic principle that the pressure caused by the weight of the column of air above the observer decreases as the observer rises in altitude. However, the relationship between pressure and altitude is not constant because air is compressible. Additional factors, although not quite as important, which influence the density of air are temperature and humidity.

The precise *surveying altimeter* (Fig. 5-27) is an improved version of the old aneroid barometer. It is remarkably sensitive to changes in atmospheric pressure and its operation is simple. The graduation of the dial into feet makes possible direct readings of altitude. The surveying altimeter is used primarily to determine differences in elevation between points, one of which is a benchmark or station of known elevation.

Since altimeter surveying is dependent on the measurement of air pressure, all those conditions that affect air density, other than change of elevation, must be considered if the most satisfactory results are to be obtained. In other words, the assumed pressure-altitude relationship prevails only under certain standard conditions.

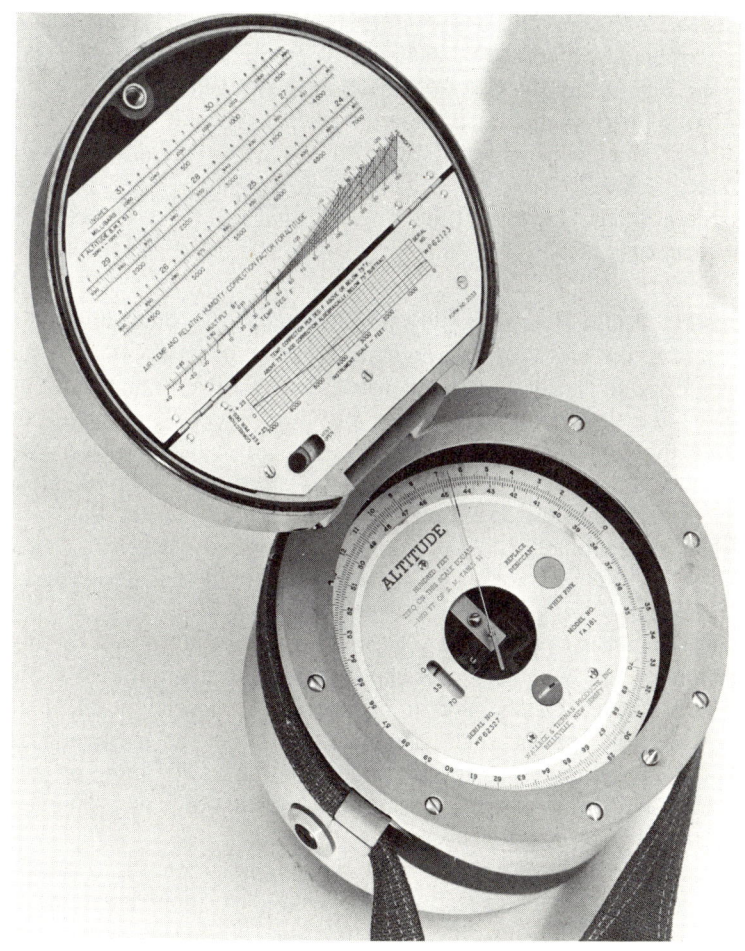

Fig. 5-27. Surveying Altimeter. *Wallace & Tiernan, Inc.*

If the survey is performed when these conditions do not exist, corrections must be applied to the observed differences of elevation.

Altimeter surveys can be conducted in several ways. However, only one procedure, the *single-base method*, will be explained. Two altimeters are employed with the single-base method. One altimeter remains at a point of known elevation where readings of the altimeter and a thermometer are made at regular intervals. The other altimeter, called the roving altimeter, is transported to those points where elevations are desired. Readings are made of the altimeter and a thermometer. To the observed difference of elevation is applied a correction for temperature (and sometimes humidity) and the corrected difference of elevation is combined with the elevation of the base station to give the elevation of the field station.

Barometric leveling is performed when expedient for exploratory surveys and in situations where the accuracy requirements are greatly reduced. Altimeter elevations are frequently used in the reduction to the horizontal of slope distances that were measured electronically.

5-21. Profile Leveling Profile leveling is the procedure of determining the elevations of points at regular intervals along a fixed line. Prior to the design and construction of sewers, highways, railroads, and similar projects, stakes are set at intervals, usually 100 ft, along the centerline. These 100-ft points are called *stations*. Points intermediate between full stations are termed *plus stations*. Hence, a stake situated 2400 ft from the point of beginning of the project would be identified as "24 + 00." A stake marking a stream crossing 75 ft in advance of station 24 + 00 would be designated as "24 + 75."

For engineering purposes, a ground profile is the trace of the intersection of an imaginary vertical surface with the ground surface. This profile is usually plotted on specially prepared profile paper, on which the vertical scale is much larger than the horizontal; and on this plotted profile, various studies relating to the fixing of grades and the estimating of costs are made. The field work of profile leveling provides the data for this work.

Assuming that the alinement has been fixed on the ground by setting stakes at 100-ft stations, the level party first determines, by the usual procedure of differential leveling, the height of the instrument, which has been set up conveniently near the line of stakes.

Foresight readings, called "ground rod" readings, are then taken on the ground at each stake and at intermediate "plus-station" points where there is a marked change in the ground slope. Since these ground-rod readings are used only for plotting and have no relation to the determination of benchmark elevations, they are taken to the nearest *tenth of a foot* only. Accordingly, all elevations of ground stations are computed to the nearest tenth of a foot only.

A staked centerline, profile, and form of notes are shown in Fig. 5-28, *a, b,* and *c.*

Figure 5-28*a* shows that the location of the instrument and turning-point stations are independent of the centerline. Figure 5-28*b* shows the plotted profile of the ground line, the horizontal and vertical scales being 100 ft and 20 ft per in., respectively. From the shape of the ground profile it is apparent why it was necessary to take plus-station readings to show the stream crossing and the changes in slope.

Figure 5-28*c* shows that the profile-level notes are similar to those for differential levels except that a separate column, headed G.R., is used for the ground-rod readings, and that several of these readings, depending on field conditions, are recorded between turning points.

It may be added that when the levelman is reading the rod to the nearest tenth of a foot only, he need not use the extreme care in centering the bubble that is necessary when sighting on benchmarks or turning points.

5-22. Cross-Section Leveling Cross sections are profiles taken transverse to the centerline of a project. They provide the data for estimating quantities of earthwork and for other purposes. Two classes of cross sections may be mentioned: (1) roadway cross sections and (2) borrow-pit cross sections.

Roadway Cross Sections. The procedure consists of taking profile readings at right angles to the centerline at each station along the route, and at any plus-stations where a cross section is necessary, to represent correctly the ground surface. For a roadway, the cross-section profile extends both to the right and to the left of the centerline, usually to the right-of-way fence, or at least as far as any possible earthwork will be constructed.

The notes are kept somewhat as shown in Fig. 5-29 for which it

Fig. 5-28. Profile Leveling.

Sta.	B.S.	H.I.	F.S.	G.R.	Elev.	HIGHWAY LOCATION—FONTANA TO WALWORTH	Dietzgen Level No. 22 June 15, 1968	F. R. Grant π M. Hughes Rod 10
B.M.	1.76	849.63			847.87			
14				2.2	847.4			
15				7.0	842.6			
T.P.	0.63	838.66	11.60		838.03		Spike in Pole 60' Left of Sta. 14 + 20	
+40				4.9	833.8			
16				6.3	832.4			
+70				6.8	831.9			
+75				13.2	825.5		Spring Creek	
+80				6.6	832.1			
17				6.2	832.5			
18				4.5	834.2			
T.P.	9.76	846.32	2.10		836.56			
19				5.1	841.2			
+40				6.2	840.1			
+60				10.1	836.2			
20				8.5	837.8			
+40				3.5	842.8			
21				3.7	842.6			
T.P.			3.24		843.08		Root of 10-in. Oak 40' Right of 21 + 20	
	12.15		16.94		847.87			
			12.15		Check → 4.79			
			4.79					

\multicolumn{4}{l}{CROSS-SECTION NOTES}	Route 47	November 4, 1968
Sta.	B.S.	H.I.	F.S.	Elev.	Berger Level No. 4 Fair, Cool 10 in. Oak, 40'L. Sta. 412	Williamson, Inst. Lessler, Rod Rogers, Tape
B.M.	6.15	757.58		751.43	Left ₵	Right
	7.32	761.74	3.16	754.42		
					30 ft 12 ft.	12 ft 30 ft
415					$\frac{4.6}{57.1}$ $\frac{5.0}{56.7}$ $\frac{5.8}{55.9}$	$\frac{6.2}{55.5}$ $\frac{6.8}{54.9}$
+50					$\frac{6.5}{55.2}$ $\frac{7.1}{54.6}$ $\frac{7.2}{54.5}$	$\frac{7.6}{54.1}$ $\frac{8.2}{53.5}$
416					$\frac{4.0}{57.7}$ $\frac{4.2}{57.5}$ $\frac{4.5}{57.2}$	$\frac{4.8}{56.9}$ $\frac{5.4}{56.3}$
417					$\frac{2.8}{58.9}$ $\frac{3.0}{58.7}$ $\frac{3.1}{58.6}$	$\frac{3.6}{58.1}$ $\frac{4.0}{57.7}$
T.P.			4.10	757.64		

Fig. 5-29. Field Notes for Cross-Section Leveling.

is assumed that readings extend 30 ft to the right and to the left of the centerline. Each ground-rod reading is recorded on the right-hand page as the numerator of a fraction; the denominator is the elevation of the point, found by subtracting the numerator from the H.I. The differential levels for determining the heights of instrument are recorded as usual on the left-hand page.

Figure 5-30 shows the manner in which the cross-section data are plotted and also the relation of the proposed roadway to the ground line. The same scales are usually used both vertically and horizontally.

From such cross sections the areas may be calculated, or determined by planimeter, and volumes computed as described in Art. 7-16.

Borrow-Pit Cross Sections. For borrow pits, for some contour maps, and for some other purposes, it is desirable to take ground-rod readings at the corners of squares of regular dimensions as shown in Fig. 7-8.

In this case, the rod readings are usually designated by some simple system of rectangular coordinates and recorded as are profile ground-rod readings. For example, starting from some assumed origin, the abscissas of points may be designated by numbers *1, 2, 3,* etc., and ordinates by letters *a, b, c,* etc. Accordingly, the ground-rod readings may be designated as *a-1, a-2, b-1, b-2,* etc.

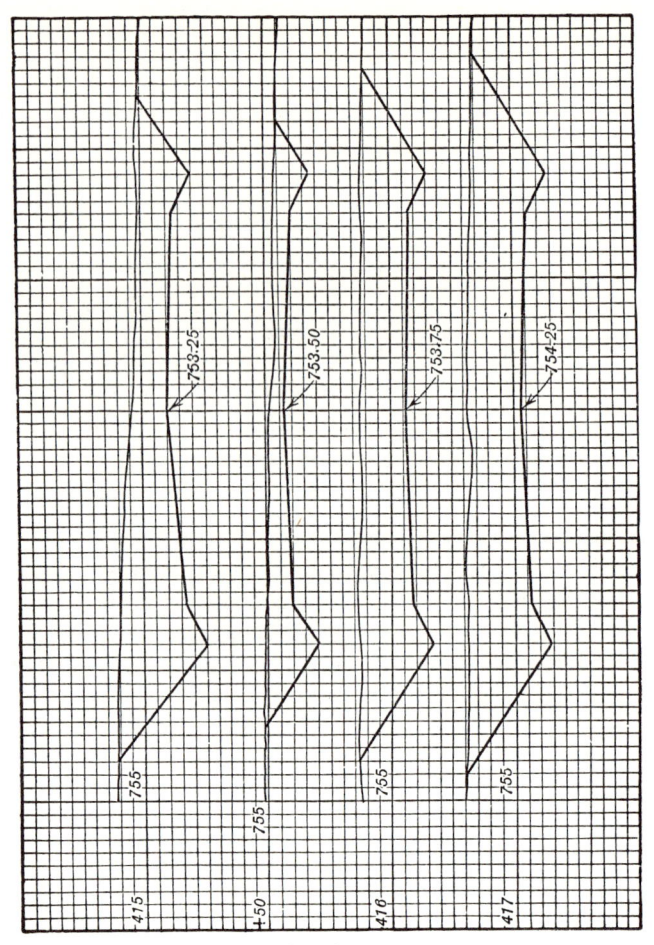

Fig. 5-30. Plotted Cross Sections.

PROBLEMS

5-1. Complete the record (on page 159) of a line of differential levels, determine the elevation of B.M. 37, and make a page check.

5-2. A peg test of a dumpy level was made over a distance of 200 ft. With the instrument midway between stakes A and B, the reading on A was 6.47; on B, 5.14. With the instrument beside B, the backsight on B was 4.82 and the foresight on A was 6.14. Is the line of sight inclined upward or downward and how much per 100 ft?

5-3. An automatic level was tested over a distance of 250 feet to determine

ELEVATION MEASUREMENTS

Sta.	B.S.	H.I.	F.S.	Elev.
B.M. 36	3.37			416.22
TP	6.14		8.21	
TP	5.08		4.17	
TP	11.10		6.52	
TP	4.32		5.91	
TP	8.89		0.32	
B.M. 37			1.15	

the inclination of the line of sight when the circular bubble was centered. With the level midway between two points A and B, the reading on A was 5.205; on B, 4.365. With the instrument beside A, the backsight on A was 4.190 and the foresight on B was 3.365. Is the line of sight inclined upward or downward and how much per 100 ft?

5-4. A determination of the sensitiveness of the bubble tube of a wye level was made. With the rod at a distance of 300 ft, a reading of 5.320 was obtained. After moving the bubble through five divisions, the rod reading was 5.155. Find (a) the sensitiveness of the level tube per 2 mm graduation, and (b) its radius of curvature.

5-5. Calculate the uncertainty in the reading of a level rod at a distance of 240 ft if the bubble is off center by one-fourth of a division in a 20" level vial.

5-6. What is the error in the reading of a level rod which is 6 in. out of plumb at a height of 11.2 ft when

(a) A backsight of 10.50 is observed?
(b) A foresight of 1.05 is observed?

5-7. A line of differential levels was run from B.M. 105 (elev. 578.43) to B.M. 122 whose preliminary elevation was determined to be 941.86. It was then discovered that the 11.0-ft rod was 0.015 ft too long with the error uniformly distributed over the entire length of the rod. Calculate the revised elevation of B.M. 122.

5-8. Assume now that the rod of Problem 5-7 was not long but was short by 0.010 ft because of wear of the shoe or foot piece. What change will be effected in the elevation of B.M. 122?

5-9. A line of differential levels was executed with an automatic level down a long slope for a distance of 7.6 miles. Commonly accepted rules for leveling were violated by carelessly permitting uniform backsights of 100 ft and foresights of 300 ft. Following completion of the levels, a peg test revealed the line of sight was inclined downward 0.008 ft per 100 ft. If the preliminary or observed difference of elevation was −802.41 ft, find the corrected difference of elevation. Do not neglect curvature and refraction effects.

5-10. A line of levels was run through a tunnel beneath the Mississippi River at New Orleans. All benchmarks and turning points were located in the crown or roof of the tunnel and the level rod was inverted when readings were obtained. Complete the following record. Datum is mean sea level.

Sta.	B.S.	H.I.	F.S.	Elev.
B.M. 57	10.02			5.87
	11.43		9.16	
	9.78		8.42	
	10.14		7.20	
	4.29		5.89	
B.M. 58			11.03	

5-11. With reference to Fig. 5-25, the following data are given: $D = 1072.85$ ft, $\alpha = +14°15'$, $i = 4.70$ ft, and $o = 6.45$ ft. Calculate the difference of elevation.

5-12. With reference to Fig. 5-26, the slope distance AB was measured electronically and found to be 6010.50 ft. The zenith distance at point A was $84°05'$. Calculate the difference of elevation. Do not neglect curvature and refraction.

5-13. Complete the record of profile leveling on page 161 and show the page check.

5-14. The 1929 mean sea-level datum at San Francisco corresponds to a reading of 8.60 ft on the tide staff there. The City of Oakland datum intersects the staff at 11.59 ft. If the mean sea-level elevation of a point is 1.46, what is its elevation relative to the Oakland datum?

REFERENCES

1. Karren, Robert J., "Recent Studies of Leveling Instrumentation and Procedures," *Surveying and Mapping,* Vol. 24, No. 3, pp. 383–397, American Congress on Surveying and Mapping, Washington, 1964.
2. Rayner, W. H., and M. O. Schmidt, *Surveying—Elementary and Advanced,* Chapter 22, D. Van Nostrand Co., Inc., Princeton, N.J., 1957.

ELEVATION MEASUREMENTS

Sta.	B.S.	H.I.	F.S.	G.R.	Elev.
B.M. 41	3.26				416.02
5 + 00				5.2	
6 + 00				4.9	
TP	0.56		5.89		
7 + 00				8.5	
7 + 60				9.2	
8 + 00				11.7	
9 + 00				7.5	
B.M. 42			6.81		

3. Ropes, Gilbert E., "Vertical Control of the Great Lakes," *Journal of the Surveying and Mapping Division,* American Society of Civil Engineers, Vol. 91, No. SU1, pp. 35–49, April 1965.
4. United States Geological Survey, *Topographic Instructions,* Book 2, Chapters 2E1–2E5, Washington, D.C., 1966.

6 Transit Traverse

6-1. Introduction In order to provide a framework of survey points whose horizontal and vertical positions are accurately known, basic horizontal and vertical control surveys are executed. The results of such surveys are the horizontal coordinates and the elevations of the points. These position data are vitally necessary for detailed or dependent surveys and for all kinds of comprehensive mapping and charting programs. The fundamental network of points whose horizontal positions have been accurately determined is called the *horizontal control*.

Horizontal control may be established either by *traverse* or *triangulation*. Any traverse consists basically of a series of lines, whose lengths and directions are known, connecting points whose positions are to be determined. Triangulation is a method for extending horizontal control, which requires few length measurements and numerous angle measurements. The survey stations are points on the ground that define the vertices of triangles forming parts of quadrilaterals or chains of triangles. The horizontal angles at each station are measured, and the lengths of the triangle sides are determined through the application of trigonometry by means of successive computations through the chain of triangles from a side of known length. This side, the *baseline,* is directly measured in the field.

For surveys of limited extent, the traverse is most satisfactory. The triangulation method is generally preferred in rough or hilly terrain where numerous promontories are available for station sites that are easily intervisible and where it would be difficult or impossible to conduct taping measurements. However, the availability and use of electronic distance measuring devices may nullify the advantages that triangulation traditionally possessed with respect to traverse. The choice of method is not dependent upon the required accuracy of results, since high-quality horizontal control data can be obtained by both types of survey. Relative economy will usually dictate the kind of survey method.

Traverses are classified and identified in a variety of ways, viz., according to the methods employed, the quality of the results achieved, the purpose served, and the configuration or form of the connecting lines. The angular measurements of transit traverse are conducted with optical theodolites as well as vernier transits. The treatment of the subject in this chapter assumes that the linear measurements are made by taping or by electronic means.

Transit traverse, especially *transit-tape traverse,* is the most common and fundamental surveying operation. The principles and practices of transit traverse are common to topographic, hydrographic, city, and land surveys and to route surveys for highways, railroads, and pipelines. The calculation of traverses, including the determination of the rectangular coordinates of survey points, will be one of the principal topics of Chapter 7. Traverses of subordinate quality, such as stadia traverse, and traverses designed to satisfy specific engineering requirements, such as those along the centerline of a proposed highway, will be discussed in Chapters 14 and 15, respectively.

In addition to the treatment of transit traverse and a brief explanation of some other transit surveying operations (see Fig. 6-1), the

Fig. 6-1. Extending Centerline of Sewer. *Chicago Metropolitan Sanitary District*

scope of this chapter will include an expanded discussion of the importance of field notekeeping, some considerations of safety in the field, and a look at field communications. It is presumed that the student by this time has performed various basic angular and linear measurements in his instructional program and that the former topics will be much more meaningful to him now.

TRAVERSE

6-2. Traverse Configuration The geometric form or configuration of a traverse is one of the most common, but not necessarily the most significant, ways of classifying it. In general, however, a traverse is classified either as a closed traverse or an open traverse.

A *closed traverse* is one that begins and ends at the same point or at points whose horizontal positions are known. These two types of closed traverses are known, respectively, as *loop traverses* and *connecting traverses*. A loop traverse forms a continuous closed circuit as shown in Fig. 6-2. A typical example of this kind of traverse

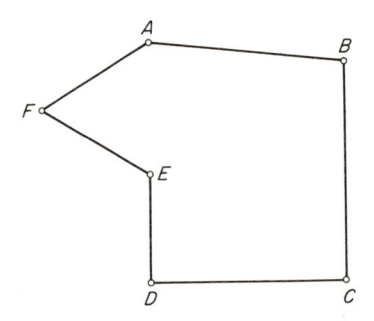

Fig. 6-2. Loop Traverse.

is the perimeter survey of a tract of land. Such a traverse is executed in order to obtain the data needed to describe adequately the parcel and ascertain its area. A connecting traverse begins and ends at widely separated points whose horizontal positions have been previously determined by a survey of at least equal accuracy and preferably of higher accuracy. The horizontal positions of the termini are held fixed in the calculation and adjustment of the connecting traverse shown in Fig. 6-3.

An *open traverse* will begin at a point of known or assumed position and will terminate at a station whose relative horizontal position is unknown. No means are available, therefore, for calculating a clo-

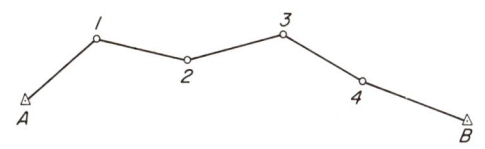

Fig. 6-3. Connecting Traverse.

sure in position and of thus assessing the true quality of the traverse operation. Traverses of this type are frequently used when making the preliminary survey for a highway.

6-3. Route Selection The route to be taken by a traverse depends upon whether it is to locate existing points or to establish points in conformance with some specific plan. The former situation is exemplified by the survey of a farm whose boundary markers have been recovered. Their relative positions are now to be determined by traverse. The latter situation refers to the execution of traverse which is to serve as the basic framework for a mapping project as described in Chapter 14.

All traverse routes should either form closed loops or begin and end at points whose positions have been fixed by control surveys of higher order. Ordinarily, the route will follow highways and railroads so as to facilitate transportation and the execution of accurate measurements. The routes should be carefully planned so that the data obtained by the traverse will satisfactorily and economically serve the purposes for which the traverse was executed. This means, in general, a suitable distribution of well monumented, properly described, and satisfactorily positioned stations.

In the case of some traverses, especially those related to the fixing of centerline positions for highways and railroads, the general directions of the traverse lines may be predetermined by such factors as the need to avoid difficult terrain, cemeteries, and residential areas. Also, various obstructions, including large and valuable trees, may exercise considerable influence upon the location of the traverse lines.

With the advent of aerial photography in mapping work and various types of engineering studies, transit traverse began to serve a very important role in providing control for such photography. Hence, today transit-traverse surveys are used with increasing frequency as control for photogrammetric procedures, as will be explained in Chapter 12. This makes quite evident the need for the traverse engineer to have a set of aerial photographs to assist him in selecting and identifying suitable locations for the traverse stations along the general route to be followed.

6-4. Stations Each one of the points in a traverse where the transit is set up and the angular change in direction is measured is termed a *traverse station*. However, not all of such points are monumented. The nature of the traverse will govern whether its stations are permanently monumented or only temporarily marked. If the traverse is an extension of the national horizontal control network and its stations were set by a federal surveying and mapping agency, they would be monumented. Upon the other hand, if the traverse constitutes the centerline of some route of transportation or communication, only a relatively few stations would remain durably marked upon the completion of construction operations.

Wood stakes are most frequently used as temporary markers for traverse stations. They are usually 1-in. by 2-in. and at least 18-in. long, with one end sharpened. They are driven flush with the surface of the ground and have a tack on top to mark the exact position of the point. The term *hub* is commonly applied to such a stake. Semipermanent markers may be iron pipe or steel rod and heavy wood hubs which have been treated with a preservative to inhibit rot. Points on concrete surfaces may be defined by chisel marks, and those on bituminous surfaces by nails driven through small metal disks called *shiners* or pieces of colored flagging. To provide for the protection and recovery of a hub, *guard stakes* may be used. They are driven obliquely so that the top is over the hub. The identification of the hub can be written on the guard stake.

All survey stakes, even the most durable hubs, are vulnerable to disturbance during construction operations and may have to be replaced. Any survey point for that matter, whether wooden hub or metal tablet in concrete, which is to have continued usefulness should be referenced to other points so that it can be quickly and accurately restored to its correct position. Such auxiliary points or *reference marks* are used to re-establish a traverse station if the marker has been moved or destroyed. The perpetuation of key survey markers is an extremely important matter to both survey engineer and technician and a topic to which the attention of the student will be directed from time to time in this text.

A typical way of referencing a survey point is depicted in Fig. 6-4. Recovery of the point is effected by describing arcs of the recorded

Fig. 6-4. Referencing a Survey Point.

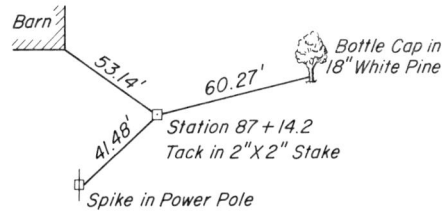

radii about the three reference points as centers. Note that the documented distances must be horizontal and that the reference marks should be located to provide a strong intersection of the arcs. Whenever possible, the marks should be within one tape length of the survey station.

In the case of transit traverse that is to serve as the framework for national mapping programs and the long-time needs of engineers and surveyors, permanent markers are commonly set at intervals not exceeding one-half mile in urban districts, or more than approximately 2 miles in rural areas. Station sites should be selected with a view to minimizing the possibility of the marker being disturbed by construction operations of the future. It is of the greatest importance that the traverse stations be durably monumented if they are to perpetuate the results of the survey. Hence, the marker should be substantial, adequately referenced, and carefully described. A 4-in. circular bronze tablet, appropriately inscribed (as in Fig. 6-5) and with a 3½-in. shank, when anchored in a concrete post, a

Fig. 6-5. Traverse Station. Tennessee Valley Authority

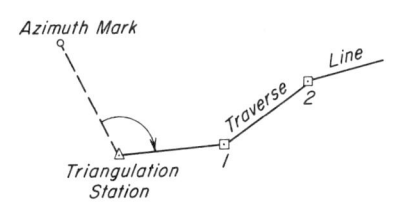

Fig. 6-6. Connection to Triangulation Station.

masonry structure, or solid rock in place, makes a very satisfactory marker.

Traverse stations, as along a line of third-order traverse (Art. 6-11), are identified by letters and numbers. The use of such numbers should not be confused, however, with the system of stationing and the employment of station numbers for points along the centerline traverse for a highway as explained in Art. 5-21.

Occasionally, it may be desirable to set the traverse stations in pairs not less than 1000 ft apart, in order to provide a reference line of known azimuth from which subsidiary or local surveys can be conveniently originated. If such a practice is not followed, one or two azimuth points should be observed at each traverse station to serve the same purpose. Such azimuth points should be objects of a permanent character that are clearly visible from the station at all times of the year. Examples of suitable points are water tanks, airway beacons, church spires, and forest lookout towers.

A typical description of a transit-traverse station is as follows:

LHT 621 (TVA, 1937; Jefferson County, Tenn.)

Located 9.6 miles southeast of Dandridge, 0.8 mile northwest of Reidtown, on the property of Bud Manning, 80 ft northeast of United States Highway 25 W, 6 ft northeast of old Reidtown Rd., 3 ft northeast of telephone pole, 60 ft north of road that connects Highway 25 W and old Reidtown Rd., 3 ft north of fence corner, and 100 ft southwest of Bud Manning's house. A standard TVA tablet set in top of concrete post and stamped "LHT 621 1938."

Reference Mark 1: 3.3 ft S. 35° W. to telephone pole.
Reference Mark 2: 27.9 ft N. 10° W. to east gate post.
Grid position (Tenn.): $X = 2,802,775.2$; $Y = 592,778.5$.
Grid azimuth to WPA 45-27-12 (distance 848.8 ft): 334°31'20".
Elevation: 1,108.459 ft.

6-5. Party Organization A transit-traverse party may consist of as few as two men and as many as eight to ten, depending upon the personnel policies of the engineering firm or the governmental surveying agency performing the work. Another important factor affecting the size of the party is whether the distance and angle measurements are performed simultaneously or at different times.

The field work is generally divided into two basic activities, distance and angle measurement. The methods of operation are highly variable and depend upon the personnel available, the manner of determining distance, i.e., electronically or by taping, and the desired accuracy. Heavy timber cutting and surveying along a route relatively inaccessible to vehicles are other factors influencing the size of the traverse party. Despite its flexibility the typical traverse party will be headed by a chief who is usually a graduate engineer. Working under his direction will be recorders, instrumentmen, rodmen, and axmen whose duties are self-evident.

6-6. Connections to Existing Control If a traverse is to fulfill its basic function of providing horizontal control, it must begin at a point of known position. Furthermore, in the case of connecting traverses, the line must end at a point whose coordinates are fixed. At the termini of the connecting traverse the chief of party must be certain that the survey points to which the traverse has been tied or connected have been properly identified through their descriptions and that they have not been disturbed. Whenever possible, even an open-ended traverse should start at a point whose position is known.

Although new traverse lines frequently begin at established traverse lines of equal or higher quality, they are sometimes connected to an existing triangulation system (see Art. 9-2). The purpose of the connection is to provide a means for introducing both relative horizontal position and direction into the traverse. In Fig. 6-6 the transit is set up on the triangulation station and the horizontal angle is turned from the azimuth mark to the first traverse point. The azimuth mark is a monumented point or a well-defined feature like a church spire at least one-fourth mile away. The azimuth to it from the triangulation station is known. Distance measurements along the traverse begin at the triangulation station.

6-7. Linear Measurements The lengths of traverse courses are determined by either taping or electronic measuring procedures. The purpose for which the traverse is being executed, the equipment available, the nature of the terrain, and the desired accuracy of the results will effect the choice of distance-measuring methods.

For traverses of moderate and lower accuracies, as later defined in this chapter, taping is the most common way of making linear measurements. This is particularly so for highway and railroad centerline surveys where the 100-ft tape is most useful in establishing stationing. For the most part, therefore, it is assumed that traverse courses are measured with a tape. A few summarizing remarks will be made concerning taping and the specific role of electronic measurements will be indicated.

(*a*) *Taping.* For routine traverse work such as in ordinary land surveys, a single measurement of each course is adequate. As the requirements for accuracy increase and assurance is desired that no blunders are made, double determinations are made of the lengths. The taping is conducted forward and backward and the allowable maximum descrepancy between the two results is specified as some fraction, such as 1/7500, of the length. From time to time, the tape should be compared with a master tape or some standard of length. The manner of conducting the taping, applying the corrections, and evaluating the results is substantially the same as that described in Chapter 2. The importance of good taping cannot be overemphasized. It is the major factor affecting the accuracy of transit-tape traverse and it is the most important element influencing the productivity of the traverse party and, hence, the cost of the traverse survey.

(*b*) *Electronic Measuring.* When the accuracy requirements of a traverse operation are strict, the use of electronic distance-measuring devices may be dictated. Also, for surveys of ordinary accuracy, considerations of economy may indicate that any of the instruments described in Chapter 4 shall be used rather than a tape.

Electronic traverse methods have virtually superseded all transit-tape procedures that were once widely utilized by the U.S. Geological Survey for establishing horizontal control. Private engineering firms, state highway departments, and land surveyors are making increasing use of the Tellurometer, Geodimeter, and Electrotape on traverse and related operations. There is no practicable limit as to the lengths of lines other than those stated by the manufacturer and imposed by the topography. The minimum length of the traverse course is that for which the instrumental error becomes an excessively large proportionate part of the length of the line as explained in Chapter 4. Furthermore, it is manifestly uneconomical to make and reduce an electronic measurement if the distance can be easily and satisfactorily measured with a tape in a fraction of the time. Under special circumstances, such as in the case of traverse surveys through busy city streets, it may be very hazardous to tape. The advantages there of electronic distance measurements are obvious.

6-8. Angular Measurements The horizontal angle measured at a traverse station serves basically to express the difference in the directions of the two lines at that point. Although any angle between the lines will satisfy this purpose, several kinds of angles have been widely used. Traverses are sometimes designated, in a superficial way, by the type of angle that was measured. The most common types are *deflection angle, angle-to-right,* and *interior angle.*

The deflection angle was defined in Art. 3-3. This angle can be turned either to the right or left of the preceding line extended. When the deflection angle is very small, the transitman is likely to make a mistake in denoting the direction in which it was turned. For this reason, the use of this angle is probably decreasing. For a loop traverse the algebraic sum of the deflection angles should equal 360°.

EXAMPLE 6-1: A loop traverse has been executed and deflection angles were measured as shown in Fig. 6-7. The bearing of line *AB*

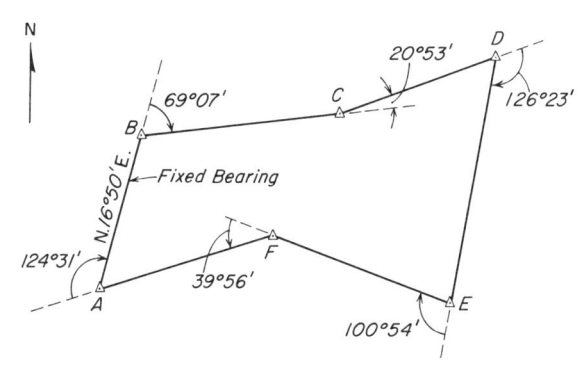

Fig. 6-7. Deflection Angle Traverse.

is N. 16°50′ E. Calculate the best bearings for the remaining sides.

The angular error of closure of the traverse will be the difference between the algebraic sum of the angles and 360°. The algebraic sum is 420°55′ − 60°49′ or 360°06′. Hence, the closure is 0°06′. It is distributed by subtracting 0°01′ from the angles at points *A, B, D,* and *E* and adding 0°01′ to the other angles. The bearing of *BC* then becomes N. 85°56′ E. The student should calculate all other bearings and utilize the check mechanism of using the revised angle at point *A* and the bearing of line *FA* to see if the computed bearing of line *AB* is the same as the fixed value.

The angle-to-right has very largely supplanted the deflection angle, especially for open traverses. At the survey point occupied by the transit or theodolite, a sight is taken toward the last station occupied and the clockwise angle is turned to the next traverse point. This and related procedures have led to the practice of referring to these points as the *rear station,* the *occupied station,* and the *forward station.* Typical field notes are shown in Fig. 6-8.

Angle-to-Right Record				Wild T-1 No. 3785	Oct. 27, 1968
Station Occupied	From To	Circle ° ′	Angle ° ′	Windy Temp. 45°F	R.P. Smith – Transit W.J. Rockford M.T. Bruns } Rodmen
38	37 39	0 00.0 17 28.8 34 57.8	17 28.9		
39	38 40	0 00.0 121 18.1 242 36.2	121 18.1	Traverse along County Highway H from Elkhorn to Tibbits.	
40	39 41	0 00.0 183 14.1 6 28.3	183 14.15	RPS-16: RR spike in bituminous pavement, west side of road, opposite entranceway to farm of J.P. Hollingsworth; 0.8 mile South of Tibbits.	
41	40 42	0 00.0 201 15.7 42 31.6	201 15.8		
42	41 RPS-16	0 00.0 195 27.9 30 55.6	195 27.8		

Fig. 6-8. Record of Angles-to-Right.

If the interior angles of a loop traverse are measured, their sum should equal $(n - 2)\,180°$, where n is the number of courses.

EXAMPLE 6-2: The interior angles of the loop traverse shown in Fig. 6-9 were measured. The azimuth of line AB is fixed at $42°16'$ (reckoned from north). Calculate the azimuths of the remaining lines.

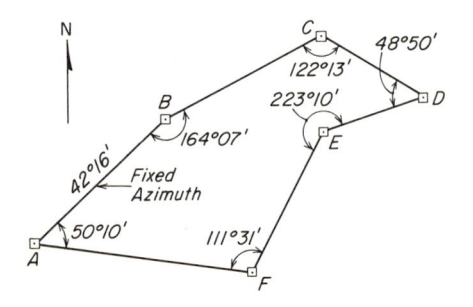

Fig. 6-9. Interior Angle Traverse.

The angular error of closure of the traverse will be the difference between the sum of the angles and $(n - 2)\,180°$. The actual sum is $720°01'$ and the theoretical sum is $720°00'$. Hence, the closure is $+0°01'$. In the absence of other information and presuming the angles were measured with a one-minute transit, a single correction of $-0°01'$ is applied to the angle flanked by the shortest sides. Hence, the corrected angle at point E is $223°09'$.

The azimuth of BC becomes $58°09'$. The student should calculate all other azimuths and employ the check available by calculating the azimuth of AB using the azimuth of FA and the angle at point A to see if the fixed azimuth of AB is obtained.

The determination of vertical angles or zenith distances is a part of the program of angular measurements at a traverse station. Such angles are used to reduce slope distances to the horizontal and, in some cases, to calculate differences of elevation between the stations.

Although angular measurements on traverse operations are conducted in substantially the manner already described in Chapter 3, a few supplementary remarks are appropriate.

The use of very short courses should be avoided because of the inherent weakness of the angles through which the azimuth must be carried.

The *range pole,* already mentioned in Arts. 2-5 and 3-13, is widely employed as a target in the measurement of horizontal angles. It is manufactured of fiber glass, steel, or wood in lengths of from 6 to 8 ft, and with an outside diameter of about 1 in. It can be held by the rodman or secured in a vertical position in the head of a tripod.

In order to promote speed and increase accuracy on traverse operations, special targets like those in Fig. 6-10 are used. The target assembly includes a tribrach with three-screw leveling head, circular bubble, and optical plummet. This arrangement permits the targets and theodolite heads to be interchangeably mounted on the tripod. Thus, only a single setup of the tripod is required at all stations. Provision is made for artificial illumination of the target.

Some surveying organizations utilize flashing signal lamps as targets on traverses of the highest quality.

6-9. Angular and Linear Accuracies In all surveying operations, it is desirable that a logical relation be maintained between the quality of angular and linear measurements if the most economical results are to be obtained. This relationship is shown in Fig. 6-11,

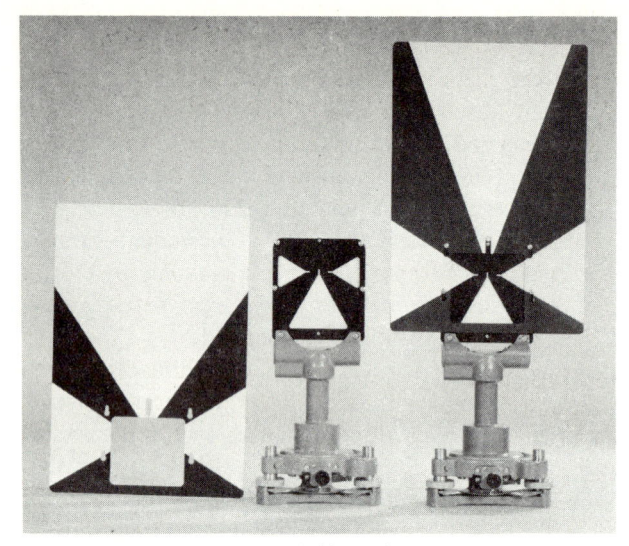

Fig. 6-10. Traverse Targets. *Wild Heerbrugg Instruments, Inc.*

where it is supposed that point B is to be located with respect to point A, and that this requires the measurement of a distance D, and an angle α. Each measurement is subject to error, and let these be represented by E_d and E_a, respectively. The linear displacement of B with respect to A, due to the angular error E_α is E_a and is given by the product $D \times \tan E_\alpha$. If the two sources of error are consistent, then E_d will be equal to E_a.

An error in distance is expressed as a ratio of the error to the

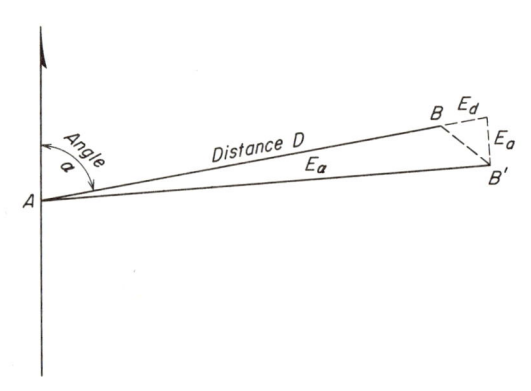

Fig. 6-11. Angular and Linear Accuracies.

distance measured and is reduced to a fraction whose numerator is unity. Hence, if a distance has been measured with a precision such that the error is 1/1000, and if an angle is to be measured with a corresponding precision, the permissible error will be such that the tangent of the angle E_α will also be 1/1000 or 0.001. Since the sines or tangents of small angles may be considered to vary directly with the angles, and since the tan of 01' is 0.0003 (nearly), then it is evident that, if the accuracy of a measured distance is 1/1000, the corresponding accuracy of a related angle is 0.001/0.0003 = 03'. Similarly, if a distance is measured with an accuracy of 1/10,000, related angles should be measured with an accuracy of 0.0001/0.0003 = 20''.

In general, it can be said that there will be consistency between linear and angular measurements if the relative error in distance equals the angular error in radians.

6-10. Checking Traverses The quality of traverse operations is dependent upon the accuracy of the angular and linear measurements. Whenever possible, checks should be imposed on these measurements in order to disclose blunders and to prevent the accumulation, beyond permissible limits, of accidental and systematic errors.

For a loop traverse the geometric check on the sum of the angles is easy to apply. The computed closure, as explained in Chapter 7, is indicative of the general quality of the traverse but provides no assurance that there are no systematic errors in distance measurement. On more important traverses of this type, it is desirable to double angles and measure courses both forward and backward.

In the case of connecting traverses executed between points of known horizontal position of higher quality, the calculated closure at the terminal point offers an excellent basis for the assessment of accuracy. It is very common, however, to introduce at regular intervals on a long traverse astronomic checks on the azimuths of the lines. The subject of astronomic determination of the bearing or azimuth of a line is treated in Chapter 10. Figure 6-12 portrays the initial part of a traverse line beginning at triangulation station Jones and extending eastward. Angles-to-right have been measured. The true (or astronomic) meridians through Jones and station 4 are shown, as well as the shifted position of the true meridian through Jones at station 4. Even though all measured angular values are perfect, it is apparent that the astronomically derived bearing of line 4-5 will exceed the calculated value by the amount, θ. The

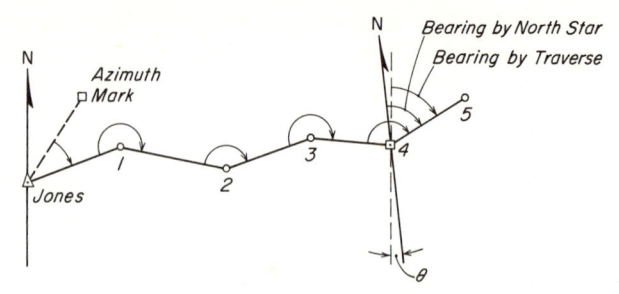

Fig. 6-12. Convergency of True Meridians.

quantity, θ, is the angular convergency of the true meridians. Convergency can be evaluated by means of the expression

$$\theta'' = 52.13L \tan \phi \qquad (6\text{-}1)$$

where L is the east-west distance in miles between two points in mean latitude, ϕ. For a complete treatment of convergency, see Art. 360 of Ref. 2.

The foregoing remarks indicate that on both connecting and open-ended traverses the application of angular checks by astronomical methods requires that proper allowance be made for convergency of meridians when comparing azimuths at a check point.

EXAMPLE 6-3: In Fig. 6-12, traverse station 4 is 8.35 miles east of triangulation station Jones. (In Chapter 7 this distance will be termed a departure distance.) The true bearing of line 4-5 as computed from the traverse angles is N. 44°10'20" E. The bearing as determined by a sight on the North Star is N. 44°16'50" E. The mean latitude of points Jones and No. 4 is 36°43½' N.

Calculate (a) Convergency
(b) Preliminary azimuth descrepancy
(c) Azimuth closure

(a) $\theta'' = 52.13$ L tan ϕ
$\theta'' = (52.13)(8.35)(.74605)$
$\theta'' = 324$
or $\theta = 0°05'24"$

(b) Preliminary azimuth descrepancy $= 0°06'30"$

(c) Azimuth closure $= 0°06'30" - 0°05'24" = 0°01'06"$

6-11. Grades of Traverse Brief consideration will be given to the nature of angular and linear errors as they affect the accuracy of traverse surveys. Ordinarily, angular errors are mainly accidental in character. Hence, they will accumulate as the square root of the opportunities for making them, or as the square root of the number of traverse points, as mentioned in Art. 1-16. On the other hand, the most significant linear errors are apt to be systematic. Therefore, the accuracy of the horizontal position of any point is likely to be affected much more by the systematic linear errors than by the accidental angular errors. Accordingly, the most common method of expressing the limit of error for traverse closure is that employing a ratio, such as 1/10,000, in the same manner as in the case of assessing the precision of taping.

The specifications shown in Fig. 6-13 are those used by the federal

	First Order	Second Order	Third Order
Number of azimuth courses between azimuth checks not to exceed............	15	25	50
Astronomical azimuth: probable error of result.................................	0.5 sec.	2.0 sec.	5.0 sec.
Azimuth closure at azimuth check points not to exceed.........................	2 sec. $\sqrt{N}$ or 1.0 sec. per station	10 sec. $\sqrt{N}$ or 3.0 sec. per station	30 sec. $\sqrt{N}$ or 8.0 sec. per station
Distance measurements accurate within.....	1 in 35,000	1 in 15,000	1 in 7500
After azimuth adjustment, closing error in position not to exceed...............	0.66 ft. $\sqrt{M}$ or 1 in 25,000	1.67 ft. $\sqrt{M}$ or 1 in 10,000	3.34 ft. $\sqrt{M}$ or 1 in 5000

N is the number of stations for carrying azimuth
M is the distance in miles

Fig. 6-13. Specifications for Traverse.

surveying and mapping agencies and other related organizations. Special attention is directed to the specifications for third-order work.

6-12. Traverse Control Data The results of third-order traverse with which this chapter has been mainly concerned are the computed coordinates of the transit stations and frequently, also, additional intermediate points. Such position data are invaluable as control for national topographic mapping. They can be used for a wide variety of other engineering purposes also, such as providing a framework of reference to which extensive highway surveys can be tied. As a matter of fact, a vast mileage of highway centerline surveys are executed every year which quite closely approximate in accuracy third-order traverse.

Third-order traverse data, particularly when positions are expressed in state coordinates as explained in Chapter 11, will be found useful in perpetuating private land boundaries, in providing lines of known azimuth with which local surveys can be oriented, and in making possible the application of checks on the accuracy of long route surveys whose termini are tied into traverse points of known position.

Much third-order traverse has been executed by such governmental agencies as the Tennessee Valley Authority, Corps of Engineers, and the U.S. Geological Survey. The best general source of information as to horizontal and vertical control in a given area is the appropriate *geodetic control diagram* which portrays in map form (1° of latitude by 2° of longitude) all work done by the U.S. Geological Survey and the U.S. Coast and Geodetic Survey in the specific region. Geodetic diagrams can be ordered from the Coast and Geodetic Survey, Washington, D.C., and the detailed leveling and traverse data desired can be subsequently obtained from the agency that performed the work.

6-13. Other Transit Surveys In addition to its use on basic traverse operations, the transit plays an important role on a multitude of other surveys. Its application to topographic, land, pipeline, railroad, and similar surveys will be described in later chapters. However, the function of the transit in the *location of detail* deserves explanation here.

Virtually all transit surveys include the location of certain natural or artificial features, termed details, with respect to the transit or survey lines. For example, when a traverse is executed along the route of a new highway, buildings, utilities, and all other existing features must have their positions determined relative to the traverse. A very common way of effecting the necessary tie measurements is shown in Fig. 6-14. The corners, nearest the transit line, of the rectangular building are located by means of stationing and a taped offset distance.

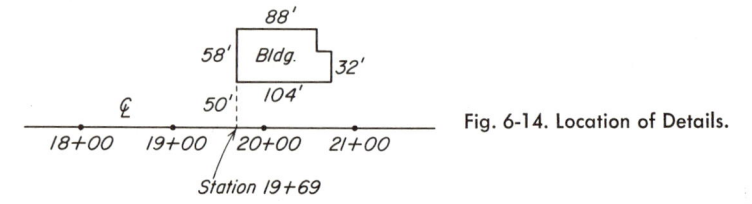

Fig. 6-14. Location of Details.

The method of radiation employing a transit and a tape is particularly useful when a number of features can be located from a central position. With the transit occupying a station of known position and with an initial backsight along a line of known azimuth, the horizontal angles between the various features are noted and the distances are measured. If the points are not close enough for a tape to be used conveniently, an electronic distance-measuring device can be utilized to great advantage. However, vertical angles also must be measured in order to convert slope distances to the horizontal.

FIELD NOTEKEEPING

6-14. Importance Recording the measurements of a surveying project in a field book is a most important task. Despite the meticulous refinement with which a survey may have been executed, its objectives will not be attained if the documentation is deficient in quantity, defective in quality, illegible, or ambiguous. The field notes of a survey represent the original record of what was found and what was done. When the notekeeping is closely supervised or performed by the party chief himself, its quality may be a significant indication as to the diligence, thoroughness, and care with which the survey was conducted. However, the neatest of notes may document major blunders in measurement or have serious informational inadequacies.

The field notebook record is particularly important in certain kinds of preliminary engineering investigations, such as for highways and major structures. Initial decisions affecting planning and design are based to a large extent on boundary, topographic, and other surveys. Even if such surveys are complete and accurate, poor notekeeping will portray an incorrect picture of the project situation and influence the deduction of erroneous conclusions.

The importance of high-quality documentation (see Fig. 6-15) of a surveying operation cannot be overemphasized. Although it is true that automatic data processing equipment, such as digital electronic computers, are having an effect on notekeeping as well as computation procedures, the use of field notebooks and the need for skillful recorders will continue in the foreseeable future.

6-15. Field Notebooks The engineer's field book is generally used for recording surveying data. The most common page size is

Fig. 6-15. Recording Notes. *Tennessee Valley Authority*

4⅝″ × 7½″. Bound and loose-leaf types are available. They are ruled horizontally with six vertical columns on the left page, and cross lines and a single vertical red line on the right page. A single page number is placed in the upper right-hand corner of each double page.

The selection of a field book should be based primarily on considerations of permanency of the written record. The book must be durable in all kinds of weather and should be able to withstand repeated handling and the aging effects of time for the many years it may be in a filing repository. The book should be strongly bound so that rough field usage will not break the binding, and the lines on opposite pages should be in correct register. A leather-bound field book best satisfies these requirements.

Loose-leaf books are convenient when performing work involving the progressive transfer of notes from field to office. However, the sheets are easily lost or misplaced or the recorder may fail to place an identifying caption on each one.

The duplicating notebook provides for detaching the original leaf while the carbon copy remains in the book.

When a field notebook is issued, the party chief should enter information as to its ownership on the outside and inside of the book, reserve the first 2 or 3 pages for an index, and paginate the entire volume. Certain other mechanical details may be significant. A few are listed as follows:

1. Make clear and legible entries with a well-sharpened, medium-hardness (3-H or 4-H) black pencil.

2. Letter all entries.

3. Make no erasures of observed data. Erroneous entries should be struck out by a single line and the corrected value entered above the original value.

4. Make liberal use of sketches but draw lines with a straight edge.

5. Indicate the general orientation of field drawings with a meridian arrow.

6. Use abbreviations and symbols to promote conciseness but be certain their meaning is known to the office engineer.

7. Keep the index up to date.

8. Do not hesitate to use narrative to explain more clearly some significant aspect of a surveying project.

9. Acknowledge the original source of any data used in initiating or closing a survey. Cross-reference important information.

10. Record data in its original observed form. For example, if deflection angles were measured on a traverse, do not mentally transform to equivalent angles-to-right and record these.

6-16. Fundamental Requirements Uniformity in the format and recording of field survey notes is important, particularly on a given project or within an organization. In a textbook of surveying it is manifestly impossible to present recommended note forms for documenting all kinds of surveying operations. However, it is practicable to name the fundamental requirements that all good field notes should satisfy. With a minimum of comments, they are

1. *Neatness*—neatness and orderliness will promote clarity.

2. *Completeness*—every pertinent item of information should be recorded. Placing notes on loose slips of paper with the intention of

transcribing them later into the book is bad practice. Relying on the memory is equally hazardous. It may be impracticable to return to the job site for a single significant item of overlooked information.

3. *Clarity*—all entries and statements must admit of only one interpretation—the correct one.

4. *Legibility*—illegible entries reflect discredit on the notekeeper and are costly to the engineering firm because the data are numerically uncertain. A return trip to the project area may be necessary.

5. *Accuracy*—there is no substitute for data that fails to meet the standards prescribed for the survey.

6. *Integrity*—the notes should reflect the personal integrity of the chief of party and his recorder. The data should be honestly acquired and not be molded or altered in any way in order to satisfy stated accuracy requirements.

6-17. Memorandum of Survey Sometimes a memorandum of survey is prepared to identify, describe, and summarize a field survey. This report will facilitate the subsequent review, evaluation, and recovery of the basic survey work. The memorandum will include information as to the party chief, survey accuracies, horizontal and vertical ties, field book numbers, equipment used, records of survey monuments, and other details. Monumentation records, sometimes supported by photographs, are invaluable.

Indexing, filing, and storage of completed field notebooks in a central office is an essential aspect of the maintenance of survey records, but this subject lies beyond the scope of this book.

FIELD SAFETY

6-18. General Considerations Attention to safety is an important consideration in the conduct of all surveying operations. At every level of responsibility from project engineer to rodman there must be a clear recognition of the hazards of surveying whether on a busy highway, in a transformer yard, on top of a triangulation tower, or in a remote area having dangerous terrain features. It is the primary responsibility of the chief of party to recognize the obscure as well as the more apparent hazards on a job site and to train his subordinates to be alert at all times, to avoid dangerous practices and situations filled with peril, and to know what to do in case of trouble. Each survey party should have a first-aid kit and some personnel trained in first-aid measures.

6-19. Promoting Safety Safety is the concern of every member of a survey party. All employees should observe sensible safeguards in handling equipment, dealing with a climatically hostile environment, and in working under hazardous conditions. This treatment cannot deal with such special situations as survival in the Arctic, helicopter transport to isolated mountain peaks, or snake bite in the desert (see Fig. 6-16). Only the more common hazards, especially those associ-

Fig. 6-16. Use of Leggings in Snake-Infested Areas. *Tennessee Valley Authority*

ated with engineering construction and along highways, will be mentioned.

It is essential because of noise, dust, and movement of all kinds of equipment to be especially vigilant around construction activities. Make it obvious to operators of cranes (see Fig. 6-17), trucks, dozers, and other mobile equipment that you are working in the vicinity.

When working near others, carry range poles and rods vertically against the body so that another's head or eyes will not be struck if you suddenly turn. Avoid guiding a stake with your hand while another person is driving it with a sledge hammer. Permitting a tape to slide rapidly through the hands may cut them. Sheath cutting tools (see Fig. 6-16), like the machete, when not in use or put them in a safe place. When cutting brush, be at least 10 ft away from the

Fig. 6-17. Transitman on Construction Survey. *Dravo Corp.*

nearest man. Use metal tree climbers in order to ascend trees and poles.

Protective headgear (see Fig. 6-18) should be worn whenever there is danger of head injury from falling objects or impact with stationary features. When walking up steep and rocky slopes or out of excavations, do not climb directly behind another man. A spill by the leading man or a rock loosened by him could cause trouble for those following.

Do not enter manholes unless there is assurance that dangerous gases are not present. Provide a suitable barrier around the manhole if the cover has been removed. Do not throw any kind of a tape across electric wires.

When working along a highway, use signs and flags to notify motorists of your presence, but do not depend on such markers to slow down everybody. All party personnel must be continually alert and ready to leap off the road. The wearing of jackets or vests of distinctive colors, such as fluorescent red or orange, will increase visibility to motorists. Painting of tripods in contrasting color bands is another safety measure. In general, an effort should be made to select

Fig. 6-18. Surveying in Sewer Tunnel. *Chicago Metropolitan Sanitary District*

a time for the survey when the traffic is slack and the party chief should minimize the time spent on the traveled roadway by working along offset lines on the shoulders or sidewalks. The safest solution in especially hazardous situations is the protection of a law enforcement officer.

FIELD COMMUNICATIONS

6-20. Hand Signals In all surveying operations it is essential that the various members of the party are able to communicate with each other. This is particularly necessary in transit traverse and other surveying activities which may widely separate the personnel. Survey party members, particularly the instrumentman and the rodmen and tapemen, generally communicate by hand signals over distances beyond conversational limits or where noise seriously interferes with conversation. Prolonged shouting should be avoided, whether on college campus or on the job. It identifies the beginner.

A few of the more common hand signals are as follows:

Give a Foresight. When the transitman desires a foresight he signals the head tapeman by holding his arm straight above his head and waving it in a circle about a vertical axis.

Set a Hub. When the head tapeman desires to set a transit point, he holds his range pole in a horizontal position above his head and waits until the transitman gives an "all right" signal. The range pole is then brought to a vertical position and lined in for a stake.

Set a Tack. When a stake has been driven and the head tapeman desires to set a tack, he holds the point of his pole on the stake and waves the top of the pole slightly to the right and left until the transitman answers with an "all right" signal. The point is then lined in on the stake.

Right or Left. When lining in a range pole the transitman should give definite signals that can be readily interpreted, e.g., he may signal to the right with one arm and to left with the other.

When showing signals, the tapeman should remember that objects are seen by contrast. Hence a taping pin or dark pencil will be seen

Fig. 6-19. Two-Way Radio Communication. *Bureau of Land Management*

best in front of a white background, but a white or yellow pencil or the white bands on the range pole will be seen best against a dark background.

All Right. Both arms are extended horizontally and the forearms waved vertically. This signal may be given by any member of a party.

6-21. Radio Where distances are too great for good visual identification, portable radio equipment may be used. It is a standard accessory on electronic distance-measuring devices. Two-way radio communication systems (see Fig. 6-19) have been found very helpful in coordinating and accelerating surveying operations on extensive route engineering projects. They also provide the means for a central office to keep in constant touch with several field parties and to expedite the dispatch of personnel to a new assignment.

PROBLEMS

6-1. A loop traverse originated and ended at point A. The bearing of line, AB, is fixed at S. 42°16½'W. Points C, D, and E lie westward from line AB. The following interior angles were measured with a 30" engineer's transit:

Sta.	Angle
A	89°15'00"
B	111°42'30"
C	97°36'00"
D	102°14'00"
E	139°15'00"

Distribute the angular error of closure and calculate both bearings and azimuths of the remaining traverse sides.

6-2. The tabulated deflection angles in a loop traverse were measured with an optical theodolite. The azimuth of line AB is known to be 345°16'10". Distribute the angular error of closure and calculate the azimuths of the remaining traverse sides.

Sta.	Angle
A	115°02'25" R.
B	92°52'15" R.
C	47°18'30" L.
D	121°47'05" R.
E	77°36'20" R.

6-3. The angles-to-right of an open-ended traverse are as follows:

Sta. Occupied	From	To	Angle
A	Az. Mark	1	50°27'
1	A	2	220°14'
2	1	3	178°36'
3	2	4	225°39'
4	3	5	155°10'

If the azimuth from station A to its azimuth mark is 310°11', calculate the azimuths of the remaining traverse courses.

6-4. Calculate the angular convergency of the true meridians passing through two points whose average latitude is 40° if the departure distance between the points is

(a) 1.40 miles
(b) 100 ft

REFERENCES

1. Pafford, F. William, *Handbook of Survey Notekeeping,* John Wiley & Sons, Inc., New York, 1962.
2. Rayner, W. H., and M. O. Schmidt, *Surveying—Elementary and Advanced,* Chapter 20, D. Van Nostrand Co., Inc., Princeton, N.J., 1957.

7 Computations

7-1. Introduction The practice of surveying comprises both field and office operations. The field work includes primarily the procurement of data and the layout of construction. The office work deals with the computations that are necessary to convert the observed measurements into such form as the purpose of the survey may require. For example, one of the important objectives of most land surveys is the determination of the area of the tract.

The scope of this chapter will encompass a brief treatment of the major principles of good computing practice and the detailed calculation and adjustment of a loop traverse followed by the presentation of the basic concepts of area and volume determination. It is assumed that the principal calculating device available to the student is the electric desk calculator. However, the versatility and greater capabilities of the electronic desk calculator in surveying computations are mentioned.

The words, calculations and computations, will be considered synonymous. The term, computer, has been widely applied in surveying to indicate both the calculating device as well as the person performing the calculations. Here, however, when applied to a calculating device, computer will signify a high-speed digital computing mechanism.

GENERAL

7-2. Basic Considerations Neatness and uniformity in method are as essential in computing practice as they are in field notekeeping. An orderly arrangement with logical sequential development of the solution not only aids the computer but also facilitates the work of the checker.

Most engineering and surveying organizations have designed standard computation forms both for general purposes and for specific problems. In very wide use is the 8½" ×11" printed sheet divided by faint blue lines into one-quarter inch squares. Such ruling is of great assistance in promoting neatness and legibility. An ample margin is maintained along the left edge of the sheet for permanent binding and space is provided at the top for all essential information. The latter would include a general subject caption identifying the problem, date, names of computer and checker, and the source of the original data. Each sheet is further identified as, for example, "sheet 3 of 7 sheets."

The results of all engineering calculations are considered provisional until they have been verified for correctness. Various forms of checks are discussed later.

7-3. Computational Aids The most essential computational aid in the operation of an electric desk calculator is a table of natural trigonometric functions. Although the use of 8-place sines, cosines, and tangents provided by such tables as Peters (see Ref. 2) is unjustified for most calculations in ordinary surveying, the ease with which the table can be entered is a matter of considerable advantage. Peters, for example, tabulates these trigonometric functions for every second of arc. Hence, the desired trigonometric function of an angle measured to a fraction of a minute by either a transit or theodolite can be quickly obtained. The computer should use only as many places as are appropriate in relation to the precision of the linear measurements.

An adding machine is a very desirable computing accessory if a great deal of addition and subtraction is involved and the desk calculator does not provide a printed record of the entries.

A slide rule, although generally of limited application in survey calculations, is very useful in determining taping corrections, adjusting traverses, and otherwise performing multiplication and division in those situations where the result need not be expressed to more than three or four significant places. The 10-in. slide rule will provide solutions to three significant places.

Nomograms are special graphs that are designed to aid in the rapid solution of certain surveying problems. They are available, for example, to obtain the angular convergency of two meridians.

Logarithms can be employed to make all surveying calculations. If a desk calculator is not available, logarithms must be used.

For field purposes a pocket-sized, hand-driven calculator known as the Curta (see Fig. 7-1) has been developed. It is an 8-oz miniature, all-purpose computing device.

Various commonly used trigonometric formulas are summarized in Table XXII.

Fig. 7-1. Hand Calculator. *Tennessee Valley Authority*

7-4. Desk Calculators The introduction of mechanical calculating machines, initially hand driven, produced a major revolution in methods of engineering computation. The use of logarithms has virtually ended and logarithmic tables of the trigonometric functions have been replaced by tables of the natural values of the trigonometric ratios. In the modern survey office the electric desk calculator itself is being superseded in part by a superior computing device, the electronic calculator which will be discussed in Art. 7-21.

Electric desk calculators are of various types, designs, and capabilities. The older models performed the operations of multiplication and division in steps, digit by digit. Subsequent designs incorporated an automatic division capability. Today (1969) the surveying organization can make its selection of desk computing devices from a wide array of rotary and printing calculators having various notable features. No attempt can be made to individually describe the makes available. Some of the most impressive operating characteristics are instant multiplication, automatic accumulation of multipliers and quotients, automatic decimals and zeros, and automatic squaring. Another feature is a memory element that can store the answer to any calculation for use in subsequent computations. The capacities of the electro-mechanical calculators vary somewhat. Both listing and totaling capacities are given in terms of the numbers of digits or columns of figures used to express an entry or an answer. The modern electric desk calculator, such as shown in Fig. 7-2, represents a major advance in speed, accuracy, and ease of office computations. It has increased the productivity of office personnel and thereby decreased the cost of surveying calculations.

Fig. 7-2. Electric Desk Calculator. S.C.M. Corporation

7-5. Significant Figures The significant figures in a number are the digits whose values are known. They are identified as those digits, proceeding from left to right, beginning with the first nonzero digit and ending with the last digit of the number.

Some illustrative examples are as follows:

(a) 541.6800 has seven significant figures.
(b) 50.0006 has six significant figures.
(c) 0.00058 has two significant figures.
(d) 0.006200 has four significant figures.
(e) 8.000050 has seven significant figures.
(f) 51.0 has three significant figures.

[7-5] COMPUTATIONS

The subject of significant figures is highly important both in the field work and office computations of surveying. Since neither the measurements or the quantities mathematically deduced from them can be exact, it is essential to use the appropriate number of significant figures to express a final result. Utilizing more significant figures than the precision of the field measurement warrants is apt to convey a misleading impression as to the quality of the end product of the survey. Some quantities, such as the number of degrees about a point, are mathematically exact. The fraction ¾ may be an exact number if it is written as 0.75 or 0.75000 with no distinction between these two expressions. Numbers obtained by counting are exact.

The precision with which a measurement is conducted largely determines the number of significant figures with which to express the quantity. However, one more significant figure than seems justified is frequently used. As an illustration, consider the taping of a line over rough terrain. Although the recorded distance is 857.86 ft, it should be obvious that the fifth digit is somewhat uncertain. It, nevertheless, is retained. If the measurement is definitely uncertain by five or more units in the fifth figure, it would be better practice to express the result with four significant figures or to the nearest 0.1 ft.

In order to add further clarification to the concept of significant figures, the following rules may be helpful:

(a) All nonzero digits are significant.
(b) Zeros at the beginning of a number merely indicate the position of the decimal point. They are not significant.
(c) Zeros between digits are significant.
(d) Zeros at the end of a decimal number are significant.

When performing addition (or subtraction), the result should be expressed in a logical manner. Consider the summation of the following three measured segments of a survey line:

$$\begin{array}{r} 24.2 \text{ ft} \\ 468.46 \\ \underline{156.} \\ 648.66 \text{ ft} \end{array}$$

The sum cannot be properly expressed closer than to the nearest foot, because one of the component quantities has been measured to the nearest foot only. The sum should be expressed, therefore, as 649 ft.

When performing multiplication (or division) the number of significant figures in the product cannot be greater than the number of significant figures in the factor having the fewest significant figures.

For example, consider the combined operations of multiplication and division indicated by

$$\frac{5.27 \times 838 \times 51.3781}{52 \times 62.581076}$$

The result should be expressed to two significant figures, the number of significant digits in the term, 52, which has the fewest significant digits. If the calculations extend over several steps, it is common practice to retain one more digit than is necessary until the final result is attained. The answer then is expressed with the correct number of significant figures.

In Art. 6-9 it was noted that in surveying operations involving both angular and linear measurements it is desired that the relative error in distance should equal the angular error in radians. Although the rates of change of the trigonometric functions depend upon the function and the size of the angle, the following general guides, as applied to average size angles, will be helpful in determining how many places (significant figures) in the trigonometric function should be used in typical calculations:

(a) For a 01' error in angle, use four places.
(b) For a 10" error in angle, use five places.

7-6. Rounding Off When there are more significant figures in a quantity than are required, the number is rounded off to the number of places that are needed. When rounding off, it is advisable to adopt a definite system so that the laws of chance operate and equitable results are obtained. A specific rule also permits the checker to perform his task more speedily.

If the result is to be expressed to n significant figures, the nth figure should be retained as is if the figure following it is less than 5 in the $(n + 1)$th place. If the digit following the nth significant figure is greater than 5 in the $(n + 1)$th place, the nth figure should be increased by one unit. When the $(n + 1)$th digit is a 5, round off to the nearest even digit in the nth place.

The following examples illustrate the usual rules:

(a) 6746.589 to five significant places is 6746.6.
(b) 837848 to four significant places is 837800.

(c) 468.767 to five significant places is 468.77.
(d) 468.762 to five significant places is 468.76.
(e) 468.755 to five significant places is 468.76.
(f) 468.745 to five significant places is 468.74.

7-7. Checks In prior considerations of field operations, emphasis was placed on the great need to be constantly vigilant against the introduction of blunders or mistakes. The same comments apply to office calculations. It is particularly important that data be correctly entered into the desk calculator and the machine results be correctly transcribed to the computation form. Faulty use of trigonometric tables is not uncommon.

In an engineering office the results of routine calculations are generally verified by the work of a checker who uses the original computation sheets. The most effective check would be an independent calculation by a second person preferably using different formulas.

Approximate checks to uncover the presence of large mistakes can be made sometimes with the slide rule. Graphical methods serve the same purpose.

When a check computation apparently reveals errors or mistakes in the original calculation, it is clearly essential to confirm the correctness of the check computation before accepting its results.

In all calculating work, particularly that performed by the surveying student, it is always advisable for the computer at the conclusion of a problem to ask himself *whether the result appears to be reasonable.*

TRAVERSE COMPUTATIONS

7-8. Rectangular Coordinates In surveying practice it is customary to define the position of a point with reference to two lines that intersect each other at right angles at some selected position. The plane rectangular coordinates of a point are the distances to the point from such a pair of mutually perpendicular axes. The distance from the X-axis is the Y-coordinate, and the distance from the Y-axis is the X-coordinate. In the United States, the X-axis is assigned the east-west direction and the Y-axis is assigned the north-south direction. This convention, however, is not universal. The opposite is true in Europe.

In American practice x-coordinates increase eastwardly and y-

coordinates increase northwardly. Frequently such coordinates are termed *East Coordinates* and *North Coordinates,* respectively. In order to avoid the generation of negative coordinate values, the origin ($x =$ zero, $y =$ zero) is placed sufficiently far to the south and west of the survey area. Systems of state-wide plane rectangular coordinates are discussed in Chapter 11.

The use of rectangular coordinates is the most convenient method of expressing the horizontal positions of survey points. The coordinates of a point uniquely define its position relative to any other point located in the same system.

Coordinates are widely employed for many purposes, including the plotting of maps and the calculation of land areas.

7-9. Forward and Inverse Problems The principal objectives of rectangular coordinate calculations in surveying have been traditionally the solutions to the forward and inverse problems.

Figure 7-3 illustrates the *forward problem.* The coordinates of

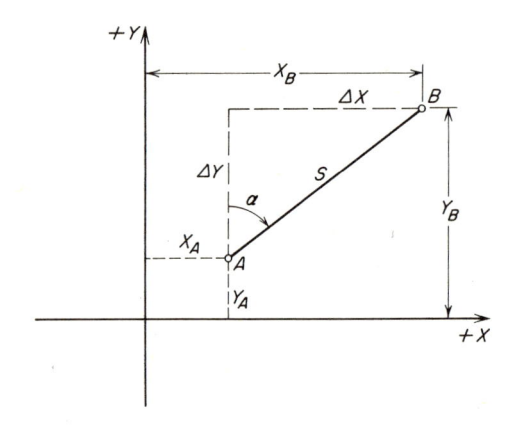

Fig. 7-3. Forward and Inverse Problems.

point A are known as well as the azimuth and length of AB. The coordinates of B are required.

It is apparent that the desired quantities are given by the following equations:

$$X_B = X_A + S \times \sin \alpha \qquad (7\text{-}1)$$

and

$$Y_B = Y_A + S \times \cos \alpha \qquad (7\text{-}2)$$

The *inverse problem* is that of determining the length and azimuth of AB when the rectangular coordinates of points A and B are known. The solution is provided by the equations:

$$\tan \alpha = \Delta X/\Delta Y \tag{7-3}$$

and

$$S = \Delta X/\sin \alpha \tag{7-4}$$

or

$$S = \Delta Y/\cos \alpha \tag{7-5}$$

7-10. Latitudes and Departures The terms, latitude and departure, are widely used in rectangular coordinate calculations of surveying. They are defined as follows:

The *latitude* of a line is its projection on the reference meridian.

The *departure* of a line is its projection on the east-west line perpendicular to the reference meridian.

It is readily evident that latitude as here defined is not the same as geographic latitude.

The basic expressions for calculating latitude and departure are

$$\text{Latitude} = \text{length} \times \text{cosine bearing angle} \tag{7-6}$$

$$\text{Departure} = \text{length} \times \text{sine bearing angle} \tag{7-7}$$

Latitudes are North or positive for lines having a northerly bearing and South or negative for lines having a southerly bearing.

Departures are East or positive for lines having an easterly bearing and West or negative for lines having a westerly bearing.

These concepts are portrayed in Fig. 7-4 where β denotes the bearing angles.

Latitudes are also termed *"northings"* and Y-differences; departures are similarly called *"eastings"* and X-differences.

7-11. Traverse Computation The calculation of a loop traverse, including the determination of the enclosed area, embraces the execution of the following instructions:

1. Make any necessary correction to the observed course lengths including reduction of slope distances to the horizontal.
2. Distribute the *angular error of closure* and calculate the bearings.
3. Compute the latitudes and departures.
4. Find the *linear error of closure* and the *relative error of closure*.

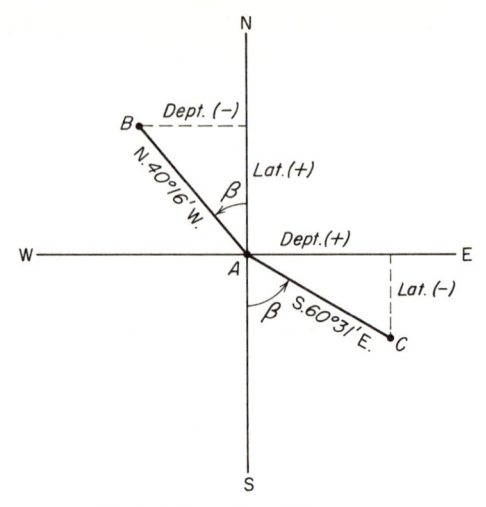

Fig. 7-4. Latitude and Departure.

5. Balance or adjust the survey.
6. Compute the coordinates.
7. Compute the area.

These steps will now be explained with reference to the calculations portrayed in Fig. 7-5. The corrected distances and bearings are tabulated therein. Point 1 has the arbitrary coordinates, $x = 0$ and $y = 0$. The line, 1–2, has a bearing of N. 16°50′ E. The calculation of latitudes and departures is not shown. They were determined with a table of natural trigonometric functions and a desk calculator.

In any loop traverse the algebraic sum of the latitudes (Σ Lats.) and that of the departures (Σ Depts.) should be zero because the survey originates and closes back on the same point. However, because of the inevitable errors in linear and angular measurements these criteria will never be exactly satisfied. The calculations depicted in Fig. 7-5 indicate that Σ Lats. is 0.03 ft and Σ Depts. is 0.30 ft.

Step 4 entails first the computation of the *linear error of closure*. This can be considered as the hypotenuse of a triangle of error whose legs are the Σ Lats. and Σ Depts. as portrayed in Fig. 7-5. It is evaluated as follows:

$$\text{Linear error of closure} = \sqrt{(\Sigma \text{ Lats.})^2 + (\Sigma \text{ Depts.})^2} \qquad (7\text{-}8)$$

The *relative error of closure* provides a better assessment of the

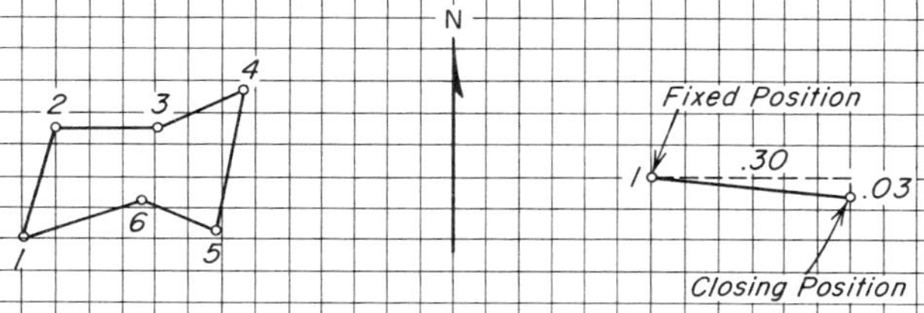

[7-11] COMPUTATIONS

TRAVERSE COMPUTATIONS

Sheet No. 1 of 1
Traverse No. 52
Project No. B12
Place St. Joseph, Ill. Date 10/17/68
Datum Arbitrary Field Books 472
Computed by G.P. Roll
Checked by M.T. Leicht
Remarks: Angular error of closure (−0°03′) distributed equally among all angles.

Traverse adjustment by Compass Rule.

Station	Bearing	Distance (ft.)	Departures East	Departures West	Latitudes North	Latitudes South	Balance Dept.	Balance Lat.	Coordinates x	Coordinates y
1	N.16°50′E.	354.51	102.66		339.32		+102.60	+339.33	0.00	0.00
2	N.85°56′E.	318.54	317.74		22.59		+317.69	+22.59	102.60	339.33
3	N.65°00′E.	274.17	248.48		115.87		+248.44	+115.87	420.29	361.92
4	S.11°23′W.	469.65		92.69		460.41	−92.76	−460.40	668.73	477.79
5	N.67°42½′W.	246.57		228.14	93.53		−228.17	+93.53	575.97	17.39
6	S.72°18½′W.	365.01		347.75		110.93	−347.80	−110.92	347.80	110.92
1		2028.45	668.88	668.58	571.31	571.34	0.00	0.00	0.00	0.00

Linear Error of Closure = $\sqrt{(.03)^2 + (.30)^2} = 0.30$ ft

Relative Error of Closure = $\dfrac{.30}{2028} = \dfrac{1}{6800}$

Fig. 7-5. Traverse Calculation.

quality of a traverse than the linear error of closure. Obviously, a traverse 5.0 miles long developing a linear closure of 4.60 ft will be of greater precision than a traverse only 2.5 miles long that has the same closure. It is common practice, therefore, to calculate the relative error of closure, which is merely the linear error divided by the length of the traverse. Of course, both quantities must be in the same units. The result is expressed in the form of a ratio with unity as the numerator. Accordingly, the relative error of closure in Fig. 7-5 is 1/6800. The denominator is usually rounded off in the manner indicated. This result indicates, on the average, that one foot of error was generated every 6800 ft of traverse.

Some comments are desirable with respect to traverse calculations indicating a very large relative error of closure. If it is believed the field work was done with reasonable care, a careful check should be made of the computations to uncover blunders. Sometimes latitudes and departures, as well as sines and cosines, are interchanged, a decimal point is misplaced, or the addition is faulty. If no mistakes can be found, it is necessary to return to the field and make new measurements of suspected distances and angles.

7-12. Traverse Adjustment After the relative error of closure has been determined and its value satisfies the specifications governing the quality of the survey, the traverse must be *balanced* or adjusted. This operation refers to the distribution in an equitable and logical manner of corrections to the latitudes and departures so that their algebraic sums are made equal to zero. Such a procedure will make the traverse a mathematically closed figure. In some cases the experienced surveyor will prefer to make adjustments to the latitudes and departures in a manner that may seem rather arbitrary but is based on recollections of the relative difficulty in measuring certain courses and angles. As an example, a line taped over very rough terrain will have its latitudes and departures corrected more than those of courses over level ground.

In general, it is desirable to utilize an adjustment process that is both economical and logical. One widely employed procedure is that known as the *compass rule*. It assumes that the quality of angular and linear measurements is approximately the same and that the corrections to the latitudes and departures vary as the length of the course.

The compass rule states that the correction to the latitude (or de-

parture) of a course is to the total error in latitudes (or departures) as the length of the course is to the length of the traverse. With reference to course 1–2 the correction to the latitude is calculated as

$$\frac{C_l}{.03} = \frac{354.51}{2028.45} \quad \text{or} \quad C_l = 0.01 \text{ ft}$$

and the correction to the departure is

$$\frac{C_d}{.30} = \frac{354.51}{2028.45} \quad \text{or} \quad C_d = 0.06 \text{ ft}$$

Generally, the course and perimeter lengths shown above need not be expressed to more than three significant figures and a slide rule provides ample accuracy. The corrections must be applied in an appropriate way. In the case of the calculations of Fig. 7-5, the sum of the south latitudes exceeds the sum of the north latitudes. Therefore, corrections to south latitudes are minus and to north latitudes plus. The proper sign must prefix the corrected latitudes in the columns containing the balanced quantities. The final latitude of line, 1–2, is plus or North. As a precautionary measure the algebraic sums of the balanced latitudes and departures should be determined and the result clearly entered on the form. Usually, one latitude or departure may have to be arbitrarily changed by .01 ft in order to nullify rounding-off errors. A connecting traverse between two points of fixed rectangular coordinates will be adjusted in a modified manner by the compass rule in Chapter 11.

The execution of step 6 in the traverse computation comprises the calculation of the coordinates of the survey points. The coordinates of a traverse station must be known or they must be assumed as in Fig. 7-5. The other coordinates are found by the successive algebraic addition of the balanced latitudes and departures to the coordinates of the preceding point. Hence,

$$Y_2 = Y_1 + \text{latitude of } 1\text{-}2$$
$$X_2 = X_1 + \text{departure of } 1\text{-}2$$

A check on the arithmetic is obtained if the coordinates of point 1 as determined from point 6 are the same as the original given values.

AREA COMPUTATIONS

7-13. Area by Coordinates One of the principal purposes for executing land surveys is to acquire the data needed to determine the area of the tract. Usually, a traverse is run along the perimeter of the parcel and the office calculations proceed in the manner already outlined. Since the coordinates of the corner points are thus already available, it is particularly convenient to use them to compute the area.

The procedure for calculating the area within any closed plane figure bounded by straight lines can be expressed as follows:

Rule. The area is equal to one-half the algebraic sum of the products of each ordinate multiplied by the difference between the two adjacent abscissas, always subtracting the preceding from the following abscissa.

This rule can be readily derived by summing algebraically the areas of the trapezoids formed by projecting the traverse courses upon a reference meridian to the west of the tract. In applying the foregoing rule to surveying practice, the terms ordinate and abscissa are supplanted by the corresponding coordinates, *North* and *East*.

With these substitutions made, using the letters N and E to indicate the coordinates, the rule may be applied by means of the following arrangement. The coordinates for each vertex are written in the form of a fraction, of which the numerator is the ordinate (N) and the denominator is the abscissa (E). Also, the series of fractions thus written is enclosed between vertical dashed lines. Now the first numerator N_1 is to be multiplied by the difference between the two adjacent denominators, E_2 and E_6, always subtracting the preceding abscissa E_6 from the following E_2. To indicate this operation the denominator of the last fraction to the right, E_6, is written outside the dashed line to the left of the first fraction. Likewise, the denominator of the first fraction E_1 is written outside of the dashed line to the right of the last fraction. The completed arrangement follows:

$$\overline{E_6} \left| \frac{N_1}{E_1} \quad \frac{N_2}{E_2} \quad \frac{N_3}{E_3} \quad \frac{N_4}{E_4} \quad \frac{N_5}{E_5} \quad \frac{N_6}{E_6} \right| \overline{E_1}$$

The area is now given by the equation:

$$A = \tfrac{1}{2}[N_1(E_2 - E_6) + N_2(E_3 - E_1) + N_3(E_4 - E_2) \\ + N_4(E_5 - E_3) + N_5(E_6 - E_4) + N_6(E_1 - E_5)]$$

In order to determine the area within the traverse of Fig. 7-5, the fractions and calculations are arranged as follows:

| 0.0   339.33   361.92   477.79   17.39   110.92 |
| 347.80 | 0.0   102.60   420.29   668.73   575.97   347.80 | 0.0

+	−
0.0 × (−245.20) = 0	0
339.33 × 420.29 = 142,617	
361.92 × 566.13 = 204,894	
477.79 × 155.68 = 74,382	
17.39 × (−320.93) =	5,581
110.92 × (−575.97) =	63,887
421,893	69,468
−69,468	
2) 352,425	
176,212 sq ft = 4.05 acres

7-14. Area of Irregular Tracts Areas with irregular or curved boundaries are usually measured by establishing a baseline conveniently near and by taking offsets at regular intervals from the baseline to the boundary. Three methods are most commonly used, namely, (1) the trapezoidal method, (2) Simpson's one-third rule, and (3) the coordinate method.

Figure 7-6, *a* and *b*, represents two types of irregular areas, the first with an irregular boundary and the second bounded by the circular curve of a street or a roadway property line. Offsets h_1, h_2, etc., have been measured from the baseline to the boundary at regular intervals, *b*.

1. *Trapezoidal Method.* If the ends of the offsets in the boundary line are assumed to be connected by straight lines, a series of trapezoids is formed, the bases being the offsets and the altitudes being the common distance *b*. Accordingly, the area of the first trapezoid is $\frac{b(h_1 + h_2)}{2}$, of the second it is $\frac{b(h_2 + h_3)}{2}$, etc. Summing these up, we have for the total area, *A*, the following equation, in which *n* equals the number of offsets:

$$A = b\left[\frac{h_1 + h_n}{2} + (h_2 + h_3 + \cdots + h_{n-1})\right] \qquad (7\text{-}9)$$

EXAMPLE 7-1: Find the area of Fig. 7-6*a* if the common interval

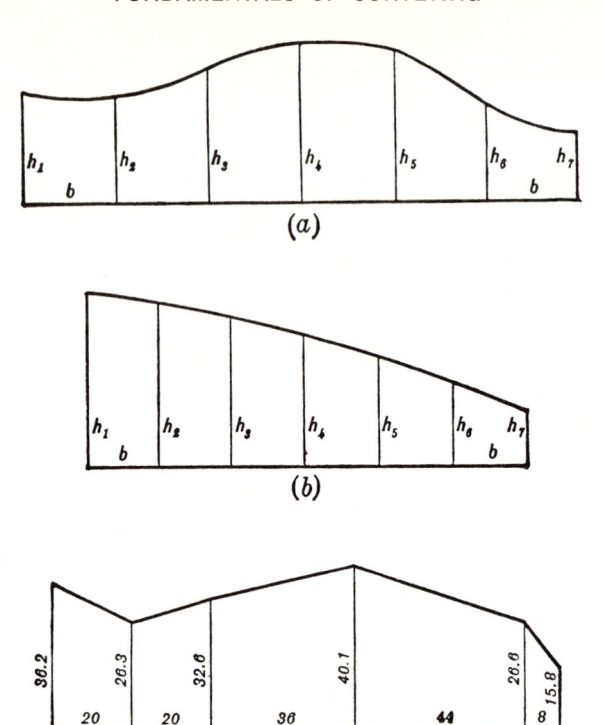

Fig. 7-6. Irregular Tracts.

is 25 ft and if the offsets are 29.6, 28.2, 34.3, 41.5, 39.6, 27.2, and 18.4 ft, respectively.

$$A = 25 \left(\frac{29.6 + 18.4}{2} + 28.2 + 34.3 + 41.5 + 39.6 + 27.2 \right)$$

$$= 4870 \text{ sq ft}$$

2. *Simpson's One-Third Rule.* Simpson's one-third rule may be applied to areas similar to those illustrated in Fig. 7-6a and b, where the offsets have a common interval, b, and provided an odd number of offsets is taken. The rule may be stated as follows: *The area is equal to one-third of the common interval between offsets, multiplied by the sum of the first and last offsets, plus two times the sum of the other odd offsets, plus four times the sum of the even offsets;* or, if n equals the number of offsets,

$$A = \frac{b}{3}\bigg[h_1 + h_n + 2(h_3 + h_5 + \cdots + h_{n-2})$$
$$+ 4(h_2 + h_4 + \cdots + h_{n-1})\bigg] \quad (7\text{-}10)$$

This rule is based on the assumption that the curve passing through the ends of the first three offsets is a parabola; the same is true for the curve through the ends of offsets 3, 4, and 5, and through the ends of offsets 5, 6, and 7, etc. It is supposed that this series of parabolic curves will approximate the boundary line more closely than do straight lines, and so yield a more accurate value for the area.

EXAMPLE 7-2. Find the area of Fig. 7-6b if the common interval is 20 ft and the offsets are 44.3, 42.0, 39.4, 34.6, 28.7, 22.3, and 14.6, respectively.

Since there is an odd number of offsets, the rule can be applied to the entire area as follows:

$$A = \frac{20}{3}\bigg[44.3 + 14.6 + 2(39.4 + 28.7) + 4(42.0 + 34.6 + 22.3)\bigg]$$

$A = 3938$ sq ft

3. *The Coordinate Method.* If the boundary of an area is such that offsets are best taken at irregular intervals, as shown in Fig. 7-6c, the area may be calculated as a series of separate trapezoids, or by the coordinate method. The latter method has been explained in previous articles.

EXAMPLE 7-3. Find the area of Fig. 7-6c with dimensions in feet as shown. Taking the origin at the foot of the left-hand offset, the arrangement of coordinates and computation are as follows:

$$\frac{\vert 0}{128.0\vert 0} \quad \frac{36.2}{0} \quad \frac{26.3}{20.0} \quad \frac{32.6}{40.0} \quad \frac{40.1}{76.0} \quad \frac{26.6}{120.0} \quad \frac{15.8}{128.0} \quad \frac{0\vert}{128.0\vert 0}$$

$$\begin{aligned}
36.2 \times 20.0 &= 724. \\
26.3 \times 40.0 &= 1052. \\
32.6 \times 56.0 &= 1825.6 \\
40.1 \times 80.0 &= 3208.0 \\
26.6 \times 52.0 &= 1383.2 \\
15.8 \times 8.0 &= 126.4 \\
&\ 2\overline{)8319.2} \\
&4159.6
\end{aligned}$$

The relative merits of the above methods may be compared as follows:

1. The trapezoidal method is simplest and is sufficiently accurate for most areas, provided proper care is taken in the measurements. It may be noted (Fig. 7-6a) that the calculated area will be slightly too large for trapezoids where the boundary line is concave upward, and too small for trapezoids where the line is concave downward. Thus, this method is more accurate where the boundary line consists of segments of contrary flexure.

2. Simpson's rule is more laborious to apply than the trapezoidal rule, but it is more accurate for all conditions. It is especially applicable to figures like the one shown in Fig. 7-6b.

3. The coordinate method is best for offsets at irregular intervals.

7-15. Area by Polar Planimeter The planimeter is an instrument by means of which the area of a plotted, closed figure may be determined directly by tracing the perimeter and reading the result from the scale.

The polar type of this instrument is illustrated in Fig. 7-7. Its essential features are an anchor point or pole P, a tracing point T, and a roller R, which has a graduated scale on a drum.

The two arms connecting these points are movable about the connecting pivot. In the older type, the tracing arm is of fixed length and capable of reading areas in one unit only, usually square inches. In another type, Fig. 7-7, the tracing arm is adjustable in a sleeve, to read areas in different units, corresponding to the scale of the map. The arms are usually adjusted so that one revolution of the roller measures an area of 10 sq in.; the scale is divided into 100 parts, and the vernier reads to 1/10 of a scale division, or 1/1000 of 10 sq in. = 0.01 sq in.

With some instruments there is provided a flat metal bar with a needle point at one end and a hole drilled through the bar at a distance from the needle point equal to the radius of a circle of 10 sq in. area. With the needle point pressed into the drawing paper and the tracing point inserted in the hole, an area of 10 sq in. can be quickly and accurately traced, and, if necessary, the tracing arm can be precisely adjusted. If the instrument is of the fixed-arm type, the tracing arm cannot be adjusted, but a constant can be determined by which all results are to be multiplied, to determine correct values.

If a proving bar, described above, is not provided, a square of

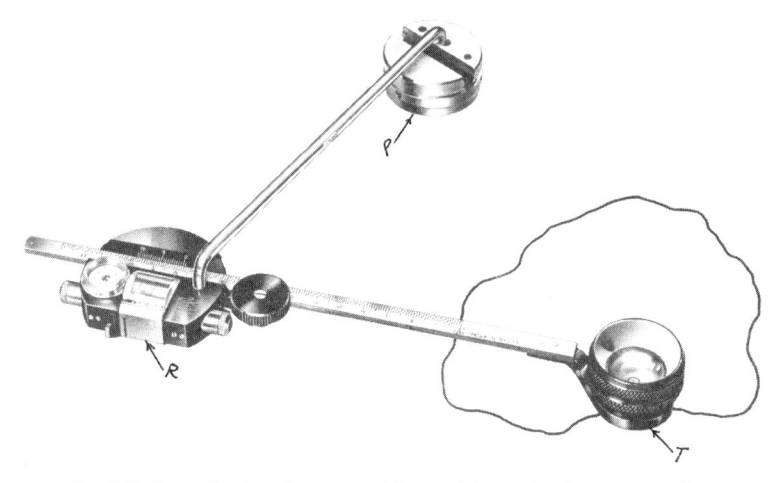

Fig. 7-7. Optical Polar Planimeter. *National Surveying Instruments, Inc.*

known area can be carefully drawn and the perimeter traced a few times, either to adjust the tracing arm or to find the instrument's constant.

In use, the pole is placed in any convenient position *outside* the area and the tracing point is placed at some initial point of the perimeter to be traced. The scale is then read by means of the index and vernier provided on the roller frame. This reading of the scale is recorded as the initial reading. The tracing point is then moved carefully around the perimeter, during which process the roller will both turn and slide, until the initial point is again reached. The scale is again read, and recorded as the final reading. The difference between the two readings is a measure of the area within the perimeter traced. If the direction of the tracing point has been clockwise, the result is positive, and, if counterclockwise, it will be negative. Two or more determinations of each area should be made to provide a check and to secure a more accurate result.

Because of backlash in the instrument, no attempt should be made to set the fixed-arm planimeter at zero. The optical planimeter, however, is quickly and accurately set to zero by depressing a button.

EXAMPLE 7-4: A fixed-arm polar planimeter is used to determine the area of a parcel of land shown on a map whose scale is 1 in. = 400 ft. The initial reading is 5678 and the final reading is

7215. If one unit in the fourth place equals .01 sq in., find the land area in acres.

Map area = (7215 − 5678).01 = 15.37 sq in.

$$\text{Land area} = \frac{(15.37)(160{,}000)}{43{,}560} = 56.5 \text{ acres}$$

If the area is too large to be included by the tracing arm for one position of the pole, the area can be divided into the requisite number of subdivisions.

Areas are often desired on cross-section or profile paper on which the horizontal and vertical scales are not the same; but this causes no difficulty, since such areas are always proportional to the product of horizontal and vertical dimensions.

EXAMPLE 7-5: The areas of plotted highway cross sections are determined with an optical planimeter whose adjustable arm is set so that one unit in the fourth place equals .015 sq in. The horizontal and vertical scales of the cross sections are 1 in. = 10 ft and 1 in. = 5 ft, respectively. The mean difference of readings of several runs around the figure is 1280. What is the end area?

$$\text{Area} = (1280)(.015)(10 \times 5) = 960 \text{ sq ft}$$

When carefully manipulated, planimeter results are surprisingly accurate. The percentage of error decreases as the size of the area increases, one reason being, of course, that the area increases with the square of the perimeter. The mean of two determinations should be correct within 1% for small areas and may easily be within 0.1% for areas of 20 sq in. or more. This precision is sufficient for such purposes as determining drainage areas, cross section and contour areas for earthwork, and reservoir areas and volumes. Its precision and facility in operation make this a most useful instrument for determining plotted areas of any shape whatsoever.

VOLUME COMPUTATIONS

7-16. Average-End-Area Method In engineering projects, volumetric quantities are usually determined by finding the areas of parallel sections and multiplying their mean value by the perpendicular distance between them. This procedure is called the method of *average end areas*.

This method is usually used for finding earthwork volumes for the construction of highways, railways, drainage ditches, canals, etc. In such projects the cross-section areas are usually determined at intervals of 100 ft. Volumes are calculated by the average-end-area formula

$$V = \left(\frac{A_1 + A_2}{2}\right)\frac{L}{27} \tag{7-11}$$

in which V is the volume in cubic yards, A_1 and A_2 are the areas of the end sections in square feet, and L is the distance between the sections in feet.

If the ground is uneven or changes slope abruptly, intermediate sections are taken such that the errors in resulting volumes will not be serious. Theoretically, this method is not exact unless the two end areas are equal, but except as indicated in the following article, the resulting errors are insignificant.

7-17. Prismoidal Formula When the ground surface is such that the two end areas are widely different in area, or when high precision is desired, as in the cases of rock quantities in excavation and of volumes of concrete structures, the average-end-area method is not sufficiently exact and the volumes are calculated as prismoids.

A *prismoid* may be defined as a solid having parallel, plane bases and sides that are plane surfaces. Its volume is given by the equation

$$V = \frac{L}{27}\left(\frac{A_1 + 4A_m + A_2}{6}\right) \tag{7-12}$$

in which V is the volume in cubic yards; L is the perpendicular distance between the bases; A_1 and A_2 are the end sections; and A_m is the area of a mid-section.

It should be noted that the area of a mid-section will not ordinarily be the same as the mean of the two end areas, but must be computed from the linear dimensions which are the average of those of the end sections.

The application of the prismoidal formula to earthwork sections is somewhat laborious since the mean dimensions of the end sections must be found before the area of the mid-section can be computed. These computations are simplified by computing a correction to be applied to the volume as calculated by the average-end-area method of the previous article. This correction is $C_v = 0.31(C_1 - C_2)(W_2 - W_1)$, in which C_v is the prismoidal correction in cubic yards for an earthwork section 100 ft long, C_1 and C_2 are the center cuts or fills at the end sections A_1 and A_2, and W_1 and W_2 are the corresponding distances between slope stakes at these sections.

Slope stakes are discussed in Chapter 15. The prismoidal correction is applied to the average-end-area volume with the sign indicated by the calculation for C_v. However, normally, it is subtracted.

7-18. Borrow Pits In constructing the earthwork for roadways, levees, etc., it is frequently necessary to excavate material from areas adjacent to the project to form the embankments. Such excavations are called *borrow pits*. They may be quite irregular in shape, and the volume of material is usually determined by first dividing the area into squares of suitable size and then taking level readings at the corners of the squares both before and after the work of excavation. From these data the volume may be computed, the volume removed from any square being calculated as the average of the heights at the four corners times the area. Some corners are common to more than one square, thus *d*, in Fig. 7-8, is common to one, *e* is common

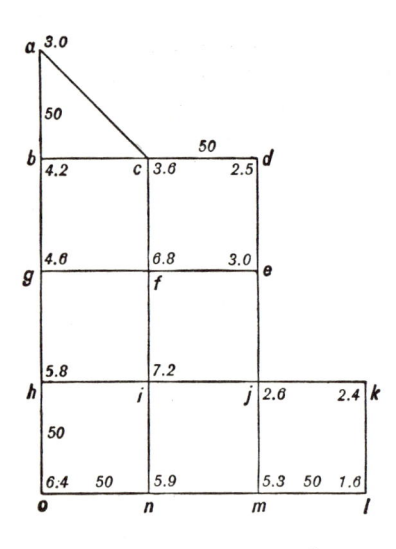

Fig. 7-8. Borrow Pit.

to two, *j* is common to three, and *f* is common to four squares. Hence, the work may be simplified somewhat if each corner height is multiplied by the number of squares to which it is common, and the sum taken and averaged, to find the average cut of the whole area. By this method the volume included within a group of squares is given by the following equation:

$$V = \frac{A}{4 \times 27}(h_1 + 2h_2 + 3h_3 + 4h_4) \qquad (7\text{-}13)$$

in which A is the area of one square in square feet, h_1, h_2, h_3, and h_4 are the corner heights common to one, two, three, and four squares, respectively.

The total area may include additional triangles or trapezoids, which must be computed separately.

The volume indicated in Fig. 7-8 may be computed as follows:

h_1	h_2	h_3	h_4
$b = 4.2$	$c = 3.6$		$f = 6.8$
$o = 6.4$	$g = 4.6$	$j = 2.6$	$i = 7.2$
$l = 1.6$	$h = 5.8$		
$k = 2.4$	$n = 5.9$		
$d = 2.5$	$m = 5.3$		
	$e = 3.0$		
$\overline{17.1}$	$\overline{28.2}$	$\overline{2.6}$	$\overline{14.0}$

$$V_1 = \frac{2500}{4 \times 27}[17.1 + (2 \times 28.2) + (3 \times 2.6) + (4 \times 14.0)] = 3178$$

$$V_2 = abc = \frac{2500}{2 \times 3 \times 27}(10.8) \qquad\qquad = \underline{167}$$
$$\phantom{V_2 = abc = \frac{2500}{2 \times 3 \times 27}(10.8) \qquad\qquad = } 3345 \text{ cu yd}$$

OTHER TOPICS

7-19. Cutoff Line A wide variety of miscellaneous calculations could be presented but only two will be mentioned. Suppose it is desired to compute the bearing and length of the line, 1–4, in Fig. 7-5.

From Eq. 7-3

$$\tan \alpha = \frac{\Delta X}{\Delta Y} = \frac{668.73}{477.79} = 1.39963$$

$$\alpha = 54°27.3'$$

Hence, the bearing is N. 54°27.3′ E.
Then, from Eqs. 7-4 and 7-5, respectively,

$$S = \frac{\Delta X}{\sin \alpha} = \frac{668.73}{.813659} = 821.88 \text{ ft}$$

and

$$S = \frac{\Delta Y}{\cos \alpha} = \frac{477.79}{.581342} = 821.87 \text{ ft}$$

The length could be found also from

$$S = \sqrt{(\Sigma \text{ Lats.})^2 + (\Sigma \text{ Depts.})^2}$$

7-20. Auxiliary Traverse Sometimes in executing a land survey it is very difficult to measure directly along the property lines because of obstructions. The lengths and bearings of these lines can be calculated from an auxiliary traverse whose stations are situated in the clear and conveniently close to the property corners. This situation is depicted in Fig. 7-9. Angle and distance ties are carefully

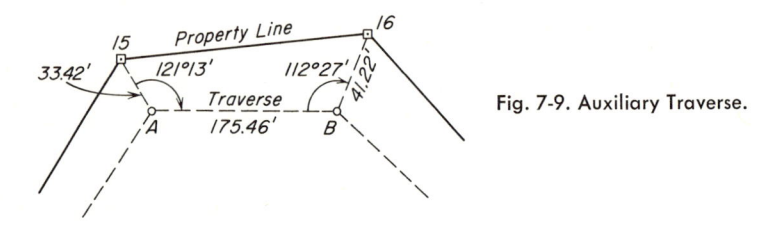

Fig. 7-9. Auxiliary Traverse.

made from each transit hub to the nearby corner. The calculation of the traverse will yield coordinates for the hubs and from such values the coordinates of the corners are determined. The inverse problem is then solved to obtain the lengths and bearings of the property lines.

7-21. Electronic Calculator The science of computing has been tremendously advanced by the introduction and progressive improvement of various electronic computing devices. These machines span a wide spectrum from the electronic desk calculator to the expensive, large-storage, high-speed computers needed to solve the most intricate problems of engineering and science. The term, *electronic calculator,* as used here refers to any one of the several desktop solid-state computers, which play a role midway between that of the electric desk calculators and the highly sophisticated computer centers.

The use of an electronic calculator to solve surveying problems not only enables the work to be performed in a fraction of the time required by a rotary calculator, but it decreases the opportunities for making mistakes. There is no need to look up and record the functions of angles. The electronic calculator figures the value of the desired function in a fraction of a second.

The data for a given problem and the appropriate set of instructions for effecting a solution are fed into the machine and the results are usually obtained in the form of a typewritten record. The calculator shown in Fig. 7-10 has a keyboard for the control of instruc-

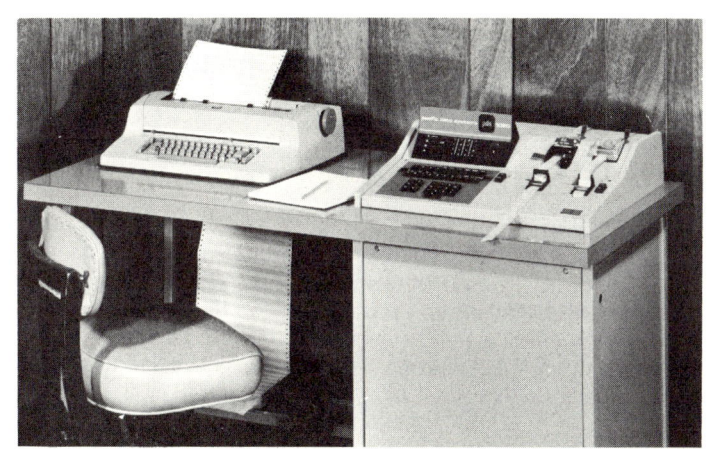

Fig. 7-10. Electronic Calculator. *Keuffel & Esser* Co.

tions to the machine. The programs are on punched paper tape rolls. Devices of this general type occupy the space of a small office desk, are highly portable, and utilize a standard 115v AC outlet. Operation can be learned in a few hours.

Perhaps the most cogent way of expressing the capabilities of the typical desktop electronic calculator is by comparing its computation of the traverse of Fig. 7-5 with that by electric calculator. Following the introduction to the machine of the coordinates of the starting point and the bearings and lengths of all courses the device will calculate and print out in a fraction of a minute the linear error of closure, the balanced coordinates of the points, and the area. Other surveying computing programs, such as those for determining earthwork volumes, solving the inverse problem, determining the coordinates of line intersections, and calculating horizontal and vertical curves, are available.

PROBLEMS

7-1. The following data were obtained in executing a loop traverse:

Course	Distance (ft)	Interior Angle
AB	417.26	80°46'
BC	219.78	132°00'
CD	374.63	112°41'
DE	318.25	122°48'
EA	551.40	91°42'

The bearing of AB is fixed at S. 75°00' W. Note that the first angle is at point A and that the traverse lies to the west of line AE.
Perform the following operations:

(a) Distribute the angular error of closure.
(b) Calculate latitudes and departures.
(c) Calculate the linear and relative errors of closure.

7-2. Balance the traverse of Problem 7-1 by the compass rule and calculate the coordinates of the points. The coordinates of station A are $x = 9256.41$ ft and $y = 8476.29$ ft.

7-3. Calculate the area within the traverse of Problem 7-2 to the nearest 0.01 acre. In order to most easily accomplish this, use new coordinates deduced from assumed values of point A or any other point and the balanced latitudes and departures.

7-4. The following data refer to a loop traverse:

Course	Distance (ft)	Azimuth
AB	2538.1	209°37'
BC	3923.7	96°01'
CD	4973.5	357°46'
DE	3698.5	269°26'
EA	2630.0	151°43'

The coordinates of point A are $x = 10,000.0$ ft and $y = 10,000.0$ ft.

Perform the full sequence of steps in the calculation of this traverse including the determination of the area to the nearest 0.1 acre.

7-5. A traverse was run around a tract of land with the following results:

Course	Distance (ft)	Azimuth
AB	236.34	126°30'
BC	212.36	195°43'
CD	182.78	174°47'
DE	313.10	301°10'
EF	164.38	42°04'
FA	243.34	2°05'

(a) Perform a complete traverse calculation. The coordinates of point A are $x = 2178.42$ ft and $y = 1831.07$ ft. Arbitrary coordinates can be chosen when determining the area. Calculate it to the nearest 0.01 acre.

(b) Find the length and bearing of the line FB.

7-6. An open-ended traverse, $ABCD$, was executed in the cultivated area skirting a heavy forest, through which the line AD passes. There is need to mark out on the ground the line AD whose ends are not intervisible.

The results of the traverse are as follows:

Point	Distance (ft)	Deflection Angle
A		
	478.20	
B		60°20′ R.
	692.15	
C		93°45′ R.
	781.40	
D		

Calculate (a) the angle, to nearest 01′, at A needed to define the direction of AD with respect to AB, and (b) the length of AD to the nearest 0.01 ft.

7-7. The four corners, $EFGH$, of a small tract of land have been recovered and a traverse of the perimeter is desired. However, an old stone fence and several large trees make measurements along the boundaries very difficult. Therefore, an auxiliary traverse, $ABCD$, is conducted within the borders of the parcel and distance and direction ties are obtained from the traverse stations to the property corners as follows:

Line	Bearing	Length (ft)
AB	S. 89°58′ E.	296.40
AE	N. 20°00′ W.	34.20
BC	S. 43°20′ W.	333.90
BF	N. 60°20′ E.	16.90
CD	S. 80°21′ W.	215.60
CG	S. 73°00′ E.	27.60
DA	N. 27°24′ E.	314.20
DH	S. 36°30′ W.	15.70

Calculate the following:

(a) The coordinates of each transit point and property corner. Balance traverse by compass rule. Point D has the coordinates $x = 200.00$ ft and $y = 200.00$ ft.

(b) The length and bearing of each side of the parcel and tabulate them. Express lengths to 0.01 ft and bearings to nearest 01′.

(c) The area of the parcel to the nearest 0.01 acre.

216   FUNDAMENTALS OF SURVEYING

7-8. In order to determine the area of a tract of land bordered by the high-water line of a sinuous stream, offsets from a transit line were measured at regular intervals of 25 ft as follows: 20.8, 16.7, 21.5, 29.3, 31.0, 25.1, 15.7, 18.0, and 23.2. Find the area by Simpson's one-third rule.

7-9. Find the area of the tract of Problem 7-8 by the trapezoidal rule.

7-10. The area of an irregular parcel of land is to be determined from the following field record of offsets from a transit line, AB:

Distance from Point A (ft)	Offset (ft)
0	14.0
15	22.5
35	12.7
70	26.3
105	29.8
120	12.0
155	5.2

Find the area by the method of coordinates.

7-11. A fixed-arm polar planimeter is used to measure the area of a highway cross section that has been plotted with a horizontal scale of 1 in. = 20 ft and a vertical scale of 1 in. = 5 ft. The mean difference for several runs between the initial and final readings was 1036 with one unit in the fourth place being equal to 0.01 sq in. Find the end area to the nearest square foot.

7-12. A planimeter was used to determine the area of a tract of timber shown on an aerial photograph whose scale is 1 in. = 600 ft. If the initial reading was 3781 and the final reading (going clockwise) was 5859 and the unit's place has a value of 0.01 sq in., find the area to the nearest acre.

7-13. The end areas along a proposed drainage ditch are as follows:

Station	End Area (sq ft)
16 + 22	85
17 + 00	110
17 + 81	260

Find the total volume (nearest cubic yard) of excavation between 16 + 22 and 17 + 81.

7-14. The end areas along a proposed levee are as follows:

Station	End Area (sq ft)
37 + 00	596
38 + 00	1242
38 + 80	978

Find the total volume (nearest cubic yard) of excavation between 37 + 00 and 38 + 80.

REFERENCES

1. Montgomery, C. J., "Impact of Electronic Calculators on Survey Computations," *Surveying and Mapping,* Vol. 25, No. 1, pp. 49–64, American Congress on Surveying and Mapping, Washington, D.C., 1965.
2. Peters, Jean, *Eight-Place Tables of Trigonometric Functions for Every Second of Arc,* Chelsea Publishing Co., New York, 1963.

8 Errors and Adjustments

8-1. Introduction Only the briefest mention was made of the subject of errors in Chapter 1. Under the assumption that the reader has by now performed typical field measurements of the types explained in the intervening chapters, it should be evident to him that all measurements of physical quantities are inexact. It is most appropriate, therefore, to expand on the earlier treatment of errors and introduce related topics.

A broader and more penetrating look will be taken at the uncertainties of surveying data and various terms will be defined. Particular attention will be directed to the distribution of accidental errors and the use of probable error as a measure of precision. The difference between accuracy and precision will be carefully delineated, and the manner in which systematic and accidental errors accumulate will be illustrated. Since many surveying dimensions are obtained indirectly by calculation, attention will be directed to the error of computed quantities.

Finally, the necessity and the basis for making adjustments of observed data will be indicated, including a consideration of weighted measurements.

The foregoing rudiments of error analysis will continue to be utilized in succeeding chapters of this book and should suffice to assist the surveyor in making assessments of the reliability of his measurements.

8-2. Errors The difference between an observed or computed value of a quantity and the ideal or true value is termed the error. Errors are of various kinds, depending upon how they originate and propagate.

Surveying measurements are affected by two general classes of errors, viz., *systematic errors* and *accidental errors*. Systematic errors are those whose magnitudes and signs are directly related to the conditions surrounding the measurements. They conform to known physical laws and are susceptible of mathematical determination. Changes in conditions are accompanied by corresponding changes in the magnitude, and sometimes the sign, of the resulting error. A systematic error that is the same in both magnitude and sign throughout a series of measurements is sometimes called a *constant error*. Examples of constant errors are the index error of the vertical circle of a transit and the error associated with the use of a tape that is consistently too long.

Systematic errors are effectively eliminated by evaluating them and making the proper correction or by employing a suitable measurement program. Although it is said that a systematic error can be removed by computation or by the use of proper instrumental techniques, there will generally remain a small residual error, which is unknown in sign and magnitude.

Systematic errors accumulate directly as the number of opportunities for making them. For that reason, they are sometimes termed *cumulative errors*. For example, a 300-ft steel tape which is 0.09 ft short introduces an error of +0.09 ft each time it is used.

Accidental errors are those errors for which it is equally probable that the sign of the error is plus or minus. The algebraic sign and the magnitude of an accidental error are matters of chance and are governed by the law of probability. It is not possible to apply a correction to nullify an accidental error. In a series of measurements accidental errors are compensative in their total effect, i.e., they tend to partially cancel themselves. For that reason they are sometimes referred to as *compensating errors*. Frequently they are termed *random errors*. They accumulate as the square root of the number of opportunities for making the error, as mentioned in Art. 1-16 where this principle was called the law of compensation. An example of an accidental error would be that caused by failure to read correctly the rod when running a line of differential levels. If, for example, the estimated error in reading the rod is ±0.004 ft at a distance of 200 ft and there are 11 setups between benchmarks, the total estimated error from this source alone is $.004 \times \sqrt{22} = \pm 0.019$ ft.

Mistakes or *blunders* should not be confused with errors. They are irregular in their occurrence, obey no mathematical law, and are relatively large in size. They result from haste, carelessness, and ignorance. They must be eliminated by careful work and by periodic checking. Examples are transposition of digits in recording an observation, misreading the tape when a fractional distance is measured, and taking a sight with the transit on the wrong range pole when measuring an angle.

Sometimes errors are classified as to source or cause. Hence, *natural errors* are those associated with environmental factors, such as temperature, humidity, and wind. *Instrumental errors* are those due to imperfections in the construction or adjustment of the instrument. *Personal errors* are illustrated by failure to perceive when a target is bisected by the vertical sight line of a transit or which line on a vernier matches that on a scale. Natural and instrumental errors are usually systematic, but personal errors tend to be random.

Two other terms that are widely used to indicate the quality of measurements are discrepancy and uncertainty.

A *discrepancy* is merely the difference between two measurements of the same quantity. A small discrepancy indicates that no mistakes were made and that accidental errors were small. It offers no indication of the existence or magnitude of systematic errors. For example, the forward and reverse taping of a baseline 1000 ft long may produce a discrepancy of only .04 ft, but neglect of the slope and temperature corrections may make both measurements highly faulty.

The *uncertainty* in a measurement is a rather loose expression of the difference between the true value and the measured quantity. Hence, it might be said that the uncertainty of an angle is ±15″. Sometimes it is used synonymously with standard error (see Art. 8-5).

8-3. Precision and Accuracy If two separate series of measurements are made of the same quantity and the individual observations of the first set show more scatter or are more discordant than those of the second set, it is quite logical to conclude that the measurements of the second series were conducted with more care or under more favorable circumstances than those of the first series. It can be said that the *precision* or the degree of conformity of the measurements of the second series is greater. However, precision, which denotes relative or apparent nearness to the truth, must not be confused with *accuracy,* which expresses correctness of the results or absolute nearness to the truth.

It will be helpful to keep in mind that precision relates to the degree of refinement in the performance of an operation or in the statement of a result. It is a term that is best associated with the quality of execution of a measuring task, whereas accuracy refers to the quality of the result.

The accuracy of the product of a measurement program is always somewhat affected by systematic errors. However, such errors, if undetected, would not prevent a series of observations from being closely grouped or from indicating a high precision. For example, consider that two similar tapes of different length characteristics have been standardized. The report on tape A was erroneously applied to the results of several measurements of a baseline with tape B. The agreement between the measurements was excellent, but the accuracy of the average length was seriously prejudiced.

In general, it can be said that discordant measurements indicate a lack of precision, but precise measurements may or may not be accurate.

Consider also the following example. Suppose the *true* length of a line is 710.325 ft. Party C measured the line and found it to be 710.35 ft. Party D also measured the line with somewhat greater refinement and recorded the length as 710.367 ft. It can be concluded that the measurement by Party C is more accurate but less precise than that by Party D.

8-4. Distribution of Accidental Errors Suppose that a fairly large number of careful measurements are made of some quantity. The mean of these measurements is the *most probable value*. For the proof of this the references at the end of the chapter should be consulted. In order to determine the errors of the individual measurements, it is assumed that the most probable value is the true value of the quantity. Then the error of each separate measurement is taken to be equal to the difference between the measurement and the most probable value. This difference is termed a *residual* or *deviation*.

The residuals of like sign and within certain ranges of values are then counted and plotted as rectangular coordinates with the magnitude of the error along the horizontal axis, plus on the right and minus on the left of the origin, and with the number of errors of that size, or the frequency of occurrence, plotted upward on the vertical axis.

A smooth curve is then drawn, connecting the points as closely as possible. This curve, shown in Fig. 8-1, portrays the following characteristics of accidental errors:

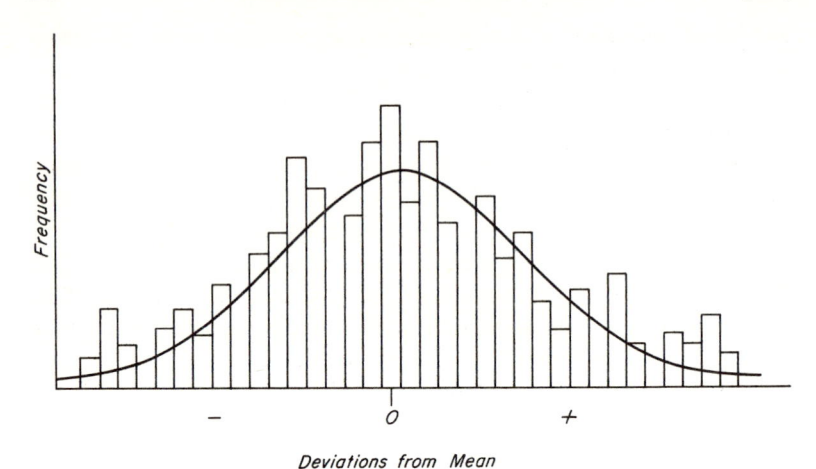

Fig. 8-1. Distribution of Accidental Errors.

1. Positive and negative errors of the same magnitude occur with equal frequency.
2. Small errors occur more frequently or are more probable than large ones.
3. Very large errors seldom occur.

It should be recalled that the number of measurements was fairly large. The smaller the number of measurements, the less closely the residuals will conform to define a representative curve. If, however, it be assumed that an unlimited number of observations are made, the mathematical principles of probability permit a theoretical curve to be derived. Such a curve, shown in Fig. 8-2, is called the *probability curve* and has the following properties:

1. The total area under the curve is unity.

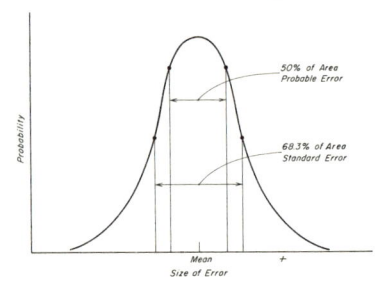

Fig. 8-2. Probability Curve.

2. Each ordinate expresses in per cent the probability of occurrence of an error whose magnitude is given by the abscissa.
3. The sum of the probabilities is 100%.
4. In a specific range of error the area beneath the curve between the designated limits (plus to minus) indicates the relative probability of occurrence of errors in such a range.

8-5. Measures of Precision In order to compare the relative quality of various sets of physical measurements of the same quantity it is convenient to calculate a numerical index that is expressive of the precision of the observations. Two widely used indices are the standard error and the probable error.

The *standard error,* also termed the *standard deviation* and the *mean square error,* is utilized in the interpretation of biological, sociological, psychological, and related physical data and to a limited extent in the assessment of surveying observations. The *probable error* has been employed to indicate the precision of surveying measurements for many years.

The defining equations are as follows:

$$E_s = 0.6745\sqrt{\frac{\Sigma v^2}{(n-1)}} = 0.6745\sigma_s \tag{8-1}$$

and

$$E_m = 0.6745\sqrt{\frac{\Sigma v^2}{n(n-1)}} = \frac{E_s}{\sqrt{n}} = 0.6745\frac{\sigma_s}{\sqrt{n}} \tag{8-2}$$

where E_s is the probable error of a single observation in a series of observations, E_m is the probable error of the mean of all the observations, v is a residual, n is the number of observations, and σ_s (sigma) is the standard error of a single observation.

The significance of probable error and standard error will be clarified by further examination of Fig. 8-2. The probable error of an observation in a series is the middle one of all the errors (or residuals) when they are arranged in numerical order. Since the number of errors greater than the probable error is the same as the number less than it, the probability of an error exceeding the probable error is equal to that of an error being less than it because the total probability is unity. Hence, the chances are equal that an error taken at random from the series will be greater or less than the probable error. The limits of the probable error are indicated in Fig. 8-2 by ordinates equidistant from the vertical axis and enclosing an area beneath the probability curve equal to 0.50. This means that there is a 50% likelihood that the error of any measurement will fall between $+E$ and $-E$. Hence, the probable error may be defined as that quantity which, added to or subtracted from the most probable value, fixes the limits within which there is an even chance that the true value of the measured quantity must lie.

With respect to the standard error or sigma, the analogous interpretation is that there is a 68.3% certainty that the error will be within the limits defined by $+\sigma$ and $-\sigma$.

The manner of calculating the probable error and the standard error will be illustrated by the following example.

EXAMPLE 8-1: In carrying a line of levels across a river, a series of ten readings was taken by observer A on a level rod held on the distant benchmark. At a later time observer B, with a different instrument and under different weather conditions, took ten readings on the same benchmark. The two series of readings and the calculations of probable error resulting from the use of the appropriate parts of Eqs. 8-1 and 8-2 follow:

OBSERVER A

Rod	v	v^2
5.36	0.00	0.0000
5.43	0.07	0.0049
5.27	0.09	0.0081
5.38	0.02	0.0004
5.37	0.01	0.0001
5.31	0.05	0.0025
5.32	0.04	0.0016
5.41	0.05	0.0025
5.39	0.03	0.0009
5.36	0.00	0.0000
5.36 = Mean	±0.36 = Σv	0.0210 = Σv^2

$$E_s = 0.6745\sqrt{\frac{0.0210}{9}} = \pm 0.033 \text{ ft}$$

$$E_m = \frac{\pm 0.033}{\sqrt{10}} = \pm 0.010 \text{ ft}$$

OBSERVER B

Rod	v	v^2
6.51	0.00	0.0000
6.37	0.14	0.0196
6.59	0.08	0.0064
6.56	0.05	0.0025
6.51	0.00	0.0000
6.50	0.01	0.0001
6.41	0.10	0.0100
6.65	0.14	0.0196
6.52	0.01	0.0001
6.49	0.02	0.0004
6.51 = Mean	±0.55 = Σv	0.0587 = Σv^2

$$E_s = 0.6745 \sqrt{\frac{0.0587}{9}} = \pm 0.055 \text{ ft}$$

$$E_m = \frac{\pm 0.055}{\sqrt{10}} = \pm 0.017 \text{ ft}$$

The value $E_s = \pm 0.033$ ft signifies that, in the series of readings taken by observer A, for any single reading (taken at random), say 5.38, there is an even chance that the true value lies within the limits 5.38 ± 0.033 ft.

The value $E_m = \pm 0.010$ ft applies to the mean or most probable value, 5.36 ft, and signifies that, in the series of readings taken by observer A, there is an even chance that the true value of the measured quantity lies within the limits 5.36 ± 0.010 ft.

The corresponding values of sigma for observer A are

$$\sigma_s = \pm 0.049 \text{ ft} \quad \text{and} \quad \sigma_m = \pm 0.015 \text{ ft}$$

It should be noted that neither the probable error nor the standard error can be applied as a correction to observed values. Both carry the double sign, plus or minus.

The term, probable error, should not be construed to mean the most probable or most likely error. Probable error is best understood simply as an index or a measure of precision.

The presence of undetected systematic errors in the various observations of a series would not affect the calculation and expression of the precision, but the arithmetic mean of all the measurements obviously would be an inaccurate value.

8-6. Propagation of Accidental Errors In surveying practice both direct and indirect measurements are made. A direct measurement is one in which the value of the quantity is determined by measuring the quantity itself. An indirect measurement is one requiring the computation of the desired quantity from measurements on other quantities. For example, the horizontal distance between two points can be calculated from the measured slope distance and the observed difference of elevation.

The fundamental principles used to evaluate the error of a quantity computed from indirect measurements will be presented. Derivations are to be found in the literature cited at the end of the chapter.

(*a*) *Error of a sum.* The probable error of the sum of the independent measurements whose probable errors are $E_1, E_2, \ldots, E_n$, respectively, is

$$E_{\text{sum}} = \sqrt{E_1^2 + E_2^2 + \cdots + E_n^2} \tag{8-3}$$

Equation 8-3 and the equations that follow are valid for probable error, standard error, or any other specified error or uncertainty.

When a series of similar quantities, such as tape lengths, are involved and the accidental error E of transferring each tape length to the ground is considered to be the same, a special case of Eq. 8-3 arises and

$$E_{\text{series}} = \sqrt{E^2 + E^2 + \cdots + E_n^2} = E\sqrt{n} \tag{8-4}$$

Equation 8-4 is the mathematical expression of the law of compensation mentioned earlier.

EXAMPLE 8-2: A distance of 3600 ft is to be measured with a 100-ft tape. Consider that the accidental error of measuring one tape length is ± 0.005 ft. The total accidental error will be $(.005)\sqrt{36} = \pm 0.030$ ft.

(*b*) *Error of a product.* If $U = XY$,

$$E_U = \sqrt{X^2 E_Y^2 + Y^2 E_X^2} \tag{8-5}$$

EXAMPLE 8-3: The dimensions of a rectangular city lot are 123.30 ft and 48.30 ft, with uncertainties of ± 0.04 ft and ± 0.03 ft, respectively. What is the uncertainty in the area? By Eq. 8-5 the uncertainty is

$$E_U = \sqrt{(123.30)^2(.03)^2 + (48.30)^2(.04)^2} = \pm 4.1 \text{ sq ft}$$

(c) *General equation.* The fundamental equation expressing the relationship between the accidental errors of the measurements and the accidental error of the final result is

$$U = \sqrt{\left(\frac{\partial U}{\partial X}\Delta X\right)^2 + \left(\frac{\partial U}{\partial Y}\Delta Y\right)^2 + \left(\frac{\partial U}{\partial Z}\Delta Z\right)^2 + \cdots + \left(\frac{\partial U}{\partial N}\Delta N\right)^2} \quad (8\text{-}6)$$

where U is a function of the measured quantities X, Y, and Z.

EXAMPLE 8-4: The difference of elevation h between the ends of a traverse course is to be computed from the measured slope distance, $L = 837.30 \pm 0.10$ ft and the vertical angle, $\alpha = 3°20' \pm 30''$. What is the uncertainty in the difference of elevation if the estimated uncertainties in L and α are as indicated?

$$h = L \sin \alpha$$

By Eq. 8-6,

$$\Delta h = \sqrt{\left(\frac{\partial h}{\partial L}\Delta L\right)^2 + \left(\frac{\partial h}{\partial \alpha}\Delta \alpha\right)^2}$$

where

$$\frac{\partial h}{\partial L} = \sin \alpha = 0.058$$

$$\frac{\partial h}{\partial \alpha} = L \cos \alpha = (837.30)(.998) = 835.63$$

$$\Delta \alpha = (30)(.00000485) = \pm 0.00015 \text{ radian}$$

and

$$\Delta L = \pm 0.10 \text{ ft}$$

Hence,

$$\Delta h = \sqrt{(.058 \times .10)^2 + (835.63 \times .00015)^2} = \pm 0.12 \text{ ft}$$

8-7. Weighted Measurements Thus far it has been assumed that all the measurements have been made under the same conditions and that they are of equal quality. Sometimes, however, one observation of a series may be more reliable than another. Such an observation should exert greater influence upon the calculation of the results. The degree of reliability is commonly termed the *weight* of the measurement. This is merely the relative value of that observation to the others of a series. It is expressed as a number and, being strictly relative, may be multiplied by any factor provided that all the others in a series are multiplied by the same quantity.

The general expression for a weighted mean, $\overline{M}$, of n observations M_1, M_2, M_3, ..., M_n having the respective weights w_1, w_2, w_3, ..., w_n is

$$\overline{M} = \frac{w_1 M_1 + w_2 M_2 + w_3 M_3 + \cdots + w_n M_n}{w_1 + w_2 + w_3 + \cdots + w_n} \quad (8\text{-}7)$$

The assignment of weights is largely a matter of judgment based on experience and on a knowledge of the field conditions at the time the various observations or measurements were made.

When calculating the mean value of some quantity from two or more sets of observations it is logical to give consideration to the calculated precision of each of the sets or series. The weights are taken to be inversely proportional to the square of the probable error (or the standard error), or

$$w_1/w_2 = E_2^2/E_1^2 \quad (8\text{-}8)$$

EXAMPLE 8-5: A line was carefully taped on two different days. The mean of the measurements on the first day was 1278.562 ft, and the calculated probable error of the mean was ±0.012 ft. On the second day the respective quantities were 1278.551 ft and ±0.021 ft. What is the weighted mean length of the line?

$$\frac{w_1}{w_2} = \frac{(.021)^2}{(.012)^2} = \frac{7^2}{4^2} = \frac{49}{16}$$

Hence,

$$\overline{M} = \frac{(1278.562)(49) + (1278.551)(16)}{49 + 16} = 1278.559 \text{ ft}$$

8-8. Adjustments Suppose that in the course of a survey the three horizontal angles of a triangle have been measured with a 30″ transit and their sum fails to equal the theoretical amount, viz., 180°. Before calculating the lengths of the triangle sides, it is both customary and essential to adjust the three measured angular values by applying equal corrections to them. Furthermore, consider the matter of adjustment of benchmark elevations. Benchmarks A and B are ten miles apart and have had their elevations determined by second-order levels. A connecting line of third-order levels (see Fig. 8-3) is now run between these benchmarks and several intermediate

Fig. 8-3. Running Third-Order Levels. *U.S. Geological Survey*

benchmarks are set. Quite obviously a closure will be developed when terminating the levels at benchmark B. Therefore, the provisional elevations of the intermediate benchmarks must be adjusted.

The procedure of equitably and logically making corrections to the original data is termed *adjustment*. The results are termed the adjusted or corrected values. Such an adjustment as that of the interior angles of a single small triangle is performed very easily. Others are highly complex. The student has thus far adjusted the interior angles of a loop traverse in order to make the sum of the angles equal to the theoretically correct amount. He has also balanced or adjusted the original computed latitudes and departures of such a traverse in order to make their algebraic sums zero. Two other adjustment calculations will be now illustrated, as follows:

1. Figure 8-4 shows the manner in which the results of a line of third-order levels between two points of fixed elevation are tabulated. The closure error is distributed among the several sections of the line in proportion to their lengths in order to obtain the correction or adjustment.

2. Figure 8-5 depicts four traverses intersecting at a common point. It is necessary to determine the best values of the coordinates of P. A satisfactory solution can be effected by finding a weighted mean for each of the desired coordinates. The weight assigned to the values obtained from each of the traverses is taken to be inversely proportional to the length of the traverse. It is assumed, of course, that all traverses are of comparative quality and originated at points

230 FUNDAMENTALS OF SURVEYING [8-8]

State: Illinois
County: Champaign
Line: Urbana to Glover
Computed by: WTZ *Date:* 6-15-68
Checked by: RTJ *Date:* 6-18-68

Level Book	Benchmark	Dist. (miles)	Summation B.S.	Summation F.S.	Diff. Elev.	Correction	Adjusted Diff. Elev.	Adjusted Elev.	Remarks
15-1968	USGS "G-1"							722.210	Fixed
	PBM 7	0.8	56.012	53.068	+2.944	+.006	+2.950	725.160	
	TBM 1	1.1	90.513	92.517	−2.004	+.009	−1.995	723.165	
	PBM 8	0.7	54.222	60.127	−5.905	+.006	−5.899	717.266	
	TBM 2	0.9	63.319	70.811	−7.492	+.007	−7.485	709.781	
	TBM 3	1.2	49.106	52.616	−3.510	+.010	−3.500	706.281	
	PBM 9	0.7	36.515	35.109	+1.406	+.006	+1.412	707.693	
	TBM 4	0.9	57.819	67.313	−9.494	+.007	−9.487	689.206	
	PBM 10	1.1	97.111	100.106	−2.995	+.009	−2.986	695.220	
	PBM 11	0.8	35.689	34.123	+1.566	+.006	+1.572	696.792	
	TBM 5	1.0	67.822	74.120	−6.298	+.008	−6.290	690.502	
	TBM 6	0.7	41.103	39.115	+1.988	+.006	+1.994	692.496	
	PBM 12	0.6	32.719	43.778	−11.059	+.005	−11.054	681.442	
	USGS "G-2"	1.0	56.189	58.734	−2.545	+.008	−2.537	678.905	Tie
		11.5	738.139	781.537	−43.398	+0.093			Closure = 0.093 Allowable closure = 0.170

Fig. 8-4. Tabulation of Third-Order Levels.

[8-8] ERRORS AND ADJUSTMENTS

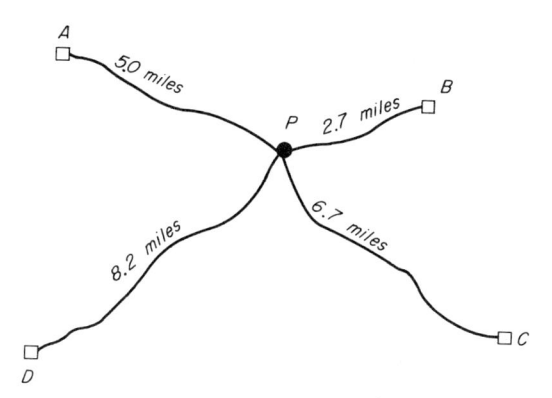

Fig. 8-5. Junction-Point Traverse Adjustment.

whose coordinates are dependably known on the same common coordinate system.

The procedure for calculating the coordinates of P is shown in the following tabulation:

From Point	Coordinates (ft) North (y)	Coordinates (ft) East (x)	Length, L (miles)	Weight $\frac{1}{L}$
A	32,548.9	47,685.3	5.0	0.20
B	32,547.3	47,686.6	2.7	0.37
C	32,548.1	47,686.9	6.7	0.15
D	32,549.5	47,684.7	8.2	0.12
				0.84

The weighted mean of the last two figures in the north coordinate of P is:

$$\frac{(8.9 \times 0.20) + (7.3 \times 0.37) + (8.1 \times 0.15) + (9.5 \times 0.12)}{0.84} = 8.2$$

Similarly, the weighted mean of the last two figures in the east coordinate is:

$$\frac{(5.3 \times 0.20) + (6.6 \times 0.37) + (6.9 \times 0.15) + (4.7 \times 0.12)}{0.84} = 6.1$$

Hence, the adjusted coordinates of P are:

$$N = 32,548.2 \text{ ft} \qquad E = 47,686.1 \text{ ft}$$

A final comment must be made concerning the matter of adjustments. The mechanism of adjustment will not improve the quality of the observed data to the extent of guaranteeing that a final correct result will be attained. Adjustment will remove inconsistencies between the observations but cannot make up for deficiencies in them. It is exceedingly important that the engineer-surveyor secure data that are satisfactory for the use to which they will be put. No amount of mathematical manipulation will produce accurate results from raw data that are inherently faulty.

PROBLEMS

8-1. The uncertainty in centering the bubble of a level is estimated to be 1/8 of a division of the vial. The sensitiveness is 16″. Calculate the uncertainty in the difference of elevation between two benchmarks 4.0 miles apart if the sight distances are maintained at 150 ft.

8-2. It is estimated that the instrumentman was able to obtain the rod readings of Problem 8-1 with an uncertainty of ± 0.003 ft. Calculate the uncertainty in the difference of elevation due to this source of error.

8-3. Combine the two separate accumulated uncertainties of Problems 8-1 and 8-2 and find the total estimated uncertainty in the observed difference of elevation.

8-4. The various measurements of the length, in feet, of a survey line are as follows: 827.461, 827.508, 827.512, 827.493, 827.468, and 827.452. Calculate the probable error of the mean. Also, express in the form of a ratio with unity in the numerator.

8-5. An angle was carefully measured 10 times with an optical theodolite by observers A and B on two separate days. The calculated results are as follows:

Observer A	Observer B
Mean = 42°16′25.2″	Mean = 42°16′20.4″
$E_m = \pm 3.2''$	$E_m = \pm 1.6''$

Determine the best value of the angle.

8-6. A long baseline was divided into 4 sections and each section was carefully measured a number of times. The calculated probable errors, in feet, of the mean lengths of each section were as follows: ± 0.018, 0.043, 0.031, and 0.020. Determine the probable error of the length of the baseline.

8-7. The sides of a rectangular field measure 1247.7 ft and 627.0 ft with estimated uncertainties of ± 0.7 ft and ± 0.5 ft, respectively. Determine the area and the uncertainty of its value.

8-8. Four traverse lines, one of which, AP, was double taped, were

ERRORS AND ADJUSTMENTS 233

executed to a junction point P. The calculated coordinates of P were as follows:

Via Route	Distance (miles)	Coordinates (ft) x	Coordinates (ft) y
A	2.20	27,891.32	31,431.07
B	4.35	27,892.00	31,431.32
C	1.60	27,892.05	31,430.98
D	5.95	27,891.76	31,430.90

Find the adjusted coordinates of P. (Note that the weight accorded the coordinates of P deduced from line AP should be calculated from an effective distance equal to one-half the actual distance or 1.10 miles.)

8-9. The difference of elevation between two points was determined by trigonometric leveling. The slope distance was measured electronically and found to be 1468.72 ft and the zenith distance was $83°14'20''$. The respective uncertainties in these two quantities are ± 0.05 ft and $\pm 15''$. Calculate the difference of elevation and the uncertainty in this value.

8-10. Determine the weighted mean of the following measurements of an angle:

$62°27'38.5''$ weight 2
$62°27'41.3''$ weight 1
$62°27'47.0''$ weight 3
$62°27'40.2''$ weight 2

REFERENCES

1. Barry, B. Austin, *Engineering Measurements*, John Wiley & Sons, Inc., New York, 1964.
2. Beers, Y., *Theory of Errors*, Addison-Wesley Publishing Co., Inc., Reading, Mass., 1958.
3. Kissam, Philip, *Surveying for Civil Engineers*, Chapter 21, McGraw-Hill Book Co., New York, 1956.

9 Triangulation

9-1. Introduction Triangulation is a method of surveying to determine the horizontal positions of points on the surface of the earth. It is a very efficient mode for making surveys over extensive areas because it avoids the necessity of measuring the lengths of all lines that enter into a survey. A triangulation system basically consists of a configuration of triangles all of whose angles have been carefully measured but very few of whose sides have been directly measured. Sides whose lengths are actually measured are known as *bases* or *baselines*. The survey points or triangulation stations are located at the vertices of the triangles. By the use of the measured angles and bases, the lengths of all other sides in a connected system can be successively determined by trigonometry. Furthermore, if the horizontal coordinates of one point are known as well as the azimuth to another station, it is possible to derive the coordinates of all other points and the azimuths of all other lines.

Triangulation surveys are classified according to accuracy into several orders or categories. Also, such surveys may be considered to be of two types: plane, in which the curvature of the earth's surface is ignored, and geodetic, in which the true figure and size of the earth are taken into account. *Geodetic triangulation* requires extremely precise instrumental equipment and observational techniques and offers an excellent example of one of the several activities of geodetic surveying mentioned in Art. 1-4. *Plane triangulation* has usefulness only over areas of limited extent, but its concepts are particularly relevant, therefore, to this textbook whose scope is largely restricted to plane surveying practice. However, the treatment of triangulation principles provided in this chapter will include an introduction to the national system of geodetic triangulation, the methods employed, the results obtained, and the means by which the local engineer-surveyor can identify and procure needed triangulation data.

The basic system of horizontal control for defining the geographic positions of points within the United States and serving as the authority for all subordinate surveys is the fundamental *horizontal control network* of traverse and triangulation that was begun by the U.S. Coast and Geodetic Survey and is continually being expanded by that agency. Ultimately, as the network becomes more densely developed, the engineer executing a local control survey in most parts of the nation will have, within a few miles, a survey point of accurately known position.

The principle of the triangulation method is not of recent origin. Tycho Brahe, the Danish astronomer, is commonly credited with having conducted the first survey of this type. In 1578–79 (see Ref. 3) he connected Denmark and Sweden with a network of triangulation across the intervening sea. The operation is particularly remarkable when it is realized that it antedated the development of the telescope by Lippershey in 1607.

The horizontal control surveys of the United States were begun during the early part of the nineteenth century at several points, mostly along the coasts, and existed initially as separate surveys. Some examples of such detached surveys are the early triangulation in New England, a portion in the vicinity of St. Louis, Missouri, and separate surveys near San Francisco and San Diego, California. By 1900 all of the detached surveys were welded together into a single national coordinated survey system.

9-2. Triangulation Systems A cardinal principle in the extension of horizontal and vertical control is that of initially establishing master frameworks of reference, such as networks of high-order triangulation and levels, respectively, and subsequently subdividing such networks into secondary and tertiary systems of successively lower quality. This principle is sometimes termed "working from the whole to the part." The horizontal positions of the salient triangulation stations and the vertical positions of the key benchmarks are determined with the highest accuracy. The intervening country is then progressively subdivided by arcs of triangulation (and lines of traverse) and lines of levels through the employment of methods of lower precision until the desired density of control is attained. The reason for executing the initial surveys with the highest precision is to restrict the accumulation of errors that would otherwise take place. The reverse process can be easily imagined to produce large and serious errors. In essence, then, an accurate basic control survey will assure that all subsidiary and detail surveys will properly fit together. Triangulation (and also traverse) of requisite accuracy performs this role with respect to horizontal position. Any system of triangulation consists of a series of connected triangles which adjoin or overlap each other. Figure 9-1 illustrates four systems of triangulation.

A *single chain of triangles* is a rapid and economical system for covering a narrow strip of terrain as, for example, a river valley. It is not as accurate as other systems, and baselines must be introduced

(a) Chain of Triangles *(b) Quadrilaterals* *(c) Central Point Figures*

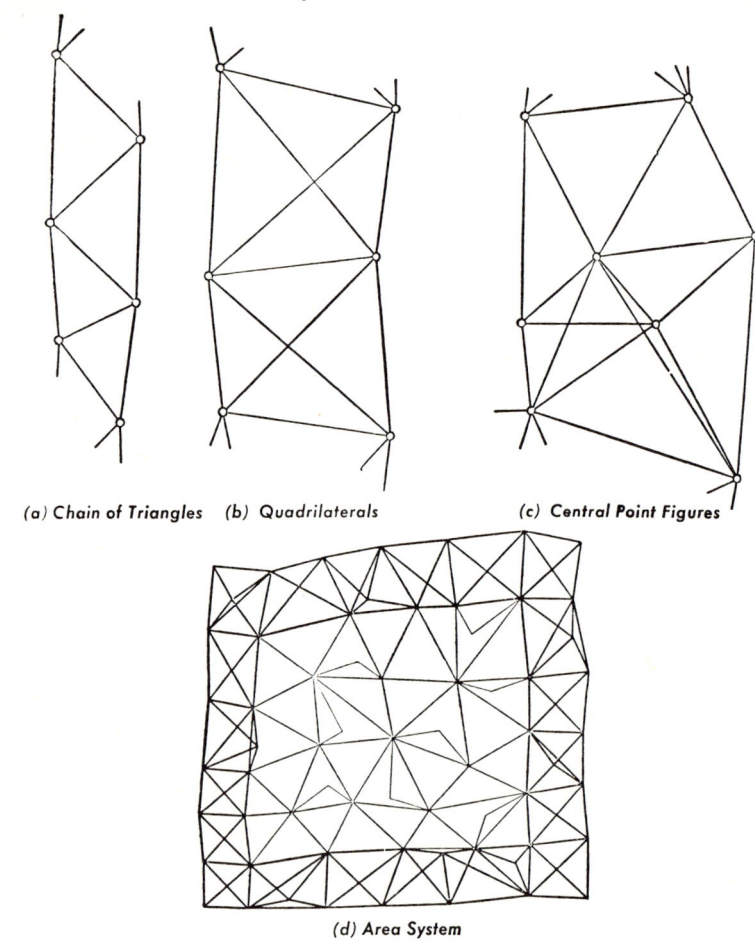

(d) Area System

Fig. 9-1. Triangulation Systems.

frequently if the accumulation of errors is not to become excessive. It is important in this system that no small angles (say less than 20°) shall be permitted. High quality triangulation systems never contain single triangles as units in a chain of figures.

Quadrilaterals afford an excellent system since the computed lengths of the sides can be carried through the system by different combinations of sides and angles, and thus the accuracy of the results is increased and checks on all computations are frequently obtained. Most of the major arcs of triangulation in this country consist of chains of quadrilaterals.

If a wide area is to be covered with a relatively dense distribution of points, such as would be the case for a large metropolitan triangulation survey, *central-point figures* are used.

The *area system* originated when it became necessary to provide triangulation control in the areas bounded by the principal arcs of triangulation.

All stations in any triangulation system are ordinarily occupied in order to measure the angles of each triangle. Additional points useful for control purposes may have their positions determined by two or more intersecting lines of sight from the main scheme stations. Such intersection points may be water tanks, radio masts, lookout towers, industrial chimneys, and other tall and conspicuous objects.

The lengths of triangulation lines vary over a broad range. On the principal arcs, lengths of 100 miles are not uncommon. Accompanying the development of the triangulation system in the United States has been a progressively closer distribution of survey points so as to satisfy better the requirements of state and municipal governments, and engineers and surveyors in private practice. This desire to meet local needs generally has shortened the average triangle side so that many are from 5 to 15 miles long. The execution of an independent triangulation survey across a wide stream in order to obtain data for the construction layout (see Fig. 9-2) of a bridge may entail distances of only a few thousand feet.

9-3. Classification of Triangulation The basis of classification of horizontal control surveys is the accuracy with which the length and azimuth of a line of the triangulation or traverse are determined. Triangulation surveys are of widely different accuracies, depending on the extent and purpose of the survey. The generally accepted standards of accuracy for *first-, second-,* and *third-order triangulation* are

Fig. 9-2. Bridge Layout Survey. *Dravo Corp.*

given in Fig. 9-3. These standards are achieved through the use of instruments and procedures that are consistent with the desired order of accuracy.

The principal criterion for judging the quality of triangulation is the relative base-to-base length discrepancy. This discrepancy between the measured length of a baseline and its length as computed through the system from the next preceding base shall not be greater than 1 part in 100,000 of the length of the base for first-order, class I work; 1 part in 20,000 for second-order, class I; and 1 part in 5000 for third-order. These are minimum requirements. In general, most length closures are much better than the prescribed requirements.

The basic horizontal control network for the nation consists of arcs of first-order, class II triangulation. Eventually the entire country will be spanned by such arcs at spacings of 60 miles and with the areas between the arcs in the approximate shape of squares.

Second-order, class I triangulation comprises systems of triangles that subdivide the area between the first-order arcs, and, together with second-order, class II triangulation, will provide an eventual distribution of approximately one station of either first- or second-

… TRIANGULATION …

	First Order			Second Order		Third Order
	Class I	Class II	Class III	Class I	Class II	
Principal use	Urban surveys, scientific studies	Basic network of United States	State, county, private	Area networks and supplemental cross arcs in national net	Coastal areas and inland waterways. Supplemental and cadastral	Topographic
Base measurement						
Actual error not to exceed	1 part in 300,000	1 part in 300,000	1 part in 300,000	1 part in 300,000	1 part in 150,000	1 part in 75,000
Probable error not to exceed	1 part in 1,000,000	1 part in 1,000,000	1 part in 1,000,000	1 part in 1,000,000	1 part in 500,000	1 part in 250,000
Triangle closure						
Average not to exceed	1″	1″	1″	1.5″	3″	5″
Maximum seldom to exceed	3″	3″	3″	5″	5″	10″
Closure in length						
Discrepancy between computed length and measured length of base, or adjusted length of check line, not to exceed	1 part in 100,000	1 part in 50,000	1 part in 25,000	1 part in 20,000	1 part in 10,000	1 part in 5,000

Fig. 9-3. Specifications for Triangulation.

order control in each 50 square miles of area. Second-order, class II triangulation is represented by the lightweight lines in Fig. 9-1d.

Third-order triangulation is extended from triangulation of higher order and is used principally as the control for plotting charts and maps. However, it can be used also for a wide variety of engineering and construction surveys. Primary attention is directed to this grade of triangulation. Higher-quality triangulation is thoroughly treated in Ref. 1.

9-4. Reconnaissance All triangulation, even that of limited extent, is usually preceded by a preliminary field study, called the *reconnaissance,* to select the best station sites. Some of the criteria in determining the location and distribution of stations are *intervisibility* and *strength of figure.*

The intervisibility of stations must be established before the angle observing program is begun. In some cases a visual test of the lines upon the preliminary visit to the station site is all that is required. In most situations it will be helpful to obtain data from a topographic map as to the character of the intervening terrain, the ground elevations at the stations, and the approximate length of the line. If a map is not available, the elevations can be obtained by barometric or trigonometric leveling, and various expedients can be utilized to ascertain the length of sight. Profile studies can then be conducted to determine whether the obstructions are cleared by the line of sight and proper allowance is made for earth's curvature and atmospheric refraction. The combined effect of *curvature and refraction* is expressed by

$$h = 0.574M^2 \qquad (9\text{-}1)$$

where h is in feet, and M is the length of the sight in miles.

Even in flat country where the stations may be a considerable distance apart, towers may be needed to overcome the effect of earth's curvature. The computed tower heights are generally increased by at least 10 ft to avoid sights that graze the ground or water surfaces where refraction effects are uncertain, and to allow for errors due to uncertainties in elevations and distances of the profile data.

In Fig. 9-4 the target at station B must be 10 ft + (0.574 × 100) = 67.4 ft above the water. If both the instrument at A and the target at B are to be elevated, the necessary height above the water surface for each is 10 ft + (0.574 × 25) = 24.4 ft.

[9-4]   TRIANGULATION   241

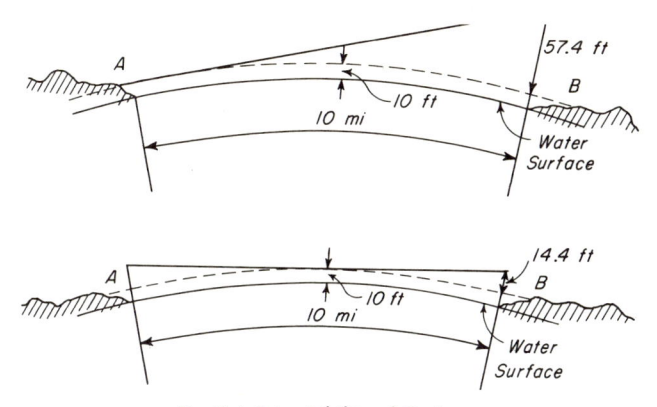

Fig. 9-4. Intervisibility of Stations.

Strength of figure refers to the effect of the shape of a triangle upon the accuracy with which the length of a side can be calculated. In any system of triangulation the lengths of the triangle sides are computed by the law of sines. The starting data are the measured length of a line, called the *base,* and the horizontal angles at the vertices of the triangles. Since, for a given change in the angle, the sines of small angles change more rapidly than those of large ones, it is evident that the percentage error in the computed side of a triangle will be larger if the side is opposite a small angle than if it is opposite a larger angle. It is to be assumed that the accuracy with which an angle is measured is independent of its size.

In the two triangles, *ABC* and *DEF* of Fig. 9-5, let the side *b* equal the side *e*. Also, let it be supposed that the sides *c* and *f* are to be computed, and that the opposite angles, 60° and 15°, respectively, have been measured with the same accuracy—namely, one minute of arc. Since the lengths of *c* and *f* are directly proportional to the

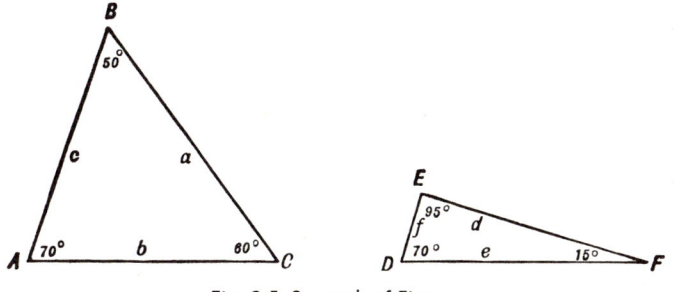

Fig. 9-5. Strength of Figure.

sines of the opposite angles, it follows that the percentage error in the computed length of a side will be the same as the relative uncertainty in the sine of the angle. An examination of a table of natural sines will reveal that the error in the computed length of side c opposite the 60° angle will be 0.017%; whereas the error in the computed length of side f opposite the 15° angle will be 0.10%, or six times the percentage error in the side c.

Although the foregoing analysis indicates the great importance of well-shaped figures in a triangulation network, it is not to be concluded that all small angles are undesirable. In computing a given side in any triangle there are always two *distance angles*—namely, the angle opposite the known side and the angle opposite the unknown side. In the triangle ABC, if it is supposed that the side b is known and the side c is to be computed, then the distance angles are 50° and 60°, respectively. Because of this condition there is always one angle in a triangle that has no effect in computing a given side. Hence, this noneffective angle can be very small and still the accuracy of the computed length will in nowise be affected. For example, in the triangle DEF, if the sides e and d are the known and unknown sides, respectively, then the distance angles are 95° and 70°. Accordingly, the small angle 15° has no effect on the computed result.

9-5. Signals and Towers Various kinds of targets may be used depending upon the quality of the triangulation, the length of sight, the conditions of visibility, and other circumstances.

A satisfactory target for daylight observations for lines up to only a few miles long may be a straight mast with alternate sections of it painted in contrasting colors. The pole must be vertical, centered accurately over the station, and adequately guyed with wire. A pennant at the top of the mast will be helpful to triangulation observers in finding the target.

In Fig. 9-6 is shown a daylight fluorescent type of target, which is suitable for moderate lengths of sights.

Practically all important and high-quality triangulation angle measurements are conducted at night because of the superior observing conditions at that time. Electric signal lamps make excellent targets.

Although triangulation stations are most frequently situated on the tops of hills or promontories, it is usually necessary to erect a tower to elevate the instrument or the target, or both, to assure intervisibility. Such towers are of various designs, heights, and materials, but all consist of two separate and independent structural elements, each having its own foundation, so that the movements of the observer and the recorder are not transmitted to the theodolite. A tower of moderate height is shown in Fig. 9-7.

Fig. 9-6. Triangulation Signal. C & R Mfg. Co.

Fig. 9-7. Triangulation Tower. Parsons, Brinkerhoff, Quade & Douglas

9-6. Station Markers Much of the value of a triangulation survey depends upon the permanence and recoverability of the *station markers*. It cannot be too strongly emphasized, therefore, that all horizontal control survey stations of third or higher order be adequately monumented. A monumented station is one that has a durable marker which is properly stamped or identified. The station must be described in a manner that will assure its positive recovery any time in the future.

The following comments refer particularly to the practice of the U.S. Coast and Geodetic Survey.

Most triangulation stations are monumented with a bronze disc or tablet of distinctive design which is set in a concrete post. Occasionally such discs are placed in holes drilled in outcropping rock or massive masonry structures and grouted into place with cement. A typical triangulation tablet is shown in Fig. 9-8. Where the soil con-

Fig. 9-8. Triangulation Marker.

ditions permit, a subsurface marker should be set in a small mass of concrete with its face about 4 ft (or sufficient to extend below the frost line) below the surface of the ground and with a layer of sand 3 to 4 in. thick between it and the bottom of the concrete post in which the surface marker is embedded. The practice of double monumentation is of particular value in areas subject to cultivation.

As a further safeguard against the loss of a station, it is necessary that there be set at least two *reference marks* each consisting of a disc bearing an arrow set to point toward the station. The true bearings and horizontal distances to the reference stations from the main station should be measured and recorded.

Another important station accessory is the *azimuth mark* which can be used to furnish an azimuth to local surveyors and engineers initiating a survey at the main station. Each triangulation station usually has an azimuth mark situated not less than 1000 ft away and in such a location that it is visible from the ground instrumental setup at the main station.

The following is a typical description of a U.S. Coast and Geodetic Survey triangulation station:

Name of Station, WILSON; State, Wisconsin; County, Eau Claire

About 14 miles east of Fall Creek, 10 miles northeast of Augusta, approximately 14 miles north of Fairchild and situated near the center of section 28, T27N, R5W near the Wilson fire lookout tower.

To reach from the Augusta State Bank in Augusta, go north and east on County Route G, 12.5 miles to where it turns north and a dirt road continues east. Proceed east on dirt road 1.25 miles until it turns north. The Wilson lookout tower is about 500 ft to the southeast.

Station mark, a standard disk in a concrete monument stamped "WILSON 1934" is 233.4 ft southeast of the center of the lookout tower, 38 ft northwest of a 15-in. oak tree, and about 200 ft southwest of the southeast corner of a fenced pasture. Monument projects 4 in. above ground surface.

Reference mark 1, a standard disk marked "WILSON No. 1 1934" set in concrete is situated toward the northeast 125.04 ft.

Reference mark 2, a standard disk marked "WILSON No. 2 1934" set in concrete is situated under the Wilson lookout tower and 233.48 ft toward the northwest. The distance between the reference marks is 230.31 ft.

Azimuth mark, a standard disk set in concrete and stamped "WILSON 1934" is situated on the property of E. A. Shambaugh, about 350 ft northeast of his house, near the crest of a ridge 100 ft south of the east-west fence on the north side of the field and about 200 ft west of the north-south fence. Monument is flush with surface.

9-7. Angle Measurement The direction theodolite is the principal instrument used in the measurement of angles in triangulation. Its horizontal circle remains fixed as the telescope is pointed on various signals and the circle is read for each direction. The difference between successive directions equals the value of the included angle.

A broad range of optical-reading direction theodolites suitable for triangulation are commercially available. All these instruments possess essentially the same basic characteristics. A representative type, the Wild T-2 depicted in Fig. 9-9, will be described. This instrument is satisfactory for the measurement of angles on all triangulation and traverse except that of the highest accuracy. Hence, it is well adapted to meeting the instrumental requirements of second- and third-order horizontal control surveys.

After the instrument is properly leveled with the single plate-bubble tube and the three leveling screws and a pointing has been made, the procedure for reading a direction is briefly as follows:

1. The black line on the inverter knob is set in a horizontal position so that the horizontal circle is visible in the reading microscope.

2. The knob (optical micrometer) for coincidence setting is turned until in the middle of the field of view the graduation lines of the upper half-image of the circle appear to be in perfect coincidence with those of the lower half-image (see Fig. 9-10).

3. The reading is obtained by counting the number of 10′ intervals from 285° to 105°. This is five intervals, which represents 50′. The additional minutes and seconds are given by the micro-

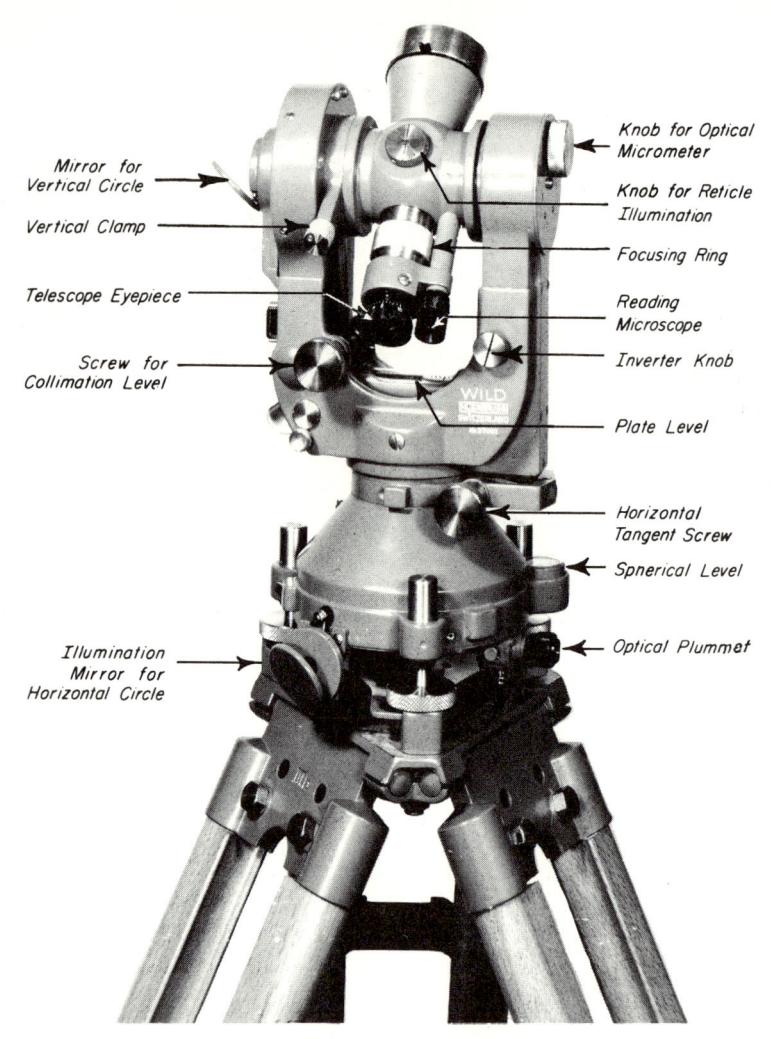

Fig. 9-9. Wild T-2 Theodolite. *Wild Heerbrugg Instruments, Inc.*

meter drum reading of 1′54.6″. Hence, the complete reading is 285°51′54.6″.

The vertical circle is read in a similar manner. The angular quantity obtained is, however, the zenith distance or the complement of the altitude rather than the altitude itself.

The following observing procedure is recommended for the Wild T-2 theodolite when used on triangulation. First, any suitable main

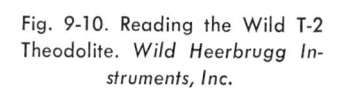

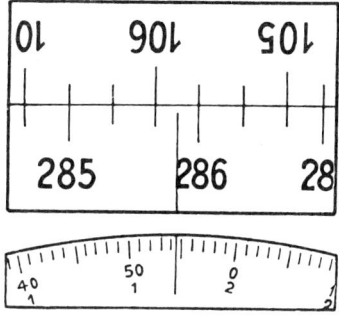

Fig. 9-10. Reading the Wild T-2 Theodolite. *Wild Heerbrugg Instruments, Inc.*

scheme line is selected as the initial or reference line. With the instrument level and the telescope in the direct position the observer sets the circle for the first position. The use of different circle settings for each of several initial pointings ensures that all parts of the circle will be used in the determination of the various angles.

The observer then makes a pointing on the initial station, reads the direction, and proceeds to move around the horizon in a clockwise manner, making pointings upon and reading directions to all stations. The telescope is then reversed and directions are observed to all the same stations in a counterclockwise manner. This program of observations constitutes one *position*. The number of positions, each of which requires a different initial setting, varies with the required accuracy of the survey.

9-8. Base Measurement Baselines whose lengths are determined by direct taping are used infrequently on third-order triangulation because of the general policy of beginning and closing subsidiary triangulation along lines of adjusted higher-order triangulation. Furthermore, with the introduction of electronic distance measuring devices, most baselines can be measured more economically, at no sacrifice in accuracy, with such devices.

However, all distances measured with such instruments are uncertain by a fixed amount of unknown sign, such as ±0.03 ft, plus a fraction of the distance (see Art. 4-11). Thus, for a short baseline the proportionate error in the length may be intolerably large. Hence, baselines may be taped when they are too short to be measured electronically with the requisite accuracy, and, of course, when an electronic device is not available.

The base tape used on modern triangulation work is preferably made of an alloy of nickel and steel. Such tapes are known as *Invar, Nilvar,* or *Lovar* tapes. Their coefficient of expansion is only about 1/30 of that of steel. This is of marked advantage when the field temperature conditions are variable, for an error of only 0.5°F in the mean temperature of a steel tape alone introduces an error of 1 part in 300,000 of the measured length. Such an error is not permitted by the specifications for higher-order baselines.

The first task of a base measuring party is to clear the selected site for the base and set carefully on line at the proper intervals, depending upon the length of the tape and the number of intermediate supports, the 4-in. × 4-in. wood posts over which the taping is conducted. If the posts are carefully lined in with a transit, any resulting errors due to alinement will be negligible.

The front end of the tape should be marked with a sharp pencil or with a finely pointed awl on a small strip of copper tacked to the top of the wood posts. The measurement between two adjacent posts should be made with the tape in contact with the tops of the posts.

In order to determine the temperature of the tape while the base is being measured, thermometers attached to both front and rear ends of the tape are read for each measured tape length.

The data for determining the slope or grade correction are usually obtained by running a line of profie levels over the tops of the posts. An inspection of the inclination corrections in Table X will be useful in deciding what accuracy is necessary in such leveling.

In general, incorrect tension or sag is not a significant source of error on baseline taping since the tape is used in the field under the same conditions of tension and support as when it was standardized in the laboratory. However, undetected variations in tension, although insufficient to elongate the tape, may introduce an error due to sag which can be considerable. Base tapes for important triangulation work should be calibrated by the U.S. Bureau of Standards, Washington, D.C. The report for a typical test is shown in Fig. 9-11.

The ends of the baseline should be permanently marked with massive concrete monuments having metal tablets in their surfaces. It is best to construct a protective barricade around the point, as shown in Fig. 9-12. Also depicted therein is a portable device, called a taping tripod or *buck,* which was used to support the tape during the base measurement.

9-9. Baseline Reduction Calculations After the field measurement of a base has been completed, it is necessary to calculate and

U.S. DEPARTMENT OF COMMERCE
NATIONAL BUREAU OF STANDARDS
WASHINGTON, D.C. 20234

NATIONAL BUREAU OF STANDARDS

REPORT OF CALIBRATION
on
100-Foot Iron-Nickel Alloy Tape

Maker: Keuffel & Esser Co. NBS No. 10247
Maker's No. 5582

Submitted by

University of Illinois
Department of Civil Engineering
Urbana, Illinois 61801

This tape has been compared with the standards of the United States. The horizontal straight-line distances between the terminal points of the indicated interval have the following lengths at 68°Fahrenheit (20°Celsius) when the tape is subjected to the horizontally applied tensions and conditions of support given below:

Supported on a horizontal flat surface:

Tension	Interval	Length
10 kilograms	0 to 100 feet	100.004 feet

Supported only at the 0- and 100-foot points with both points of support in the same horizontal plane:

Tension	Interval	Length
15 kilograms	0 to 100 feet	100.003 feet

The values given for the lengths of the indicated interval are not in error by more than 0.001 foot. This limit of error is based upon the limits imposed by the standards used for the calibration of the tape, the length of the tape interval, the character of the terminal markings, and the behavior of the tape during the period of calibration.

The comparisons of this tape were made at a mean temperature of 68.0°Fahrenheit using the centers of the ends of each graduation near the edge of the tape marked with small dots near the graduation. The coefficient of linear thermal expansion of this tape is assumed to be 0.00000022 per degree Fahrenheit (0.0000004 per degree Celsius), and is the value authorized for this tape by the Keuffel & Esser Company.

Test No. 212.21/G-37913
Date: January 23, 1967

Fig. 9-11. Baseline Tape Calibration Report.

apply various corrections to the observed lengths in order to obtain the best value of the length of the base.

These corrections, which are fundamentally the same as those described and computed in Chapter 2, are as follows:

1. Tape-length correction
2. Temperature correction
3. Grade correction

Fig. 9-12. End of Baseline. *Tennessee Valley Authority.*

The temperature correction is usually computed and applied for the entire length of a section of the baseline making use of the mean observed temperature for that section. The grade correction must be computed for each full or partial tape length.

Only in special situations, such as those involving the measurement of a partial tape length, is it necessary to consider a correction for sag. Ordinarily it is very unlikely that a correction for incorrect tension will be required. In an extended system of triangulation, particularly over regions of widely varying relief, it is necessary to reduce all baseline lengths to the plane of mean sea level. Failure to do this prevents the attainment of satisfactory closures between bases at different elevations.

In Fig. 9-13, let C be the correction to be subtracted from the measured baseline L, which has an elevation H above sea level, to reduce it to its sea-level length. Then, since for a given angle arcs are proportional to their respective radii, the following proportion can be written:

$$\frac{L-C}{R} = \frac{L}{R+H}, \quad \text{or} \quad C = \frac{LH}{R+H}$$

Since H is small compared with R, it may be disregarded in the denominator, and then

$$C = \frac{LH}{R} \tag{9-2}$$

An average value of the earth's radius may be taken as 20,906,000 ft.

EXAMPLE 9-1: A baseline has been measured as 4352.416 ft at an average elevation of 1140 ft. Find the sea-level length of this line.

$$C = \frac{(4352)(1140)}{20,906,000} = 0.238 \text{ ft}$$

Hence, the sea-level length is $4352.416 - 0.238 = 4352.178$ ft.

9-10. Angle Adjustment When an arc of triangulation consists of a chain of triangles like that in Fig. 9-1a, the adjustment of the angles involves the application to each angle of a correction equal to one-third of the triangle closure. In the case of a quadrilateral like that of Fig. 9-14, in addition to the satisfaction of the *geometric condition*,

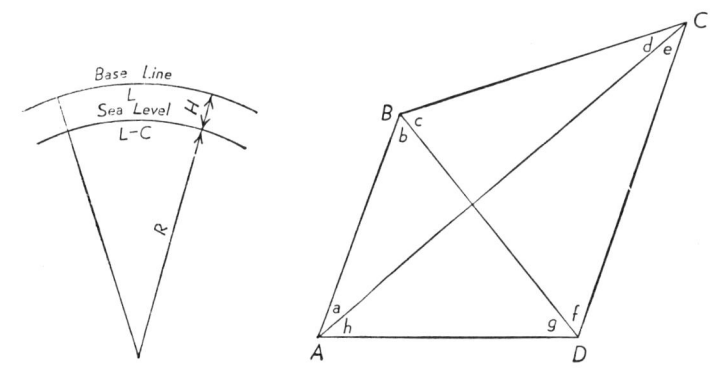

Fig. 9-13. Reduction to Sea Level.

Fig. 9-14. Adjustment of a Quadrilateral.

viz., that the sum of the angles of each triangle be made equal to 180° exactly, a *trigonometric condition* should be satisfied. This entails a secondary correction of the angles so that the distance BC as computed from the starting line AD will be the same no matter what computation route is taken through the quadrilateral. An elementary example of the geometric adjustment of the angles of a quadrilateral will be presented. An adjustment of this nature is ordinarily satisfactory for a small triangulation survey, such as a bridge crossing. A more adequate treatment of the entire quadrilateral adjustment is to be found in Ref. 2.

EXAMPLE 9-2: The angles in the quadrilateral of Fig. 9-14 have the following observed values:

	°	′	″
$a =$	29	25	34
$b =$	58	41	20
$c =$	69	36	20
$d =$	22	17	02
$e =$	30	29	25
$f =$	57	37	33
$g =$	50	35	44
$h =$	41	17	26
	360	00	24

The first step in the adjustment process is to decrease all angles by 3″ so as to eliminate the closure of +24″.

In order to achieve equality of the sums of opposite pairs of angles it is necessary that the following conditions also be satisfied:

$$a + b = e + f$$

and

$$c + d = h + g$$

Hence, using the first adjusted values, there results

$$a + b = 88°06'48''$$
$$e + f = 88°06'52''$$

Therefore, a correction of +1″ is applied to angles a and b, and a correction of −1″ to angles e and f.

Similarly, for the other pairs of opposite angles:

$$c + d = 91°53'16''$$
$$h + g = 91°53'04''$$

Therefore, a correction of −3″ is applied to angles c and d, and a correction of +3″ to angles h and g. It is to be noted that these corrections are applied to the preliminary adjusted values. Hence, for example, the final value of angle a is 29°25′34″ − 3″ + 1″ or 29° 25′32″. The adjusted angles are as follows:

	°	′	″
$a =$	29	25	32
$b =$	58	41	18
$c =$	69	36	14
$d =$	22	16	56
$e =$	30	29	21
$f =$	57	37	29
$g =$	50	35	44
$h =$	41	17	26
	360	00	00

9-11. Length and Position Calculations In the calculation of the sides of the triangles in a triangulation system one side and all three angles are known. The law of sines is used to determine the remaining sides.

EXAMPLE 9-3: Let the side AD of the quadrilateral of the preceding example have a length of 984.68 ft. Calculate the lengths of all sides and make a double determination of the length of BC by using two computation routes through the figure. A desk calculator and natural trigonometric functions are utilized.

Proceeding through the quadrilateral with triangles ADB and ABC:

$$AB = \frac{AD \sin g}{\sin b} = \frac{(984.68)(.772684)}{(.854353)} = 890.553 \text{ ft}$$

$$DB = \frac{AD \sin (a + h)}{\sin b} = \frac{(984.68)(.943894)}{(.854353)} = 1087.891 \text{ ft}$$

$$AC = \frac{AB \sin (b + c)}{\sin d} = \frac{(890.553)(.784860)}{(.379169)} = 1843.398 \text{ ft}$$

$$BC = \frac{AB \sin a}{\sin d} = \frac{(890.553)(.491292)}{(.379169)} = 1153.896 \text{ ft}$$

Then, using the route through triangles ADC and DBC:

$$DC = \frac{AD \sin h}{\sin e} = \frac{(984.68)(.659878)}{(.507375)} = 1280.648 \text{ ft}$$

$$BC = \frac{DC \sin f}{\sin c} = \frac{(1280.648)(.844559)}{(.937306)} = 1153.927 \text{ ft}$$

It is seen that the two calculated lengths for the side BC are substantially the same and a mean value would be taken as 1153.91 ft.

In order to proceed with the calculation of coordinates, assume that point A has the rectangular coordinates, $N = 10,000.00$ ft, $E = 10,000.00$ ft, and that the bearing of AD is due east. A double-position calculation of the coordinates of point B will be performed.

EXAMPLE 9-4: Calculate the rectangular coordinates of point B using the triangle ABD and the preceding data.

Station	Bearing	Distance	Cos	Sin	N (ft)	E (ft)
A					10,000.00	10,000.00
	N. 19°17′02″ E.	890.55	.943894	.330249	+840.58	+294.10
B					10,840.58	10,294.10
D					10,000.00	10,984.68
	N. 39°24′16″ W.	1087.89	.772684	.634790	+840.59	−690.58
B					10,840.59	10,294.10

Thus, it is apparent that the calculation of the coordinates of point B has been checked. They were derived from arbitrary values assigned to point A. In Chapter 11 the subject of state plane coordinates will be presented. They should be used whenever it is feasible to do so.

9-12. Network of Triangulation in the United States A vast network of triangulation of various orders of accuracy and different densities of distribution covers the entire nation. Although the major portion of the higher-order work has been executed by the U.S. Coast and Geodetic Survey, other federal agencies have assisted with the extension of the triangulation, particularly with third-order work. Such organizations include the U.S. Geological Survey, the U.S. Bureau of Reclamation, the Mississippi River Commission, the Bureau of Land Management, the U.S. Corps of Engineers, and the U.S. Forest Service.

Figure 9-15 depicts a small part of a large index map showing the first- and second-order triangulation net in the country. As of June 30, 1965, the total number of triangulation stations set by the U.S.

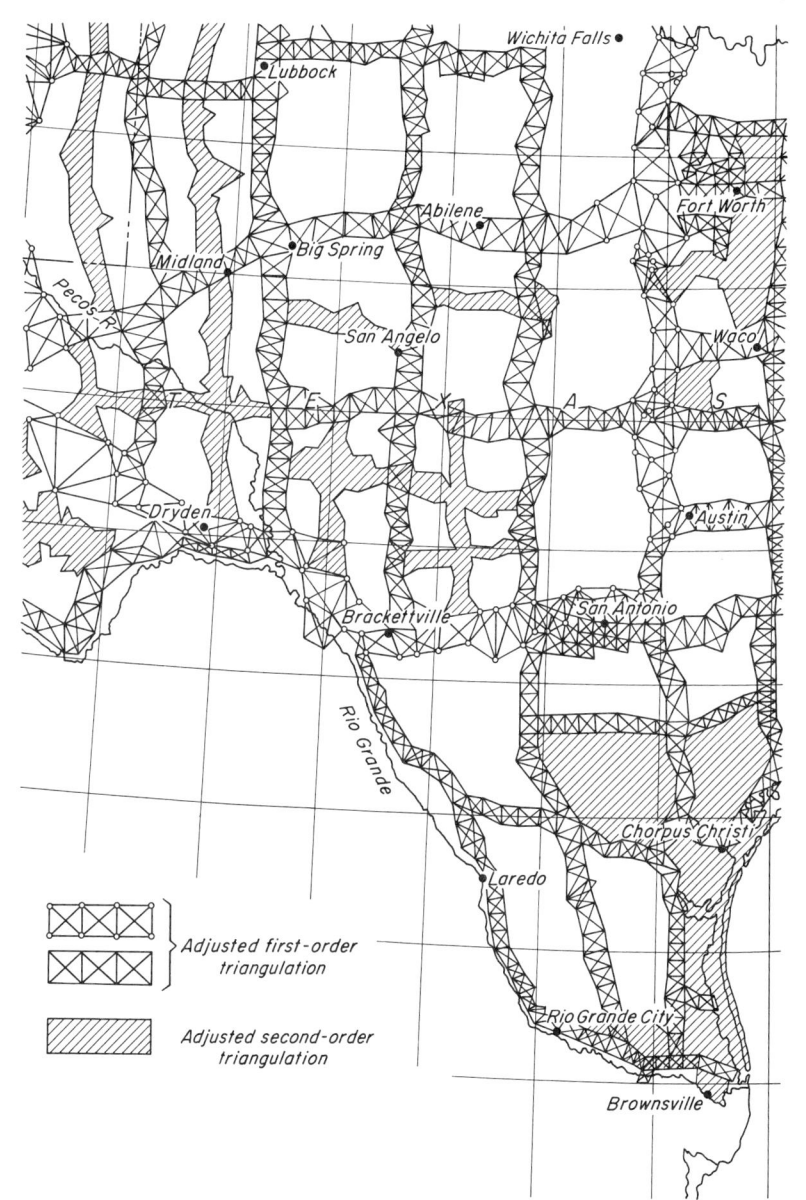

Fig. 9-15. Network of Triangulation. *U.S. Coast and Geodetic Survey*

Coast and Geodetic Survey was approximately 175,000. In the fiscal year ending June 24, 1965, a total of 1880 new stations were set. It is evident, therefore, that continuing progress is being made in expanding the nation's fundamental triangulation network.

The engineer who wishes to initiate a local survey at any point of known horizontal position, whether it be a triangulation or traverse station, should know how to secure the position coordinates of the point. In all cases, the name of the agency setting the mark will be found on the metal tablet. Additional identification of the mark can be secured by noting carefully the name or number of the station, also the year and other information that may be stamped on the disc. Having ascertained the name of the agency and having properly identified the mark, an inquiry can be directed to the organization giving this information and also the general location of the station by state, county, and proximity to the nearest city or town.

Frequently the engineer or surveyor will wish to ascertain what horizontal control is available in a given area. An initial request can be directed to the Environmental Science Services Administration, Coast and Geodetic Survey, Washington Science Center, Rockville, Maryland, for a copy of the Geodetic Control Diagram covering the particular area. These diagrams portray both horizontal and vertical control executed by the Coast and Geodetic Survey and by the U.S. Geological Survey in a format having latitudinal and longitudinal dimensions of 1° and 2°, respectively. A request for specific desired data can then be transmitted to these agencies.

Figure 9-16 shows the manner in which the positions and azimuths are published. It should be noted that geodetic (or true) azimuths are reckoned from south. The grid coordinates and plane azimuth will be explained in Chapter 11. Obviously, the description that accompanies the survey data is very important, because in its absence the engineer could not find the station he wishes to occupy.

The entire system of geographic positions is based on the *North American Datum of 1927* which is defined by a reference surface conforming to the dimensions of the *Clarke Spheroid of 1866* on which a triangulation station, *Meade's Ranch,* in central Kansas has specific values of latitude and longitude, and the azimuth of the line from Meade's Ranch to triangulation station *Waldo* has a fixed value. This system has been extended into and utilized by Canada, Mexico, and the Central and South American countries.

ADJUSTED HORIZONTAL CONTROL DATA

NAME OF STATION: THOMAS **YEAR:** 1959
STATE: Illinois **LOCALITY:** Ill. Hwy. Survey, Champaign to Kankakee
Second-ORDER Triangulation **SOURCE:** G-12097 **FIELD SKETCH:** Ill. 22-I
and Traverse

GRID DATA	COORDINATES (Feet)	PLANE AZIMUTH θ(OR Δα) ANGLE	MARK
STATE: Ill. ZONE: E CODE: 1201	x 535,920.13 y 1,300,893.44	209°01'40" + 0 04 59	AZ MK RANTOUL CHANUTE AFB SOUTH TANK OF 2
STATE: ZONE: CODE:	x y		

GEODETIC DATA	POSITION		SECONDS IN METERS	ELEVATION	
	LATITUDE:	40° 14' 19."6187 NORTH		Bench	METERS
	LONGITUDE:	88 12 16.8218 WEST		Mark	FEET

TO STATION	GEODETIC AZIMUTH (From south)	DISTANCE (Meters)
	SECOND-ORDER	
TWIN	27°14'44."86	3,022.479*
BABB RM 3	89 32 09.97	2,350.747*
BORO	217 08 55.34	3,946.086*
	THIRD-ORDER	
RNG RANTOUL RADIO RAN	202 20 31.4	2,034.31
RANTOUL MUNICIPAL STANDPIPE	205 01 39.5	9,007.38
RANTOUL CHANUTE AIR FORCE BASE S TANK OF 2	209 06 38.7	8,089.98
RANTOUL MUNICIPAL WATER TANK	213 53 03.8	10,249.95
TACAN RANTOUL RAN	221 24 30.7	8,318.92
RANTOUL CHANUTE AIR FORCE BASE SE TANK	232 40 30.7	8,743.03

*Tellurometer length

Fig. 9-16. Horizontal Control Data. *U.S. Coast and Geodetic Survey*

9-13. Uses of Triangulation Data The purpose of triangulation is simply that of accurately determining the relative positions of widely distributed survey stations. The documented coordinates of such points and the monumentation that serves to physically define them constitute the product of the triangulation survey.

The basic triangulation control network of the nation has many very important uses. First and foremost, it provides a rigid framework for many kinds of mapping and charting projects. It serves to locate national, state, and county boundaries; it perpetuates the results of private boundary surveys which are tied to it; it facilitates the correlation of local surveys and mapping programs, and it can be used to provide accuracy checks for a variety of surveys, such as highway and other route surveys.

Basic horizontal control data, whether obtained by triangulation or traverse, can be said to be an important requirement for developing the nation's resources and planning large engineering projects, such as those for flood control, navigation, and irrigation.

9-14. Trilateration The method of extending horizontal control by the traverse method has gained increased acceptance in recent years because of the economies attained through the use of electronic distance measuring devices. However, inasmuch as traversing does not have the geometric checks available in the triangulation method, the latter is preferred for the most important work; but when the observing conditions are unfavorable, the effect on triangulation operations is likely to be more severe than on traverse because traverse sights are generally much shorter. Further consideration of the relative advantages of traverse and triangulation has led to the occasional employment of a new type of triangulation in which the sides of the triangles instead of the angles are measured. This operation is termed *trilateration*. The lengths of the triangle sides can be determined by electronic distance-measuring devices in weather that would be unsuitable for triangulation. The angles can then be calculated from the sides by trigonometry. The use of such angles, a starting azimuth, and known coordinates for a point make it possible to calculate positions.

9-15. Three-Point Problem Under certain conditions it may be useful to employ the principle of the *three-point problem* to obtain the horizontal position of a survey point. The field work involves only the measurement of the horizontal angles between three objects of known position. These can be church steeples, water tanks, radio masts, and other features that were previously located by intersecting lines of sight from the main triangulation stations. The circumstances leading to the use of the three-point problem are those in which an isolated point quite remote from a traverse or triangulation station is to be located with reasonable accuracy.

[9-15] TRIANGULATION

The theodolite is merely set up over a point whose position is desired and the angles are measured. It is necessary, however, that consideration be given to the shape of the resulting figure. In the worst possible situation, viz., when the instrument station whose position is sought lies on or near the circle through the three known stations, the solution will be very weak or indeterminate.

Various analytical, semigraphical, and graphical solutions to the three-point problem are available. A typical analytical solution will be presented.

In Fig. 9-17, A, B, and C are points of known positions. The

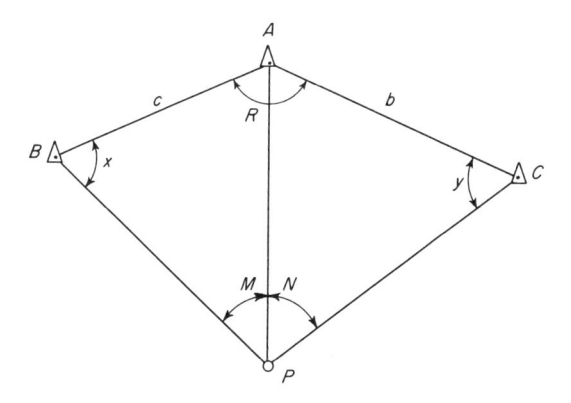

Fig. 9-17. Three-Point Problem.

angle, R, is known as well as the distances, b and c. At P, the unknown position of the instrument station, the angles, M and N, have been measured.

$$y + x = 360° - (M + N + R) = J \quad (9\text{-}3)$$

also,
$$\frac{\sin x}{AP} = \frac{\sin M}{c}$$

and
$$\frac{\sin y}{AP} = \frac{\sin N}{b}$$

whence,
$$\frac{c \sin x}{\sin M} = \frac{b \sin y}{\sin N}$$

or
$$\frac{\sin x}{\sin y} = \frac{b \sin M}{c \sin N} = H \quad (9\text{-}4)$$

now, from 9-3,

$$x = J - y$$

and, from 9-4,

$$H \sin y = \sin x = \sin (J - y)$$

or

$$H \sin y = \sin J \cos Y - \cos J \sin y$$

dividing through by cos y and rearranging,

$$\tan y = \frac{\sin J}{H + \cos J} \qquad (9\text{-}5)$$

Then, x can be found from 9-3, and both triangles, ABP and ACP, can be solved.

PROBLEMS

9-1. Find the sea-level length of a baseline for which the following data are given: length of lovar tape, 100.002 ft while supported at the end points under a tension of 15 kg at a temperature of 68°F; coefficient of expansion is 0.00000022 per degree Fahrenheit. The tape was used in the field under the same conditions of support and tension as when it was standardized. The average observed temperature was 88.6°F, and the mean of the field measurements (uncorrected) was 1005.482 ft. Profile levels over the supports gave elevations as follows:

Sta.	Elev.	Sta.	Elev.
0	99.98	6	106.67
1	97.12	7	107.95
2	98.70	8	108.58
3	101.22	9	108.69
4	102.32	10	112.51
5	104.42	+05	112.52

The partial tape length at the end of the base was measured with a steel tape fully supported.

9-2. A baseline was measured 5 times and the reduced lengths, in feet, were as follows: 1346.788, 1346.772, 1346.748, 1346.781, and 1346.761.

Determine the probable error of the mean length of the base. Also express the result in the usual ratio form.

9-3. The measured values of the angles in the quadrilateral of Fig. 9-14 are as follows: $a = 42°38'36''$, $b = 64°52'28''$, $c = 40°32'57''$,

$d = 31°56'07''$, $e = 62°00'46''$, $f = 45°29'58''$, $g = 33°31'32''$, and $h = 38°57'40''$.
Adjust these angles after the manner of Example 9-2.

9-4. The side AD of the quadrilateral of Problem 9-3 has a measured length of 2947.62 ft. Calculate the length of side BC by two independent routes through the figure.

9-5. The coordinates of point A in Problem 9-4 are $x = 10{,}000.00$ ft and $y = 10{,}000.00$ ft. The bearing of AD is S. $80°00'00''$ E. Find the coordinates of points B, C, and D.

9-6. The data for a three-point problem (see Fig. 9-17) are as follows: $R = 68°14'10''$, $M = 132°13'40''$, $N = 105°32'20''$, $c = 49{,}321.4$ ft, $b = 34{,}217.4$ ft. Find the angles x and y and the side AP.

REFERENCES

1. Gossett, F. R., *Manual of Geodetic Triangulation*, U.S. Coast and Geodetic Survey, Special Publication No. 247, Revised, Washington, D.C., 1959.
2. Rayner, W. H., and M. O. Schmidt, *Surveying—Elementary and Advanced*, Chapter 18, D. Van Nostrand Co., Inc., Princeton, N.J., 1957.
3. Skop, Jacob, "The First Modern Triangulation," *Surveying and Mapping*, Vol. X, No. 4, pp. 274–277, American Congress on Surveying and Mapping, Washington, 1950.

10 Observations for Meridian

10-1. Introduction Every survey must embody some concept of direction and be properly oriented. *Orientation* refers to the establishment of the correct relationship of a survey line to a cardinal line of direction. In Chapter 3 various kinds of bearings and azimuths were described, but no treatment of the methods for determining actual magnitudes was presented. That is the principal purpose of this chapter. The phrase, *observations for meridian,* refers to the measurements which are made to fix the position of a meridian or to deduce the angle that a survey line makes with such a reference line.

In general, original determinations of direction are either *geophysical* or *astronomical.* Measurements of the first type are made with the magnetic compass or a *gyro-theodolite.* Most engineer transits are equipped with a magnetic compass located between the standards on the upper plate. The uncertainty of a direction determination by the magnetic compass is hardly less than 15′. This has limited the use of the compass to rather rough and preliminary surveying operations, including certain mapping activities. Since the compass was used on many early property surveys, it may be helpful to have a thorough understanding of the compass in order to retrace old boundary lines.

The gyro-theodolite shown in Fig. 10-1 consists of a theodolite and a north-seeking gyroscopic device mounted on top of it. This instrument is useful when magnetic orientation is not sufficiently accurate and astronomical observations are not possible. High cost has precluded its substantial acceptance by organizations other than the military. An accuracy of ±30″ is attainable.

Astronomical determinations of direction can be made with varying degrees of accuracy, depending upon the quality of the equipment, the techniques employed, and the skill of the observer. Although

Fig. 10-1. Gyro-theodolite. Bureau of Land Management

both stellar and solar sightings can be taken to determine the true bearing of a line, the following treatment is restricted to observations upon the North Star, Polaris. It is presumed an ordinary engineer's transit or theodolite is available and that an uncertainty of not more than 30″ is satisfactory.

Astronomic observations for azimuth are made to prevent the accumulation of errors in triangulation and traverse, ascertain the true directions of land boundaries, orient radar antennae, and for numerous other purposes. In Chapter 11, it will be learned that it is occasionally necessary to make an astronomic determination of the azimuth of a line before its state plane or grid azimuth can be calculated.

Up to this time the term, true, has been used with certain expressions of direction, such as true bearing or true azimuth, to distinguish them from the corresponding magnetic values. From now on the preferred term, astronomic, will be occasionally used. True and astronomic should be considered as having identical connotations.

MAGNETIC OBSERVATIONS

10-2. Magnetic Compass The earth's magnetic field is characterized by lines of force which, except for small variations, remain fairly constant in direction. Accordingly, at any given place, the compass needle will indicate the same direction over a considerable period of time. For most places this direction will not be true north but will be either to the east or west of north, depending on the lo-

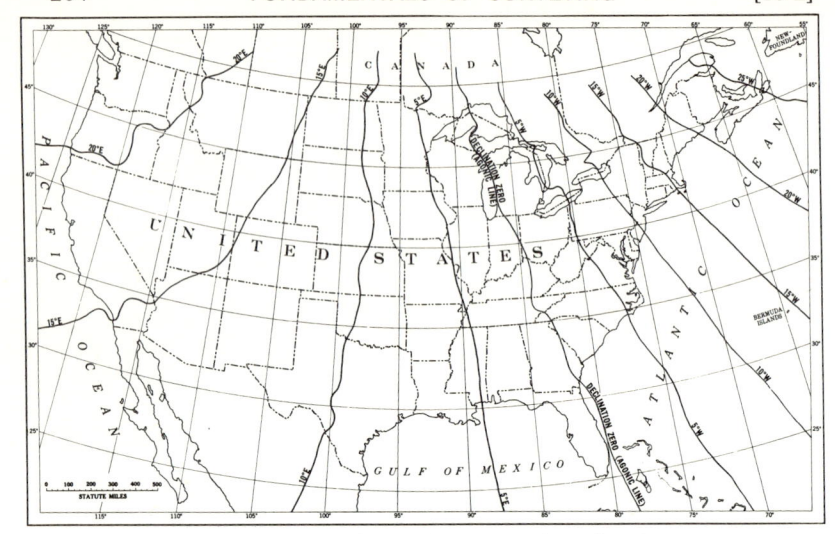

Fig. 10-2. Isogonic Chart. U.S. Coast and Geodetic Survey

cality. This angle between the direction of the needle and true north has already been defined in Art. 3-4 as the declination of the needle.

Figure 10-2 is an *isogonic chart* of the United States for 1960, which shows lines of equal magnetic declination. The position of the *agonic* line or line of 0° declination is to be noted. To the west of it the declination is east and to the east of it the declination is west.

It is to be noted that the chart is a generalized portrayal of magnetic declination and that pronounced differences from the normal pattern may exist in certain localities, such as in the iron ore regions.

The magnetic lines of force are also inclined to the horizontal, so that a needle which is balanced before it is magnetized will afterward, in the Northern Hemisphere, have its north end deflected downward. This deflection is called the *dip* of the needle.

The essential features of the compass are shown in Fig. 10-3. These include a circle graduated in quadrants from 0° to 90° both east and west from both the north and south points of the compass; a magnetized steel needle, supported on a steel pivot and jeweled bearing and counterbalanced against the dip of the needle; also a line of sight fixed with respect to the north-south points of the compass box. Since the needle dips down to the north, the counterbalance is placed south of the pivot and serves to indicate the north and the south ends of the needle. It will be noticed that, in sighting any

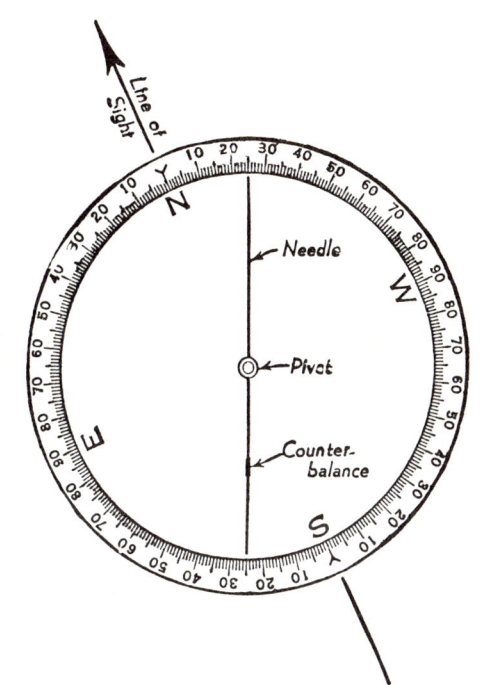

Fig. 10-3. Magnetic Compass.

point, since the graduated circle turns while the needle remains stationary on its pivot, the west and east points of the circle are interchanged, thus to give the correct bearing if the north end of the needle is read.

A clamp is provided to lift the needle from its pivot when not in use. It is important that this be done; otherwise, the needle will jar on its bearing and soon become sluggish and insensitive.

As a part of the transit, the compass box is mounted on the upper plate of the transit between the standards. The telescope, of course, provides the line of sight.

To read a bearing with the compass, the transit plate is leveled and the needle is released by its clamp and allowed to come to rest. The telescope is sighted along the line whose bearing is desired and the north end of the needle then indicates the bearing. If the south end of the needle were read, it is obvious that the back bearing of a line would be read. In Fig. 10-3 the bearing of the line of sight is found to be N. 25° W.

Since the magnetic needle is affected by any nearby iron or steel or other electromagnetic influence, it is obviously essential that no setup be made in close proximity to power lines, vehicles, hydrants, and like objects.

10-3. Declination Changes The direction of the magnetic meridian at any given place is subject to a number of variations which constitute a source of error that may or may not be serious, depending on the field conditions and the accuracy desired.

There is a *daily* variation, i.e., a slight swing of the needle of a few minutes of arc between morning and afternoon observations; there is also a small *annual* variation from month to month. These are negligible for ordinary compass work.

A *secular* variation, extending over a long period of time (about 250 years) is of greater importance because, although the variation from one year to the next is slight, it is a systematic movement which, in a term of years, attains considerable magnitude. By reason of this variation the direction of the magnetic meridian, at any given place, changes slowly (perhaps 3' per year) to the west for a long period of time and then it swings back to the east over a similar period. At the present time the swing is toward the west and, if the declination of the needle at any given place were 5° east of north 30 years ago, then, assuming a rate 3' westward per year, it would be 3°30' east now, and it will be 2° east 30 years hence. This is a matter of some importance in retracing compass surveys made many years ago.

ASTRONOMICAL OBSERVATIONS

10-4. Celestial Sphere The procedures of practical field astronomy are greatly simplified and an understanding of space relationships more easily achieved by the concept of the celestial sphere. The stars, which are at astronomical distances from the earth, are considered to be fixed on the inner surface of a sphere of infinite radius whose center is the earth. It is immaterial in this discussion whether the observer's position on the earth's surface, or the center of the earth, is considered as the center of the celestial sphere, since the radius of the earth is utterly insignificant when compared with the distances to the stars.

Figure 10-4 depicts the celestial sphere whose essential elements are defined as follows:

Great Circle. A great circle of a sphere is the trace on its surface

[10-4] OBSERVATIONS FOR MERIDIAN

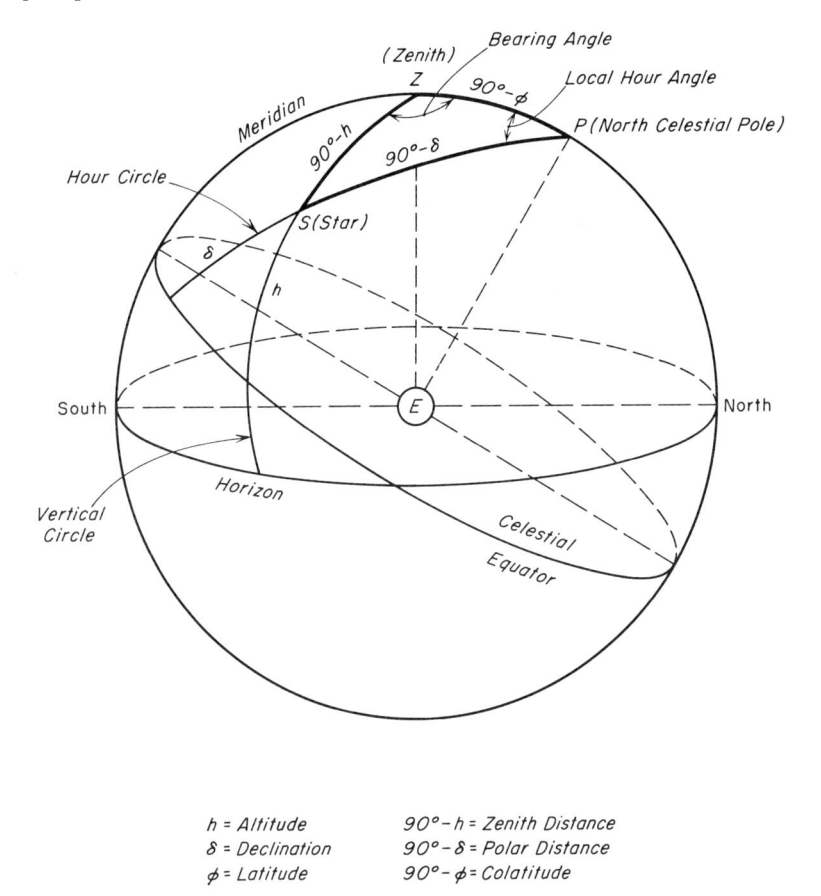

h = Altitude $90° - h$ = Zenith Distance
δ = Declination $90° - \delta$ = Polar Distance
ϕ = Latitude $90° - \phi$ = Colatitude

Fig. 10-4. Celestial Sphere.

of the intersection of a plane passing through the center of the sphere.

Celestial Poles. The north and south geographic poles are at the extremities of the polar axis of the earth. The celestial poles are situated at the points where the earth's polar axis extended intersects the celestial sphere.

Celestial Equator. The celestial equator is a great circle whose plane is perpendicular to the celestial polar axis.

Zenith. The zenith is that point on the celestial sphere where a plumb line extended upward intersects it. The opposite point is termed the nadir.

Horizon. The horizon is the great circle that is situated halfway between the observer's zenith and nadir points and whose plane is perpendicular to the plumb line. Azimuth is measured in the plane of the horizon.

Vertical Circle. A vertical circle is the great circle passing through the observer's zenith and any celestial object. Obviously it must be perpendicular to the horizon. The altitude of a heavenly body above the horizon is measured along a vertical circle.

Hour Circle. An hour circle is the great circle joining the celestial poles and passing through any celestial body. Thus, there is an hour circle for each object in the sky. Of necessity, an hour circle must be perpendicular to the celestial equator.

Meridian. The observer's celestial meridian is both an hour circle and a vertical circle, since it passes through the celestial poles as well as the zenith and nadir points.

The various angular quantities and the *PZS* astronomical triangle will be explained in subsequent articles.

10-5. Apparent Motion of Celestial Sphere The earth is a planet which completes a rotation from west to east about its polar axis in approximately 24 hours and effects a circuit about the sun in approximately 365 days.

For the purposes of astronomical observations, the earth is regarded as stationary and the celestial sphere is assumed to rotate about it from east to west. This concept of motion of the celestial sphere will be helpful in understanding the apparent movement from east to west of the celestial bodies and particularly in making clear the apparent daily rotation of the pole star, Polaris.

10-6. Time Time and the conversion of one kind of time to another are of great significance in all aspects of astronomy, since the daily motion of the stars and their positions at any instant are intimately related to time. Although several kinds of time are used in astronomy, it will be necessary to deal only with the familiar standard time in this abbreviated treatment of the azimuth problem.

If the movement of the sun from east to west is considered, it is recognized that its crossing of each meridian signalizes noon for all points on that meridian and that meridians to the east have already experienced noon while those to the west have yet to experience noon.

[10-6]   OBSERVATIONS FOR MERIDIAN   269

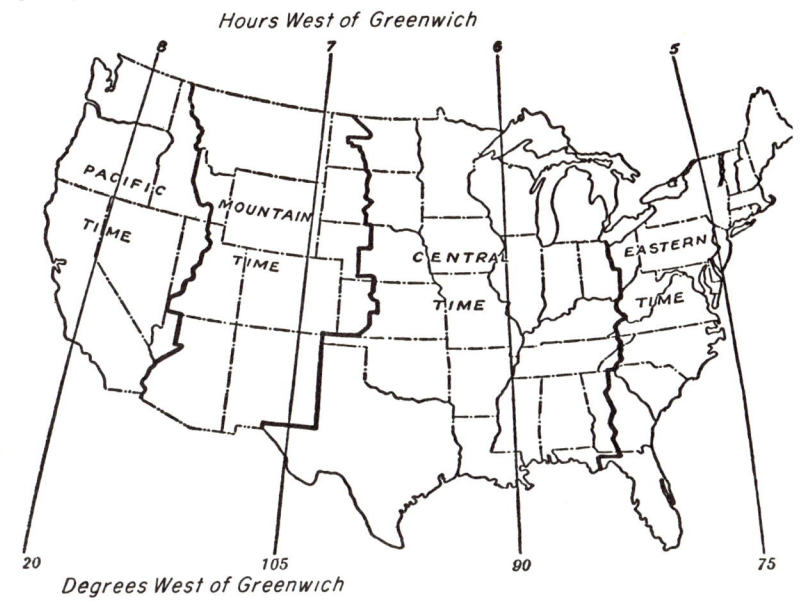

Fig. 10-5. Standard Time Zones in the United States.

Figure 10-5 shows the standard time zones that have been established in the United States. It will be noticed that the central meridians of the time belts differ by 15° of longitude, or 1 hour of time, since the sun completes an apparent revolution about the earth every 24 hours. Although the instant of noon (12 o'clock) on any observer's meridian is nominally that corresponding to the appearance of the sun exactly over the meridian, it is obvious that the use of a multiplicity of local times would result in endless confusion. Hence, all timepieces within a given time zone are set to keep the same kind of time as that which pertains to the zone's central meridian only. This kind of time is termed *standard time*.

Since the sun apparently moves from east to west, a given moment such as noon is experienced earlier in the zones to the east. For example, 12 o'clock Central Standard Time (CST) is 1 P.M. Eastern Standard Time (EST) and 11 A.M. Mountain Standard Time (MST).

The prime meridian for reckoning world longitude passes through the observatory at Greenwich, England. Thus the longitude of the central meridian of the Greenwich standard time belt is 0° or 0^h. The standard time for this zone is called *Greenwich Civil Time* (GCT) (also Greenwich Mean Time or Universal Time) and is widely used in astronomical work. GCT is obtained merely by adding to the standard time of any zone in the United States the appropriate number of hours of longitude that its central meridian is removed from Greenwich.

Daylight Time in any zone is equivalent to standard time in the zone next removed to the east; for example, 9 A.M. Central Daylight Time (CDT) is the same instant as 9 A.M. EST.

In astronomical calculations the hours of the day are numbered consecutively from 0 to 24 beginning at midnight. Thus an event which occurs at 3:00 A.M., September 1, would be recorded as 3^h00^m September 1, and an event 12 hours later would be recorded as 15^h00^m September 1. The designations of A.M. and P.M. are unnecessary and are frequently productive of serious errors in the expression of time.

EXAMPLE 10-1: Find the GCT of an observation made at 9:50 P.M., CST.

$$\begin{array}{r} 9:50 \text{ P.M., CST} \\ +12 \hphantom{:00 \text{ P.M., CST}} \\ \hline 21^h50^m \text{ CST} \\ +\ 6 \hphantom{^h00^m \text{ CST}} \\ \hline 3^h50^m \text{ GCT (on following day)} \end{array}$$

10-7. Time Signal Service The ease with which correct standard time, or the error of one's timepiece, can be obtained has contributed materially to the simplicity with which the results of astronomic observations for azimuth can be calculated. Although extreme accuracy in time is not required if azimuth values are to be correct to the nearest half minute only of arc, it is recommended as a principle of uniformity that the following time service be utilized to secure the watch correction to the nearest second.

If an ordinary short-wave radio set is available, time signals can be received 24 hours a day from Station WWV, the National Bureau of Standards' Central Propagation Laboratory in Ft. Collins, Colorado. Every second there is a tick and every 5 minutes there is voice identification of the signals. These time signals are transmitted on a frequency band ranging from 2.5 to 25 megacycles.

10-8. Latitude and Longitude The latitude and longitude of the observer furnish an expression of his position on the earth. Since values of these coordinates are required for astronomic computations of azimuth, it is desirable to define them and provide information as to how they can be obtained.

Latitude is merely the angle between the direction of the plumb line and the plane of the earth's equator. It can also be defined as the angular distance that the observer is north or south of the equator.

Longitude is the angular distance that the observer is either east or west, 0°–180°, from the Greenwich meridian. Longitude can be expressed either in terms of arc or time measure. Thus, the longitude of a point in Urbana, Illinois, is $5^h 52^m 54^s$ West or $188°13'30''$ West. Note that a careful distinction must be maintained between arc and time expressions of the same longitude because a minute of arc is not equivalent to a minute of time. The following tabulation offers a useful summary of time and arc relations:

$$24^h = 360° \qquad 1° = 4^m$$
$$1^h = 15° \qquad 1' = 4^s$$
$$1^m = 15'$$
$$1^s = 15''$$

Both latitude and longitude can be scaled from a good map. The maps or charts published by various government organizations are best for this purpose. Those most widely available are the standard quadrangle topographic maps of the U.S. Geological Survey. Also, some highway maps are now being published with geographic grids. If no reputable map is available, an inquiry for the observer's latitude and longitude may be addressed to the Director, U.S. Coast and Geodetic Survey, Washington, D.C. The location of the observer can be expressed by giving the name of the nearest post office or, in sectionized areas, the number of the section as well as the designation of township.

A simple astronomic procedure for ascertaining latitude is described in Art. 10-19.

10-9. Position of a Celestial Body In much the same manner that the geographical coordinates, latitude and longitude, are employed to define the position of an observer on the earth, astronomical coordinates are used to indicate the position of a heavenly body on the celestial sphere. These coordinates are *Greenwich Hour Angle* (GHA) and *Declination*.

The GHA of a star is the distance, measured westward, from the

Greenwich meridian to the hour circle passing through the star. The GHA can be expressed in time measure (0^h to 24^h) or in arc measure (0° to 360°). Since the motion of the celestial sphere is from east to west, while the earth remains stationary, it is apparent that the GHA of all stars constantly increases with time. This rate of increase is somewhat more than 15° per hour of time, because the celestial sphere actually completes slightly more than one complete rotation every 24 hours.

The declination of a star is the angular distance it is above or below the celestial equator. For Polaris, it is convenient to use the term, *polar distance,* which is the angular distance the star is from the North Pole.

Figure 10-6 shows, for a given moment, the GHA and declination of a star. Several publications of astronomic data have tabulations

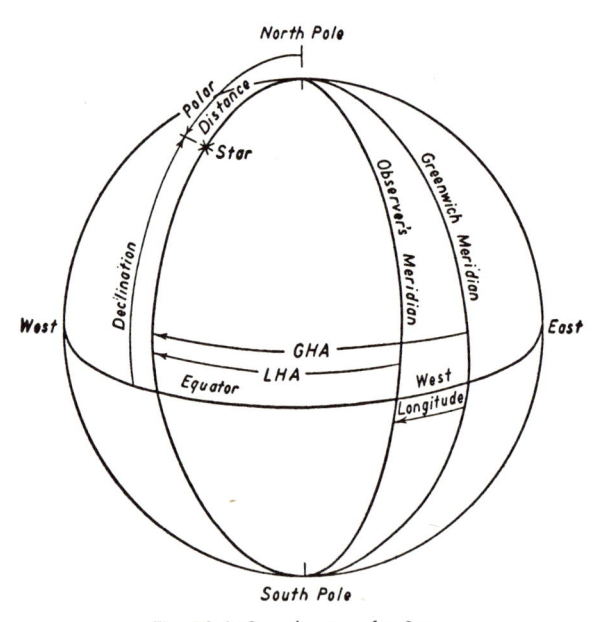

Fig. 10-6. Coordinates of a Star.

of GHA for certain of the heavenly bodies at the moment of 0^h GCT (Greenwich midnight) for each day in the year. This makes the problem of finding GHA relatively easy. It is merely necessary to convert the standard time of observation to GCT by adding the appropriate number of whole hours (the zone correction) to standard time; then, to the tabulated GHA of the star at 0^h GCT is added the increase in the GHA occurring in the elapsed time interval between 0^h GCT and the GCT of the observation.

EXAMPLE 10-2: An observation was made on Polaris at Pittsburgh, Pennsylvania, at $22^h30^m10^s$ EST on June 17, 1965. What is the GHA of Polaris?

$$\begin{array}{r} 22^h30^m10^s \text{ EST} \\ + \ 5^h \hphantom{30^m10^s} \\ \hline 27^h30^m10^s \end{array}$$

or $3^h30^m10^s$ GCT, June 18.

$$\begin{array}{r} 237°54.2' \text{ GHA, } 0^h \text{ GCT June 18 (Table I)} \\ + \ 52°41.1' \text{ Increase for elapsed time (Table II)} \\ \hline 290°35.3' \text{ GHA of Polaris at moment of observation} \end{array}$$

10-10. Local Hour Angle The *Local Hour Angle* (LHA) of a star is the distance, measured westward, from the observer's meridian to the hour circle through the heavenly body. LHA can be expressed either in time units (0^h to 24^h) or arc measure (0° to 360°). LHA is obtained from GHA merely by subtracting from it the west longitude or adding the east longitude of the local meridian, depending upon the position of the observer. Thus,

$$\text{LHA} = \text{GHA} - \text{west longitude} \tag{10-1}$$

or

$$\text{LHA} = \text{GHA} + \text{east longitude} \tag{10-2}$$

Under certain circumstances, it will be necessary to add 360° (or 24^h) to the GHA of Eq. 10-1 in order to perform the required subtraction.

In connection with azimuth observation on Polaris, it is extremely important to realize that, when the LHA is 0° to 180° (0^h to 12^h), the star is west of north; and when the LHA is 180° to 360° (12^h to 24^h), it is east of north.

A convenient expression of the hour angle position of a star is the *meridian angle* (t). It is reckoned both westward and eastward from the observer's meridian up to a maximum value of 180°. Thus, when LHA is less than 180°, t is numerically equal to LHA and has the suffix "west." When LHA is greater than 180°, t is equal to 360° less LHA and is termed the meridian angle "east."

For example,

When LHA = 37°10′ then t = 37°10′ W.
When LHA = 210°35′ then t = 149°25′ E.

10-11. The Astronomical Triangle A spherical triangle is the figure formed by joining any three points on the surface of a sphere by arcs of great circles. A particular spherical triangle having as its vertices a celestial pole, the observer's zenith, and a heavenly body is called the "astronomical triangle," and it is of much importance in engineering astronomy. See Fig. 10-4.

Figure 10-7 shows the *PZS* (pole, zenith, star) triangle as viewed

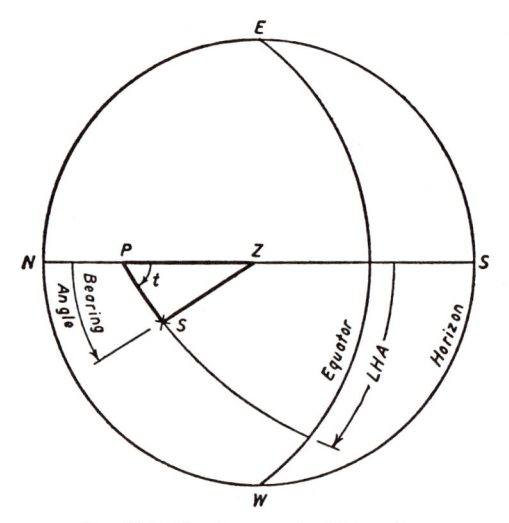

Fig. 10-7. The Astronomical Triangle.

from the observer's zenith. In the typical azimuth problem, the known quantities in the triangle are sides *PZ* (90° − latitude) and *PS* (90° − declination), and the included angle, t, at the pole. Side *ZS* equals (90° − altitude). It is frequently called the *zenith distance*. Through application of the spherical trigonometry, the required angle at *Z* can be calculated. This angle is commonly termed the bearing angle of the star and is reckoned both east and west from north. In Fig. 10-7 the star is depicted west of the meridian. It will be noticed that t is west and is numerically the same as LHA. If the star were east of the meridian, LHA would exceed 180°, t would become east, and the angle at Z would become directly the azimuth of the star.

10-12. Polaris Polaris is a celestial body of great importance to the civil engineer and surveyor. It is a fairly bright star (magnitude 2.1) situated about 1° from the north celestial pole. It can be easily identified by prolonging the line connecting the two stars in the bowl of the Big Dipper on the side most remote from the handle. It is also helpful to remember that the altitude of Polaris will be always within 1° of the observer's latitude and that there are no other stars near Polaris that are likely to be confused with it.

As viewed from a position on earth (Fig. 10-8), Polaris rotates in

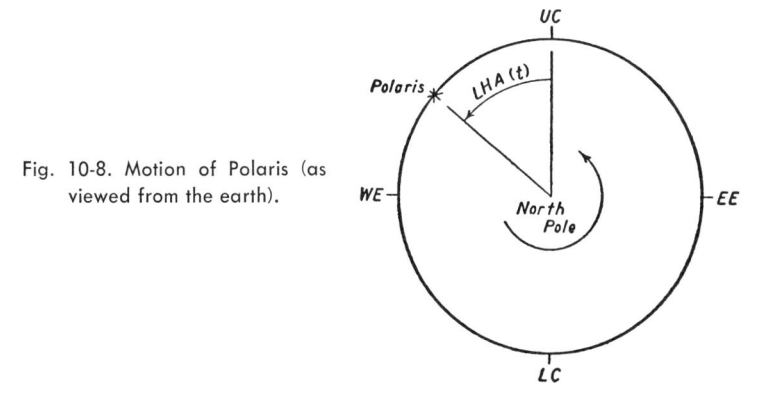

Fig. 10-8. Motion of Polaris (as viewed from the earth).

a counterclockwise direction about the north celestial pole. It makes a complete revolution approximately every 24 hours.

The instant that Polaris crosses the upper branch of the observer's meridian is called time of Upper Culmination (UC). At that moment LHA of Polaris is zero. Approximately 6 hours later Polaris reaches Western Elongation (WE). Polaris passes over the meridian again at Lower Culmination (LC) and attains its most easterly position at Eastern Elongation (EE).

From this description of the motion of Polaris it is apparent that, at the moments of elongation, the star's motion will be nearly vertical and that its bearing will change but slightly with time. At culmination, however, the rate of change of azimuth of the star with respect to time will be a maximum.

The certainty with which Polaris can be identified, its proximity to the north pole, the fact that it is constantly above the horizon, and the ready availability of specially prepared tables to facilitate the solution of the azimuth problem make it a particularly useful astronomical body.

10-13. The Azimuth Problem The typical azimuth problem consists of two basic operations. They are as follows:

1. The observation of Polaris at any time in order to measure the horizontal angle between the star and a terrestrial signal called the "mark." At the moment a pointing is made on the star, time is noted.

2. The computation of the bearing angle of Polaris at the instant of observation. This calculated angle, when appropriately combined with the measured field angle, will yield the required astronomic azimuth or true bearing of the line.

For any azimuth observation the following data in addition to the measured horizontal angles and the observed times are essential: (1) the date of the observation, (2) the name or identity of the instrument station, (3) the name of the mark, (4) the latitude and longitude of the instrument station as scaled from a map or determined from any other reliable source, (5) the watch error, if any, and (6) the kind of time kept by the watch (MST, CDT, etc.). A good sketch portraying the general position of the terrestrial line with respect to astronomic north is very desirable.

10-14. Field Procedure This treatment emphasizes the inherent advantages of observing Polaris for azimuth at any local hour angle. The facility with which the computations can be made and results of good quality obtained make the local hour angle method an entirely satisfactory procedure. It is preferred to the elongation method even for the most precise azimuth determinations of geodetic surveys.

It can be easily verified that, at the most unfavorable moment, i.e., Polaris at culmination, an error of as much as 1 minute in time causes, at a latitude of $40°$, a bearing angle error of only $0.3'$ of arc. Furthermore, the local hour angle method permits the observation to be made at any moment convenient to the observer. He is not required to wait for the precalculated instant of elongation which may take place at a most unsuitable time or when the star happens to be obscured by clouds.

The field procedure begins with setting up the transit over the station, performing the usual centering and leveling, and taking an initial sight at the mark. It will be necessary to illuminate the crosswires by shining a flashlight obliquely into the objective end of the telescope. Optical transits, however, are wired for internal illumination. It is best that the terrestrial line be at least 1000 ft long so that the objective lens does not require refocusing when sighting the star. A good signal at the mark consists of a slit cut in the side of a box or tin can in which a light is placed. The signal must be well centered, of course, over the mark. When pointing upon the star, it is very essential that Polaris appear as a tiny pinpoint of light rather than as a disc. Difficulty will be experienced in finding the star unless the objective lens is correctly focused. This can be accomplished by taking a preliminary sight on any other prominent heavenly body or upon a distant terrestrial light.

The use of either a standard 30″ engineer's transit or an ordinary optical theodolite will provide satisfactory azimuths for most local control work. With both instruments, the horizontal angles are doubled. The field procedure is as follows:

1. Set up the instrument and level it carefully.
2. Set the horizontal circle at 0° and sight the mark.
3. Sight the star and note the time.
4. Read and record the single angle.
5. Release the lower motion, invert the telescope, and sight the mark a second time.
6. Sight the star and note the time.
7. Read and record the double angle.

Typical field notes are shown in Fig. 10-9.

10-15. Ephemerides The tables in this chapter are available in more complete form in the ephemerides published by governmental agencies and by the various makers of surveying instruments. All are published annually, and a copy for the current year in which field observations are being conducted is indispensable for making the necessary reduction computations. A particularly useful, entirely adequate, and very inexpensive publication for the surveyor is the "Ephemeris of the Sun, Polaris, and other Stars," which is prepared by the Bureau of Land Management, Department of the Interior. It is available from the Government Printing Office. The abridged ephemerides provided by the instrument makers are furnished with-

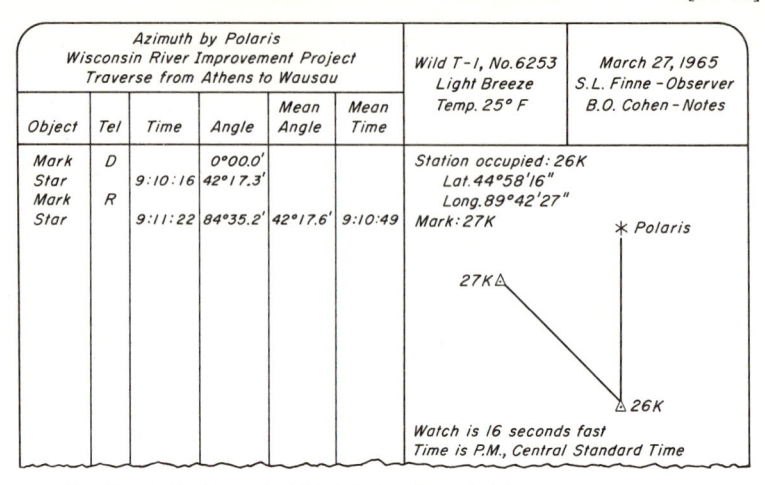

Azimuth by Polaris Wisconsin River Improvement Project Traverse from Athens to Wausau						Wild T-1, No.6253 Light Breeze Temp. 25° F	March 27, 1965 S.L. Finne -Observer B.O. Cohen - Notes
Object	Tel	Time	Angle	Mean Angle	Mean Time		
Mark Star Mark Star	D R 	9:10:16 9:11:22	0°00.0' 42°17.3' 84°35.2'	 42°17.6'	 9:10:49	Station occupied: 26K Lat.44°58'16" Long.89°42'27" Mark: 27K 27K △ △ 26K Watch is 16 seconds fast Time is P.M., Central Standard Time	✶ Polaris

Fig. 10-9. Field Notes for Third-Order Azimuth Observation on Polaris.

out charge and will generally provide all the data necessary for the computations treated here. Another suitable ephemeris is "The Star Almanac for Land Surveyors," which is prepared annually by H. M. Nautical Almanac Office and published by Her Majesty's Stationery Office, London. It can be obtained from the British Information Service, 45 Rockefeller Plaza, New York, N.Y.

10-16. Reduction Procedures Three reduction procedures for calculating the azimuth of Polaris at any hour angle will be presented.

1. A formula of wide and general application for computing the precise azimuth of any celestial body is as follows:

$$\tan Z = \frac{\sin t}{\cos \phi \tan \delta - \sin \phi \cos t} \qquad (10\text{-}3)$$

where

ϕ = latitude
δ = declination
t = hour angle
Z = azimuth of star

The quantity Z is reckoned east or west from true north.

2. For a less precise value of the azimuth a more simple formula can be used. It can be derived directly from the law of sines for spherical trigonometry as applied to the PZS triangle of Fig. 10-7. Hence, it is seen that

$$\sin Z = \frac{\sin t \sin (90° - \delta)}{\sin (90° - h)}$$

or

$$\sin Z = \frac{\sin t \sin p}{\cos h} \quad (10\text{-}4)$$

where

p = polar distance, $90° - \delta$.

h = altitude

For a circumpolar star like Polaris, angles Z and p will always be small. Therefore, we can substitute the values of the angles in seconds (or minutes) for their sines and obtain

$$Z = \frac{p \sin t}{\cos h} \quad (10\text{-}5)$$

where Z and p are both in seconds or in minutes of arc.

It will be noted that Eq. 10-5 requires the measurement of the altitude, h. This is the value after instrumental corrections and refraction are applied to the observed altitude. If a reliable measurement of altitude seems unlikely and the latitude, upon the other hand, has been well established, the following formula can be used:

$$Z = \frac{p \sin t}{\cos (\phi + p \cos t)} \quad (10\text{-}6)$$

The quantity $p \cos t$ can be obtained from prepared tables like Table VI. This quantity is the difference between the altitude of Polaris and the North Pole at any hour angle, t. Note that Table VI is designed to convert the true altitude of Polaris to the altitude of the pole which latter quantity is numerically the same as the observer's latitude as explained in Art. 10-19. In the use of Eq. 10-6 the quantity $p \cos t$ taken from Table VI is either added to or subtracted from the known latitude, depending upon whether the star is above or below the pole at the moment of observation. The position of Polaris is given by its hour angle as shown in Fig. 10-8.

3. A convenient solution of the *PZS* triangle for the azimuth of Polaris can be made with certain prepared tables. Such a solution is presented in detail in Art. 10-17.

10-17. Azimuth from Polaris at Any Time A typical reduction computation is presented for an azimuth observation made upon Polaris on May 5, 1965, at Urbana, Illinois. The given time and horizontal angle represent the mean values from a single set, direct and reverse, of observations on the star.

EXAMPLE 10-3:

Station occupied: Illinois
Latitude: 40°06.3′ N.
Longitude: 88°13′30″ W.
Observed time: 8ʰ23ᵐ20ˢ P.M., CST
Angle: mark to star (clockwise) 46°17½′

Mark: Sta. 25
Date: May 5, 1965
Observer: J. L. M.
Watch 35ˢ slow

It is required to find the true bearing of the line. The solution is as follows:

1. *Calculation of GCT*

Watch time (P.M.)	8ʰ23ᵐ20ˢ CST
Correction	+ 35ˢ
Corrected time	8ʰ23ᵐ55ˢ
Time (24-hr basis)	20ʰ23ᵐ55ˢ
Zone correction	+6ʰ
GCT	2ʰ23ᵐ55ˢ
Greenwich date	May 6

2. *Calculation of LHA*

GHA at 0ʰ GCT	195°41.4′	(Table I)
Increase in GHA	36°04.6′	(Table II)
GHA	231°46.0′	
Less west long.	88°13.5′	
LHA	143°32.5′	
t	143°32.5′ (west)	

3. *Preliminary Bearing Angle of Polaris.* A double interpolation is made from Table III which is entered with the known latitude and previously calculated LHA. This interpolation effects a mechanical solution of the *PZS* triangle for an average value, 0°56.8′, of the polar distance of Polaris for the year.

LHA	LATITUDE		
	40°	40°06.3′	42°
140°	0°47.1′	0°47.2′	0°48.5′
143°32.5′	—	0°43.6′	—
145°	0°42.0′	0°42.1′	0°43.3′

Thus, the preliminary bearing angle of Polaris is 0°43.6′.

4. *Final Bearing Angle of Polaris.* The actual polar distance of Polaris, 0°57.0′, is obtained from Table IV and affords, together with the preliminary bearing angle of Polaris, a means for entering Table V to obtain a supplementary correction. As a glance at Table V discloses, this correction is usually very small.

$$\begin{array}{ll} \text{Preliminary bearing angle} & 0°43.6' \\ \text{Correction (Table V)} & +\ \ 0.1' \\ \text{Final Bearing Angle of Polaris} & \overline{0°43.7'} \end{array}$$

5. *True Bearing of Line.* The final bearing of Polaris is combined with the measured field angle to obtain the bearing of the line. Since the LHA of Polaris is between 0° and 180°, the star is west of north. A good sketch (Fig. 10-10) depicting the position of the star

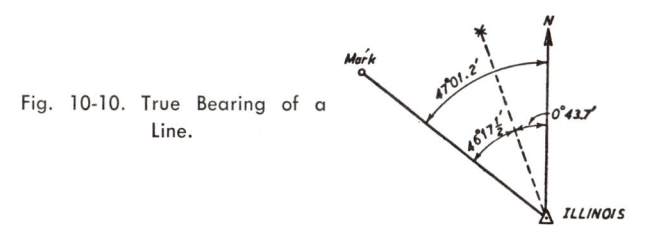

Fig. 10-10. True Bearing of a Line.

in relation to the meridian and the terrestrial line always should be drawn.

The true bearing of the line is thus found to be N. 47°01.2′ W.

Alternate solution. An alternate solution of this problem will now be effected with the use of Eq. 10-6 as follows:

$$Z = \frac{p \sin t}{\cos (\phi + p \cos t)}$$

where

$$\begin{array}{ll} p = 57.0' & \text{(from Table IV)} \\ t = 143°32.5' & \text{(as previously calculated)} \\ p \cos t = -45.7' & \text{(from Table VI)} \\ \phi = 40°06.3'\ \text{N.} \end{array}$$

hence,

$$Z = \frac{(57.0)(0.59424)}{\cos 39°20.6'} = 0°43.8'$$

10-18. Quality of Azimuth Determination The quality of an astronomic determination of azimuth is dependent upon the accuracy of the measured field angle, as well as upon the accuracy of the computed bearing angle of the star.

Consideration will be given to the effect of an error in time upon the quality of the calculated bearing angle of Polaris. Assume that an observation was made on May 1, 1965, in latitude 40°10′ N. when the hour angle of Polaris was 5°. The polar distance for the date is 0°57.0′. Assume uncertainty in time is ±1 minute.

Differentiating Eq. 10-5, we have

$$dz = \frac{p \cos t \, dt}{\cos h} \tag{10-7}$$

where p and dz are in minutes of arc and dt is in radians.

$$h = \phi + p \cos t = 40°10' + 56.4' = 41°06.4'$$

Recalling that 1 minute of time equals 15′ of arc and 1″ of arc equals 0.00000485 radian, we have

$$dt = (1)(15)(60)(.00000485) = \pm 0.00436 \text{ radian}$$

Hence, $\quad dz = \dfrac{(57.0)(.996)(.00436)}{(.753)} = \pm 0.33'$

Since the preceding observation was made near the moment of upper culmination when the azimuth of the star is changing most rapidly, it is obvious that time is not critical for azimuth observations of ordinary accuracy on Polaris.

10-19. Latitude from Polaris at Any Time Figure 10-11 portrays a section along the observer's meridian through the portion of the celestial sphere above his horizon. It is readily seen that, since the polar axis is perpendicular to the equator and the plumb line is normal to the plane of the horizon, the altitude of the pole equals

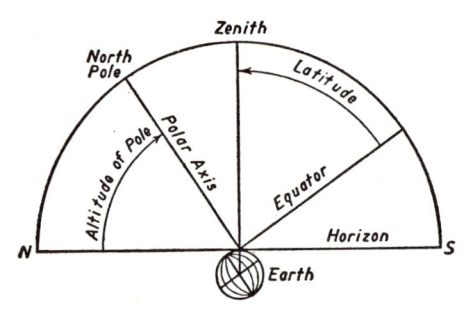

Fig. 10-11. Latitude and Altitude of Pole.

the latitude. At culmination, Polaris is above or below the pole by an amount of arc equal to its polar distance. At elongation, it is directly opposite the pole and its altitude equals the latitude. For other intermediate positions, Table VI tabulates the angular corrections which, for various values of LHA, can be applied to the observed altitude of Polaris to reduce to the altitude of the pole.

EXAMPLE 10-4: An observation for latitude was made upon Polaris at a point in northern Minnesota on July 27, 1965. The observed altitude of Polaris was 48°06½′, and the calculated LHA for the observed moment of observation was 215°20′. What is the observer's latitude?

LHA is 215°20′ or t is 144°40′ E.
48°06.5′ observed altitude
+46.4′ Reduction to pole (Table VI.)
48°52.9′ Latitude

PROBLEMS

10-1. Calculate the GCT corresponding to the following moments of standard time:

(a) 5 P.M. CST Feb. 10 (*Ans.* 23^{h}00^m Feb. 10)
(b) 23^{h}50^m EST Mar. 30 (*Ans.* 4^{h}50^m Mar. 31)
(c) 18^{h}40^m CDT Jan. 6 (*Ans.* 23^{h}40^m Jan. 6)
(d) 21^{h}16^m MST June 2 (*Ans.* 4^{h}16^m June 3)

10-2. For the year 1965 calculate the GHA of Polaris for each of the moments tabulated in Problem 10-1. (*Ans.* (a) 97°42.2′ (b) 232°55.5′ (c) 73°03.0′ (d) 287°22.1′.

10-3. If the observer is in west longitude 92°15.0′, calculate both the LHA and t for each of the preceding moments of time. (*Ans.*)

(a) LHA = 5°27.2′ t = 5°27.2′ W.
(b) LHA = 140°40.5′ t = 140°40.5′ W.
(c) LHA = 340°48.0′ t = 19°12.0′ E.
(d) LHA = 195°07.1′ t = 164°52.9′ E.

10-4. An azimuth observation was made upon Polaris on March 17, 1965, at Pittsburgh, Pennsylvania. The latitude of the station is 41°40.0′ N., and the longitude is 83°36.5′ W. The mean angle, measured clockwise, from the mark to the star is 23°14½′ and the corresponding observed time is 10^{h}43^{m}25^s P.M., EST. The watch is known to be 30^s fast. Find the true bearing of the line. (*Ans.* N. 24°20.0′ W.)

10-5. An azimuth observation was made upon Polaris on June 16, 1965, at a third-order traverse station near Milwaukee, Wisconsin. As scaled from a map, the latitude of the station is 43°07′20″ N., and the longitude is 87°58′50″ W. The mean of the angles, measured clockwise, from the azimuth mark to the star with a T-1 transit is 56°42.3′, and the corresponding observed time is 9^{h}16^{m}15^s P.M., CDT. The watch is known to be 10^s slow. Find the astronomic azimuth of the line (from south).

10-6. Reduce the field notes of Fig. 10-9 to find the azimuth (from south) of the line.

10-7. Suppose the watch correction of Problem 10-4 were neglected. Make a new calculation to determine the resulting error in the bearing angle of the star. (*Ans.* 0.1′) (Note that 30^s error in time will have the same effect on the calculation of LHA of Polaris as an error of 7½′ in longitude.)

10-8. A latitude observation was made on Polaris on November 1, 1965, at a point in central California. The observed altitude of Polaris was 37°17′, and the calculated LHA of the star was 289°10′. Find the observer's latitude. (*Ans.* 36°59′ N.)

10-9. A latitude observation was made on Polaris on May 16, 1965, at a third-order traverse station in north-central Montana. The mean altitude of Polaris as measured with a T-1 transit was 47°48.7′, and the calculated LHA of the star was 136°12.5′. Find the observer's latitude.

REFERENCES

1. Hosmer, G. L., and J. M. Robbins, *Practical Astronomy*, 4th ed., John Wiley & Sons, Inc., New York, 1956.
2. Nassau, J. J., *Practical Astronomy*, 2nd ed., McGraw-Hill Book Co., New York, 1948.

11 State Coordinate Systems

11-1. Introduction The position of a point on the surface of the earth is completely defined by stating its latitude and longitude and its elevation above mean sea level.

The necessity of relating the elevations of one survey to those of another has become so common that the desirability of a standard datum for elevations is well recognized. Accordingly, every new survey in the United States involving vertical control of any importance is based upon the 1929 Mean Sea Level Datum.

The parallel necessity of relating the horizontal positions of points in one survey to those of another by means of a common *horizontal datum* or a system of coordinates has assumed increasing importance in recent years. The number, complexity, and areal extent of modern engineering projects have made urgent the need not only to make more usable existing geodetic position data by reducing them to a convenient system of rectangular coordinates, but also to express the results of future horizontal control surveys, particularly those of third order, in terms of some standard coordinate system. Before explaining the fundamental properties of the state plane coordinate systems, it will be well to trace the steps leading to their evolution, mention the major advantages of such systems as compared with a geographic system, and state the minimum accuracies with which surveys based on state coordinates should be executed.

11-2. National Network of Horizontal Control Because of the need for perpetuating national and state boundaries, providing control for mapping and charting programs, and other purposes, the federal government has long recognized the necessity for an accurate network of horizontal control for the nation. For more than a century the U.S. Coast and Geodetic Survey has engaged in operations that determined the geodetic positions of thousands of well-monumented points in all parts of the country. Supplementing the network of first- and second-order stations of this agency have been the very extensive third-order horizontal control operations of other surveying organizations.

Formerly, it was the practice to publish and make available to the engineers of the country the positions of these survey points in terms of their spherical coordinates only, viz., latitude and longitude. The use of control data in such form required a working knowledge of geodetic surveying formulas and methods which the average surveyor and engineer did not possess. Also, these processes were too slow and tedious when compared with plane surveying procedures to justify their use in common survey practice. Hence, for many years the problem besetting the increased use of higher-order horizontal control data was that of finding a simple and practicable means for utilizing such position data for detail surveys. Prior to 1933 this could be done only through geodetic methods similar to those by which such control surveys were originally executed and extended or by setting up local systems of plane coordinates of very limited extent. However, lack of training and experience on the part of engineers and the need for special equipment and mathematical tables prevented any substantial use of geodetic control data by the first method. Some cities, notably New York City and Pittsburgh, set up tangent plane coordinate systems based on the national horizontal control system.

11-3. Beginnings of State Coordinate Systems In order to make the geodetic data of the national horizontal control system readily available in a form acceptable to engineers and surveyors, the U.S. Coast and Geodetic Survey established in 1935 what are known as the *state coordinate systems.* Each state has a separate system. The basis of each system is the mathematical projection of the earth's surface upon a surface like that of a cone or cylinder, which can then be developed into a plane. The Coast and Geodetic Survey computes and publishes the plane coordinates, on the appropriate state coordinate system, of all points for which it determines geographic positions. Hence, just as there is but one point on the surface of the earth corresponding to a geographical position expressed by latitude and longitude, so likewise there is but one point corresponding to a given pair of plane coordinates, expressed as x and y, for a particular zone of a given state. The substitution of these simple rectangular plane coordinate systems and their associated position data for the relatively unwieldy and complex geographic coordinates and geodetic computations has led to a substantial increase in the use of national horizontal control data and to a better understanding of the benefits of referencing local surveys to the federal network. Of paramount importance, however, in this connection is the fact that the engineer or surveyor who ties his carefully executed survey to the national net may easily adjust his work to the data of that net using the ordinary office procedures of plane surveying. By doing this he is able to incorporate into his survey the accuracy qualities of the parent geodetic survey.

11-4. Advantages of State Coordinate Systems In subsequent discussions of the applications of state coordinates to specific engineering surveys, the detailed advantages of using such a system of coordinates will be explained. It will suffice here to mention three major advantages of a state coordinate system. These are inherited from the national geodetic network upon which the state coordinate system depends.

1. Positive checks can be applied to all surveys to prevent the accumulation, beyond permissible limits, of errors in the measurement of angles and directions.

2. Surveys that are initiated at widely separated points, and perhaps for different projects, will have at their junction substantially the same azimuth for a given line and the same coordinates for a given point. If it were not for the errors of field measurement, the agreement in azimuth and coordinates would be exact. Thus, two such surveys can be coordinated and used to supplement each other.

3. Any station whose state coordinates have once been accurately determined may be said to be permanently located. Even though the marker itself is destroyed, its position is perpetuated by the record of coordinates and the station can be restored by careful measurements from the nearest recovered stations in the system.

11-5. Quality of Surveys The field procedures associated with the subsequent computation of state coordinates are the simple methods of plane surveying involving nothing more complicated than transit traverse followed by latitude and departure calculations. It should be emphasized, however, that wholly reliable results will be obtained only if the field work is based on first- or second-order

Fig. 11-1. Traverse Operations.
Tennessee Valley Authority

control and executed in a careful manner (see Fig. 11-1) so that third-order accuracy is obtained.

To clarify some misconceptions, it is worth mentioning that there is no special operation of surveying which is called state coordinate surveying. Any suitably accurate method of carrying distance and azimuth from a point of unquestioned state coordinate position will serve to make possible the calculation of the position of other points. Furthermore, there is no survey point that can be properly described as a state coordinate station. Defining the position of a station in the nation's horizontal control network by state coordinates is merely utilizing a mode of expressing position which is parallel to that of using geographic coordinates, because for every pair of state coordinates there is a pair of particular geographic coordinates and vice versa.

11-6. Local Plane Coordinates Engineers and surveyors have used for many years in their work a variety of unrelated and arbitrary coordinate systems. Frequently, different systems of this type are utilized in the same community or even at the same industrial site. Such coordinate systems are defined by assigning x and y values to a chosen survey point and taking either an assumed meridian or the true meridian through the initial point as grid north. These systems lack official recognition, cannot be correlated with other surveys, and pose problems associated with convergency of the meridians when extended over a large area. An isolated system is still too frequently merely an expedient choice. When ties to a recognized system, such as the state coordinate system, are easily feasible, the continued use of local systems is hardly defensible.

11-7. Map Projections The mathematical theory of map projections is too complex to be treated here, but it is thought desirable to discuss some elementary principles and present the general properties of two kinds of projections.

The surface of a sphere cannot be developed into a plane without distortion; hence, if a considerable portion of the earth's surface is to be shown on a map, the dimensions must be distorted one way or another. The character of the distortion can be controlled if the points on the earth's surface are mathematically projected upon a plane, or upon a surface (cone or cylinder) that can be developed into a plane. After such projection and development, the points will represent in the plane, i.e., on the map, with a minimum of scale distortion the correct relative positions of the corresponding points on the earth's surface.

Distortions on a map projection are negligible if they are too small to be plotted at the scale of the map. However, distortions entering into the use of rectangular plane coordinates are negligible only if they are so small as to fall within the usual limits of accidental errors in the field measurements.

In general, a plane coordinate system which is to have maximum engineering utility should have the following features:

1. The y- and x-coordinates of a survey point in the plane rectangular system should be readily obtainable from the latitude and longitude. Also, the reverse process should be equally feasible.
2. The forward and back azimuths of a line in the rectangular grid system should differ by exactly 180°.
3. The length of a survey line as calculated from the grid coordinates of its termini should be equal to the ground distance or a means be available to effect readily a transformation from one length to the other.
4. The grid azimuth of a line should be simply related to the true or astronomic azimuth of the line.

Several projections satisfying the requirements of a rectangular plane coordinate system have been devised, but only three will be described here. They may be classed according to area limitations as local systems of coordinates or as regional systems. To the first class belongs the *tangent-plane* projection and to the second belong the *Lambert* and *Transverse Mercator* projections.

11-8. Tangent Plane Coordinates The only simple way by which the results of the highly accurate geodetic surveys can be easily and readily used for controlling ordinary surveying operations is by transforming the geographic coordinates to some system of plane coordinates.

For a relatively small area, such as an average-sized city, a projection on a tangent plane gives an accuracy that is satisfactory for survey purposes. Such a projection represented the first attempt to utilize existing higher-order horizontal control by transforming geographic to plane coordinates. The projection consisted basically of the representation of points projected radially from the center of the earth to a plane tangent to the earth at a point in the general vicinity of the center of the city.

The use of such a projection was made quite simple by U.S. Coast and Geodetic Survey Special Publication No. 71, entitled "Relation Between Plane Rectangular Coordinates and Geographic Positions." This publication contains tables by means of which one can reduce geographic coordinates to rectangular coordinates on a tangent plane at sea level. This projection is limited to relatively small areas, however, because, as the distance from the origin increases, the difference between the measured length of a line and the length computed from the plane coordinates of its termini increases rather rapidly. At 40 miles from the origin, the scale error is 1 part in 20,000 and at 80 miles it is 1 part in 5000.

Despite its limitations, the tangent plane coordinate system served very well the needs of several large cities, among them Pittsburgh, Pennsylvania. However, ever since the establishment of the state coordinate systems, which have broad regional coverage, there is no longer any need to establish new tangent plane systems. Hence, it is not conceivable that any more projections of this type should be devised for American engineering practice, because they are distinctly inferior to a state plane coordinate system.

11-9. Lambert Projection The Lambert projection has different forms, but the one adopted for state-wide plane coordinates is a modified conic projection whose general characteristics may be de-

[11-9] STATE COORDINATE SYSTEMS 291

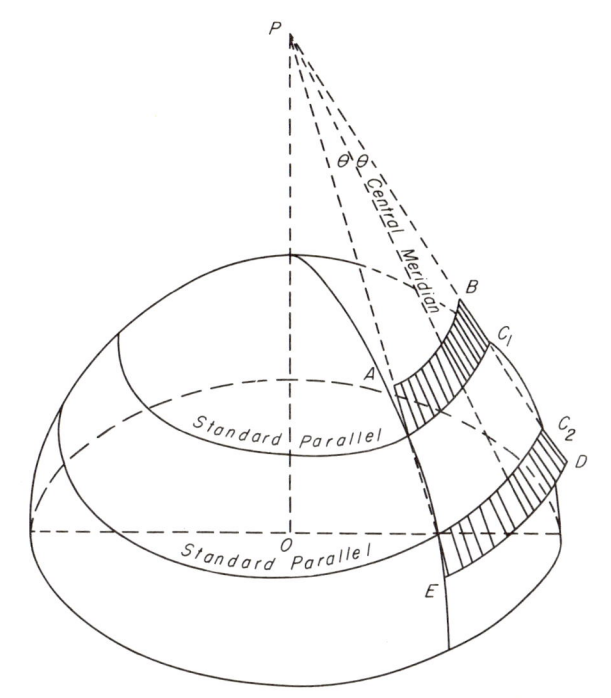

Fig. 11-2. Lambert Projection.

scribed by referring to Fig. 11-2. It consists of an imaginary cone whose axis OP is assumed to be coincident with the axis of the earth, and whose elements, one of which is PBD, cuts the earth's surface at two points C_1 and C_2. A partial frustrum of the cone is shown as $ABDE$. Two small circles, called *standard parallels,* are generated where the conical surface cuts the earth's surface.

If the frustrum $ABDE$ is developed into a plane surface, it will appear as shown in Fig. 11-3. A central meridian will have a known longitude, and the longitude of any point, as A or B, will be given with respect to the central meridian by adding (or subtracting) the angle θ, at P, between the central meridian and the element through the given point. The latitudes of the two circles C_1 and C_2 are known, and the latitude of any point as A, not on these circles, can be found from its known distance along its terrestrial meridian, north or south of the standard parallel.

The projections from the earth's surface onto the cone are made along radii from the earth's center, O.

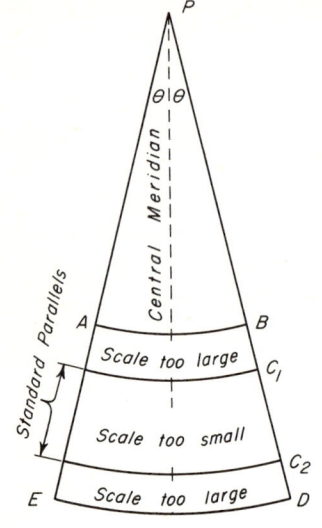

Fig. 11-3. Lambert Projection Developed into a Plane.

From the conditions stated above, it is evident that this projection has these following characteristics: (1) Since the projection cone cuts the earth's surface along the standard parallels, the longitude scale along these circles will be exact. (2) The conical surface is so nearly coincident with the earth's surface that the projection may be said to be *conformal;* i.e., both the latitude and longitude scales are so nearly exact that angles between lines on the projection are very nearly the same as the angles between the corresponding lines on the earth's surface. (3) For the zone between the standard parallels, the scale of the projection will be too small; and likewise, for the zones north and south of the middle zone, the scale will be too large. (4) The projection can be extended indefinitely in an east-west direction, without affecting the accuracy of the projection; but as the projection is extended in a north-south direction, the scales are modified and change in a rapidly increasing ratio as this dimension is increased.

11-10. The Mercator Projection The Mercator projection consists of an imaginary cylinder with its axis coincident with the earth's axis and its surface tangent to the earth's surface at the equator. The meridians on this projection are straight lines perpendicular to the baseline, or equator, and hence are parallel with each other. Since the earth's meridians converge at the pole, it is evident that the scale of the projection for east-west dimensions increases as the latitude increases.

At the equator, however, the scale is exact in all directions and the projection is therefore conformal. As the projection is extended northward (or southward) from the equator, the scale of the meridians is changed to correspond with the change in the scale of the parallels, so that this projection possesses the unique quality that a *rhumb line,* i.e., a line of constant bearing, is always a straight line. Hence, this projection is much used in navigation.

11-11. Transverse Mercator Projection The Transverse Mercator projection, which is a modification of the Mercator projection, has been designed to meet the requirements of state-wide plane coordinates for those states whose greatest dimension lies in a north-south direction. The projection may be described by reference to Fig. 11-4.

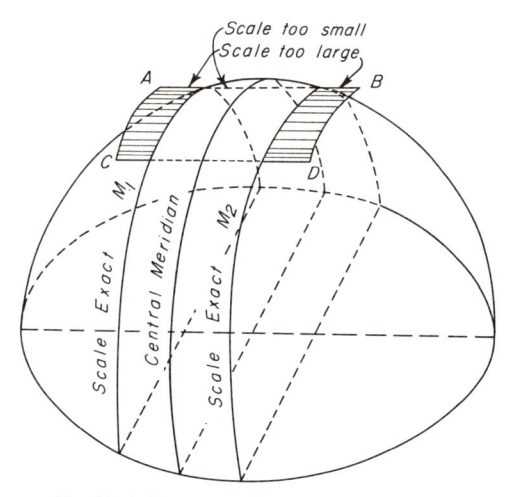

Fig. 11-4. Transverse Mercator Projection.

This projection consists of a cylinder which is turned 90° from that of the Mercator projection and which has its radius slightly reduced so that, instead of being tangent to the earth's surface along a central meridian, it cuts the sea-level surface along two parallels as M_1 and M_2. A portion of such a projection is shown as $ABCD$.

This projection has the following characteristics: (1) The scale of the projection is exact along the two parallels M_1 and M_2. (2) The scale is too large for the zones outside of the two intersecting parallels, and it is too small for the zone included between them. (3) This projection can be extended indefinitely in a north-south direction without changing the scale relations, but these relations change rapidly as the length of the projection is extended in the east-west direction.

11-12. The State Coordinate Systems As mentioned in a previous article, the errors involved in using a system of tangent plane coordinates increase rapidly as the distance from the point of tangency increases. This is a significant weakness of a coordinate system to be used for a state-wide project, such as a network of superhighways, or to connect surveys extending over several large counties. The urgency of creating state plane coordinate systems that could utilize existing geodetic data over an entire state without involving anything more complicated than plane surveying procedures led to the establishment of the state coordinate systems.

These systems are based upon the Lambert and Transverse Mercator projections. The former is used for those states whose greatest dimension lies in an east-west direction, and the latter is employed for those states whose greatest dimension lies in a north-south direction. In two states, New York and Florida, both projections are used. Almost all the states are divided into several belts or zones with each zone having its own origin and reference meridian.

State coordinate systems based on the Lambert and the Transverse Mercator projections are commonly designated as Lambert and Transverse Mercator grids, respectively.

The states and their grid systems are tabulated on page 295.

11-13. Computation of Plane Coordinates on Lambert Grid Before describing the calculation of a traverse based on state plane coordinates, it will be desirable to show the manner in which the geodetic coordinates of a survey point can be transformed into plane coordinates. Only a limited discussion of the theory underlying the computational procedure for the Lambert grid and none for the Transverse Mercator grid will be presented. However, it should not be difficult for any engineer to perform the calculations shown in this and the following article. Furthermore, since it is now the practice of the U.S. Coast and Geodetic Survey to compute and publish

LAMBERT SYSTEM

Arkansas	North Dakota
California	Ohio
Colorado	Oklahoma
Connecticut	Oregon
Iowa	Pennsylvania
Kansas	South Carolina
Kentucky	South Dakota
Louisiana	Tennessee
Maryland	Texas
Massachusetts	Utah
Minnesota	Virginia
Montana	Washington
Nebraska	West Virginia
North Carolina	Wisconsin

TRANSVERSE MERCATOR SYSTEM

Alabama	Mississippi
Arizona	Missouri
Delaware	Nevada
Georgia	New Hampshire
Idaho	New Jersey
Illinois	New Mexico
Indiana	Rhode Island
Maine	Vermont
Michigan	Wyoming

BOTH SYSTEMS

Florida	New York

the coordinates on the appropriate state system of all points for which it determines the geographic positions, it is possible for the interested engineer to make request for such data.

The transformation computation for a Lambert grid will be illustrated by reference to the Minnesota state coordinate system. As shown in Fig. 11-5, this state is covered by three overlapping zones, the North, Central, and South zones. Each zone has different axes for x and y, although all y-axes, passing through the center of the respective zone, are given an x value of 2,000,000 ft. The x-axis is placed well below the southern limit of each belt and has a value of zero feet. The geographic coordinates of a point in the North Zone will be transformed into plane coordinates.

The North Zone has for a central parallel the parallel of latitude 47°50′. Along this line the scale ratio (error of the projection) is

296 FUNDAMENTALS OF SURVEYING [11-13]

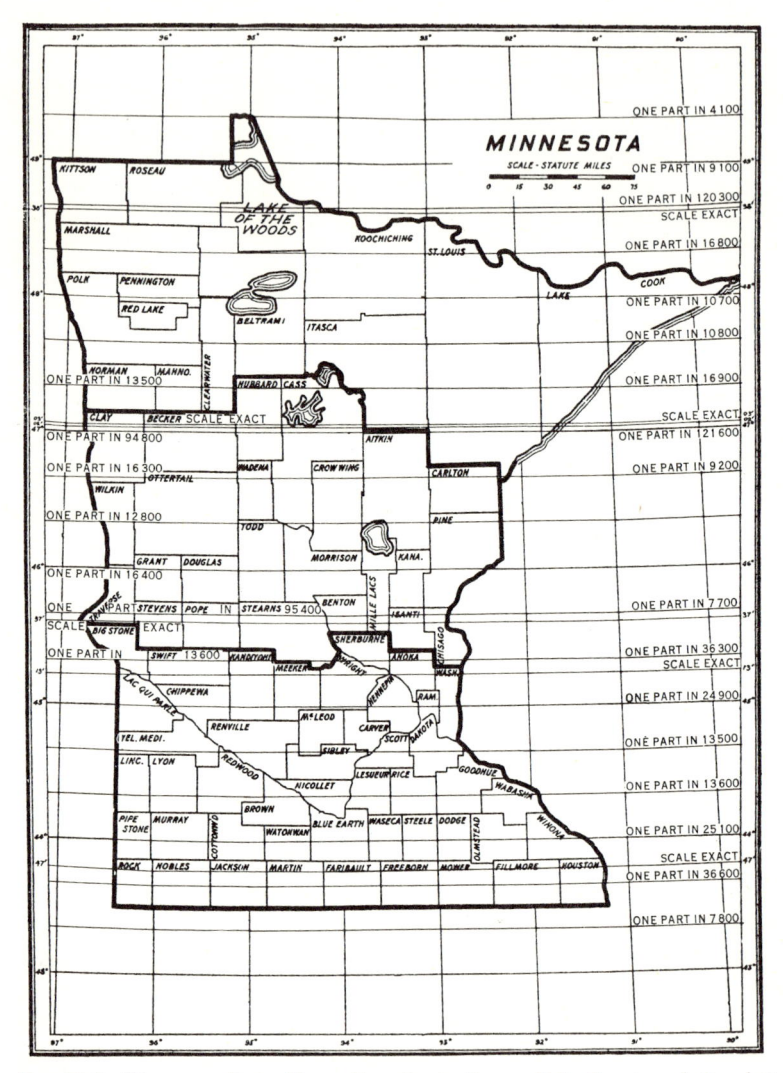

Fig. 11-5. Minnesota State Plane Coordinate Zones. *U.S. Coast and Geodetic Survey*

1 part in 10,300 parts too small. The standard parallels along which the scale is true are 47°02′ and 48°38′. The meridian of longitude, 93°06′, is the y-axis and the parallel of latitude 46°30′ defines the x-axis. Hence, the origin of coordinates for the North Zone is a point on the 46°30′ parallel situated 2,00,000 ft west of longitude 93°06′.

In Fig. 11-6 is shown the fundamental basis for transforming the

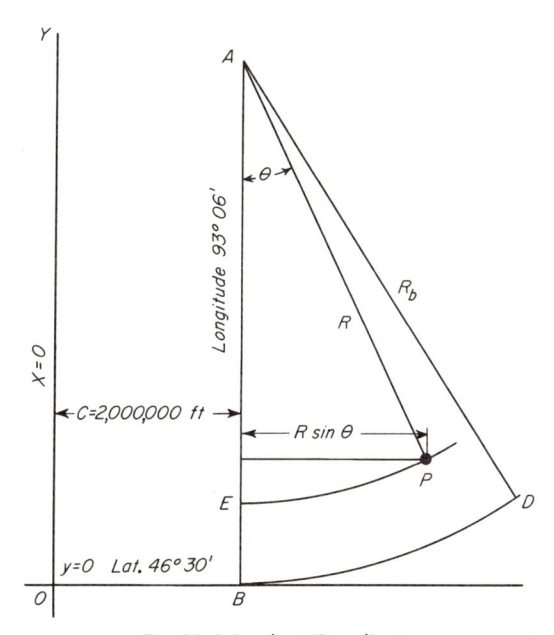

Fig. 11-6. Lambert Coordinates.

geographic coordinates of a point, P, into its Lambert grid coordinates. Point O is the origin of coordinates, and AB is the central meridian of the system. The value of the x-coordinate of the central meridian is designated as C. The point A represents the apex of the cone on which the area is projected, and the arcs PE and DB represent portions of parallels of latitude through point P and the lower extremity of the y-axis. The distance R_b is the largest latitude radius of the zone and is a constant for the zone. It is to be noted that the y-coordinate of point A, the apex of the cone, is equal to R_b. The angle, θ, is the angle of convergency between the central meridian and the meridian through the point, P. The values of R for each whole minute of latitude, and the values of θ for each whole minute of longitude are given in U.S. Coast and Geodetic Survey Special Publication No. 264, entitled "Plane Coordinate Projection Tables—Minnesota." The angle, θ, is considered positive if P is east of the central meridian and negative if P is west of the central meridian. It is to be noted that *grid north* and *geodetic north* are identical along the *central meridian*.

The expressions for calculating the Lambert coordinates of a point are then as follows:

$$x = R \sin \theta + C$$
$$y = R_b - R \cos \theta$$

A typical transformation computation will now be demonstrated. The pertinent portions of Special Publication No. 264 are shown in Tables XVI and XVII.

EXAMPLE 11-1:

Given: Station "Blackduck Tank"
Latitude 47°43'50.270". Longitude 94°32'58.240"
State—Minnesota; Zone—North
$C = 2,000,000$ ft $\qquad R_b = 19,471,398.75$

Required: Lambert coordinates, x and y.

Solution: $R = 19,022,539.81$ (Table XVI)
$\theta = -1°04'27.8621''$ (Table XVII)
$\sin \theta = -0.0187508257$
$\cos \theta = +0.9998241876$

Then, $\quad x = R \sin \theta + C$
$x = -(19,022,539.81)(0.0187508257) + C$
$x = 1,643,311.67$ ft

and

$y = R_b - R \cos \theta$
$y = 19,471,398.75 - (19,022,539.81)(0.9998241876)$
$y = 452,203.34$ ft

It is to be noted that $\sin \theta$ is negative in this problem. The term, $R \cos \theta$, is subtractive in all cases. Also, the preceding calculation makes quite obvious the need for a 10-bank computing machine as well as an expanded table of natural sines and cosines. Special Publication No. 246 of the U.S. Coast and Geodetic Survey contains tables of these functions to 10 decimal places. This and similar surveying publications are available at nominal prices from the

Superintendent of Public Documents, Government Printing Office, Washington, D.C.

11-14. Computation of Plane Coordinates on Transverse Mercator Grid The transformation computation for a Transverse Mercator grid will be illustrated by reference to the Illinois state coordinate system. As will be recalled from a previous article, the Transverse Mercator projection is often described by stating that the axis of the tangent cylinder lies in the plane of the earth's equator. This projection may be illustrated by a cylinder cutting the surface of the spheroid along two small ellipses equidistant from the central meridian of the zone. The cylinder is then considered to be cut along an element and developed into a plane.

As shown in Fig. 11-7, the State of Illinois is covered by two overlapping Transverse Mercator zones, the East Zone and the West Zone. Each zone has its own axis for y, although both axes, passing through the centers of the respective zones, are given an x value of 500,000 ft. Both zones use the same x-axis, which is situated well below the southern limit of the state and has a value of zero feet. The geographic coordinates of a point in the East Zone will be transformed into plane coordinates.

The central meridian of the East Zone is 88°20′ west longitude. Along this line the scale of the projection is 1 part in 40,000 parts too small. The lines of exact scale are parallel to the central meridian and situated approximately 28 miles (147,900 ft) east and west of it. To the east and west of these lines, respectively, the scale is too large. The parallel of latitude 36°40′ defines the x-axis. Hence, the origin of coordinates for the East Zone is a point on the 36°40′ parallel situated 500,000 ft west of longitude 88°20′.

The expressions for calculating the Transverse Mercator coordinates of a point are as follows:

$$x = x' + 500,000$$
$$x' = H \cdot \Delta\lambda'' \pm ab$$
$$y = y_o + V\left(\frac{\Delta\lambda''}{100}\right)^2 \pm c$$

where y_o, H, V, and a are quantities based on the geographic latitude, and b and c are based on $\Delta\lambda''$; also x' is the distance the point is either east or west of the central meridian.

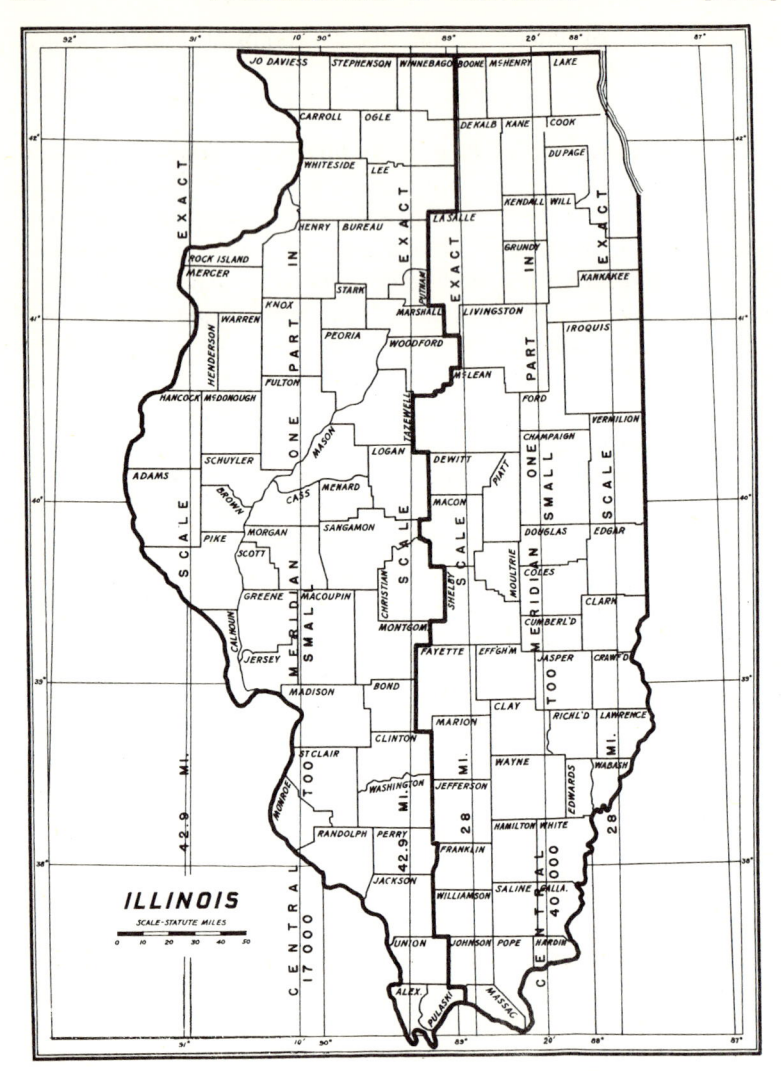

Fig. 11-7. Illinois State Plane Coordinate Zones.

A typical transformation computation will now be demonstrated. The pertinent portions of Special Publication No. 303 of the U.S. Coast and Geodetic Survey, entitled "Plane Coordinate Projection Tables—Illinois," are shown in Tables XVIII, XIX, and XX.

EXAMPLE 11-2:

Given: Station "King"
Latitude 40°43′37.202″. Longitude 88°41′35.208″
State—Illinois; Zone—East
Central Meridian: 88°20′00″

Required: Transverse Mercator coordinates, x and y.

Solution: $\Delta\lambda = -0°21′35.208″$
$\Delta\lambda'' = -1,295.208$

$\left(\dfrac{\Delta\lambda''}{100}\right)^2 = 167.756$

$H =$	76.992654	(Table XVIII)
$V =$	1.217932	(Table XVIII)
$a =$	-0.492	(Table XVIII)
$b =$	$+2.545$	(Table XIX)

$$H \cdot \Delta\lambda'' = -99,721.50$$
$$ab = -\underline{1.25}$$
$$x' = H \cdot \Delta\lambda \pm ab = -99,720.25*$$

$x = x' + 500,000 = 400,279.75$ ft

$y_o = 1,478,725.73$ (Table XVIII)

$V\left(\dfrac{\Delta\lambda''}{100}\right)^2 = 204.32$

$c = \underline{-0.04}$ (Table XIX)
$y = 1,478,930.01$ ft

11-15. Grid Azimuths The projection lines of a state plane coordinate system, whether Lambert or Transverse Mercator, comprise what is called a *grid* because all north-south lines are parallel with the central meridian and perpendicular to all east-west lines. Because of the convergency of the true or geographic meridians, it is obvious that the grid azimuth of a line will be the same as the geodetic azimuth only when the station at which the azimuth is expressed is situated on the central meridian. For all other lines in both systems, the grid azimuth will differ from the geodetic azimuth of the line. This difference becomes greater with increasing distance

---

*When ab is negative, decrease $H \cdot \Delta\lambda''$ numerically. If ab is positive, increase $H \cdot \Delta\lambda''$ numerically. Note also that since $\Delta\lambda''$ is negative because the station is west of the central meridian, x' is also negative.

of the survey station from the central meridian and is substantially equal to the angular convergency between the central meridian and the true meridian passing through the station.

Figure 11-8 shows that for stations west of the central meridian,

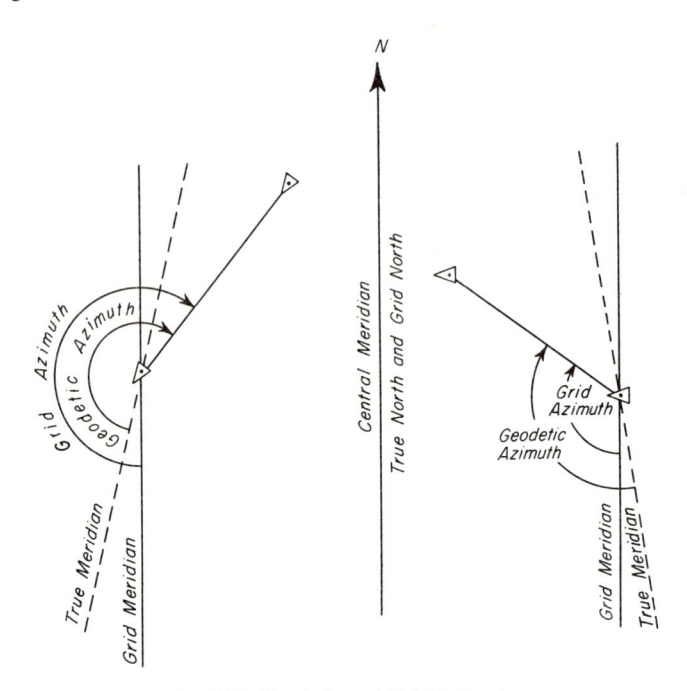

Fig. 11-8. Geodetic and Grid Azimuths.

the grid azimuth is greater than the geodetic azimuth and, for stations east of the central meridian, the grid azimuth is less than the geodetic azimuth.

In a state coordinate system, the forward and back grid azimuths of any line differ by exactly 180°. Grid azimuths are frequently reckoned from the south in keeping with the common practice in geodetic surveying.

11-16. Computation of Grid Azimuths When the grid coordinates of two stations are known, the grid azimuth of the connecting line is found by a simple computation in which the tangent of the azimuth angle is equal to the difference in the x-coordinates divided by the difference in the y-coordinates. Thus, $\tan a = \Delta x/\Delta y$. The usual sign conventions apply to this computation; and the magnitude of the azimuth, i.e., the quadrant in which it falls, can be readily ascertained by the use of a sketch.

Before the plane coordinates of a traverse can be computed, the grid azimuth of a reference line at the beginning station must be known so that the grid azimuths of each course can be determined. In some cases a distant triangulation or traverse station of known position can be sighted and the traverse initially oriented by this line of known grid azimuth. In other situations a sight can be taken on the azimuth mark, which is usually located at a distance of one-fourth to one-half mile from the triangulation station. With increasing frequency, the grid azimuth of the line to the azimuth mark is being published. If it is not, it will be necessary to convert the geodetic azimuth, which is always available, to a grid azimuth by one of the following expressions:

Lambert Systems

Grid Azimuth = Geodetic Azimuth $- \theta +$ *second term*

where θ is the familiar angle of convergency between the central meridian of the projection and the true meridian through the given station and must be used with careful regard always to sign. The second term is small and may be neglected for all situations involving third-order traverse except those in which the orientation sight is substantially greater than 5 miles.

If no azimuth mark is available, it will be necessary to determine the astronomic azimuth of the first line by solar or stellar observation and convert the astronomic azimuth to a grid azimuth by the foregoing expression.

Transverse Mercator Systems

Grid Azimuth = Geodetic Azimuth $- \Delta \alpha -$ *second term*

where

$$\Delta \alpha'' = \Delta \lambda'' \sin \phi + g$$

The second term is again negligible for most circumstances. The astronomic azimuth can be considered equal to the geodetic azimuth. The quantity, g, is obtained from Table XX and should always be applied to $\Delta \lambda'' \sin \phi$ so as to increase it numerically.

Figure 9-16 depicts the manner in which state coordinates and grid azimuths are published.

11-17. Determination of Geodetic Distance from Ground Distance Before the grid coordinates of the traverse stations of any survey can be computed, it is necessary (1) to reduce all ground distances to mean sea level to determine their equivalent geodetic distances, and (2) to convert these geodetic distances to grid distances on the plane of the state projection.

It is presumed, first of all, that the proper corrections, including those for temperature, inclination, and error in absolute length of tape, have been applied to the observed field distances to obtain the best ground distances. These, then, are to be reduced to sea level. Of course, distances can be measured electronically.

This reduction is facilitated by the use of Tables XI and XII. Table XI lists the factors by which the ground distance at various elevations is to be multiplied in order to obtain the geodetic distance. Elevation factors for intermediate elevations can be found by interpolation. Table XII lists the corrections to be subtracted from a ground distance of 1000 ft at various elevations in order to obtain the sea-level distance. In the computation of both tables the mean radius of the earth was taken as 20,908,000 ft.

To illustrate the use of Tables XI and XII, consider a ground distance of 2165.87 ft at an elevation of 2000 ft. From Table XI the sea-level distance is (2165.87)(.9999043) or 2165.66. From Table XII the correction is (2.166)(0.0957) or 0.21. This quantity subtracted from the ground distance gives the sea-level distance of 2165.66 ft.

For most third-order traverse and for higher-order traverse at elevations under 500 ft, the sea-level correction is considered insignificant. Furthermore, approximate elevations as obtained from a map are entirely satisfactory for calculating the correction. An error of 500 ft in the elevation of a traverse course will cause a proportional error in the sea-level length of the course of only 1 part in 41,800 parts.

11-18. Determination of Grid Distance from Geodetic Distance The ratio of the plane or grid distance of any line in a state plane coordinate system to the geodetic or sea-level length of the line is known as the *scale factor*. The scale factor may be found for any line by referring to the appropriate part of the U.S. Coast and Geodetic Survey projection tables for the given state. For the Lambert systems the table of scale factors is entered with the latitude. For the Transverse Mercator Systems, the entering argument is the quantity, x', which is the distance from the central meridian.

The scale factors are expressed both as a ratio and as the correction, in units of the 7th place of logarithms, to the sea-level length of the line. The sign of the logarithm and the ratio are those to be used for changing a geodetic distance to a grid distance. In the event the conversion is made in the opposite direction, the sign should be changed and the reciprocal of the indicated scale ratio used.

It is to be noted that, when the scale factor is less than 1, it is frequently more convenient to subtract the tabular value of the scale factor from 1 and apply a subtractive correction to the geodetic length. Thus, if the scale factor is 0.9999374 and the geodetic length is 5280.00 ft, the grid distance is equal to

$$(5280.00) - (5280.00)(0.0000626)$$

or

$$5280.00 - 0.33 = 5279.67 \text{ ft}$$

In order to determine the scale factors that should be used in the calculation of a traverse on either the Lambert or Transverse Mercator grid, it will be found essential to plot the traverse to some convenient scale, such as 1 in. equals 4000 ft. The data for plotting the traverse will be the known grid coordinates of the initial station and the field lengths and azimuths of the various courses.

If the Lambert system is used, it will be necessary to determine the mean latitude of each traverse course. This can be done with sufficient accuracy by scaling the distance, measured parallel to grid north, from the initial station, whose latitude is known, to the middle of each course. This distance in feet can be converted with adequate accuracy to seconds of latitude by dividing by 100. The resulting latitude difference, in seconds of arc, can then be applied to the latitude of the beginning traverse station in order to determine a latitude for entering Table XVI and finding the scale factor by interpolation.

If the Transverse Mercator System is used, a similar plot can be employed in order to determine the mean value of x' for each traverse course. It should be recalled that if the x-coordinate of the initial traverse station is 520,000.00 ft in the Illinois system—East

Zone, the value of x' is +20,000.00 ft. The scaling of east-west distances from the traverse station to the middle of each course will permit the determination of satisfactory values of x' for entering Table XXI. Since all new U.S. Geological Survey topographic maps now contain marginal ticks for drawing the grid lines of the state coordinate system, such maps will be found to provide a convenient base for making a plotting of the traverse.

11-19. Grid Factor Since the elevation factors of Table XI and the scale factors of Tables XVI and XXI are quantities by which the ground distance and geodetic distance, respectively, are multiplied in order to determine the grid distance, it will be found helpful to use the product of these factors as a means for more directly calculating the grid distance. This combined factor is called the *grid factor*.

TRAVERSE COMPUTATIONS

11-20. Preliminary Remarks The use of the principles explained in the preceding articles will now be illustrated by the calculation of a transit traverse. After giving the essential data controlling the traverse, the major steps in the computation and adjustment of the traverse will be explained.

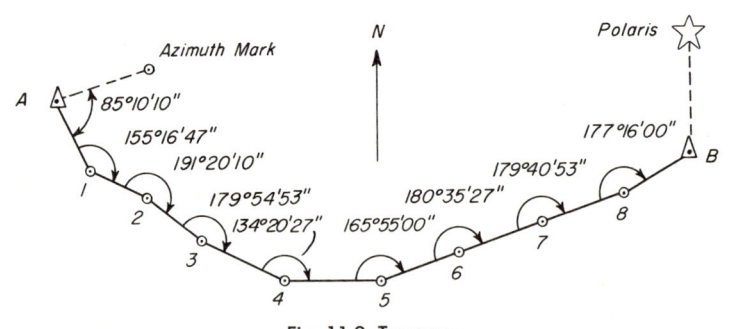

Fig. 11-9. Traverse.

11-21. Control Data The traverse shown in Fig. 11-9 was executed between two fixed points of higher-order horizontal control in western Illinois. The average elevation of the traverse above sea level is 750 ft, and the Illinois (West Zone) grid coordinates of the initial and terminal points are as follows:

[11-22]   STATE COORDINATE SYSTEMS   307

	x	y
A	461,577.68	1,556,540.53
B	474,718.81	1,554,910.45

The grid azimuth from A to the azimuth mark is 250°05′52″. At station B, an observation was made on Polaris for azimuth. The $\Delta\alpha$ angle was applied to the astronomic azimuth as explained in Art. 11-16, and the resulting grid azimuth for the line B to 8 was found to be 79°35′57″. These azimuth values (reckoned from south) are held fixed. The measured angles-to-right are shown in Fig. 11-9, and the distances are tabulated in Fig. 11-10.

Line		Corrected Field Distance (ft)	Elevation Factor (a)	Scale Factor (b)	Grid Factor (a) × (b)	Grid Distance (ft)
From Sta.	To Sta.					
A	1	754.25	0.9999642	0.9999424	0.9999066	754.18
1	2	517.12				517.07
2	3	808.11				808.03
3	4	1617.63				1617.48
4	5	982.61				982.52
5	6	3165.07				3164.77
6	7	2354.55				2354.33
7	8	3296.43				3296.12
8	B	1241.74	0.9999642	0.9999424	0.9999066	1241.62

Fig. 11-10. Computation of Grid Distances.

11-22. Computation of Grid Coordinates The computations comprise six operations as follows: (1) adjustment of the observed angles; (2) computation of grid azimuths; (3) reduction of field distances to grid distances; (4) computation of latitudes and departures; (5) computation of preliminary grid coordinates; and (6) computation of adjusted grid coordinates.

1. *The adjustment of the observed angles*—Fig. 11-11 indicates that the preliminary azimuth closure is 18″. This error is distributed equally among the 9 angles so that each angle is increased by 2″.
2. *Computations of final grid azimuths*—these are shown in Fig. 11-11, where it is noted that the corrections to the preliminary azimuths are the cumulative sums of the corrections to the measured field angles.

FUNDAMENTALS OF SURVEYING [11-22]

Sta.		Preliminary Azimuths			Azimuth			Correction for Closure		Final Azimuths		
		From Sta.	To Sta.	°	′	″	′	″	°	′	″
A	Angle	A Az.Mk. A	Az.Mk. 1 1	250 85 335	05 10 16	52 10 02	(Fixed)	02 02	250 85 335	05 10 16	52 12 04
1	Angle	1 A 1	A 2 2	155 155 310	16 16 32	02 47 49		02 04	155 155 310	16 16 32	04 49 53
2	Angle	2 1 2	1 3 3	130 191 321	32 20 52	49 10 59		02 06	130 191 321	32 20 53	53 12 05
3	Angle	3 2 3	2 4 4	141 179 321	52 54 47	59 53 52		02 08	141 179 321	53 54 48	05 55 00
4	Angle	4 3 4	3 5 5	141 134 276	47 20 08	52 27 19		02 10	141 134 276	48 20 08	00 29 29
5	Angle	5 4 5	4 6 6	96 165 262	08 55 03	19 00 19		02 12	96 165 262	08 55 03	29 02 31
6	Angle	6 5 6	5 7 7	82 180 262	03 35 38	19 27 46		02 14	82 180 262	03 35 39	31 29 00
7	Angle	7 6 7	6 8 8	82 179 262	38 40 19	46 53 39		02 16	82 179 262	39 40 19	00 55 55
8	Angle	8 7 8	7 B B	82 177 259	19 16 35	39 00 39		02 18	82 177 259	19 16 35	55 02 57
B	Angle	B B	8 8	79 79	35 35	39 57	(Fixed)		79	35	57
	Angle					18″ Closure					

Azimuth correction = $\frac{18''}{9}$ = 2″ per station

Fig. 11-11. Azimuth Computations.

3. *Reduction of field distances to grid distances*—these calculations are shown in Fig. 11-10. The scale factor experiences no substantial change in the entire length of the traverse. An average value is used.

4. *Computation of latitudes and departures*—these quantities resulting from a machine computation are shown in the appropriately headed columns of Fig. 11-12. It is to be noted that these computations were made with 8-place tables of natural sines and cosines. In most cases, 6-place tables will be adequate for the lengths of the lines encountered in third-order traverse.

5. *Computation of preliminary grid coordinates*—these are shown in Fig. 11-12 as the first values of the coordinates for each station. From the closure of −0.46 ft in the x-coordinate and +0.62 ft in the y-coordinate, the linear error of closure, 0.77 ft, and the relative error of closure, 1:19,000, are calculated.

6. *Computation of adjusted grid coordinates*—by the use of the Compass Rule, the preliminary x- and y-coordinates are adjusted. The x-correction is −0.0312 ft per thousand feet of cumulative traverse length, and the y-correction is +0.0421 ft per thousand feet. The cumulative lengths of the traverse are indicated in parentheses in the column of grid distances, and the coordinate corrections, which are easily computed with a slide rule, are tabulated in the grid coordinate columns.

Although the traverse computation is now completed, it is worth mentioning that the azimuths of all lines were affected slightly when the coordinates of the traverse stations were adjusted. Hence, if the engineer wishes to use the best azimuth of a given line, it should be computed from the final coordinates. Ordinarily, this change in the azimuth is not very significant unless the line is short and the traverse closure is relatively large.

11-23. Some Uses of State Plane Coordinate Systems The use of state plane coordinates provides all the advantages of geodetic position data from which they were derived and to which they are permanently related without introducing any of the difficulties associated with geodetic computations. State plane coordinates are conveniently determined through the application of little more than the ordinary field and office procedures of plane surveying. Yet they serve as immutable expressions of horizontal position which can be

310 FUNDAMENTALS OF SURVEYING [11-23]

Station	Grid Bearing ° ′ ″	Grid Distance (ft)	Sin Bearing Cos Bearing	Dept.	Lat.	Grid Coordinates x	Grid Coordinates y
A					(Fixed)	461,577.68	1,556,540.53
	S. 24 43 56 E.	754.18 (754)	0.41837794 0.90827303	+315.53	−685.00	461,893.21 −0.02	1,555,855.53 +0.03
1						461,893.19	1,555,855.56
	S. 49 27 07 E.	517.07 (1271)	0.75986099 0.65008559	+392.90	−336.14	462,286.11 −0.04	1,555,519.39 +0.05
2						462,286.07	1,555,519.44
	S. 38 06 55 E.	808.03 (2079)	0.61724569 0.78677046	+498.75	−635.73	462,784.86 −0.06	1,554,883.66 +0.09
3						462,784.80	1,554,883.75
	S. 38 12 00 E.	1617.48 (3696)	0.61840840 0.78585689	+1000.26	−1271.11	463,785.12 −0.12	1,553,612.55 +0.16
4						463,785.00	1,553,612.71
	S. 83 51 31 E.	982.52 (4679)	0.99426092 0.10698232	+976.88	−105.11	464,762.00 −0.15	1,553,507.44 +0.20
5						464,761.85	1,553,507.64
	N. 82 03 31 E.	3164.77 (7844)	0.99040992 0.13816003	+3134.42	+437.24	467,896.42 −0.24	1,553,944.68 +0.33
6						467,896.18	1,553,945.01
	N. 82 39 00 E.	2354.33 (10,198)	0.99178318 0.12793015	+2334.98	+301.19	470,231.40 −0.32	1,554,245.87 +0.43
7						470,231.08	1,554,246.30
	N. 82 19 55 E.	3296.12 (13,494)	0.99105775 0.13343366	+3266.65	+439.81	473,498.05 −0.42	1,554,685.68 +0.57
8						473,497.63	1,554,686.25
	N. 79 35 57 E.	1241.62 (14,736)	0.98356885 0.18053345	+1221.22	+224.15	474,719.27 −0.46	1,554,909.83 +0.62
B					(Fixed)	474,718.81	1,554,910.45

Linear error of closure = $\sqrt{(0.46)^2 + (0.62)^2} = 0.77$

Relative error of closure = $\dfrac{1}{19,000}$

x-correction = $\dfrac{-0.46}{14.736}$ = −0.0312 ft per thousand feet

y-correction = $\dfrac{+0.62}{14.736}$ = +0.0421 ft per thousand feet

Fig. 11-12. Computation of Coordinates.

used for a wide variety of engineering surveys. Some specific uses of state plane coordinates will be described briefly.

The centerline surveys for long route projects, such as a toll highway, are frequently based on state plane coordinates. The use of such horizontal control makes possible the utilization of several survey parties working in widely separated portions of the project with the positive assurance that all survey data can be correlated.

The use of state plane coordinates in extensive city and county mapping projects permits the execution of separate and detached surveys with the knowledge that, when component surveys meet, they will do so harmoniously without overlaps, gaps, or offsets at the junction points.

State plane coordinates provide a means for constructing a simple rectangular projection for the preparation of maps covering large areas and for plotting data thereon. Examples of such maps are those prepared by air mapping organizations for state highway commissions in connection with extensive highway projects. Figure 11-13 shows a portion of a table (Ref. 6) which gives values of state plane coordinates at 2½ minute intervals of latitude and longitude. This table is designed primarily to facilitate the preparation of maps based on the Transverse Mercator projection.

The exchange and use of survey and map information is greatly facilitated by the use of state coordinates. On the other hand, a multiplicity of local coordinate systems, like an excess of level datum planes, prevents the widest possible utilization of control information.

The use of state coordinates makes possible the certain recovery of any lost or destroyed survey point whether it be a property corner, an important construction control monument, or a triangulation station. The field recovery procedure consists of running a traverse from the nearest point of known position to a temporary marker in the vicinity of the lost station. A comparison of the computed coordinates of the temporary marker and those of the lost station provide a means for calculating the length and azimuth of the connecting line.

The use of state coordinates can eliminate the need for a random line in land surveys. A line between two stations that are not intervisible but can be tied to the state coordinate system can be run out directly with a minimum of clearing in brushy country.

State coordinates can be used to express the positions of important engineering structures, such as oil wells, including those in tidal

INDIANA—EAST ZONE

State Plane Coordinates for
2½-Minute Intersections (Feet)

Longitude	x	Latitude 37 45 00	y	x	Latitude 37 47 30	y
86 00 00	403	620.96	91 201.00	403	675.02	106 373.01
02 30	391	573.54	91 246.59	391	634.36	106 418.62
05 00	379	526.10	91 297.55	379	593.68	106 469.60
07 30	367	478.65	91 353.87	367	552.99	106 525.94
86 10 00	355	431.18	91 415.56	355	512.28	106 587.65
12 30	343	383.69	91 482.61	343	471.55	106 654.73
15 00	331	336.18	91 555.02	331	430.80	106 727.17
17 30	319	288.65	91 632.80	319	390.03	106 804.98
86 20 00	307	241.10	91 715.95	307	349.23	106 888.16
22 30	295	193.51	91 804.46	295	308.41	106 976.70
25 00	283	145.91	91 898.33	283	267.56	107 070.61
27 30	271	098.27	91 997.57	271	226.68	107 169.89
86 30 00	259	050.60	92 102.18	259	185.78	107 274.53
32 30	247	002.90	92 212.15	247	144.83	107 384.55
35 00	234	955.17	92 327.49	235	103.86	107 499.93
37 30	222	907.40	92 448.20	223	062.85	107 620.69
86 40 00	210	859.59	92 574.28	211	021.81	107 746.81
42 30	198	811.74	92 705.72	198	980.72	107 878.30
45 00	186	763.86	92 842.53	186	939.60	108 015.16
47 30	174	715.93	92 984.70	174	898.43	108 157.39
86 50 00	162	667.96	93 132.24	162	857.22	108 304.98
52 30	150	619.94	93 285.15	150	815.96	108 457.95
55 00	138	571.88	93 443.44	138	774.66	108 616.29

Fig. 11-13. Grid Intersection Coordinates. *U.S. Coast and Geodetic Survey*

waters, and transmission towers; and they are becoming more widely utilized in pipeline location and description as well as in the planning, construction, and expansion of large industrial plants. Here, good vertical control is likewise very important in the protection of underground utilities and process interconnections.

The number of engineering agencies that use the various state plane coordinate systems as the basis for all control surveys is increasing. These coordinate systems provide the computing and plotting bases for a wide variety of special surveys and all kinds of planimetric and topographic maps.

Occasionally, misunderstandings may develop concerning the requirements of the state enabling acts establishing and recognizing the legality of such coordinate systems. Such legislation usually makes mandatory the use of precise surveying methods in extending horizontal control from a triangulation station or a traverse point. This provision, however, applies only to court recognition of the use of state coordinates in deeds and other legal instruments. Highway and construction surveys, regardless of their quality, can be based on a state coordinate system. Naturally, their usefulness will be increased if they are, at least, of third-order accuracy. Plane coordinate projection tables are now available for every state in the nation and can be procured from the Superintendent of Public Documents, Government Printing Office.

PROBLEMS

11-1. A second-order triangulation station in northern Minnesota is situated in latitude 47°35′26.870″ and longitude 94°34′01.505″. Determine (a) the θ angle and (b) the scale factor at this station.

11-2. Compute the Lambert grid coordinates of the triangulation station of problem 11-1.

11-3. A second-order traverse station in east-central Illinois is situated in latitude 40°46′12.695″ and longitude 88°45′26.359″. Compute the Transverse Mercator (East Zone) coordinates of this point.

11-4. For the station of Problem 11-3 determine (a) the $\Delta\alpha$ angle (to the nearest second) and (b) the scale factor.

11-5. The state plane coordinates, in feet, of two survey stations, A and B, are as follows:

	x	y
A	527,946.70	1,250,583.24
B	528,198.77	1,300,758.10

Compute (a) the grid azimuth (reckoned from south) to the nearest second of the line from station A to station B, and (b) the grid distance between these points.

11-6. If the geodetic azimuth to the azimuth mark of Problem 11-1 is 27°18′11″, what is the grid azimuth?

11-7. If the geodetic azimuth to the azimuth mark of Problem 11-4 is 329°14′56″, what is the grid azimuth?

11-8. The grid length of a line as calculated from the state plane coordinates of its termini is 4010.56 ft. If the scale factor is 0.9999800 and the line is situated at an elevation of 4500 ft, find the ground length.

11-9. Using the adjusted coordinates of the line, 4–5, of Fig. 11-12, compute the final grid bearing and grid length of this line.

REFERENCES

U.S. Coast and Geodetic Survey, Washington, D.C., Special Publications:

1. *Relation Between Plane Rectangular Coordinates and Geographic Positions,* No. 71, 1932.
2. *The State Coordinate Systems,* No. 235, 1945.
3. *Sines, Cosines, and Tangents, Ten decimal places with Ten-second Intervals, 0° — 6°,* No. 246, 1949.
4. *Plane Coordinate Projection Tables—Minnesota (Lambert),* No. 264, 1952.
5. *Plane Coordinate Projection Tables—Illinois (Transverse Mercator,* No. 303, 1953.
6. *Plane Coordinate Intersection Tables (2½-minute) Indiana,* No. 332, 1955.

12 Photogrammetry

12-1. Introduction *Photogrammetry* is the science or art of obtaining reliable measurements by means of photography. Photogrammetry is subdivided into various types, but the two basic categories are *terrestrial photogrammetry* and *aerial photogrammetry*. In terrestrial or ground photogrammetry the photographs are taken with the camera mounted on a tripod, and the optical axis of the lens is usually horizontal. Aerial photogrammetry makes use of photographs that have been taken from any airborne vehicle. Such photographs may be either vertical or oblique. *Vertical photographs* are taken with the optical axis pointing vertically downward at the moment of exposure. *Oblique photographs* are obtained when the optical axis is intentionally inclined from the plumb line.

Vertical photography (see Fig. 12-1) is most commonly obtained by cameras mounted in airplanes flying a straight course but with sufficient overlap between adjacent exposures to permit subsequent stereoscopic examination of the pictures. The photographs made on a given course constitute a flight strip, and a sufficient number of strips are taken to cover a given area. The overlap in the direction of flight is called *forward lap,* and the overlap between pictures in adjacent flights is called *side lap*. The amount of forward lap and side lap is commonly specified as 60% and 30%, respectively.

Since aerial photographs are generally taken from moving aircraft, the horizontal position, elevation, and orientation of the camera are not known. Although an attempt is made to keep the camera axis vertical, a small amount of *tilt,* whose direction and magnitude are unknown, is usually present when an exposure is made. Because of these unknown factors, photogrammetric measurements have been beset with many difficulties, but these have been overcome to such an extent that aerial methods have either displaced or considerably modified the ground methods formerly used by all governmental

Fig. 12-1. Vertical Aerial Photograph. *Wild Heerbrugg Instruments, Inc.*

mapping organizations. On practically every important survey or engineering project, by either public or private agencies, aerial photographs are used in some way. For example, wide use is now being made of photogrammetry in the planning, location, and construction of highways. From aerial photographs accurate estimates are made of the costs of the right-of-way, earthwork, and structures along alternate routes, in order to obtain the most feasible and economical location.

Closely associated with photogrammetry, or *metrical photography* as it is sometimes called, is *photographic interpretation*. This subject is concerned with the determination of the nature and description of objects that are imaged on a photograph. Hence, it can be seen that photo interpretation is qualitative in character, whereas photogrammetry is essentially quantitative.

The development of photogrammetry has been intimately connected with that of the camera, photographic materials and processes, and aviation. The first known photograph was the daguerreotype, which was produced in 1839 by Daguerre. Shortly thereafter the science of photogrammetry had its genesis.

This chapter contains the rudiments of photogrammetry. Applications to map making are presented in Chapter 14.

12-2. Definitions A few fundamental definitions will be given. Some of the terms are as applicable to the horizontal photograph (Fig. 12-2) as they are to the vertical photograph.

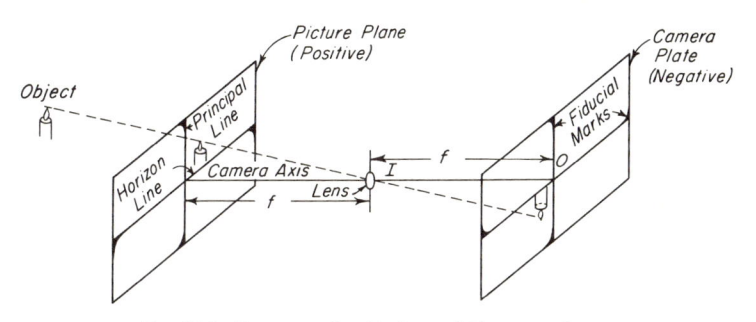

Fig. 12-2. Elements of a Horizontal Photograph.

Point of View, I, is the center of the camera lens.

Camera Axis is the line through the center of the camera lens perpendicular both to the camera plane and the picture plane.

Picture Plane is the plane perpendicular to the camera axis at the focal length distance in front of the lens. It is represented by the positive contact print or photograph taken from a plane or film.

Principal Point is the point of intersection, O, of the camera axis with either the picture plane (photograph) or the camera plane (negative).

Focal Length, f, is the perpendicular distance from the center of the camera lens to either the picture plane or the camera plate.

Fiducial Marks are index marks within the camera frame which form images on the edges of the negative. The intersection of straight lines on the photograph connecting these images fixes the principal point of the photograph.

Photograph Nadir is the point of intersection of a vertical (plumb) line through the center of the lens (at the instant of exposure) and the photograph. If there is no tilt of the camera axis when the exposure is made, the photograph nadir and the principal point will be identical.

Ground Nadir is the point of intersection of a vertical line through the center of the lens and the ground surface.

Topographic Map is a map that represents the horizontal and vertical positions of features on a portion of the surface of the earth.

Planimetric Map is a map that presents only the horizontal positions of the features portrayed. It is distinguished from a topographic map by the absence of relief representation.

12-3. Perspective Principles of Vertical Photographs Since any photograph is a perspective view, it is subject to the principles of such views whether it is a terrestrial (horizontal) or an aerial (vertical) photograph. The following principles apply to vertical photographs:

1. The photographic images of all vertical lines of objects on the ground will be radial lines which, if extended, will pass through the principal point O.

2. All parallel level lines on the ground, such as the parallel sides of a square tract of level land, will appear as parallel lines on the photograph.

The first principle stated above has the greatest significance in all photogrammetric uses of vertical photographs. Three examples will be mentioned. (1) In Fig. 12-3 the flagpole represents a vertical line

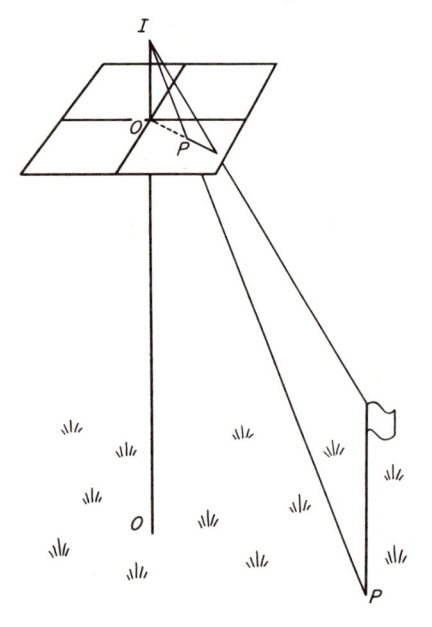

Fig. 12-3. Photographic Image of a Flagpole.

perpendicular to the picture plane, and therefore the image is a straight line which, if extended, passes through the principal point O. (2) In Fig. 12-6 are shown a vertical aerial view and the image p_h of the top of a hill P_h. The vertical projection of P_h down to sea-level datum is at P_o, and the image of this point, if it could be seen in the photograph, would appear at the point p_o. Accordingly, the image of the vertical line $P_h P_o$ is $p_h p_o$, and this line, if extended, passes through the principal point O. The length of the image d is called the *displacement* of the point p_h because of the elevation of the ground point above the datum plane. (3) In Fig. 12-4 are shown the images

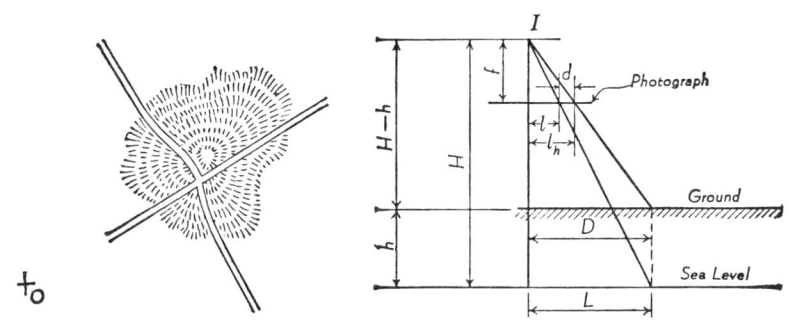

Fig. 12-4. Effect of Relief Displacement.

Fig. 12-5. Scale Relations in a Vertical Photograph.

of two roads whose alinements on the ground are straight and which intersect at right angles as they pass over a hill. The principal point of the photograph is at O. The displacement of the points on the roadway, whose direction is radial from the principal point O, is along this same direction, and therefore there is no change in the alinement of the photographic image, which remains a straight line. The points in the image of the other roadway are displaced radially from the principal point, and therefore the image of this roadway shows a convex curvature away from the point O.

12-4. Scale of a Photograph It has been shown that the images of ground points are displaced where there are variations in the ground elevation. Hence, there is no uniform scale between the many points on such a photograph; therefore, in discussing the "scale of a photograph" in this article it is assumed that the ground is perfectly horizontal and the camera axis is truly vertical.

The scale of a photograph is the ratio of a given distance on the photograph to the corresponding distance on the ground. Where English units of measure are in use, this ratio is expressed in either of two forms, which may be designated as R, the representative fraction, or S, the map scale. Thus, in Fig. 12-5 for the sea-level elevation,

$$R = \frac{l}{L} \tag{12-1}$$

in which both l and L are expressed in the same unit, or the same ratio may be expressed as

$$S = \frac{L}{l} \tag{12-2}$$

in which L is expressed in feet and l is in inches. For example, if $l = 3$ in., and $L = 4500$ ft, then

$$R = \frac{0.25 \text{ (ft)}}{4500 \text{ (ft)}} = \frac{1}{18,000}, \text{ and } S = \frac{4500 \text{ (ft)}}{3 \text{ (in.)}} = 1500 \text{ ft per in.}$$

The relationships between the scale of a photograph, the focal length f, and the height of lens H are also shown in Fig. 12-5. It is evident from the similar triangles that

$$\frac{l}{L} = \frac{f}{H}$$

and hence

$$R = \frac{f \text{ (ft)}}{H \text{ (ft)}} \tag{12-3}$$

also

$$S = \frac{L}{l} = \frac{H \text{ (ft)}}{f \text{ (in.)}} \tag{12-4}$$

Moreover, for the ground level at an elevation h above sea level, it is evident that

$$R_h = \frac{l_h}{L} = \frac{f}{H - h} \tag{12-5}$$

and

$$S_h = \frac{L}{l_h} = \frac{H - h}{f} \tag{12-6}$$

For example, if $f = 6$ in., $H = 9000$ ft, and $h = 500$ ft, then

$$R_h = \frac{0.5}{8500} = \frac{1}{17,000}; \text{ and } S_h = \frac{8500}{6} = 1417 \text{ ft per in.}$$

The latter scales are spoken of as the *representative fraction,* or the *scale,* of the photograph at the elevation h, respectively.

It is evident in Fig. 12-5 that the ground distance D is represented by two different distances on the photograph, l and l_h corresponding to the two different elevations, sea level and h above sea level. Since the ground surface is usually characterized by slopes, hills, and valleys, it is obvious that the distance on a photograph that represents a given distance on the ground varies with the different elevations, and there can be no single scale that will apply to all points appearing in the photograph. Thus, to obtain the true ground distance between points of different elevations, it is necessary to refer them to a single plane of reference called the *datum plane.* For special conditions, any assumed elevation may be used as a datum plane; but since sea level is the universal datum for ground elevations it is also commonly used as the datum for aerial photographs. Therefore, unless otherwise specified in the following pages, H will represent the height of the lens above sea level, and h will represent the elevation of a ground point above sea level in all relations dealing with scale factors.

On any given photograph, if the images of two points appear whose ground elevations are equal and if the distance between them is known, then the scale of the photograph, for the known elevation h, is readily determined by the simple relation $S = L/l$, in which L is the known distance, in feet, on the ground, and l is the distance, in inches, on the photograph. The distance L may be measured on the ground, or if a published map is available it may be possible with sufficient accuracy to scale the distance from the map.

EXAMPLE 12-1: Two road intersections, which are known to be one mile apart in rather flat terrain, are imaged on a vertical photograph. If the photographic distance is 3.22 in., what is the indicated scale of the photograph?

$$S = \frac{L}{l} = \frac{5280}{3.22} = 1640 \text{ ft per in.}$$

12-5. Number of Photographs Required Because of the overlap required in aerial photographs for mapping purposes, the net area covered by a single photograph will be that included within its full dimensions diminished by the overlap of adjacent prints.

The amount of overlap for two adjacent prints in the direction of the line of flight, called *forward lap,* is usually 60%. The distance between two principal points in a flight series is equal to the size of a print less the amount of forward lap. Thus, if the size of the print is 7 × 7 in., and the forward lap is 60%, then the distance between two adjacent principal points will be 7 in. − (7 × 0.60) = 2.8 in. Likewise, the distance between the principal points of photographs in adjacent flights is given by the size of the print perpendicular to the line of flight, less the amount of the side lap. Thus, if the print is 7 in. wide and the side lap is 25%, then the distance between two adjacent principal points will be 7 in. − (7 × 0.25) = 5.25 in. The number of photographs required, therefore, will be the number required for one strip times the number of strips.

EXAMPLE 12-2: A flight mission is to be flown under the following conditions: The area is rectangular, 15 miles by 10 miles in size; the camera negatives are 7 in. square, and the focal length is 6 in. The scale of the photographs will be approximately 1500 ft per in. The forward lap is 60%, and the side lap is 25%. How many photographs will be required?

Solution: The distance between the principal points of two photographs in the line of flight will be 7 in. − (7 × 0.60) = 2.8 in. × 1500 = 4200 ft. The total length of one flight is 15 × 5280 = 79,200 ft. Hence the number of photographs required for one flight is 79,200/4200 = 18.8 or 19. The distance between two flights will be 7 in. − (7 × 0.25) = 5.25 × 1500 = 7875 ft. The number of flights will then be (10 × 5280)/7875 = 6.7 or 7 flights. The number of photographs then is 7 × 19 = 133.

12-6. Image Displacement Caused by Ground Relief If conditions are as shown in Fig. 12-5, i.e., the photograph is truly horizontal and the ground is level, and if other sources of error are disregarded, then the photograph represents, at its proper scale, a true orthographic projection; hence it may be said to be a true map of the ground surface. Also, the photograph will have the same scale throughout the area contained within it. However, these conditions are never fully met in practice, and, since the photograph is a perspective view, any relief of the ground surface will be shown in perspective. Because of this condition, points in the photograph are said to be *displaced* from their true orthographic positions.

The displacement of an image caused by ground relief is shown in Fig. 12-6, where the image of a point P_h on a summit is shown at p_h, and the image of the vertical projection of this point to the datum plane P_o, is shown at p_o. The photographic distance $p_h p_o = d$ is the

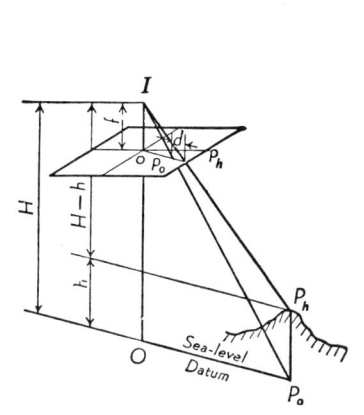

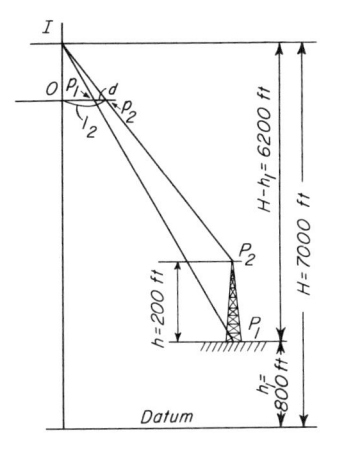

Fig. 12-6. Relief Displacement. Fig. 12-7. Image Displacement.

displacement of this image due to its elevation h above the datum.
If we let $op_h = l_h$, $op_o = l$, and $OP_o = OP_h = L$, the following equations may be written:

$$\frac{l}{L} = \frac{f}{H} \quad \text{and} \quad \frac{l_h}{L} = \frac{f}{H-h}, \quad \text{also} \quad d = l_h - l = \frac{Lf}{H-h} - \frac{Lf}{H};$$

from which

$$d = \frac{l_h h}{H} \qquad (12\text{-}7)$$

EXAMPLE 12-3: The distance from the principal point to an image on a photograph is 2.143 in., and the elevation of the object above the datum (sea level) is 850 ft. The height of lens above the datum is 7200 ft. Then $d = (2.143 \times 850)/7200 = 0.253$ in.

Another example of the use of this relation is shown in Fig. 12-7 where the images of the top and the bottom of a tower, P_2 and P_1, are shown on a photograph at p_1 and p_2. The distance to the point p_2 from the principal point O is indicated as l_2. The other values given are $H = 7000$ ft, $H - h_1 = 6200$ ft, $h = 200$ ft, and $l_2 = 2.120$ in. It is desired to find the displacement of the image of the top of the tower with respect to the image of the bottom, p_1. Following is the computation:

$$d = \frac{l_2 h}{H - h_1} = \frac{2.120 \times 200}{6200} = 0.068 \text{ in.}$$

12-7. Computed Length of Line Between Points of Different Elevations From the scale relations of Art. 12-4 it is possible to compute the correct horizontal length of a line between any two points of different elevation whose images appear on a photograph. For this problem it is necessary that the known data include the height of the lens H, the elevations of the two ends of the line, and the focal length f of the camera. The general procedure is as follows: (a) establish a system of coordinates for which the origin is the principal point, using the fiducial marks on the photograph; (b) scale the x- and y-coordinates for each end of the line; (c) compute the ground coordinates for X and Y for each of the image coordinates x and y; and (d) from the ground coordinate distances compute the length of the line.

Let A and B represent the ends of the line whose length is to be found, and a and b the images of these points. The elevations of these points are then h_a and h_b, respectively. From Eq. 12-6 the relation $S_a = (H - h_a)/f$ may be written, where S_a is the scale of the photograph for the point a, whose ground elevation is h_a above sea level. Likewise, $S_b = (H - h_b)/f$ expresses the scale for point b, whose ground elevation is h_b. Accordingly, if the x and y coordinates of the points a and b are scaled on the photograph, the corresponding coordinates of the points on the ground can be found as follows:

$$X_A = x_a \frac{H - h_a}{f}, \qquad Y_A = y_a \frac{H - h_a}{f},$$

and

$$X_B = x_b \frac{H - h_b}{f}, \qquad Y_B = y_b \frac{H - h_b}{f}$$

From these coordinates the true length of line AB may be computed, thus,

$$L = \sqrt{(X_A - X_B)^2 + (Y_A - Y_B)^2}$$

EXAMPLE 12-4:

Given: $H = 12{,}325$ ft and $f = 8.25$ in.; also the scaled coordinates of two images a and b, and the elevations of these ground points A and B. Find the true horizontal distance of the line AB.

Solution:

Point	h (ft)	$H - h$	$\dfrac{H-h}{f}$	Image	x (in.)	y (in.)	X (ft)	Y (ft)
A	720	11,605	1406.7	a	-2.174	-0.123	-3058	-173
B	915	11,410	1383.0	b	$+1.424$	-2.281	$+1969$	-3155

$$X_A - X_B = -5027, \qquad Y_A - Y_B = +2982$$
$$L = \sqrt{(-5027)^2 + (2982)^2} = 5845 \text{ ft}$$

12-8. Control for Photogrammetric Maps It would be expensive to establish the positions of a sufficient number of points by ground-control methods so that each photograph in a series could be accurately oriented in drawing the map; accordingly, graphical methods have been devised whereby the successive photographs can be oriented in the drafting room, with respect to each other and to ground features. The system of points thus fixed is termed *map control*, and a number of such systems have been devised, but the one most commonly used is known as the *radial-line* method.

12-9. Radial-Line Method The radial-line method is the most accurate means of plotting a planimetric map from aerial photographs without the use of expensive instruments. It is based on three general principles, which have been previously stated: (a) the displacements in a photograph due to ground relief are radial from the principal point; (b) images near the principal point are shown in their true orthographic positions, regardless of ground relief or tilt; and (c) the position of a point is correctly located on a map where three rays from three known points intersect.

12-10. Ground Control The *ground control* should consist of the located positions of such objects as road intersections, lone trees, bridges, and fence corners, whose images can be accurately located on the photographs. The amount of control required will be determined by the scale of the map and the accuracy required. At least three points should be located that will show in the overlap area of the first two photographs of a flight series, somewhat as shown in Fig. 12-8. It is desirable that the three points P_L, P_C, and P_R shall be distributed evenly across the photograph transverse to the line of flight.

It is advantageous to establish the ground control after the photographs have been taken, because the character and distribution of the images will indicate where the ground points should best be located. However, if the control surveys have been run previously, they can be supplemented by a few ties and extensions to suitable objects that can be selected after the photographs have been taken.

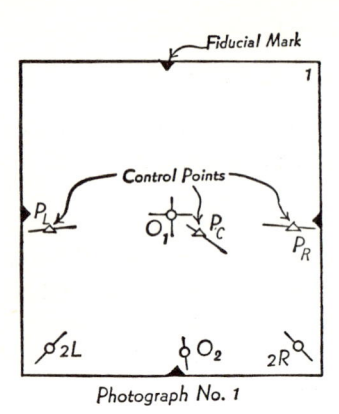

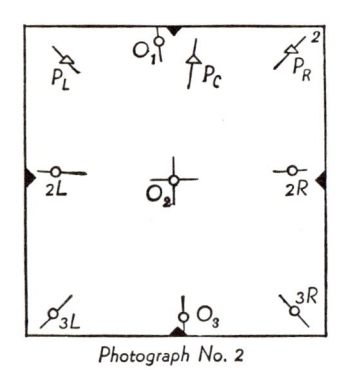

Fig. 12-8. Control Points from Overlapping Photographs.

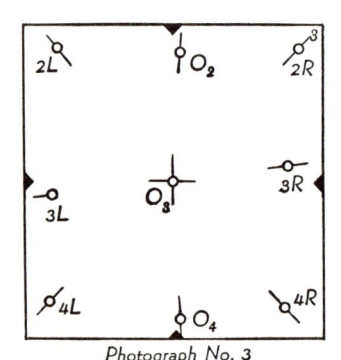

12-11. Base Map The base map can be drawn on heavy paper of high quality, on tracing cloth, or on cellulose acetate.

The coordinate grid, which has been used in computing the position of the ground-control points, is then plotted with great care on the base map, after which the positions of the control points are also plotted. The scale should approximate closely that of the photographs. Then, after the data in the photographs have been plotted to this scale, the map as a whole can be enlarged or reduced by means of a projector or by pantograph.

12-12. Plotting the Line of Flight The principal point of an aerial photograph is readily found by means of two intersecting lines drawn between the opposite fiducial marks, which appear on the margin of each photograph. For many uses of aerial photographs it is also necessary to locate on a given photograph the line of flight as determined by the plotted position of the principal points of the two adjacent pictures, one to the rear, and one forward in the direction of the line of flight. For this purpose it is necessary to transfer the principal point of one picture to the next adjacent picture. This is done with the aid of a stereoscope. The principles of stereoscopic vision and fusion are discussed in Arts. 12-17 and 12-18.

To obtain a stereoscopic view from two adjacent aerial photographs they must be oriented correctly with respect to each other, and the line of flight on the pictures must be parallel with the two lenses of the stereoscope. Then the distance between the two photographs must be adjusted until fusion occurs, when the relief in the landscape will be clearly visible.

To transfer the principal point of one photograph to the next one adjacent, it is, of course, necessary first to locate the principal point of each photograph as explained above. Then with the two pictures in their correct positions under the stereoscope, the principal point of one photograph will be seen directly and its image will be projected upon the other photograph. Then with a needle the position of the point can be transferred to the adjacent photograph.

12-13. Marking the Photographs The photographs are prepared for establishing the map control by marking on each one the radial lines and points which may be indicated by reference to Fig. 12-8, which shows the markings for the first three photographs of a flight series.

On photograph No. 1, the images of the ground-control points are identified and marked with needle points at P_L, P_C, and P_R, enclosing each within small triangles drawn with a soft, colored pencil or with ink. Short lines radial from the principal point of the photograph are drawn through each control point. The same points are also marked on photograph No. 2. The principal point and the transferred principal points of the adjacent photographs are marked on each photograph, as shown at O_1, O_2, and O_3 of photograph No. 2.

On photograph No. 1 two additional points *2L* and *2R,* called *pass points,* are selected near the left and right edges of the photograph and approximately in line with the transferred principal point of the adjacent photograph.

On photograph No. 2 images P_L, O_1, and P_R appear near the upper edge; these and the control point P_C are then marked, and also the pass points *2L* and *2R* previously selected on photograph No. 1. Again two pass points *3L* and *3R* are selected near the lower left and right corners of photograph No. 2. Short lines radial from the principal point of photograph No. 2 are then drawn through each point. This completes the marking of this photograph.

In a similar manner each succeeding photograph is marked until other ground-control points are reached.

12-14. Locating Map-Control Points After the photographs have been marked they are ready to be used in plotting the *map control.* For this purpose either tracing cloth or acetate may be used. A special acetate can be procured which exhibits very small changes in its dimensions with changing atmospheric conditions. A matte surface is also provided, which facilitates the drawing of fine pencil lines.

The plotting of the map control begins with the transfer of the plotted positions of the ground control from the base map to the acetate sheet, by pricking through with a needle. Each point is then marked as shown at P_L, P_C, and P_R, Fig. 12-9. Photograph No. 1 is then slid under the acetate and adjusted until the radial lines through P_L, P_C, and P_R on the photograph pass through the corresponding control points on the acetate. The photograph is then correctly oriented, and all radial lines upon it are traced upon the acetate, including the transferred principal point O_2.

Photograph No. 2 is then slid under the acetate and adjusted so that the lines previously drawn on the acetate pass through the corresponding points on the photograph, the lines along the line of flight, O_1O_2, always being kept in coincidence. When this orientation is completed, the rays through *3L*, O_3, and *3R* on the photograph are traced on the acetate.

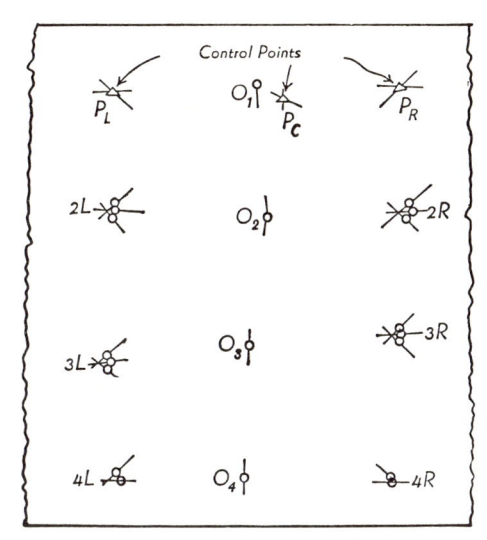

Fig. 12-9. Locating Map-Control Points.

In this manner photograph No. 3 is now placed under the acetate as explained above, and its radial lines are traced.

There will now appear three interesting rays at pass points *2L* and *2R* on the acetate. If the three rays thus drawn intersect at a point, its position is correctly located on the acetate. Usually, however, this point of intersection will not appear to coincide with the corresponding point on the photograph, because the scale of the map is not the scale of the photograph, and because of other displacements due to ground relief and tilt. This condition is shown at pass point *2L,* Fig. 12-9. Here the images of this pass point, as traced from each of the three photographs, are shown by small circles, and the true position of this point on the control map is at the intersection of the three rays.

Also, because of tilt or errors in plotting, occasionally the three rays for a given point will not intersect at a point, but will form a small triangle. In this case the position of the point is taken at the center of the triangle.

The procedure is continued until another ground-control point is reached. Here, if the map control is perfectly done, the image of the control point, as located by the intersecting rays, will coincide with the plotted position of this point on the acetate. However, this will

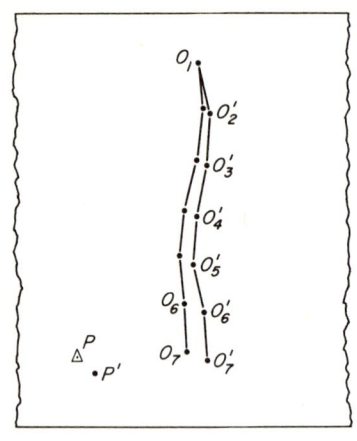

Fig. 12-10. Adjustment of Control Points.

seldom happen, and the usual condition is shown in Fig. 12-10, where P is the plotted position of the ground-control point, and P' is the position of the same point as located by the intersecting rays of the map-control plotting. In this figure the principal points of only seven adjacent photographs, as located by the map-control plotting, are shown. The distance and direction of the line PP' may be regarded as the total error of plotting the traverse O_1 to O_7; hence, the adjusted positions of the points O_2 to O_7 are found by moving each plotted point along a line parallel to the line $P'P$, a distance proportional to the distance of that point from the initial fixed point O_1. Thus O'_7 is adjusted to O_7 a distance equal to $P'P$, and O'_6 is adjusted to O_6 a distance equal to five-sixths of $P'P$, and so on. Also, the positions of all other pass points will be adjusted accordingly.

12-15. Plain Templets If a considerable area is to be plotted by the radial-line method, the procedure can be facilitated by the use of templets, which are of two kinds, namely, *plain* and *slotted*.

The plain templets are of transparent material, preferably acetate sheets, 8×10 in. in size, one for each 7×9 in. photograph. Both the ground- and map-control points are selected on each photograph as for the radial-line method and marked by needle points, each with a small ink circle around it. Then the acetate sheet is placed over it, and the position of the principal point is pricked through onto the templet. Radial lines are now drawn on the templet from the principal point to each control point previously marked on the photograph.

A base map is prepared as described previously, using a scale which is averaged as nearly as possible for the photographs of the area to be mapped.

The first templet is then laid over the base map and adjusted to the ground-control points, after which the second and third templets are also adjusted by the method previously explained; they are then fastened together by Scotch tape or by other suitable means. Each subsequent templet is added and fastened in position to the assembly of templets until the next set of ground-control points is reached. Now the ground-control points on the templets should coincide with the points on the base map; but because of the various sources of error, it will be only accidental when this coincidence is exact. The adjustment is then made by stretching or twisting the combined assembly of templets as a whole until coincidence of both ground- and map-control points has been established. The system of points that has been thus adjusted is transferred to the base map by pricking through the acetate with needles, or by laying a large transparent sheet over the templets and tracing the points upon it.

The detail of the photographs is then transferred to the base map as described in Art. 12-16.

The use of slotted templets is explained in Ref. 1.

12-16. Transferring Photographic Detail When the map control has been plotted and adjusted, the details from each photograph are then transferred to the base map by one or another of various methods. Three commonly used methods are (1) by tracing, (2) by pantograph, and (3) by projection.

If the map is to be drawn to the same scale as the base map, the detail can be transferred by tracing directly from the prints. By this method, a photograph is placed under the acetate, adjusted to the map control, and the detail is traced upon the acetate sheet. In this work the photograph will be shifted from time to time as the draftsman works from one area to another, thus always keeping the photograph closely adjusted to control points in the immediate area that is being mapped. After the detail has been drawn on the acetate sheet, it can be transferred to the base map by pantograph or by tracing over a carbon sheet. The map may then be finished by the use of the usual colors and symbols.

The pantograph may be used not only to transfer the details from the photographs but to draw the map to a scale different from that of the base map.

The best method of transferring detail from a photograph to a map is by means of a vertical reflecting projector like that shown in Fig. 12-11. Such an instrument projects the image of a photograph upon

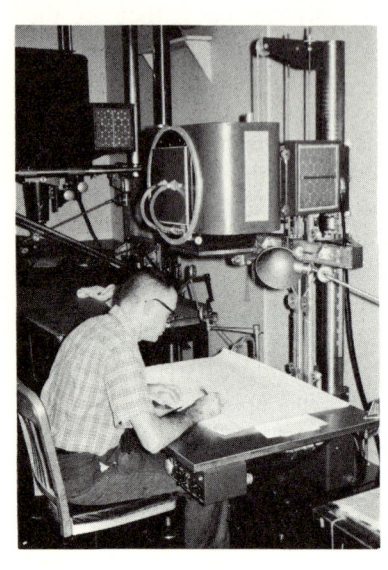

Fig. 12-11. Vertical Reflecting Projector. *Tennessee Valley Authority*

the map manuscript or base map and provides for changes in scale and for the elimination of the effect of tilt in the photograph. The image is altered so that the control points in the photograph are brought into coincidence with their plotted positions on the base map and the detail is drawn directly upon the map.

12-17. Stereoscopic Vision It is a particular phenomenon of binocular vision that the observer is able to perceive spatial relations, i.e., the three dimensions of his field of view. This perception is due partly to the relative apparent sizes of near and far objects, and to the effects of light and shade, but an important condition is the fact that a given object is viewed simultaneously with two eyes which are separated in space; hence the two rays of vision converge at an angle upon the object viewed. The angle of convergence of the two rays of vision is called the *angle of parallax,* and its magnitude has an important effect upon the accuracy with which the observer can judge the true dimensions of a given object.

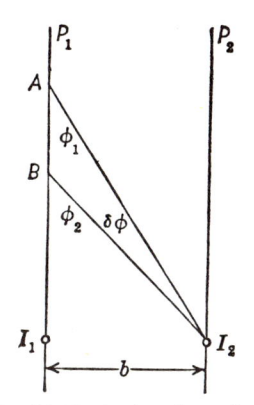

Fig. 12-12. Angles of Parallax.

Let I_1 and I_2 (Fig. 12-12) represent the positions of the two eyes of an observer, and let A and B represent two objects in the field of view. It is evident that the rays which converge upon A form the angle ϕ_1, and those upon B form the angle ϕ_2. Thus, as between two objects, that one will be judged to be nearer the observer for which the angle ϕ is the larger. It is further evident that $\phi_2 - \phi_1 = \delta\phi$, and this value is termed the *differential parallax*. This value is important since it provides a measure of the distance AB, in the line of vision, between two objects.

It is also evident that as $\delta\phi$ becomes small there is a limiting value below which the sense of stereoscopic vision is nil, and the observer is unable to judge, as between two objects, which is the nearer one. This limiting value of $\delta\phi$ for most observers is about 20″. If the distance between the observer's eyes is 2½ in., then for an angle of 20″, the rays I_1P_1 and I_2P_2 meet at a distance of about 2100 ft; at that distance or beyond, the sense of stereoscopic vision becomes inoperative, and the relative distances to objects must be judged by their apparent size or by other factors.

However, the range and intensity of stereoscopic perception can be increased in two ways—either by apparently increasing the base between viewpoints or by magnifying the field of view by the use of lenses. Some binoculars use both of these principles, having prisms that apparently spread the base *b* of vision and lenses that magnify the field. If the base is thus apparently increased 2 times, and if the lenses magnify the field 3 times, then the effect of stereoscopic perception is increased 6 times.

In aerial photogrammetry, stereoscopic vision is obtained by taking adjacent camera exposures from the moving airplane so that overlapping views of the ground are obtained. All the principles of stereoscopic vision are then applicable within the overlap area of any pair of adjacent photographs.

12-18. Stereoscopic Fusion Two simple experiments will illustrate the phenomenon of stereoscopic fusion. Near the top edge of a sheet of paper draw four large dots about as shown in Fig. 12-13*a*,

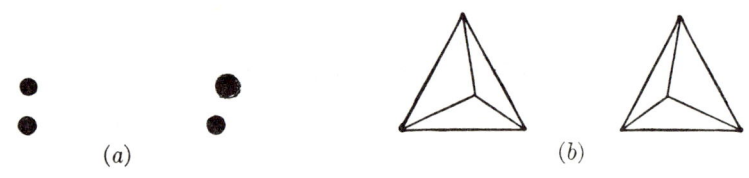

Fig. 12-13. Stereoscopic Fusion.

so that the horizontal distance between the bottom pair is slightly less than for the top pair. Hold a cardboard in the plane perpendicular to the sheet on which the dots are drawn, so that the right pair of dots is seen with the right eye, and the left pair is seen with the left eye. Then, maintaining this relation, focus the eyes on various objects at different distances behind the dots until the two pairs of dots fuse into one pair, one of the focused pairs being above the other. After a short time, as soon as the eyes have become adjusted to this fusion, one dot will appear to be nearer the observer than the other. This will be the bottom dot, if the distance between the bottom pair is the smaller.

A second drawing may be made as shown in Fig. 12-13*b*. Draw two equilateral triangles, whose centers are about 2 in. apart. For each triangle, substitute a different center, moved slightly toward the other triangle, and from each center thus chosen draw lines to the vertices of the triangles. Now observe this pair of figures in a manner similar to that described above, and when fusion of the two images has been obtained, the view will show a solid pyramid with the apex standing above the base.

In fact, with a little practice and without any screen, each pair of images shown in Fig. 12-13, *a* and *b,* may be fused, and the stereoscopic perception will be very definite. Two marginal images will be apparent when observing each pair of images, but in the attention of the observer these can be disregarded.

The accomplishment of stereoscopic fusion is aided with the use of a *stereoscope*. It is an optical instrument that helps the observer in viewing overlapping photographs to obtain the mental impression of a three-dimensional model. The simple lens stereoscope depicted in Fig. 12-14 is being used to view a pair of aerial photographs.

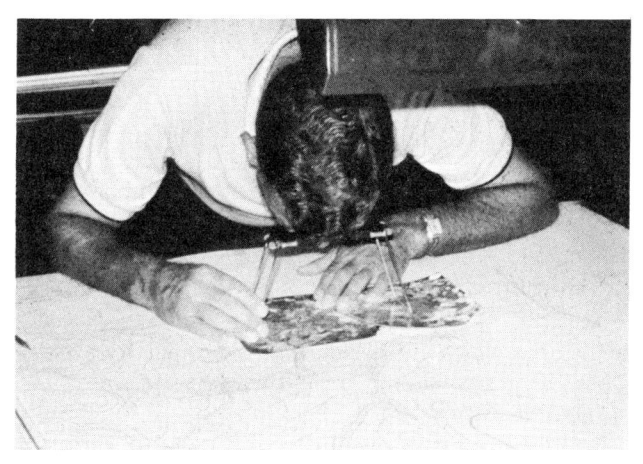

Fig. 12-14. Lens Stereoscope. *Tennessee Valley Authority*

12-19. Parallax in Aerial Stereoscopic Views The ideal conditions for obtaining aerial stereoscopic views of the ground surface are as follows: (a) two pictures are taken that overlap each other; (b) the elevation of the two camera positions, I_1 and I_2, is the same; (c) the camera axis is vertical, and therefore the picture planes (photographs) lie in the same horizontal plane. In Fig. 12-15, the two camera stations are shown at I_1 and I_2, being at the same height H above the datum. It is further assumed that the corresponding picture planes are truly horizontal and that the X-axis (principal line) of each is in the same vertical plane that contains the two

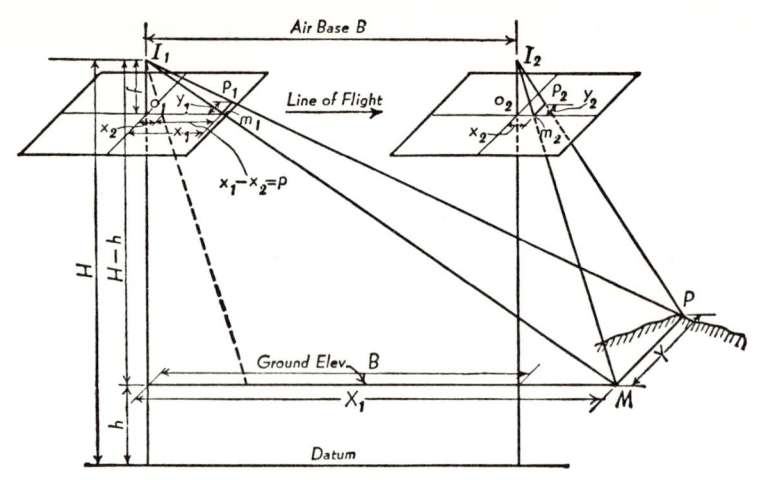

Fig. 12-15. Parallax in Aerial Stereoscopic Views.

camera stations. The line joining the two camera stations is called the *air base*.

The two images of an object P, which appear in both photographs, are p_1 for photograph I_1, and p_2 for photograph I_2. The x-coordinate of this object is x_1 for view I_1, and x_2 for view I_2, and the difference between the coordinates $x_1 - x_2 = p$ is called the *absolute parallax* of that image. It may be noted that for any stereoscopic pair of photographs for which the ideal conditions stated in the preceding paragraphs exist, the parallaxes of the images for all points having the same elevation will be equal.

It should be noted that if point P were located between the two ground nadirs, i.e., to the left of camera station I_2, then the direction of the coordinate x_2 would be to the left of o_2 in the right-hand photograph. The sign of this coordinate would be negative and, therefore, in the equation $p = x_1 - x_2$ it would be added.

An important observation should now be made; there is no parallax perpendicular to the x-axis, i.e., there is no y-parallax in a stereoscopic pair of photographs.

12-20. Space-Coordinate Equations for Aerial Stereoscopic Views
In Fig. 12-16 the relations are shown whereby the three space coordinates X, Y, and $(H - h)$ of any object with respect to the left-hand camera station are obtained from a stereoscopic pair of aerial

[12-21] PHOTOGRAMMETRY 337

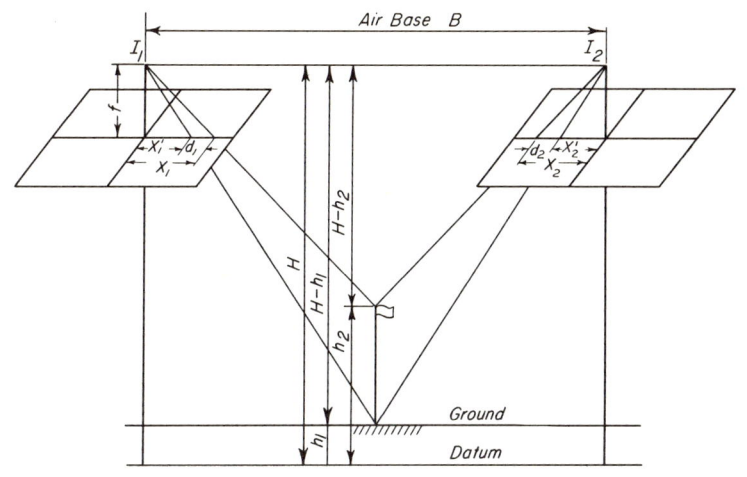

Fig. 12-16. Difference in Elevation by Stereoscopic Parallaxes.

views. From the similar triangles shown, the following equations can be written:

$$\frac{X_1}{x_1} = \frac{H-h}{f}, \text{ but } \frac{H-h}{f} = \frac{B}{x_1 - x_2} \text{ and since } x_1 - x_2 = p,$$

$$X_1 = \frac{B}{p} x_1 \qquad (12\text{-}8)$$

$$\frac{Y_1}{y_1} = \frac{H-h}{f}, \text{ but } \frac{H-h}{f} = \frac{B}{x_1 - x_2}; \text{ hence,}$$

$$Y_1 = \frac{B}{p} y_1 \qquad (12\text{-}9)$$

Also,

$$\frac{H-h}{f} = \frac{B}{p}, \text{ or } H - h = \frac{Bf}{p} \qquad (12\text{-}10)$$

These equations are of fundamental importance; they provide, for the ideal conditions stated previously, the means of determining the three space coordinates, and hence the true relative positions of all ground points within the overlap area of any adjacent pairs of photographs.

12-21. Difference in Elevation by Stereoscopic Parallaxes In Fig. 12-17 are shown the parallaxes and displacements which result

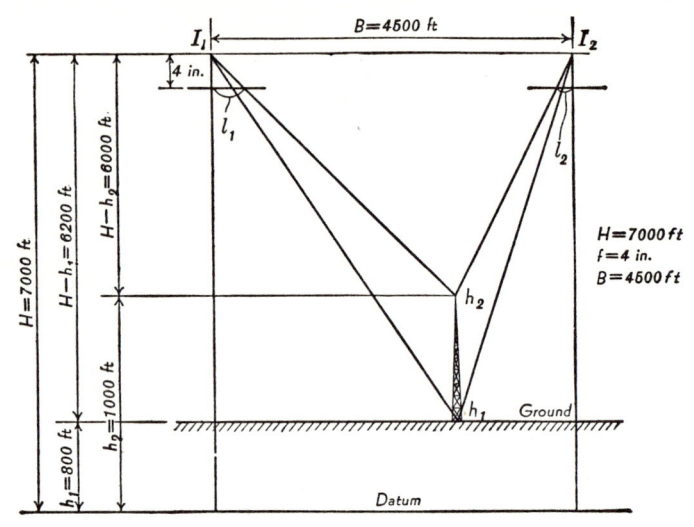

Fig. 12-17. Parallax Computations.

from the perspective rays drawn to the top and the bottom of a flagpole from each of two camera positions, I_1 and I_2. In this drawing the flagpole is assumed to be in the plane of the line of flight. If the pole were at another place not on the line of flight, anywhere in the stereoscopic view, the separate parallaxes x_1, x'_1, etc., would have different values, but the parallaxes p_1 and p_2 would be the same. In this plane of the line of flight, the parallax p_2 of the top of the pole is $x_1 - x_2$, and the parallax p_1 of the bottom of the pole is $x'_1 - x'_2$. The difference in these parallaxes is, therefore, $\Delta p = p_2 - p_1 = (x_1 - x_2) - (x'_1 - x'_2)$. In the figure the value of Δp is shown as $(d_1 + d_2)$. Thus, it is evident that the parallax difference Δp is equal to the displacements $(d_1 + d_2)$ of the top of the flagpole with respect to the bottom measured in the direction of the line of flight.

In Eq. 12-10 it is obvious that the elevation of any object above the datum can be expressed in terms of the parallax value for that point, thus,

$$h = H - \frac{Bf}{p} \qquad (12\text{-}11)$$

Therefore, the elevations of the bottom and top of the flagpole may be written

$$h_1 = H - \frac{Bf}{p_1} \quad \text{and} \quad h_2 = H - \frac{Bf}{p_2}$$

and the difference in elevations

$$\Delta h = h_2 - h_1 = \left(H - \frac{Bf}{p_2}\right) - \left(H - \frac{Bf}{p_1}\right)$$

or

$$\Delta h = \frac{(p_2 - p_1)Bf}{p_1 p_2} \qquad (12\text{-}12)$$

12-22. Parallax Computations The relations expressed in Eqs. 12-10, 12-11, and 12-12 are shown in the computations which follow (see Fig. 12-17).

Given: $B = 4500$ ft; $H = 7000$ ft; $f = 4$ in.; $h_1 = 800$ ft; $h_2 = 1000$ ft; $l_1 = 2.000$ in.; $l_2 = 1.000$ in. *Required:* p_1, p_2, d_1, d_2, and Δh.

$$p_1 = \frac{Bf}{H - h_1} = \frac{4500 \times 4}{6200} = 2.903 \text{ in.} = \text{absolute parallax for } h_1$$

$$p_2 = \frac{Bf}{H - h_2} = \frac{4500 \times 4}{6000} = 3.000 \text{ in.} = \text{absolute parallax for } h_2$$

$$d_1 = \frac{l_1 h}{H - h_1} = \frac{2.000 \times 200}{6200} = 0.0645 \text{ in.}$$

$$d_2 = \frac{l_2 h}{H - h_1} = \frac{1.00 \times 200}{6200} = 0.0323 \text{ in.}$$

$$\left.\begin{array}{l}\Delta p = 3.000 - 2.9032 = 0.0968\\ d_1 + d_2 = 0.0645 + 0.0323 = 0.0968\end{array}\right\} \text{check}$$

$$\Delta h = \frac{(p_2 - p_1)Bf}{p_1 p_2} = \frac{0.0968 \times 4500 \times 4}{3.000 \times 2.9032} = 200.0 \text{ ft}$$

12-23. Effects of Changes in Elevation and Parallax The basic equation $H - h = Bf/p$ expresses the relation between the height of lens $(H - h)$ above any object and the absolute parallax p; and Eq. 12-12 provides the means of finding the difference in elevation between any two objects whose images appear in a stereoscopic pair of photographs. By this method, however, it is necessary to measure the two absolute parallaxes p_1 and p_2 separately, and then to compute the difference in elevation, Δh. If only a few computations are desired, this method is quite satisfactory; but in mapping, where many computations are necessary, it is more expedient to make use of the difference between the two absolute parallaxes, i.e., $\Delta p =$

$p_2 - p_1$. This method makes use of instruments which measure the value of Δp directly by means of micrometer scales, and the fusion of two dots in the stereoscopic view into a so-called *floating mark* (Art. 12-26). It is not within the scope of this chapter to describe these instruments in detail, but the theory will be given so that, with a few directions, any one of the instruments may be used. Reference is made to the *parallax bar* and the *stereocomparagraph*.

The value of Δh may be derived as follows: from Eq. 12-10, Art. 12-20, $h = H - (Bf/p)$; then, by differentiation,

$$dh = \frac{Bf}{p^2} dp \qquad (12\text{-}13)$$

Again from Eq. 12-10, $p = Bf/(H - h)$, and if this value of p is substituted in Eq. 12-13 then

$$dh = \frac{H - h}{p} dp \qquad (12\text{-}14)$$

In this equation $(H - h)$ represents the height of lens above a given ground point, and p is the absolute parallax for the images of the same point in the stereoscopic pair of photographs. This equation states the rate of change in the computed elevation of any ground point corresponding to a small change in the absolute parallax of the point.

If it is assumed that this rate of change remains constant between the two points h_1 and h_2, then the difference in elevation between these points would be computed as follows:

$$\Delta h = \frac{H - h_1}{p_1} \Delta p \qquad (12\text{-}15)$$

in which p_1 is the absolute parallax of the low point h_1, and Δp is the difference in the parallaxes $p_2 - p_1$. But it is known that the relation between dh and dp is not a constant ratio, and therefore when applied to finite values, Eq. 12-15 is an approximate one. This can be proven by the example of the previous article.

Given: $H - h_1 = 6200$ ft, $p_1 = 2.903$ in., and $\Delta p = 0.0968$ in. Then $\Delta h = (6200 \times 0.0968)/2.903 = 206$ ft. The true difference in elevation $h_2 - h_1 = 200$ ft and, accordingly, the error caused by using Eq. 12-15 is 6 ft. For most conditions where the instruments mentioned may properly be used, the error caused by the approximation is not important and may be neglected.

Equation 12-15 is further simplified by making the following assumptions: (1) The approximate quantity H' is assumed to be the average height of the plane over the average elevation of the terrain in the stereoscopic view; and (2) the quantity p_1 is replaced by the quantity b, which is the mean value of the two map air bases; i.e., the mean of the photograph distances between the conjugate pairs of principal points. This latter approximation assumes that the average elevation of the nadir points below the two positions of the plane will closely approximate the average elevation of the terrain. Equation 12-15 then becomes

$$\Delta h = \frac{H'}{b} \Delta p \qquad (12\text{-}16)$$

EXAMPLE 12-5: The average height of the plane over the terrain in a stereoscopic view is estimated to be 5900 ft. The average of the two map air bases is 3.14 in. The measured difference in parallax between two points, h_1 and h_2, is 0.075 in. Then the computed difference in elevation is $\Delta h = \dfrac{5900}{3.14} \times 0.075 = 141$ ft.

12-24. Mosaics A *mosaic* is an assembly of photographs of a given area, matched as nearly as possible and pasted upon a background to form a photographic representation of an area.

For some purposes where the character of the physical features in the terrain is more important than dimensions scaled from a map, a mosaic is more useful than a map.

Because of the large displacements of points near the edges of photographs, best results will be obtained where a large amount of overlap is maintained. Then the edge portions can be trimmed away and only the central areas, comparatively free from displacements, are used. Also, for the same reason, single-lens pictures usually yield a better mosaic than multiple-lens pictures.

Mosaics are classified as either *controlled* or *uncontrolled,* depending on whether or not some kind of control is used in building the mosaic.

An uncontrolled mosaic is suitable where only a limited area is to be portrayed. No ground-control points or other maps are used in the assembly of the prints. A central photograph is chosen, trimmed, and pasted to the base. The other pictures are then matched as closely as possible and pasted in place, until the mosaic is completed.

When some adherence to scalable dimensions is desired, the positions of the prints may be controlled either by ground measurements or by a reliable existing map. Such control must be plotted at the average scale of all the photographs as nearly as can be determined. A better procedure is to rephotograph the pictures and bring them all to the same scale.

Furthermore, where a general slope affects a considerable area, as along the side of a valley, it has been found practicable by rectifying photography to obtain a photograph that represents the sloping area as though it were horizontal. Dimensions scaled from such photographs are practically free from displacements caused by relief.

After photographs have been thus prepared, they may be assembled by either of the methods explained in previous articles.

It must be added that, because of irregularities in the features of the terrain, no mosaic can be regarded as suitable as an accurate planimetric, or topographic, map for purposes of engineering designs involving linear dimensions.

Fig. 12-18 shows a mosaic being constructed.

Fig. 12-18. Mosaic. *Tennessee Valley Authority*

12-25. Photographic Mission Because of the effects of tilt and variations in the height of the lens on the resulting photographs, it is important that the pilot keep the airplane on an even keel and at a constant elevation while exposures are being made. The use of gyroscopes and spirit-level bubbles on the cameras has not proved to be satisfactory, and thus reliance must be placed on the services of well-trained pilots and stable planes capable of operating at high altitudes.

It is also important that each flight shall be straight so that the overlap between adjacent strips shall be uniform and no gaps occur.

Many factors must be considered in the design of a camera suitable for photogrammetric mapping, and underlying all of these is the resulting cost of the photography and the map. Since the cost is largely determined by the number of photographs required, every possible means is used to increase the ground area that can be included within a single exposure, the print of which will be suitable for drawing the map. The principal factors that must be considered are (a) the focal length of the lens, (b) the definition, and (c) the distortion in the photographs.

Scale Factors. The scale of a photograph is determined by the height (H) of the camera and the focal length (f) of the lens; also the area represented by a photograph of given size will be proportional to H and inversely proportional to f. Theoretically, then, to reduce the cost of the survey, f should be reduced and H should be increased. But if f is much reduced, the angle of the field of view becomes so wide that distortions near the edge of the picture are too great. Also, if H becomes too great the images are too small to be defined or recognized. Thus, there are limitations with regard to each factor, and the best design is that which keeps within proper limits with regard to each factor, and which yields satisfactory results at a minimum cost.

Maps are drawn at widely different scales, but, disregarding military surveys, the height at which an airplane can operate satisfactorily is limited between perhaps 5000 and 20,000 ft; hence a considerable range of focal lengths is required from about 4 in. to 20 in.

For topographic mapping it is important that ground parallaxes shall be as large as possible, and these will be obtained best with short-focal-length lenses, whereas longer focal lengths are more suitable for planimetric maps and are sometimes necessary for maps at large scales. Values commonly used are 5¼ in. and 8¼ in., respectively.

Definition Factors. Since airplanes are moving rapidly when exposures are made, it is essential that the shutter speeds shall be fast enough to prevent blur. These speeds, which vary from 1/25 to 1/500 sec, will depend also upon the height of the airplane.

The scale of photographs is frequently increased by enlargement from contact prints, but the amount of increase is limited by the size of the emulsion grains. The finer the grain, the better is the definition; but since the larger grains produce a faster film, a compromise must be made which will yield satisfactory definition together with the necessary shutter speeds.

The resolving power of a lens has a direct effect on the definition obtained. This quality in a lens is that which renders apparent the separateness of very small images. Thus, a good lens will show as two separate images what a poor lens will show as a single, merged, indefinite image. The resolving power of a lens for an aerial camera should be definitely specified and tested.

Distortion Factors. There are many distortion factors that affect photographic images, some of which are inherent in the film and film base rather than in the camera.

Ordinary film base will shrink during the developing and drying process, but low-shrink materials are now available which have reduced this source of distortion.

It is important that the film shall be perfectly flat and precisely in the focal plane of the lens at the moment of exposure. This is accomplished by means of a true surface platen against which the film is pressed, either by a glass plate or by air pressure caused by a vacuum acting through holes in the platen. Obviously, the faulty action of either of these mechanisms will cause distortions in the negatives.

The fiducial marks on the frame of the plate-holder, which fix the position of the principal point on each photograph, must be precisely set and maintained.

Shown in Fig. 12-19 is a precise aerial camera (WILD RC-8) as it appears in the aircraft. Cameras of this type are fully automatic in operation. The accessories include an intervalometer which regu-

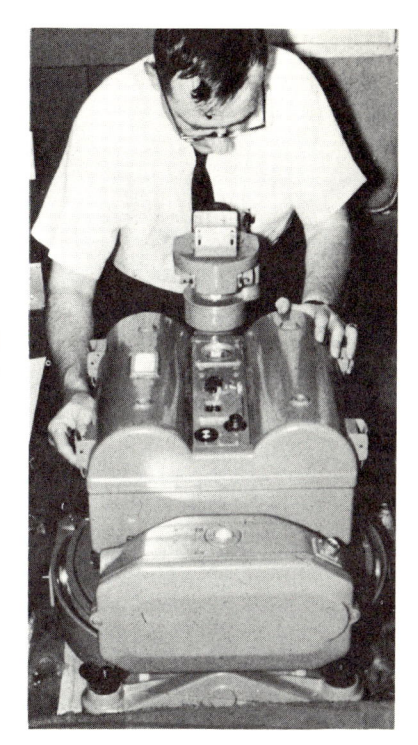

Fig. 12-19. Wild RC-8 Aerial Camera. *Tennessee Valley Authority*

lates the interval between exposures, and a viewfinder which assures proper orientation of the camera with respect to the line of flight.

12-26. Parallax Bar Article 12-23 states that the parallax bar is an instrument with which the difference in parallax, Δp, may be measured with some precision. Such an instrument is illustrated in Fig. 12-20 and is discussed briefly in the following paragraphs.

The measurement of Δp is made by means of a micrometer which moves (parallel with the line of flight) one tiny dot with respect to the other. Each of the two dots is in the center of a separate glass disc, both of which are mounted on a bar. When these two dots are viewed properly under a stereoscope they fuse into a single dot called the *floating mark*. As the right-hand dot is moved toward the left one, the floating mark appears to move vertically upward toward the ground or below the ground surface. Also, as the right-hand dot is moved to the right, the floating mark will appear to move verti-

Fig. 12-20. Parallax Bar and Mirror Stereoscope. *Fairchild Camera and Instrument Corp.*

cally downward. Hence, if the floating mark is apparently placed on the ground at a known elevation and the micrometer scale is read, and is then moved to another point of unknown elevation and the micrometer turned until the floating mark again apparently rests on the ground surface, the difference in the two micrometer readings is a measure of Δp, and therefore a measure of the difference in elevation between the two points.

The parallax bar is an instrument of relatively low precision and can hardly be considered capable of drawing a reliable map. It is more useful as an instrument for measuring spot elevations, for viewing the character of a terrain, and for making rough reconnaissance measurements. It is sometimes provided with binocular lenses to magnify further the field of view.

12-27. Stereocomparagraph The stereocomparagraph, shown in Fig. 12-21, is a stereoscopic plotting instrument which combines the features of the parallax bar, stereoscope, and tracing arm in one instrument and is capable of measuring spot elevations and of drawing planimetric features and form lines with greater precision than can be done with the parallax bar apparatus. Thus, the instrument provides a comparatively inexpensive and rapid means of constructing with some accuracy either a planimetric or a topographic map from vertical aerial photographs. The instrument is designed to be attached to a standard drafting arm.

Fig. 12-21. Stereocomparagraph. *Fairchild Camera and Instrument Corp.*

With this instrument under good conditions, spot elevations can be measured with an average error of about one five-hundredth of the height of lens. Thus, if the height of lens is 10,000 ft, the error in the measured elevation of a point would be ±20 ft. The instrument possesses the advantage that it can be operated by comparatively inexperienced draftsmen, and many operators can work simultaneously for the rapid completion of a project.

12-28. Kelsh Plotter and Wild A-7 Stereoplotter The Kelsh plotter is a precision mapping instrument. The optical and mechanical properties of this instrument permit very definite control of the floating mark, which results in highly accurate maps (Fig. 12-22).

A number of still more elaborate and universally applicable photogrammetric plotting instruments have been designed. One of these is the Wild A-7 Stereoplotter (Fig. 12-23), designed and manufactured by Wild Heerbrugg Instruments, Inc., of Heerbrugg, Switzerland. Instruments of this kind create a view of a floating mark in a true spatial model of the terrain photographed. Either vertical, oblique, or horizontal pictures may be used. The movements of the

Fig. 12-22. Kelsh Plotter with Digital Read-out System. *Tennessee Valley Authority*

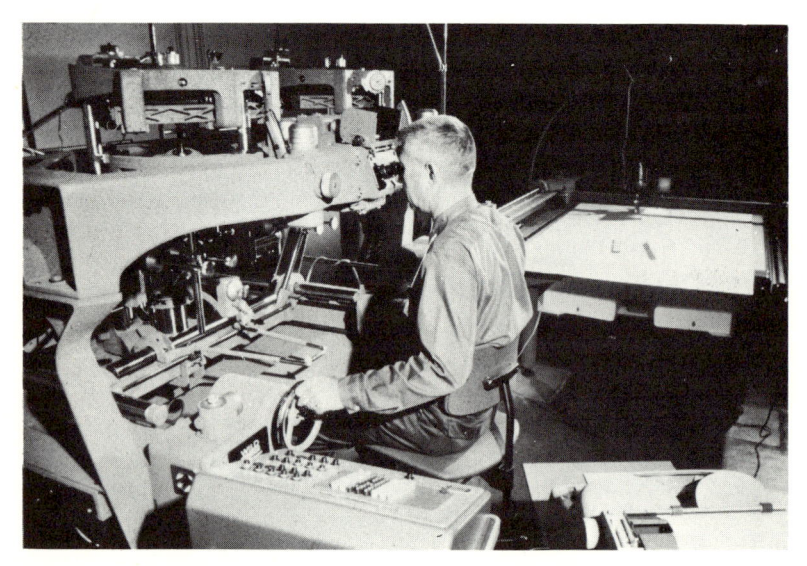

Fig. 12-23. Wild A-7 Stereoplotter. *Tennessee Valley Authority*

floating mark in the three dimensions of the model are effected by three controls simultaneously by the operator. The movements of the mark are translated by extremely precise mechanical devices to the drawing table where the map is drawn to the desired scale.

The two instruments described above may well be used in combination. The A-7 instrument may be used to extend both the vertical and horizontal map control (bridging) between ground-control stations. Then the photographs, thus controlled, can be placed in the Kelsh plotter, where the map is drawn. Thus, several relatively low-cost plotters may be used in combination with the A-7 to speed the work and reduce the total cost.

PROBLEMS

12-1. (a) If the representative fraction of a photograph is 1/15,200, what is the scale?
  (b) If the scale is 1 in. = 1820 ft, what is the representative fraction?

12-2. If the focal length of a lens is 8¼ in., at what height must the airplane fly in order to yield photography for which

  (a) $R = 1/28,000$
  (b) $S = 1500$ ft/in.

12-3. The scaled distance between the images of two points on an aerial photograph is 4.17 in. The ground distance between these points which are at substantially the same elevation is 5120 ft. What is the scale of the photograph?

12-4. Calculate the scale of aerial photography taken at an altitude of 7200 ft above mean sea level with a camera having a focal length of 6.02 in. if the average elevation of the terrain is 750 ft.

12-5. An aerial survey is to be made of an area 60 miles long by 24 miles wide. The scale is to be 1 in. = 2000 ft, the photographs are 9 in. by 9 in., forward lap is 60%, side lap is 30%, focal length of lens is 6.50 in., and average elevation of terrain is 1000 ft. Find

  (a) flight altitude above mean sea level
  (b) ground area in acres covered by each photograph
  (c) number of photographs per flight strip
  (d) number of strips.

12-6. Calculate the displacement of a point because of relief under the following conditions: flight altitude above average terrain is 5200 ft, height of the feature is 650 ft, and distance of its image from principal point of photograph is 1.952 in.

12-7. The image of the top of a radio mast is 4.781 in. from the principal point of a vertical aerial photograph and its measured displacement is 0.356 in. If the exposure was made at an altitude of 10,200 ft above the terrain, what is the height of the mast?

12-8. The average height of an aerial camera above the terrain is 7500 ft and the air base is 4.07 in. If the difference in parallax of two features is 0.118 in., find the difference in elevation between them.

REFERENCES

1. American Society of Photogrammetry, *Manual of Photogrammetry*, 3rd ed., George Banta Publishing Co., New York, 1966.
2. Moffitt, F. H., *Photogrammetry*, 2nd ed., International Textbook Company, Scranton, Pa., 1968.

13 Land Surveying

13-1. Introduction *Land surveying* deals with land boundaries. Hence, it is sometimes termed *boundary surveying*. It is one of the four major categories of surveying and mapping designated by the American Society of Civil Engineers (see Art. 1-17).

A land boundary is a line of demarcation between adjoining tracts of land. It is usually marked on the ground by various kinds of monuments placed specifically for that purpose. A boundary line between privately owned parcels of land is generally called a *property line* or a *lot line*.

A broader and somewhat more classical term embracing all kinds of land surveys is *cadastral surveying*. It is derived from the old Roman "cadastre" which is an official register of the quantity, value, and ownership of real estate. Basically cadastral surveys create, mark, define, and re-establish land boundaries. The term is used primarily to designate surveys of the public lands of the United States as described in later articles of this chapter.

Land surveying comprises the determination of the location of land boundaries and the drawing of plats portraying the subdivision of tracts into smaller parcels. It includes the preparation and interpretation of land descriptions for incorporation in leases, deeds, and other legal instruments. It sometimes embraces the design of streets, sewer and water lines, and of minor drainage structures. Certain surveys (subsequently termed engineering surveys in this text and treated in Chapter 15) such as those required in the planning, design, and construction of highways, railroads, bridges, and other engineering works are not construed to be land surveys. However, since building and right-of-way lines must be carefully delineated before construction operations commence, the role of land surveying is seen to be an important supporting one.

The practice of land surveying probably began as early as 2500 B.C.

Numerous references are made to land surveying problems in Holy Writ (see Ref. 1). The first county surveyor in the United States was George Washington. Abraham Lincoln executed numerous land surveys in central Illinois before entering politics. Today land surveying is a widely recognized professional activity whose practice is regulated by law in many states.

This chapter will provide a brief introduction to the basic concepts of rural and urban land surveys and describe the United States Public Lands Survey System. Because land surveys, their sufficiency and quality, are often subject to judicial review, some consideration will be given to certain legal aspects of boundary surveying.

GENERAL

13-2. Boundaries The function of boundaries is to define areas of jurisdiction. They serve as lines of division ranging from international and state boundaries to simple lot lines in an urban area. Obviously only boundaries that are well marked and described can properly serve the purpose for which they were established.

Property boundaries are needed to define the areal extent of estate rights and its obligations; they are essential in the maintenance of good will and cordial relations in community life. Reliable delineation of such boundaries is a primary function of land surveying. Indefinite property lines can be a source of dispute and controversy especially to resident landowners.

Many of the difficulties associated with poorly defined boundaries and conflicting descriptions of properties can be traced back to inadequacies in the original surveys. These troubles have been further compounded by frequent subdivision, disappearance of markers, blunders in retracement surveys, and the use of old descriptions that were not modified to reflect later surveys.

13-3. Land Transfers Ownership of movable and immovable property by the individual is usually regarded as one of the most fundamental of human rights. *Real property* or land is one of man's most precious possessions. Whether used as a homesite or a means to produce goods, it is traditionally considered as the most important type of property.

The gradual development of individual landownership was not uniform throughout the world. Generically, all real property was owned by the government or sovereign. When the American colonies were established or unappropriated land was settled, title to the land was secured either through grant or purchase from the state. It is to be emphasized that in all cases valid title to land derived from the authority of organized government, whatever its form.

Title to land is the legal basis for ownership. Such titles may be conveyed or transferred in various ways. The most common method of conveying interests in real estate is by a written instrument known as a *deed*. There are various kinds of deeds but all embody a description of the land whose transfer is involved.

Documentary evidence of land transfers is kept in a public depository, such as a county registry of deeds. It is customary to maintain here an alphabetic index, by years, of the names of the sellers (*grantors*) of property and of the purchasers (*grantees*), together with copies of the deeds. The ready accessibility of such records to the public and particularly to land surveyors is a major factor in the perpetuation of boundaries.

Litigation over property lines has stemmed not only from inadequate descriptions but also from legal entanglements of title. The mere execution of a deed to real estate does not invest title in the grantee unless the grantor owns the property. It is necessary to study the chain of title to the property from the time the United States government made the first conveyance. For a consideration title companies will make an examination of the history of ownership of a lot or tract of land and issue a policy guaranteeing the owner's title. The holder of a mortgage on a property is particularly interested in title insurance.

13-4. Kinds of Land Surveys Land surveying deals with the measurement, establishment, and description of the boundaries of real property. Land surveys are made for many specific purposes, such as locating on the ground described boundaries, obtaining data for a deed description, and in connection with wills, mortgages, leases, tax assessments, and condemnations for public use. In general, all land surveys are classified as being either *original* or *resurveys*. They are described as follows:

1. Original surveys are executed to define the size, shape, and relative location of a tract of land whose general boundaries as evidenced by occupancy and use and as delineated by such landmarks as rivers, fences, walls, and trees have been generally ac-

cepted by adjacent owners or adjoiners. It is necessary to mark or monument the corners of the property and to determine the lengths and bearing of the boundaries as the prelude to preparing a description. Original surveys are also executed in order to create new (smaller) parcels of land from a given tract. Such surveys, frequently termed *subdivisional surveys,* are made to subdivide a tract of land according to some plan. The U.S. Public Lands Surveys and private subdivision surveys are examples.

2. Resurveys are executed for the purpose of locating the boundaries of parcels of land already described by existing documents. Such retracement surveys are essential prior to conveyancing by deed from one party to another. It is particularly necessary to ascertain whether driveways are wholly within the boundary lines and whether there are any encroachments upon the property from buildings or structures on the adjoining lots.

Resurveys of satisfactory quality can be both difficult and time-consuming. Frequently there will be no single and unique solution. Honest differences in the evaluation of the same evidence as to line location by two equally proficient and licensed land surveyors can lead to dissimilar conclusions. Final judgment is rendered by the courts.

13-5. Rural and Urban Land Surveys Sometimes land surveys are categorized as being either rural or urban. The legal considerations affecting both types are essentially the same, although there may be differences in the technical aspects occasioned by the size of the tract, its location and topography, survey equipment used, methods employed, corner material utilized, accuracy of the results, and other factors. The problems associated with original and resurveys in both rural and urban settings will be detailed in subsequent articles. Only a few broad comments will be made here.

In general, any land survey will utilize the concepts and measurements of a traversing operation because the lengths and directions of the defining perimeter lines will be involved. When the property lines are obstructed, an auxiliary traverse is executed nearby and sufficient tie measurements are made to the corners so that the lengths and bearings of the boundaries can be calculated as already mentioned in Art. 7-20. Quite obviously there will be drastic changes in the field procedures between a survey of a large and remotely situated tract (see Fig. 13-1) of uncertain economic value and that

LAND SURVEYING

Fig. 13-1. Rural Land Survey. U.S. Geological Survey

Fig. 13-2. Lot Survey. Houston Lighting & Power Co.

of a small parcel of land (see Fig. 13-2) on which a multimillion dollar building is to be erected in the downtown area of a large city. In the latter situation the search for nearby defining property marks will be most thorough, the measurements, although not extensive, exceedingly precise, and the establishment of lot and building line positions highly accurate.

The terms *municipal surveying* and *city surveying* have connotations that are usually not the same as those of urban land surveying. Both city surveying and municipal surveying refer to comprehensive mapping programs culminating in the production of large-scale topographic maps, which are extremely important in city planning. These terms are also employed to indicate various construction or layout surveys, such as those for new streets and public utilities. A cadastral-related aspect of the city surveys is the location of all street lines in view of the fact that such lines represent the boundaries of public property. An unusually complete treatment of city surveying is to be found in Ref. 2.

U.S. PUBLIC LANDS SURVEYS

13-6. Historical In the American colonies the manner of acquiring title to the land from established government authorities was far from uniform. The boundary lines of private property were vague and consisted principally of such natural features as tidewater shorelines, streams, highways, fences, trees, and stones. The tracts thus bounded were irregular in shape except for subdivisions within the limits of towns. There was no general system to serve as a control for the positions of these boundaries and, when they became obliterated with the passage of time and the construction of public improvements, it was often difficult to restore them. Furthermore, the deed descriptions of such land parcels were complex and subject to many errors of interpretation and identification. These undesirable conditions had already prevailed for some time over most of the eastern territory of the nation when the *U.S. Public Lands Survey System* was created by an ordinance passed by the Continental Congress on May 20, 1785. The objective of this legislation was to prevent a repetition of the colonial disorder, confusion, and litigation connected with landownership and to effect a reliable and simple system of describing and identifying land. The ordinance complemented policies of land disposal which were inaugurated in order to stimulate settlement and encourage internal improvements. It was the instrument of law providing for the extension of what is loosely termed the *rectangular system* over the entire *public domain*. This is the vacant land held in trust by the federal government for the people. In 1960 more than 720,000,000 acres still remained in the public domain.

The U.S. rectangular system was never applied to the original thirteen colonies because of the impracticability of a change in land descriptions where so many boundaries would be affected.

The Bureau of Land Management, an Interior Department agency formed in 1946 through the consolidation of the General Land Office, created in 1812, and the Grazing Service, is responsible for surveying the public lands and is charged with the management, leasing, and disposal of the vacant public lands.

13-7. U.S. Public Lands Survey System Although the regulations for the subdivision of the public domain have been modified from time to time, the general plan has remained substantially the same. A synopsis of the main features of the rectangular system will be provided. A detailed account is available in Ref. 3.

The basic scheme provided for

1. The establishment of primary axes designated as *principal meridians* and *baselines* passing through an origin called the *initial point* as shown in Fig. 13-3. Each of the thirty-four principal merid-

[13-7]   LAND SURVEYING   357

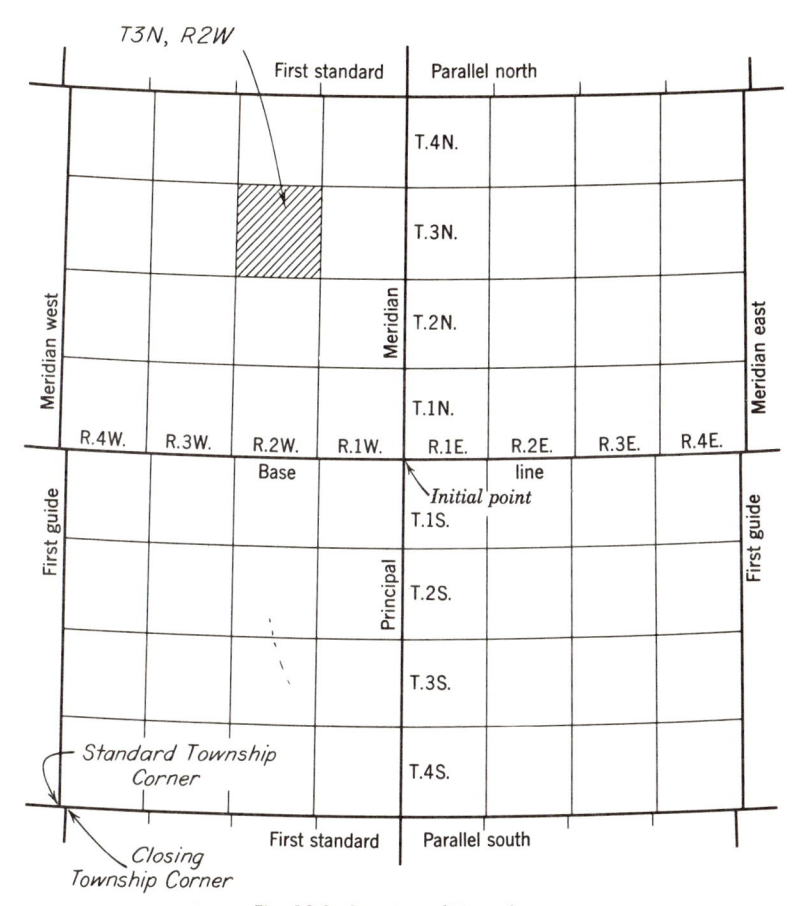

Fig. 13-3. Creation of Townships.

ians governing the entire rectangular survey system of the public lands has been named in some distinctive manner, such as the Third Principal Meridian, which governs surveys in much of Illinois. The geographic extent of the surveys originating from a given initial point is depicted in Fig. 13-4.

2. The establishment of secondary axes known as *guide meridians* and *standard parallels* or *correction lines* at intervals of 24 miles east and west of the principal meridian and at similar intervals north and south of the baseline, respectively.

3. The subdivision of the 24-mile quadrangles (see Fig. 13-3) into *townships*. A row of townships extending north and south is

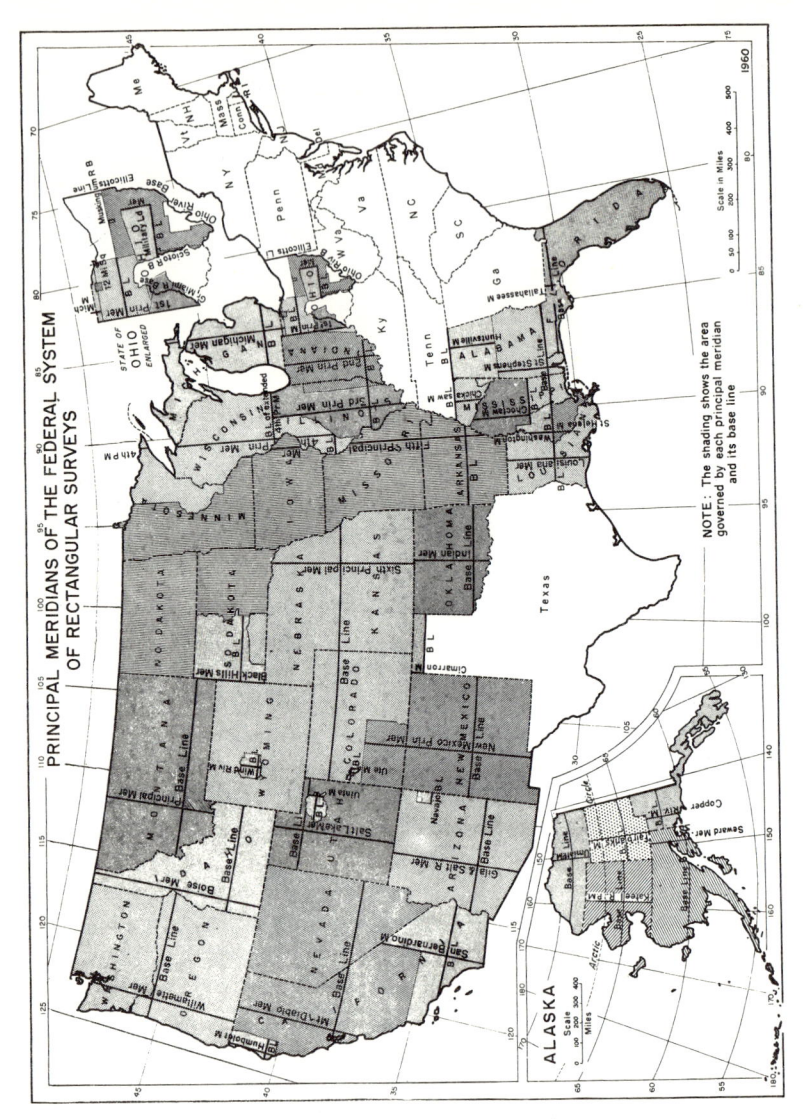

Fig. 13-4. Principal Meridians. *Bureau of Land Management*

termed a *range;* a row extending east and west is called a *tier.* A township is identified by the number of its tier and range and by the designation of its principal meridian.

The township is the basic unit of the rectangular system. A regular township is approximately 6 miles on a side and, therefore, has a nominal area of 36 square miles. In an effort to minimize the effect of convergency of meridians, *standard township corners* were set. These control the meridional subdivision of the land situated between each standard parallel and the next one to the north. *Closing township corners* were established on the baselines and standard parallels where the converging meridional lines from the south intersected them.

4. The subdivision of the townships into 36 units, known as *sections,* by running parallel lines through the townships from south to north and from east to west at distances of one mile. For a detailed explanation of this procedure, Ref. 3 should be consulted. In general, the process of subdividing the township was such as to throw the effect of convergency of meridians and of surveying errors into the northern tier and western range of sections, as indicated in Fig. 13-5.

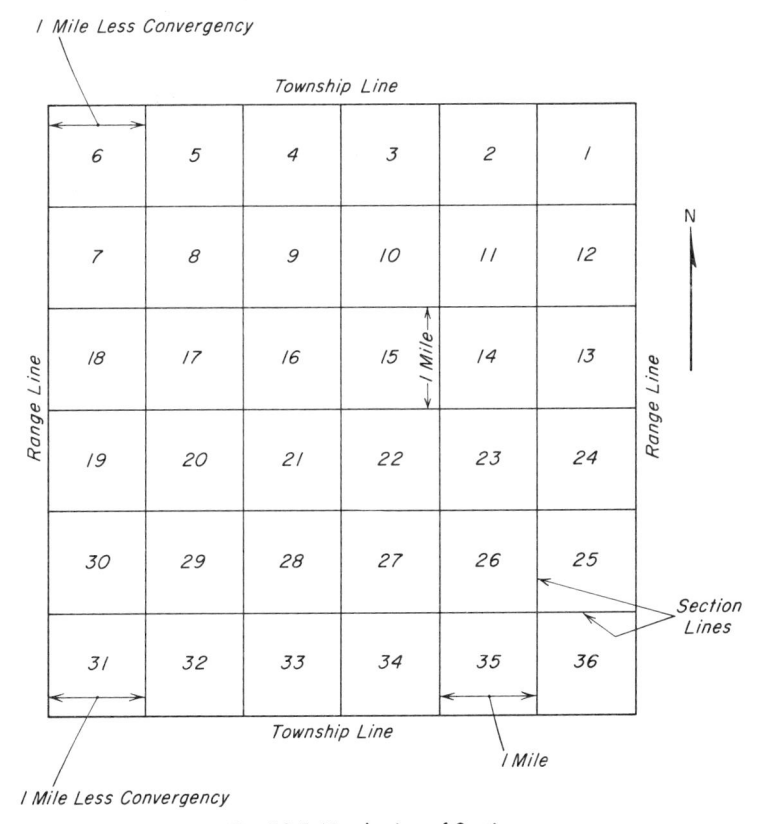

Fig. 13-5. Numbering of Sections.

The regulations prescribed suitable monumentation of the *section* and *quarter-section corners* along all lines. Quarter corners were nominally in most cases at the mid-point of the section lines.

With this step in the subdivisional process the role of the Federal surveyor ended. The subdivision of the section was performed by the private land surveyor.

5. The subdivision of regular sections, containing nominally 640 acres, into quarter sections of 160 acres each by straight lines connecting the quarter-section corners on opposite boundaries. The corner at the intersection of the quarter lines is known as the *interior quarter corner*. The 40-acre parcel (see Fig. 13-6) results from the

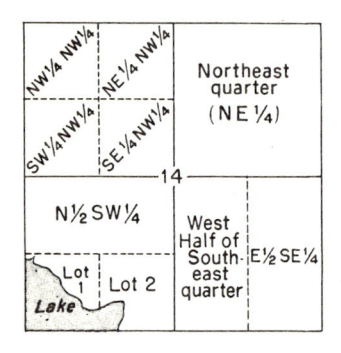

Fig. 13-6. Aliquot Parts of a Section. *Bureau of Land Management*

subdivision of the quarter sections into quarter-quarter sections by the mid-point procedure. Furthermore, the quarter-quarter sections may be subdivided into 10-acre units by continuing this procedure.

The subdivision of sections which are not regular, such as those adjacent to the north and west boundaries of a normal township, is explained in Art. 13-14.

13-8. Extent of the Rectangular System The territory of the United States that has been or is being surveyed under the rectangular system includes thirty states, as shown in Fig. 13-4. Also shown are the several principal meridians with their corresponding baselines. The shading shows the areas governed by each principal meridian and its baseline.

The rectangular system has been modified, to some extent, in each of these states by grants from Congress to individuals, grants to

Indian tribes, or claims to mineral lands; and in the southern and western states by early grants from the governments of France and Spain. Many of the Indian treaties included grants of land to individuals, both American and Indian; also, many individual property rights had become established, either by purchase or occupation, in the states west of the Mississippi prior to the date of the Louisiana Purchase. The United States system has never been applied in the state of Texas, although a modified rectangular system has been used by that state in subdividing tracts as they existed when it was admitted to the Union.

No single description of the United States public lands system will apply to the many states where it has been used, because the system has undergone many changes since it was instituted in eastern Ohio. From time to time the U.S. General Land Office and its successor, the Bureau of Land Management, have issued instructions under which the work of subdivision has been done, but many changes have been made as the work progressed. These instructions in the early years were in the form of circulars and letters to the deputy surveyors, and it is now impossible to recover all of them, although careful search has supplied many. In more recent years the U.S. General Land Office and the Bureau of Land Management have published "Manuals of Instruction" which have prescribed the procedure to be used in the western states where subdivision work has been, and still is, in progress. But it should be noted that the manner of establishing the land lines in practically all of the middle-western and southern states was very different from that described in any recent manual.

13-9. Benefits of the Rectangular System The use of the rectangular system ensures that every aliquot part of a section, whether it contains 1¼, 2½, 5, 10, 40, or 160 acres, has a definite description that is not duplicated. If the parcel of land is described with reference to the correct section, township, range, principal meridian, and state, it will be the only tract in the entire public domain with that particular description. Also, the description will readily indicate to anyone familiar with the system the general location of the parcel.

The simplicity of describing the individual tracts has made the U.S. rectangular system one of the most efficient and practical means ever devised by any nation for land identification. It has reduced the confusion and litigation over land titles, aided in the maintenance of good land title records, and stimulated the orderly settlement of the public domain.

13-10. Instruments and Methods The measurements of directions and distances in the early surveys were very crude. The magnetic compass, Fig. 13-7a, was the universal instrument used for

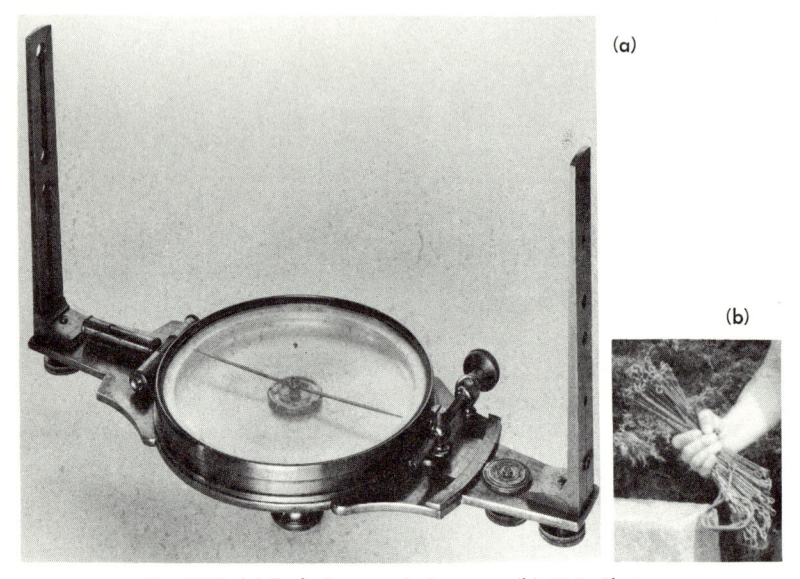

Fig. 13-7. (a) Early Surveyor's Compass; (b) 66-Ft Chain.

establishing the directions of all lines, including the principal meridians and baselines, until William Burt invented the solar compass (1836). At about the same time the transit, in something like its present form, came into use.

Of course, from the beginning, the attempt was made to establish important lines as true meridians or parallels, by taking astronomical observations and finding the declination of the needle, but these observations were crude and the procedure used to prolong these lines was imperfect so that even the principal meridians of the early surveys were not straight lines on the ground.

In running out lines with the compass, peep sights were used and a foresight, determined by a compass bearing, was taken on a range pole or some landmark ahead. The chainmen then measured the distance and the compass was moved to that point, where another foresight was taken and the work continued. Thus, no backsights were used; and, if a tree was found to be on line, the instrument was carried past it, set up by estimation on the line produced, and a new foresight taken. It is evident that any line run by this procedure would not be a straight line, but would have a small angle introduced at every instrument station.

Hence, by this procedure, no line would be established as a straight line for a distance greater than the segment between two adjacent corners; and since, from an early date, the laws prescribed quarter-section corners to be set on all lines at half-mile intervals, it may be said that for all land lines established prior to 1894, when the use of the compass was abolished, a section of land is not a square tract having four sides, but it is an irregular tract having eight sides.

There is no definite date when it may be said that the telescope came into general use, or when the use of backsights was introduced to establish straight lines. The different surveyors employed different instruments and methods, except as these were prescribed by law, or by government instructions, so that at the same time, one surveyor might be using a Burt solar compass, another might be using a transit, and another might be using a peep-sight compass for doing the same kind of work.

The chain used on the early surveys was two "poles," "perches," or "rods," 33 ft long, composed of 50 links. All recorded distances, however, were expressed in units of a four-pole chain, or 66 ft (see Fig. 13-7*b*). It was made of wire, and the many links exposed many wearing surfaces so that the length of the chain increased, perhaps as much as a half-foot in a season. A means of adjusting the length of the last link was provided, and comparisons with a "standard" chain were prescribed, but the crude methods used to deal with the errors of measuring are indicated by the poor instructions which governed the field work.

The Manual of Instructions of 1902 uses the phrase "field chains or steel tapes," which is the first evidence of the use of the steel tape in the work of the government land surveys.

13-11. Conditions Affecting Early Surveys Permissible limits of error in the field measurements of the public lands were prescribed from a very early date, but at first these were vague and indefinite. Subsequent instructions became more specific, providing limits of error for the lengths and directions of the various lines run. The fact that the measurements actually made were often in error far beyond the limits prescribed arose from many conditions other than limitations of the instruments or procedure in use, some of which should be briefly mentioned:

(a) *Land Values.* The common allotment of land to early settlers on the public lands was a "homestead" of a quarter section, 160 acres, at a cost of $1.25 per acre, including certain requirements as to improvements to be made within a specified time. Accordingly, it was believed that high accuracy in the field measurements was not warranted by such a low land value.

(b) *The Contract System.* From the beginning, haste was imposed principally by the condition that the surveys were made under contracts at a specified sum per mile. Different rates were paid depending on the importance of the lines (i.e., whether they were section lines or range lines, etc.) and on the character of the terrain (i.e., whether it was open, wooded, flat, hilly, or swampy). At all times, however, the financial return to the contracting surveyor depended on the speed with which the lines could be run.

(c) *Supervision.* All contracts were executed under oath as to the completeness and accuracy of the results, but supervision and inspection was, at first, totally lacking and was inadequate until about 1880.

It may be added that unsatisfactory work often resulted from pressure for haste exerted by settlers who occupied unsurveyed land and were anxious to have their boundaries fixed; that Indians frequently were hostile; and that competent surveyors were not available in sufficient number for the vast amount of work to be done.

However, having referred to the many sources of error in the original surveys, it should be said that much of the work was surprisingly well done, and in view of the adverse conditions, those surveyors who, over wide areas, achieved such excellent results, have well earned the gratitude and respect of their successors.

13-12. Field Notes and Plats The importance of keeping an accurate and intelligible record of the field measurements of the public land surveys was recognized from the beginning, and careful and explicit instructions have always governed this work. Of course, the notes returned did not always conform to specifications; and either because of adverse field conditions, or because of shoddy or fraudulent work of the surveyors that the lack of supervision and inspection permitted them to do, the actual present field measurements often vary widely from those shown in the notes or on the accompanying plats. However, the original record as shown in the field notes and on the plats is constantly required by present surveyors as they attempt to retrace the lines established by the original surveys; and many serious mistakes are now made by inexperienced surveyors who fail to make proper use of these sources of information. The official government township plats prepared by the Federal land surveyor were generally drawn to a scale of 1 in. equal to 40 chains and showed section boundaries, subdivisions of sections, meander lines (see Art. 13-13), and other information, as well as a limited amount of topography.

Upon the completion of the public land surveys in a state, the original field notes and township plats were transferred to the state. Copies of such notes and plats can be obtained either from the state official, usually the Secretary of State, or from the Director of the Bureau of Land Management, Washington, D.C.

13-13. Meander Lines The traverse of the margin of a permanent natural body of water is termed a *meander line*. Such lines were run as nearly as possible to conform to the mean high-water line. They served the purpose of providing data from which to calculate the areas of tracts of land made fractional by bodies of water. The conditions for and methods of running such lines were prescribed by the General Land Office. In general, they follow the margins of lakes whose areas are as large as 25 acres, and of streams whose right-angle widths are 3 chains or more.

It may be noted here that, although meander lines are shown on the plats and are used to calculate the areas of land lots, they do not constitute property lines. The ownership of property bordering on bodies of water is discussed under the subject of riparian rights (see Art. 13-32).

13-14. Subdivision of Sections Upon the plat of all regular sections the boundaries of the quarter sections are shown by straight lines connecting the opposite quarter-section corners. However, the sections bordering the north and west boundaries of the township, except section 6, are subdivided into two regular quarter sections, two regular half-quarter sections, and four fractional quarter-quarter units usually designated as lots (see Fig. 13-8). In section 6 the plan

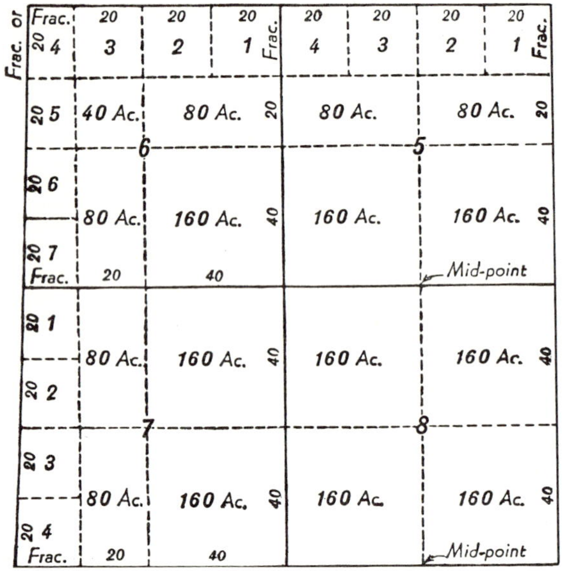

Fig. 13-8. Subdivision of Fractional Sections.

of subdivision will show one regular quarter section, two regular half-quarter sections, one regular quarter-quarter section, and seven fractional quarter-quarter units. The purpose of such subdivision is to place the excess or deficiency in measurements against the north and west boundaries of the township. It is incumbent upon the local surveyor to set the corners defining the subdivisions of a section in the same relative position they would have occupied, as indicated on the plat, had they been set by the federal surveyor.

13-15. Kinds of Corners The term "corner" has two meanings: (1) it refers to the point fixed on the ground by measurement along a line from another established point, or by the intersection of established lines; and (2) it refers to the physical object or marker, which serves as a more or less permanent monument at the given point. In the following paragraphs of this article the term has the first meaning.

Marking Corner. A marking corner is the point established by the survey measurements as the actual location of one corner of a regular tract of the subdivisional system. Such corners are designated by many different terms, depending on the location of each one in the system, such as a quarter corner, section corner, township corner, closing corner etc.

Witness Corner. Where a corner fell in a body of water or other place where it was impracticable to fix the corner itself, a witness corner was set on each one of the survey lines leading to the inaccessible corner and at a known distance from it. Thus, the corner itself was fixed by means of the auxiliary witness corners.

Meander Corners. Every meander line began at one, and ended at another, point on a line of the regular subdivisional system. At each of these two points a meander corner was established. Also, as the meander line was extended, if it intersected any of the regular subdivisional lines, a meander corner was established at each intersection. Of course, by coincidence, a marking corner or a witness corner might also serve as a meander corner.

13-16. Corner Materials The marking of the original section corners included such a variety of materials and methods that it is quite impossible to make any statements regarding them that will correctly describe them for more than a small territory and those set during a short period of time. Regarding these markers, the many Surveyors General issued their own instructions and these varied from time to time. However, it should be said in this connection that the original corner marker is incontestable evidence that fixes the position of a corner. Accordingly, the most satisfactory restoration of the original corners requires the surveyor to know the instructions under which these corners were established and to consult the field notes returned from the original survey.

These instructions were elaborated in later years as experience proved the need for more complete and permanent evidence of the corner locations. Subsurface materials of stones, or broken crockery or charcoal were placed; on the prairie, mounds of earth, or mounds of stones, and pits were made in specified forms to mark the corners. In all cases, the surveyor was instructed to record in his notes the material and method used to mark each corner.

The quality and manner of marking the corners have gradually improved, and the 1930 Manual of Instructions prescribed iron posts, 3 ft long, filled with concrete and having a metal cap (see Fig. 13-9) on which the identifying letters and numerals are stamped. If the corner falls in a roadway, a suitably marked stone is buried as a subsurface mark, and a witness corner is established nearby, outside the roadway.

Fig. 13-9. Land Corner. *U.S. Corps of Engineers*

Fig. 13-10. Douglass Fir Bearing Tree. *Bureau of Land Management*

13-17. Corner Accessories In the attempt to establish the position of a corner as permanently as possible, other objects in the immediate vicinity were used to evidence the location of the corner itself. Such objects are called *corner accessories* and these may be such natural or artificial objects as were used for the corner materials, described above. Of course, they were given specified markings, and careful descriptions were entered in the field notes.

In timbered areas, bearing trees (see Fig. 13-10) described above were universally used. Clearance of the land for lumber or for cultivation, forest fires, and other agencies have destroyed many of these trees, all of them in some regions, but the skillful surveyor may often discover slight bits of evidence which re-establish beyond doubt the position of the original corner.

13-18. Perpetuation of Land Corners From ancient times the evidence of land ownership has consisted of physical objects on or in the ground that mark the corners or boundaries of the tract owned. That condition still prevails. A legal title deed may confer ownership of the land described in the deed, but that which determines what is really owned is the physical evidence on the ground. This condition makes it absolutely necessary for the purchaser to see the physical objects that mark his boundaries if he is to know exactly what land he is buying.

This necessity of maintaining the identity of land corners is beset with many difficulties resulting from weather, building and grading operations, and other agencies that may disturb or destroy evidence of the original location of a corner or a boundary. Perpetuation of corners is one of the major problems confronting the Bureau of Land Management. With the expansion of the national horizontal control network of triangulation and traverse stations and the increasing use of state plane coordinates, it is becoming more feasible to fix the position of land boundary markers by suitable ties to such stations (Fig. 13-11). Even though the marker is destroyed, it can be replaced by means of its known coordinates. It should be understood, however, that no horizontal control survey whether its results are expressed in terms of state plane coordinates or not can have, in itself, cadastral qualities. There are certain definite legal rules, as described in Art. 13-26, which govern the restoration of lost or obliterated section and other corners. However, when the corner monument is missing because of destruction or decay, it should be apparent that the accurately known state plane coordinates of the corner may become the best available evidence of the position of the original corner, and thus provide the most satisfactory means for its restoration or recovery. This does not mean that the use of state coordinates in land conveyancing implies the abandonment of the older methods of describing land, such as by metes and bounds or by parts of a section. State coordinates can be used, however, to supplement the other kinds of land descriptions.

In order to protect the monuments established by the government in surveying the public lands, Congress on March 4, 1909, enacted legislation providing for a penalty for the unauthorized alteration or removal of such monuments. The Act provides:

Whoever shall willfully destroy, deface, change, or remove to another place any section corner, quarter-section corner or meander post, on any Government line of survey, or shall willfully cut down any witness tree or any tree blazed to mark the line of a Government survey, or shall willfully deface, change, or remove any monument or bench mark of any Government survey, shall be fined not more than $250, or imprisoned not more than six months, or both.

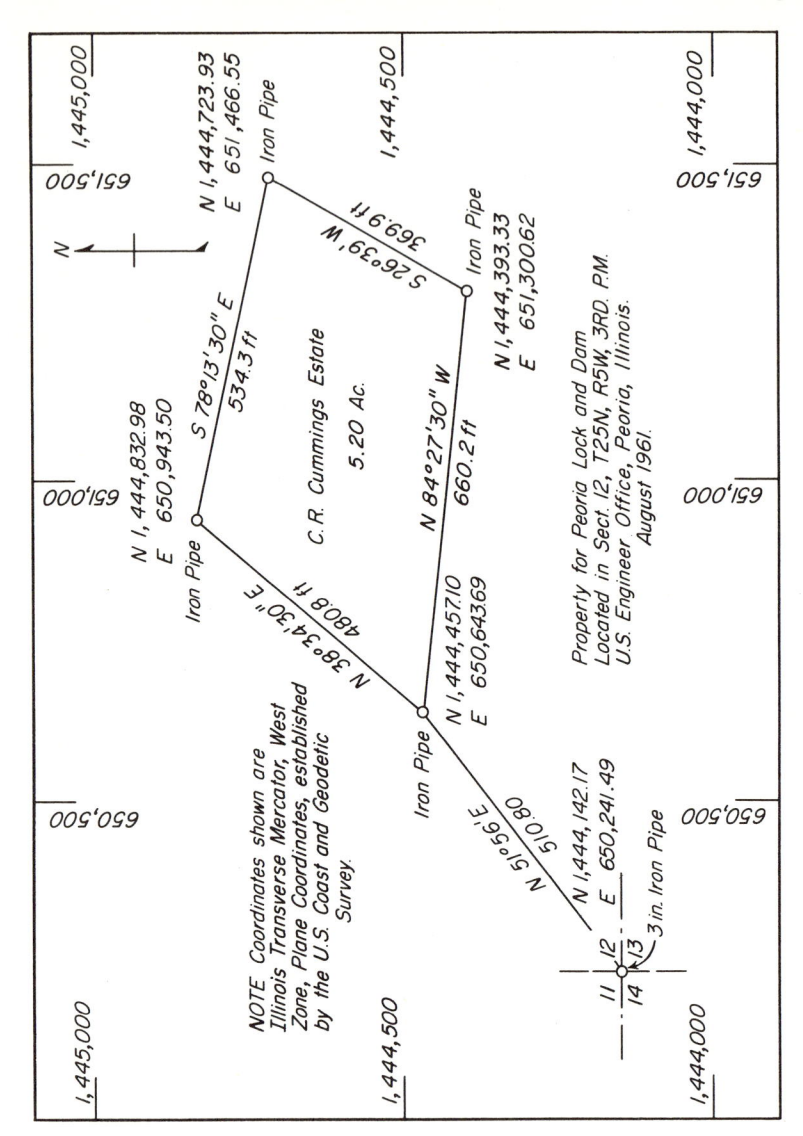

Fig. 13-11. Property Corners Fixed by State Plane Coordinates.

13-19. What Present Surveys Reveal In the preceding pages something has been said of the conditions under which the original surveys were made. Some of these conditions may be listed in summary here: (1) the instruments used, i.e., the compass and two-pole chain, rendered highly accurate measurements impossible; (2) the contract system placed upon the surveyor the incentive for speed rather than accuracy; (3) the lack of training on the part of many surveyors rendered them incapable of interpreting or properly applying the instructions which were intended to govern their work; (4) the lack of supervision or inspection permitted inaccurate and fraudulent surveys to stand.

An important part of the land surveyor's work now is the retracing and restoration of the original lines and corners of the government surveys. In some areas where those surveys have been made in more recent years this work can be done with considerable ease and satisfaction, but in many of the central and southern states where the original surveys are more than a hundred years old, this work is difficult and requires a thorough knowledge of the procedure used in the original work, and resourcefulness and good judgment in dealing with whatever evidence can now be found as to the location of the original corners.

As some indication of what the surveyor may expect to find, a few examples of conditions revealed by present surveys are given below. These are extreme examples, but they will serve to emphasize the fact that the present surveyor must expect to find that (1) no present measurement will agree with that shown on an original plat; (2) undetected mistakes are likely to be present anywhere in the original measurements and in the recorded notes; (3) no line can be regarded as being straight for more than one-half mile; (4) whereas all quarter corners, except those in the north and west sections of a township, are intended to be equidistant between adjacent section corners, such is frequently not the case.

Case I. Figure 13-12 shows two plats of the same section, the first as returned by the deputy surveyor in 1884, and the second as returned by a later survey by the General Land Office.

Case II. Figure 13-13 shows a quarter corner in Wisconsin misplaced 10 chains west of its correct position. However, the original quarter corner was found in place with two bearing trees to witness it, so that the owner of section 27 concluded he could not contest the line in court.

13-20. General Rules Governing Surveys of Public Lands The following list of general rules, which are the result of congressional legislation, is taken from the Manual of Instructions (1947), Ref. 3.

372　　　　　　　FUNDAMENTALS OF SURVEYING　　　　　　[13-20]

(A) As returned by the Deputy Surveyor in 1884

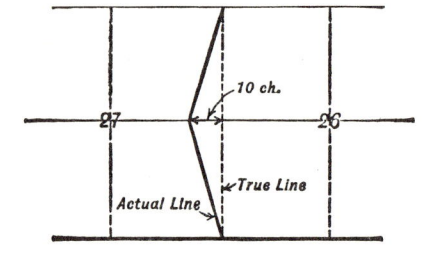

(B) As returned by the General Land Office in 1934

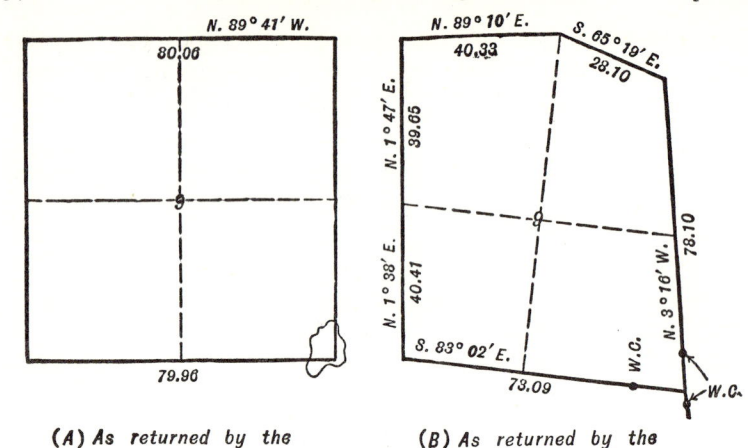

Fig. 13-12. Two Plats of Sec. 9, T. 64 N., R.1W., 4th P.M. in Minnesota

First. That the boundaries and subdivisions of the public lands as surveyed under approved instructions by the duly appointed engineers, the physical evidence of which survey consists of monuments established upon the ground, and the record evidence of which consists of field notes and plats, duly approved by the authorities constituted by law, are unchangeable after passing of the title by the United States.

Second. That the physical evidence of the original township, quarter section, and other monuments must stand as the true corners of the subdivisions which they were intended to represent, and will be given controlling preference over the recorded directions and lengths of lines.

Third. The quarter-quarter section corners not established in the process of the original survey shall be on the line connecting the section and the quarter-section corners, and mid-way between them, except on the last half-mile of section lines closing on the north and west boundaries of the township, or on other lines between fractional or irregular sections.

Fourth. That the center lines of a regular section are to be straight, running from the quarter-section corner on one boundary of the section to the corresponding corner on the opposite section line.

Fig. 13-13. Quarter Corner Misplaced 10 Chains West of the Correct Location.

Fifth. That in a fractional section where no opposite corresponding quarter-section corner has been or can be established the center line of such section must run from the proper quarter-section corner as nearly in a cardinal direction to the meander line, reservation, or other boundary of such fractional section, as due parallelism with section lines will permit.

Sixth. That lost or obliterated corners of the approved survey must be restored to their original locations whenever it is possible to do so. Actions or decisions by surveyors, Federal, State, or local, which may involve the possibility of changes in the established boundaries of patented lands, are subject to review by the State courts upon suit advancing that issue.

RURAL AND URBAN LAND SURVEYS

13-21. Land Descriptions The purpose of a deed to real property is to convey ownership of a tract of land from one party to another and to evidence title in the owner. The purpose of the description which must be a part of the deed is to furnish the information that is necessary to identify the boundaries of that particular tract on the ground, both at the time the conveyance is made and at any future time. It is evident, therefore, that the proper composition of a deed description requires a knowledge both of the law and of surveying, and something of these requirements is indicated in the following paragraphs.

Land descriptions may be grouped into four general kinds as follows: description by reference to (1) natural objects and adjoiners, without numerical data, (2) metes and bounds, (3) the public lands system, and (4) urban subdivisions.

This classification is not a definite one, for many descriptions will include characteristics of two or more of these classes, but it will serve the purpose of indicating the principal features of the various kinds of descriptions.

The *first class* includes those descriptions that refer to such natural objects as a tree, the centerline of a road, the thread of a stream, etc., or a boundary line may be described by giving no information except the names of the adjoining owners.

Such descriptions may contain some numerical data that may aid in fixing the boundaries, but in descriptions so imperfectly drawn the numerical data are likely to be so faulty as to be worthless. No reputable surveyor would return a description of this kind now, but in early times, when land values were small, such descriptions were thought sufficient. The following is an example:

Beginning at a stone in the highway leading from Portsmouth to Springfield, and on the east fence line of land of Levi Brown; thence east by the highway to land of James Green; thence north by land of Green to Spring Creek; thence westerly along Spring Creek to the east fence of land of Levi Brown, thence by land of Brown to the point of beginning; containing 22.40 acres, more or less.

The *second class* includes all pieces of land not of the first class and not regular tracts of the public lands system or an urban subdivision. The description begins by carefully describing the point of beginning and then giving the distance and bearing of each course from one corner to the next, in consecutive order around the tract. The marker at each corner is also described. These are called "metes and bounds" descriptions, of which the following is an example:

Beginning at a point, marked by an iron pin, 584.10 feet North of the S.E. corner of the N.W. ¼ of the S.E. ¼ of Sec. 2, T. 19 N., R. 8 E., 3rd P.M.; thence North 607.86 feet to the center line of Bloomington Road, marked by a cross cut in the pavement; thence in a Northwesterly direction along the center line of said Bloomington Road, 422.41 feet to another cross cut in the pavement; thence South 817.47 feet to an iron pin; thence East 366.74 feet to the point of beginning; containing 6.00 acres.

The *third class* includes all regular tracts within the areas covered by the U.S. public lands system. Thus, a 40-acre tract may be described as: "N.E. ¼, of N.W. ¼ of Section 9, Township 64 North, Range 1 West of the Fourth Principal Meridian."

The *fourth class* includes all regular parts of urban subdivisions. Thus, a tract may be described as "Lot No. 9, Block 4, in Hills and Dales Addition to the Town, now City of West Lafayette, Indiana."

13-22. Requirements of a Valid Description The essential requirements of a valid land description are that it shall be clear, accurate, and brief.

The great advantage of the United States public lands system is that it makes possible the use of descriptions so excellent in these essentials. The identification on the ground of a given tract may not be a simple matter, but that condition does not detract from the excellence of the system insofar as the description of any regular tract is concerned. The same advantages apply to regular lots of urban subdivisions. The land surveying difficulties which apply to such areas, either rural or urban, arise from other sources, which are discussed elsewhere, but the descriptions of these tracts can hardly be improved.

Faulty land descriptions are likely to occur where any irregular tract must be described by "metes and bounds," and some of the considerations which apply to such descriptions will be discussed in the following paragraphs.

Clarity. A land description should be so clear that it is subject to but one, and that the correct, interpretation; this condition should apply not only at the time the description is written, but at any future time. Therefore, the writer must keep the point of view that his description must, if followed explicitly by the surveyor, mark out the correct boundary on the ground either now or at any later time.

So much confusion has arisen from the lack of clarity in deed descriptions that many legal principles have become established as a necessary means of giving the correct interpretation to faulty descriptions. Some of these principles are discussed in the next article.

A first requirement to this end is the accurate use of words and phrases. For example, in giving directions or bearings, the word "north" should be used to indicate due north only, i.e., the direction parallel with the reference meridian of the survey; whereas, "northerly" may indicate any direction in the first or fourth quadrants. Also, the terms, "at right angles to" or "parallel with" should refer only to lines between which the angles are 90° or 0°, respectively. Other examples might be given, but it is evident that only confusion can result from the incorrect or inexact use of words or phrases.

Another source of confusion and ambiguity in land descriptions results from the incorrect use of punctuation marks. Of course, these are often misplaced in copying and recording, but great care should be taken in the sentence structure and phrases used so that the correct meaning will depend as little as possible upon the aid of punctuation marks.

A correct plat is always helpful and sometimes necessary for clarity in descriptions. Hence, it is desirable that a correct plat should accompany any resurvey, and reference made to it in the description, whereby it becomes a part of the deed (see Fig. 13-14).

In order to fix the location of a tract, it is frequently necessary to

Fig. 13-14. Plat of Rural Tract.

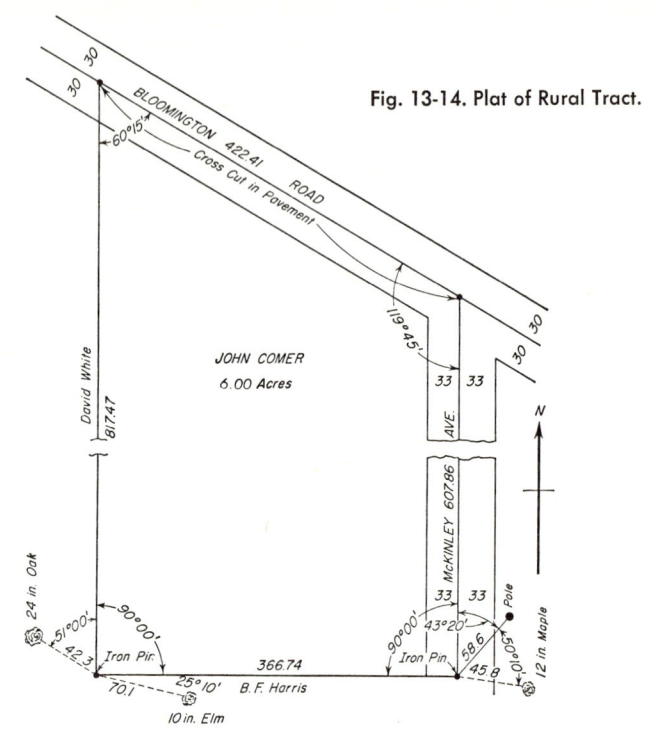

County of Champaign } s.s.
State of Illinois.

I, Godfrey Sperling, County Surveyor of the County of Champaign and State of Illinois, do hereby certify that, at the request of John Comer, I have caused a survey to be made of the following described tract of land: commencing at a point 584.10 ft North of the S.E. corner of N.W. ¼ of S.E. ¼ of Sec. 2, T.19N., R.8E., 3rd P.M., thence North 607.86 ft to the centerline of Bloomington Road, thence in a Northwesterly direction along the centerline of said Bloomington Road 422.41 ft; thence South 817.48 ft; thence East 366.74 ft to the point of beginning; containing 6.00 acres, and marked the corners thereof as shown on the accompanying plat.

Dated this 6th day of May A.D. 19____ (Signed) *Godfrey Sperling*
County Surveyor

refer to some established monument, road, or street line in the near vicinity. Care must be taken that such monument or line is definite, and identifiable upon the ground, and that the "point of beginning" of the survey shall refer to a corner of the tract surveyed and not to the monument to which the tract as a whole is referred.

Accuracy. A correct survey should, of course, precede the writing of any metes and bounds description. Any attempt to scale dimensions shown on such a plat, for the purpose of writing a present description, is sure to result in error and confusion.

For example, assume the following conditions: a city lot is shown on a plat as being 100 ft long and 50 ft wide, its length being in a north and south direction; a previous deed has conveyed out the west 12 ft of this lot; and the actual width of the lot on the ground is 49 ft.

The plat indicates the remainder of the lot as being 38 ft wide, but, as a matter of fact, it is only 37 ft wide. It is evident that it would be incorrect for the lawyer, or grantor, to describe this remainder as the "east 38 ft of said lot" because by the previous deed the remainder is only 37 ft wide, and since a deed cannot convey that which is not owned; therefore with respect to the 1-ft strip in question, the present deed is impotent. This remaining tract should properly be described as "the whole of said lot, except the west 12 ft."

This example also shows the necessity for a present survey, if the grantee is to be saved the disappointment of subsequently finding that his lot is only 37 ft wide instead of the platted 38 ft.

Surely, the grantee should be as greatly concerned that the description of his tract is an accurate one as he is that his legal title is a clear one. A lawyer understands the necessity for a valid abstract of title, but he frequently does not understand the condition that a valid description must depend on an accurate field survey.

The distances and directions of a metes and bounds description should be accurate and complete so that the error of closure of the boundary can be computed. This condition will ensure that no essential data have been omitted and will indicate the precision of the survey, as regards the accidental errors involved. It should in no way depend upon the data contained in the descriptions of adjoining tracts.

Brevity. Brevity is essential in land descriptions because brevity enhances clarity, and it reduces expense and mistakes in subsequent records. It is for this reason that a plat is essential as a part of a description, for it conveys in small space and in legible form information that otherwise would require many words.

In summary, it may be repeated that the final test of any description is whether or not the particular tract described can be satisfactorily identified on the ground.

13-23. Interpretation of Deed Descriptions In making a land survey the surveyor must refer to the deed description, and in many cases the description is faulty because of omissions of essential data or because of conflicting calls. Court decisions have established a few general rules which should govern the surveyor in making his interpretation.

1. The best interpretation is that which most plainly and completely gives effect to the intentions of the parties to the deed, as revealed by all of the evidence available.

2. As regards conflicts between the calls of a description, the order of precedence is as follows: (a) a natural corner or boundary, as a tree or a stream, will stand as against an artificial boundary, as a stake, stone, or a fence; (b) an artificial corner or boundary, if it can be identified, will stand against other conflicting calls as to direction, distance, or area; (c) the corner or line of an adjoiner, if it can be definitely identified, will control over calls for direction or distance or area; (d) in case there is a conflict between the boundary dimensions and the calculated area, the former will prevail over the latter, assuming of course that the boundary dimensions are consistent with the evidence as to the corner monuments.

3. If a description is faulty by reason of any obvious errors, or omission of essential data, the attempt is made to render it valid rather than void. Thus, if a dimension is incorrect by a full tape length, or if the length or bearing of one side of a field has been omitted, and if otherwise the description and evidence on the ground are satisfactory, the obvious omission or mistake will be corrected and the deed will be held valid.

4. If two interpretations are possible, the one that favors the purchaser will be taken.

13-24. General Scope of a Land Survey Land surveys differ in setting, purpose, and complexity and it would be impossible to describe here all phases of even the most representative types. Certain broad procedural principles, however, can be presented which apply to most property surveys. The experience and judgment of the land surveyor will indicate the manner of application of such principles.

The scope of a land survey comprises three basic steps as follows:

1. *Preliminary Study.* This phase involves the collection, study, and interpretation of all available data of record, including old field notes,

subdivisional maps, descriptions of adjoining parcels, legal documents, and other information. In the case of private surveying practice, this step should be preceded by a conference with the client to determine the nature and extent of the professional surveying service desired and to make an estimate of its cost.

2. *Field Survey.* This operation comprises a search for monuments, the delineation of lines of possession, the location of encroachments, and the execution of a closed traverse with all corners durably monumented.

3. *Preparation of Plat and Deed Description.* The plat, which is a special form of a map, portrays the tract boundaries and depicts other information essential to the description and identification of the property.

13-25. Search for Corners At the present time, every survey requires the location, if possible, of existing corners. This may be a simple matter if the work of the previous surveyor was well done and if the lapse of time has not been great. Unfortunately, both of these conditions are frequently adverse, and it requires careful and thoughtful procedure to establish the position of a corner for which much of the original evidence is missing.

The following paragraphs discuss the matter of searching for boundary markers. Although the term corner has been applied particularly to monuments set in the course of subdividing the public lands, it has here the broader connotation of being any kind of property marker.

Some fundamental definitions of corners with special reference to the U.S. rectangular system are as follows:

An *obliterated corner* is one at whose position there are no traces of the marker but whose location may be recovered by the acts of competent surveyors, the testimony of witnesses, or the use of some acceptable record evidence.

A *lost corner* is a survey point whose position cannot be determined in any way except by reference to one or more interdependent corners.

The general principles that affect the legal significance of monuments set in the subdivision of the public domain and the manner of restoring lost corners are stated explicitly in the first, second, and sixth rules governing the resurveys of the public lands as provided by the various acts of congress and reproduced in Art. 13-20.

In general, while searching for any kind of property marker the land surveyor is required: (1) to know the procedure used in the original surveys, (2) to provide himself with all data, i.e., original survey notes and plats, and the records of more recent surveys in the vicinity of the obliterated corner, and (3) to exercise good judgment and discernment in discovering and evaluating all possible evidence pertaining to the obliterated corner.

The kinds of evidence that may be used to restore a corner in rural regions include: (a) the corner itself, (b) accessories, (c) fences, (d) roads, and (e) living witnesses.

Of course the corner itself is the best evidence, and every reasonable effort should be made to find it. If it was a wooden stake, care must be taken that this evidence be not destroyed or overlooked. Having determined by preliminary measurements, as nearly as may be, the probable location of the corner, the search is begun cautiously by slicing off the surface material. If the stake has rotted, its position may be indicated by the rotted wood or discolored soil; or, sometimes in firm soil, the hole formerly occupied by the stake will be found plainly marked. If either of these evidences can be found, the position of the corner is as well recovered as though the stake itself were in place. If it appears that the soil has filled-in over the corner, this condition is usually made apparent by the different quality and color of the filled material. If the corner object is a stone, an excellent tool with which to search is a rod, about ¼ in. in diameter, 3 or 4 ft long, having an oval bead or ferrel, slightly larger than the rod, fitted over one end and a suitable handle at the other end. This rod can be pushed down vertically through 3 or 4 ft of nearly any kind of soil, including a packed roadway, to find a buried stone. By systematic procedure, a considerable area can be investigated much more efficiently by this means in a short time than can be done with a spade.

The records of any bearing trees or other reference objects should be found in the notes of the original survey or on the plats of other previous surveys. In this search, the descriptions of adjacent property should be investigated, since these might contain indispensable information. If any bearing trees or other accessories can be found, often the corner can be restored as satisfactorily as though the corner mark itself had been found.

Fences offer good evidence as to the location of a corner, since it is probable that the fence was built on the line when the corners were apparent. However, fences must be looked on with suspicion, for, in early times, a mutual agreement was frequently made whereby farmer *A* would build a line fence a rod or two over on *B*'s land, and thus secure the use of this strip of land in return for the expense of building the fence. After the lapse of years, the parties who had knowledge of this arrangement may have died or moved away and their successors left in ignorance of the true location of the property line. Or, the fence may not have been a property line at the time it was built and no attempt was made to place it on the line. However, if no such evidence of this kind can be found, it is a reasonable assumption that a fence marks the original land line.

A corner which is known to have been located in the center of a road will naturally be looked for on the centerline between existing fences. But again, the history of the fence lines must be carefully investigated to determine whether or not they have been moved since the corner was placed. Frequently when a road is widened, the change will be made on one side only, and hence the centerline between fences no longer marks the original property line.

Living witnesses frequently can give valuable information as to the location of an obliterated corner. It often happens that men living in the vicinity were present when the corner was set, or have seen it on subsequent occasions, and can recall its exact location. There is a wide difference, however, in the ability of persons to remember the locality of a place they have previously seen, and considerable allowance must be made for this fact. In any case, such evidence must be supported by other evidence, but it may be of great assistance in relocating a corner.

The kinds of evidence used to search for a corner within city limits are much the same as those in rural regions, although conditions are quite different. Building operations and street improvements effect many changes that destroy evidence as to the location of property lines; for this reason every engineer should be most careful to set and reference any corner so as to render it as permanent as possible. Lot corners are likely to be covered with filled-in material that is usually of a quality and color different from the original subsoil. Fences are seldom on the lot lines and frequently are not parallel therewith; also, pavements and sidewalks are often not parallel with adjacent property lines. All these and many other conditions make it necessary that the work be done by a competent and experienced surveyor if satisfactory results are to be obtained.

13-26. Rural Surveys Most surveys of rural property are resurveys or retracements. Proficiency in satisfactorily re-establishing boundaries in their original locations is not easily attained. It is necessary that the land surveyor not only have adequate records of former surveys but that he be able to interpret them correctly. He should be familiar with land surveying practices in the given area, possess the tenacity needed to make thorough field explorations, and be judicially minded as he assesses conflicting field evidence. He should be familiar with the decrees of the local courts in settling boundary disputes.

Some of the difficulties encountered in retracing old boundary lines are due to

1. Faulty original measurements
2. Indefinite point of beginning
3. Indefinite meridian
4. Obliterated corners
5. Inadequate descriptions
6. Enmity between adjacent property owners
7. Errors in transcribing data from field books
8. Inability to traverse along boundaries

Resurvey procedures will be considered for (a) a parcel described by metes and bounds, and (b) the restoration of a lost corner of the U.S. rectangular system.

Parcel Described by Metes and Bounds. When a tract of land is described by a recital of the bearings and lengths of all boundaries, it is said to be defined by metes and bounds. This method of land description is still widely employed, especially in the nonsectionized eastern states. An adequate description of this type starts at a point of beginning that should be monumented and referenced to nearby fixed objects. The direction and length of each course is given in a consistent direction back to the point of beginning. The type of bearing used should be stated. The markers placed at each corner must be described to aid in their recovery and identification. In the re-establishment of the boundaries it is assumed that deed descriptions are at hand, both of the property itself and of adjacent properties;

also, that one or more corners are in place. Beginning with this evidence, the surveyor retraces the existing boundaries, as nearly as may be, measuring the distances, angles, and bearings. These measurements, when compared with those of the deed descriptions or plats, will not agree exactly but will provide a proportion which, it is reasonable to suppose, will apply to the remaining boundaries to be restored. Likewise, a comparison between present and old bearings will indicate a variation, which will serve as a guide in establishing the directions of the boundary courses. By the use of these proportionate measurements the entire boundary is retraced, setting temporary corners as the work proceeds. If the location of these temporary corners seems to be consistent with all evidence at hand, the corners are then established as the permanent corners. However, it is possible that much conflicting evidence will remain after the temporary corners are set, and this is to be harmonized by a further study of all the evidence and by secondary proportionate measurements. In conformance with these adjustments, the final corners are set and carefully referenced. A traverse is then run measuring all the distances and angles, from which the error of closure and the area are calculated. A new description of the boundary is written and a plat prepared to complete the survey. Such a plat is shown in Fig. 13-14.

Restoration of a Lost Corner in the U.S. Rectangular System. Resurveys in sectionized areas of the nation may require the restoration of lost corners. Restoration by proportionate measurement as subsequently described should be employed only as a last resort, i.e., when all other measures have failed to indicate the position of a desired corner. This procedure requires a thorough understanding of the manner in which the original surveys were executed, and it would be impossible here to describe in detail the many conditions that may be encountered. Because of the complications that may arise, the engineer is cautioned about attempting this work without full information regarding the original surveys. However, the procedure for one or two simple cases will be given to indicate the general principles involved.

1. *Single Proportionate Measurement.* Suppose that in Fig. 13-8 the quarter corner on the south boundary of section 7 is lost and the westerly fractional distance of the original survey was 38.42 chains; also that the two section corners are in place, and that the present

measurement between them is found to be 5183.6 ft. The original measurement was 5175.7 ft (78.42 chains), and the lost corner was originally placed at a recorded distance of 2640.0 ft from the east section corner. Let the west, center, and east corners be represented by letters A, B, and C, respectively. Then the relations to be established are as follows:

$$\frac{BC \text{ (new)}}{AC \text{ (new)}} = \frac{BC \text{ (old)}}{AC \text{ (old)}} \quad \text{or} \quad BC \text{ (new)} = \frac{2640.0}{5175.7} \times 5183.6$$

$$= 2644.3 \text{ ft}$$

Accordingly, quarter corner B is to be restored by the measurement of 2644.3 ft from east corner C. This is called a single proportionate measurement.

2. *Double Proportionate Measurement.* Let it be supposed that the section corner between sections 15, 16, 21, and 22, as illustrated at O, Fig. 13-15, is lost, and that the corners at A, B, C, and D are

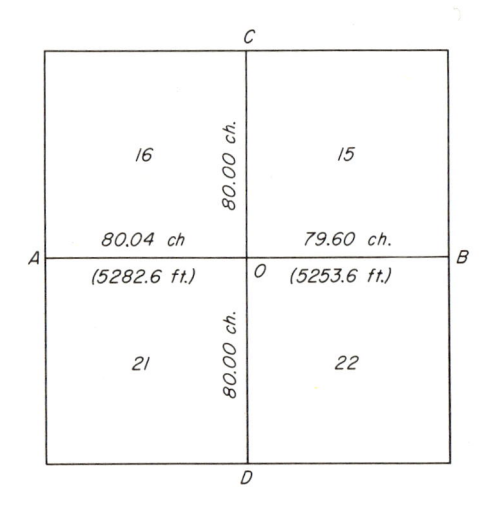

Fig. 13-15. Double Proportionate Measurement.

in place. Then, for the purpose of illustration, it may be supposed that the four quarter corners on the interior section lines are lost. Also, assume that the original measurements are as shown and that the present measurements of distances AB and CD are 10,545.1 ft and 10,571.8 ft, respectively. As above, the relations between the new and old measurements are:

$$\frac{AO \text{ (new)}}{AB \text{ (new)}} = \frac{AO \text{ (old)}}{AB \text{ (old)}} \quad \text{or} \quad AO \text{ (new)} = \frac{5282.6}{10,536.2} \times 10,545.1$$

$$= 5287.1 \text{ ft}$$

Let a temporary stake O' be set at this point.

The original position of O was midway between C and D and, hence, it should be restored in the same relationship, and the distance $CO = 10,571.8/2 = 5285.9$ ft. Let a temporary stake O'' be set at this point. It is evident that the temporary stake O' marks the longitude of the corner, and O'' marks its latitude. Hence, the corner is restored at the intersection of a north-south line passed through O' and an east-west line passed through O''. This procedure is termed double proportionate measurement.

In establishing the lines of the original government surveys it was required that the precision of the measurements should be greater on the controlling lines of the larger units than for the lines which subdivided those units. This higher precision not only was specified in the statutes under which the work was executed, but was recognized in the pay which the surveyor received. Thus, a higher price per mile was paid for establishing a baseline than for a township line; also, a higher rate was paid for a range line than for a section line. Accordingly, it is required in restoring corners that lines which were intended to be more precise than others shall have greater consideration and effect.

For example, suppose the southeast corner of section 12, Fig. 13-5, is to be restored. This corner lies on a range line which was intended to be a straight line for 6 miles and was required to be a more accurate line than the connecting section lines. Hence, to restore this corner, single proportionate measurements are taken between the nearest recoverable corners along the range line, and any conflicting evidence contributed by adjacent section lines will have no effect.

Likewise, a corner on a township line, the northwest corner of section 3, for example, will be replaced by single proportionate measurment along the township line, and any conflicting evidence contributed by adjacent section lines will have no effect.

However, the northeast corner of section 1 is a corner common to four intersecting lines of equal precision, and, hence, this corner will be restored by double proportionate measurements from the nearest recoverable corners in all four directions. Similarly, an interior section corner, as the northwest corner of section 15, will be restored by double proportionate measurements from the nearest corners in all four directions.

It is a general principle, therefore, in restoring corners, that if the lines which meet at a given corner are of different degrees of precision, single proportionate measurements are used, but if the lines are all of the same importance, double proportionate measurements are used.

13-27. Urban Surveys As a city expands, the outlying land is subdivided into lots according to a specific plan. Such subdivisional surveys represent an important function of the land surveyor in urban areas.

In general, the owner or subdivider engages a land surveyor who proceeds to execute a boundary survey of the tract and also a topographic survey in order to delineate the relief and locate the governing features of the terrain. A detailed plan of the subdivision showing all streets, blocks, lots, and their significant dimensions, as well as such other pertinent information as easements and building setback lines, is submitted to the appropriate planning body and municipal authority for approval.

Upon the completion of the layout surveys and the adequate monumentation of all property lines the drawing or plat is recorded by the county recorder and officially becomes part of the public records. For a thorough introduction to platting laws and the characteristics of subdivision plats the reader is advised to consult Ref. 1.

Frequently the urban land survey involves the retracement of the boundaries of a single city lot for title insurance and other purposes. Following the completion of the field work, the surveyor may prepare a plat similar to that shown in Fig. 13-16.

13-28. Survey for a Deed It will be helpful to summarize the general procedure for making a land survey. A parcel of land, either rural or urban, is to be conveyed and a deed description is needed. The procedure is somewhat as follows:

1. The descriptions of the tract to be surveyed and of adjacent tracts are obtained and carefully examined.

2. The corners of the tract to be sold are established by resurvey if necessary, as explained in previous articles.

3. A resurvey of the tract is made in which the following data are obtained: (a) the length of each side of the parcel, (b) the angles

Plat
of
Lot 118 in University Heights Addition to City of Urbana

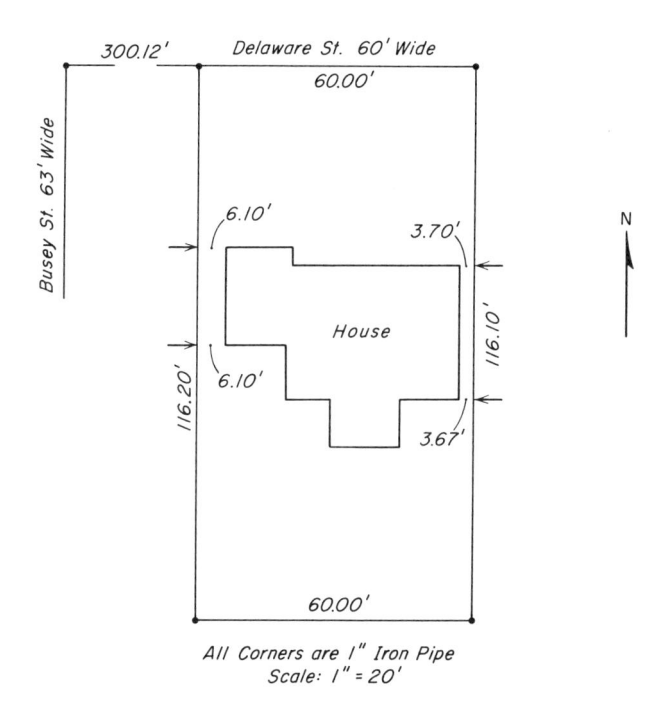

All Corners are 1" Iron Pipe
Scale: 1" = 20'

Signed and Sealed this 10th Day of July, 1968

(SEAL)

John S. Doe
Licensed Land Surveyor, Ill. 8256
Urbana, Illinois

Fig. 13-16. Plat of City Lot.

at each corner, (c) the kind and position of accessories at each corner, (d) the calculated bearing of each side, referred to true north if possible, (e) the names of adjacent owners.

4. From these data the area is calculated, a plat is drawn, and a new description is written.

The plat generally will show the following; (a) the length of each side, (b) the angle at each corner, (c) the calculated area of the tract, (d) the kind of markers set and the references to the accessories at each corner, (e) the names of adjacent property owners, (f) the positions of buildings, roads, sidewalks, or other permanent objects that would help to perpetuate the property lines, (g) a title and meridian, (h) a description of the property lines, (i) a surveyor's certificate.

13-29. Excess and Deficiency Every land survey, either urban or rural, requires the retracement and the remeasurement of property lines and, because of the inherent errors in all observations and possible mistakes, the recent measurements never agree exactly with the original; therefore, there is always some excess or deficiency to be adjusted. The discrepancy between a recent and a previous measurement may be small and the adjustment a simple one; but frequently the discrepancy is considerable, and the adjustment complex, especially where the line has been divided into a number of segments. In such cases the aid of established legal principles is helpful.

The general rule is well stated by the Supreme Court of the State of Nebraska as follows:

On the line of the same survey, and between remote corners, the whole length of which is found to be variant from the length called for, it is not to be presumed that the variance was caused from a defective survey of any part, but it must be presumed, in the absence of circumstances showing the contrary, that it arose from imperfect measurement of the whole line, and such variance must be distributed between the several subdivisions of the line in proportion to their respective lengths.

This rule has a common application where urban property is subdivided into lots and blocks, with dimensions shown on a plat and where a given parcel is described by its lot number.

The rule applies to all of the lots in a given block, even though one at the end may have a frontage dimension different from all of the others. It is sometimes supposed that such a lot, being irregular, should be considered a remnant and be assigned all of the excess or deficiency found within the block. But, unless the dimension of such a lot is omitted, or other modifying conditions apply, it is held that any discrepancy shall be apportioned among all of the lots within the block.

The rule applies also where a tract described as containing a given acreage is subdivided by stating the number of acres in each subdivided tract. Thus, if an owner wills his farm of 160 acres to his two sons, giving to one the north 60 acres and to the other the south 100 acres, and if a survey shows the whole tract to contain 180 acres, it will be divided between the sons in the proportion of 60 to 100, each receiving his proportionate share of the discrepancy.

The rule also applies commonly to the relocation of lost corners in the U.S. rectangular system, where proportionate measurements are to be made between the nearest adjacent existing corners.

The general rule that any excess or deficiency shall be apportioned throughout the tract in which it occurs does not apply in some cases. Four such cases are stated below.

1. If all of the lots in a block are given dimensions except one. For such a case it is plainly evident that the irregular lot was intended to include whatever remnant was left.

2. If streets intervene between the blocks of a subdivision and if the streets are monumented and have been used for some time, any discrepancy found in one block may not be apportioned among other blocks across street lines, because street lines used by the public soon become fixed.

3. If a tract of land has been partitioned by separate deeds by metes and bounds at different times. In this case, the tract last conveyed will be apportioned any excess or deficiency that may be found.

4. If erection of buildings and other acts of occupation show that ownership has been claimed according to platted dimensions.

13-30. Adverse Possession *Adverse possession* is a legal term which applies to the condition where the property line between adjacent owners and the titles of the adjacent properties are fixed by occupation and use of the land, as opposed to the descriptions of the properties in the deeds of ownership. Thus, it may be supposed that *A* and *B* are owners of adjacent tracts of land where a common boundary line has the same description in each of the deeds of titles vested in both *A* and *B*. However, when a fence was built, supposedly along this line, it was built over on *B*'s land, say 10 ft. Then if *A* has occupied all of the land up to the fence, believing it was his own, for the statutory terms of years (20 years in Illinois), and meeting other necessary conditions, he has acquired title to the land thus affected; and the fence, erroneously placed, has become the true boundary. This means of transfer, by occupation, of ownership and title from *B* to *A* is termed *adverse possession*.

Conditions under which title is acquired by adverse possession are as follows: occupation must be (a) actual, (b) open and notorious, (c) exclusive, (d) hostile, and (e) each of the foregoing conditions must be continuous for the statutory period fixed by the state. Each of these terms needs some further definition.

Actual occupation means that written or oral statements claiming ownership are not valid. There must be actual and visible evidence of occupation on the ground.

Open and notorious occupation can be evidenced in different ways, such as a fence, cultivation of the ground, erection of buildings, posted boundary markers, etc.

Exclusive possession means that it cannot be shared with anyone. Thus, a driveway used in common with someone else could not be used as evidence for a claim under adverse possession.

Hostile possession does not necessarily imply ill will, but it means there must be no agreement between adjacent owners, or knowledge by the claimant of the true conditions. Accordingly, he will consider as a trespass any undue entry upon the property he claims.

Continuous occupation requires that there shall be no lapses or gaps during the statutory period of possession. Thus, occasional acts of ownership will not be sufficient to establish title, nor can two or more different persons at different times exercise the rights of ownership. A possible exception to the last requirement is that in which the title to property passes through successive owners; i.e., if the property is sold to a new owner and he meets all the other conditions of adverse possession, his period of occupancy may be added to that of his predecessor. This procedure is called *tacking* and may extend through two or more adverse claimants to meet the statutory period necessary to acquire title. However, there must be no gaps between the successive periods of occupation.

In reviewing the conditions necessary to effect ownership under adverse possession, it is evident that this change of ownership is conditioned entirely on the bona fide intent and belief of the claimant. Thus, in the case of the two owners A and B mentioned above, it was supposed that each one believed the fence to be in its correct location. It frequently happens, however, that both owners of adjacent properties are not certain about the correct location of the boundary line. In such a case, if there is a mutual understanding between them that the boundary fence in place may not be the true line, but they agree to use it as the boundary, then as long as these conditions exist neither owner can claim ownership to the fence under adverse possession. But if at a later time A sells his property to another person C, who is not informed of the uncertain location of the fence and who believes it to be the true boundary, then C can claim ownership and title to the fence under adverse possession if he occupies his land according to all of the necessary conditions stated above. In the latter case the fence has become the true boundary line.

As between the public and individuals, it is a general principle that, although the government, either federal or state, may acquire title by adverse possession, the reverse is not true, i.e., the statute does not run against the state. Hence, individuals or private corporations cannot acquire public property by means of adverse possession.

13-31. Highways and Streets An engineer or surveyor frequently has occasion to deal with public improvements, such as pavements, sidewalks, and sewers, to be constructed in streets or highways, or he is required to establish property lines along a highway or street. He should, therefore, be fully informed of the conditions under which title is held for such property.

It is a common principle of law that title to property bordering on a highway or a street extends to the center of such highway or street unless there exists an explicit restriction. Such a restriction might be imposed by a metes and bounds description, or by the ordinance or statute under which the street or road was laid out.

Some statutes and city ordinances provide for no greater interest in a street by the public than an easement or right to its use as a thoroughfare, and title remains in the owners of the adjacent lots. Also, a street or a highway opened by condemnation proceedings may be governed by statutes that provide for an easement only by the public. Accordingly, when the street or highway is abandoned and the public can show no further need to use it as a thoroughfare, the right to use it reverts to the adjacent property owners.

However, some ordinances and statutes prescribe that title to property used as streets shall be held by the city, and similarly, that title to the right-of-way for improved highways shall be vested in the state.

It should be noted that a highway may become established through use only, and without having been formally laid out. If the public becomes accustomed to travel along a roadway until its use is deemed a necessity and if its use has been continued for a statutory period of time, the roadway becomes as permanently established as by any other means. This procedure has many phases similar to the principle of adverse possession, although the period of time is usually much less for the case of fixing a roadway than for establishing title by adverse possession.

13-32. Riparian Rights Riparian rights are those vested in an owner by the condition that his property borders on the bank or shore of a stream or body of water. Such rights are often of great value, as when they endow owners with many shore privileges, including the right to construct docks and wharves. Because of these values occurring from riparian ownership and because of the irregular and changeable nature of such boundary lines, it is important that the surveyor be well informed concerning the laws and customs that govern the establishment and maintenance of these lines within the state in which his survey is located.

13-33. Riparian Boundaries *Nonnavigable Streams.* In dealing with streams as boundaries, the procedure depends frequently upon whether or not the stream is "navigable." In early instructions of the U.S. Land Office regarding the running of meander lines, it was prescribed that a river was to be considered as "navigable" if the water surface had a width of three chains. Obviously, the width of the water surface for a given stream varies widely with the stage of flow; accordingly, it is quite impossible in many situations to say whether a stream is to be regarded as navigable or nonnavigable. Perhaps there is no better interpretation at present than to regard a stream as navigable if it was meandered in the original survey, and as nonnavigable if it was not meandered.

Where a nonnavigable stream serves as a boundary line, it is common law that ownership extends to the thread, or center, of the stream, but the courts differ somewhat in defining the term "thread of stream." This is sometimes defined as the line midway between the usual shorelines, regardless of the position of the main channel, but the more common interpretation defines it as the center of the main channel.

Navigable Streams. The riparian rights of an owner of property along a navigable river are fixed largely by the statutes of the state in which the property is situated. In regard to the extent of such boundary lines, laws vary from state to state between three limitations, namely: (1) to the high-water mark, or bank, (2) to the low-water mark, or (3) to the center of the stream.

The "high-water mark" or "bank" may be defined as the line where "the presence and action of water is so common and usual, and so long continued in all ordinary years, as to mark upon the soil of the bed a character distinct from that of the banks."

The low-water mark may be defined as the line to which a river usually recedes at its lowest stage, unaffected by drought.

It should be added that meander lines of the public lands system are not property lines, and, unless they are specifically designated as such in a deed description, they do not limit riparian boundaries.

Ponds and Lakes. The limits of property bordering on a pond or lake are fixed by the laws of the state in which it is situated. In a few states, ownership is limited by the shoreline, but in most states, if a pond or a lake is nonnavigable, title to property bordering on it extends to the center of it. If the land tract includes all of the pond or lake, the bed is included in the title. However, if a metes and bounds description uses the expression "along the east shore of said pond" or "thence by the edge of the lake," ownership extends to the shore only.

If a lake is navigable, title of riparian owners extends to the shore, or to the center, according to the laws of the state in which it is located.

13-34. Alluvium, Reliction, and Accretion *Alluvium* may be defined as the change in the location of a shoreline by the gradual and imperceptible deposition of soil so as to increase the area of the contiguous land.

Reliction refers to the gradual and imperceptible recession of a shoreline due either to the rising of the shore or to the subsidence, or drying up, of the body of water.

Accretion results in either of the above cases, and it is the general rule that accretion so gained is lawful and the boundary of property so affected will change with the movement of the shoreline.

Such accretion, however, has its adverse counterpart. If gradual and imperceptible erosion takes place, the riparian owner will suffer a loss by the change in the shoreline. Likewise, if the shore subsides, or the body of water rises, gradually and imperceptibly, the contiguous property will be inundated and the owner's acreage will thereby be diminished.

13-35. Avulsion *Avulsion* refers to the sudden and perceptible change in a shoreline, due usually to a river changing its course during a flood or freshet. It is the general rule that avulsion effects no changes in property lines. Thus, if a river changes its course, or if its banks are suddenly eroded during a flood, property lines will not be disturbed thereby. This rule applies to river boundaries between states as well as to private property lines.

13-36. Extension of Riparian Boundary Lines The surveyor is frequently required to extend the land lines of riparian owners in accordance with their rights. This problem may be simple or complex, depending upon conditions, and calls for good judgment and an understanding of the procedures that have been established in similar cases.

(a) *Rivers.* Where property rights extend to the center of a river, the general rule is to prolong each boundary line to the high-water mark or bank and then extend it in a direction perpendicular to the centerline of the river. Thus the west boundary of Lot 1, Fig. 13-17, is prolonged to A and then extended to B, direction AB being perpendicular to the centerline of the river.

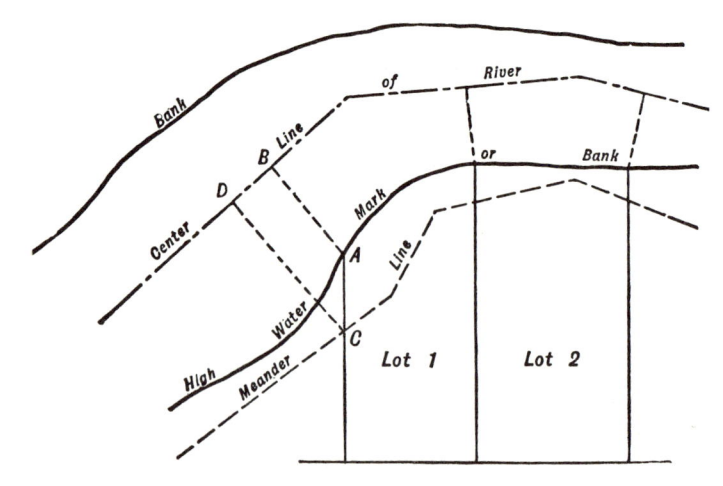

Fig. 13-17. Extension of Property Lines at a River.

In the public lands system when a bank has been meandered, it is sometimes held that the boundary line should be extended from its intersection with the meander line to the centerline of the river, as from C to D, but most court decisions are contrary to this view and specify the bank as the proper place at which to change the direction of the boundary line.

(b) *Accretion.* Where boundary lines are to be extended to include an area added by accretion, many complex situations arise, but a general rule is indicated by Fig. 13-18, where A to F repre-

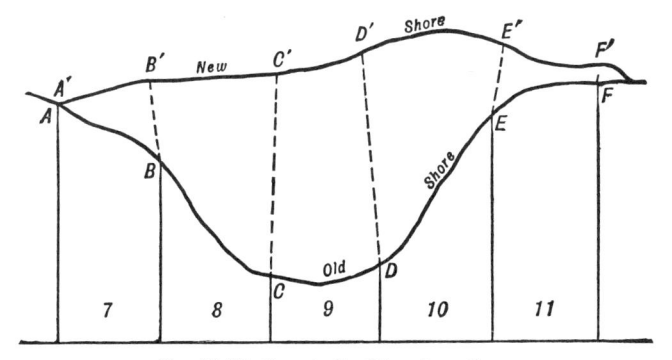

Fig. 13-18. Lines to Partition Accretion.

sents the old shoreline, and A' to F' the new line to be apportioned to lots 7 to 11. This is done by dividing the new shoreline so that each new lot is to the whole new shoreline as each old lot line is to the whole old shoreline. Thus, distances AB, BC, CD, DE, and EF are measured, and the distance A' to F' is found. Then $A'B'$ is to $A'F'$ as AB is to AF; parts $B'C'$, $C'D'$, $D'E'$ and $E'F'$ are determined in a similar manner.

(c) *Lakes.* When property rights include the bed of a lake and where the shape of the lake provides a fairly definite axis, or centerline, the boundary lines are extended from the shore perpendicularly to the centerline (Fig. 13-19). But if the lake is circular in shape or has a round area as in the figure, then the boundary lines are extended from the shore to the center of the rounded part. The courts have sometimes called this the "pie-cutting" method of subdivision.

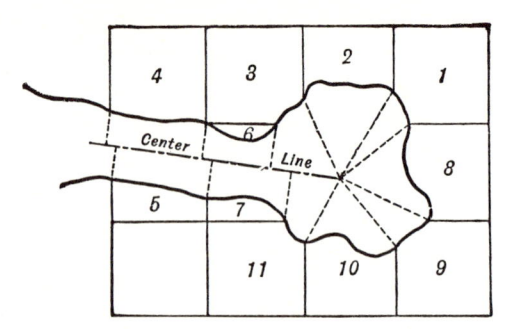

Fig. 13-19. Extension of Riparian Lot Lines for a Lake.

13-37. Legal Authority of the Surveyor The surveyor should understand clearly that, in the event of a dispute as regards the location of a corner or property line, he has no judicial authority. If he is a competent surveyor he will gather the evidence and interpret it in such a manner as usually to bring the parties to agreement out of court. However, if no agreement can thus be reached, the surveyor has no authority to impose a settlement. This is strictly the prerogative of the court, and the surveyor serves only as an expert witness in the case.

Also, to avoid unnecessary litigation and to prevent mistakes in present surveys arising from ignorance of the legal principles involved, the surveyor is required to be well informed in the laws relating to land surveys within the state and jurisdiction where the property is located. This fact is evidenced by the preponderant weight in examinations for the land surveyor's license which is given to the legal aspects of surveys as compared with the physical measurements required.

13-38. Liability A surveyor or a civil engineer belongs to a learned profession and, although that condition brings him certain privileges, it also imposes the responsibility of performing his duties with competence and honesty. If he is negligent in his work he may be held liable for damages suffered by his client. His responsibilities in this respect are similar to those of a doctor or a lawyer. Thus, if he is employed to make a survey for a building site and is told the character of the building to be erected, and his survey is carelessly done without regard to the circumstances, he may be held liable for damages to his client resulting from an erroneous location of the building.

He is not required to achieve absolute accuracy in his results, but he is obliged to exhibit that degree of care, prudence, and judgment which may properly be expected of a surveyor or a civil engineer under similar circumstances.

13-39. Registration of Land Surveyors In order to safeguard the public interest, the practice of land surveying in many states is limited to persons who have demonstrated their professional competency. The general requirements that must be satisfied are graduation from an accredited engineering curriculum followed by at least four years of approved land surveying experience and the successful passage of a two-day examination.

REFERENCES

1. Brown, C. M., and W. H. Eldridge, *Evidence and Procedure for Boundary Location,* John Wiley & Sons, Inc., New York, 1962.
2. Surveying and Mapping Division, *Technical Procedure for City Surveys,* Manual No. 10 (Revised), American Society of Civil Engineers, New York, 1963.
3. U.S. Bureau of Land Management, *Manual of Instructions for the Survey of the Public Lands of the United States,* Government Printing Office, Washington, D.C., 1947.
4. U.S. Bureau of Land Management, *Restoration of Lost or Obliterated Corners and Subdivision of Sections,* Government Printing Office, Washington, D.C., 1963.

14 Cartographic Surveying

14-1. Introduction *Cartography* may be defined as the science and art of expressing graphically, by means of maps and charts, man's knowledge of the earth's surface and its varied features. *Cartographic surveying* is, therefore, that branch of surveying that includes all the operations leading to the production of topographic maps and nautical charts.

The predominant characteristic of a *topographic map* is that it portray by some means the shape and elevation of the terrain. The most common way of representing land forms is by contour lines. In addition to depicting relief, topographic maps show drainage features and the works of man. A variety of symbols and specific colors are utilized to promote clarity of cartographic expression.

A *nautical chart* delineates the submarine topography of a given water area by means of depth curves and portrays other significant marine features, as well as those of the shore area.

Before maps or charts can be drawn, the information to be shown on them must be obtained from a survey with such detail and accuracy as is appropriate to the scale on which they are to be published. These surveys are termed *topographic* and *hydrographic surveys*. Principal emphasis here will be placed on topographic mapping procedures because topographic maps, including their preparation and use, are more relevant to engineering students than nautical charts.

This chapter will deal largely with the field methods associated with the preparation of large-scale, special-purpose topographic maps. It will provide, also, an introduction to the general-purpose mapping program of the federal government, point out the wide availability of published topographic maps, and indicate their great utility in engineering practice.

14-2. Uses of Topographic Maps A topographic map is merely a means to an end. Its purpose is to assist mankind in coping with his environment. Only its use justifies its creation. Aside from their utility in the practice of civil engineering, topographic maps are useful for many general, economic, technical and administrative planning purposes, and they assume added importance and become vitally necessary in time of war. Some of the uses to which topographic maps are put are described briefly:

1. National security—topographic maps are essential for both offensive and defensive military operations.

2. Development of mineral resources—topographic maps are a primary requirement in the exploration for minerals, oils, and solid fuels, and their quantitative evaluation.

3. Highway development—topographic maps are basic tools in the planning and design of modern turnpikes, freeways, and expressways.

4. Protection of public health—topographic maps are extensively used in connection with problems of water supply, water distribution, stream pollution, mosquito control, etc.

5. Irrigation, drainage, and water power—topographic maps are needed to show the extent and configuration of the irrigable lands. In the reclamation of land by drainage, the extent and slope of the land to be drained and the location of ditches can be more readily determined with the use of topographic maps. The study of waterpower resources and the planning and design of dams, aqueducts, and transmission lines require a detailed knowledge of the topography of the area under study.

6. Industrial development—topographic maps are useful in the selection of a suitable location for new industrial plants and housing facilities.

7. General—topographic maps are essential to other activities, such as airport design, soil erosion control, forest protection, flood control, and the location of television transmitting towers.

Finally, it is to be emphasized that any topographic map that is to be most useful must be available in advance of any development which depends on that map for basic information. In far too many cases the mapping program is not completed until after the project is well advanced. This delay obviously renders ineffective the principal function that maps are intended to serve.

REPRESENTATION OF RELIEF

14-3. Contours The most distinctive characteristic of a topographic map that differentiates it from other maps is that it portrays the configuration of the land surface. Various devices have been employed to express topography, but the most important one is the contour.

A *contour* is a line connecting points having the same elevation. It may be considered as the trace of the intersection of a level surface with the ground. The shoreline of a body of still water is an excellent illustration of a contour.

Contour interval is the vertical distance, or difference in evelation, between two adjacent contours. The contour interval in Fig. 14-1 is 10 ft.

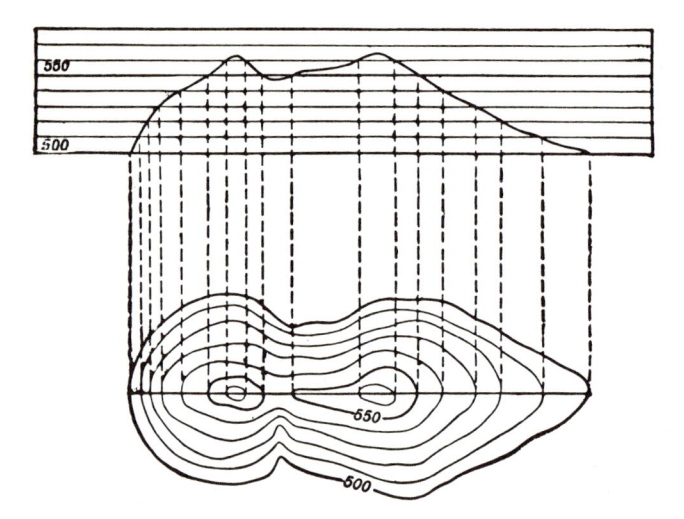

Fig. 14-1. Contours and Ground Profile.

The contour was probably first introduced in connection with sea soundings by the Dutch surveyor Cruquius in 1729. Its use as applied to terrain representation was initially suggested by Laplace in 1816, and it is considered today to be superior to all other topographic symbols for engineering purposes.

The basic datum for the vertical location by contours or depth curves of ground or marine features on maps and charts, respectively, is furnished by tidal planes of reference. For topographic maps, the most satisfactory plane is mean sea level. Hence, contours express height above (with rare exceptions) that datum.

For the soundings and depth curves of a nautical chart, a reference surface must be employed that incorporates a measure of safety to the principal user, the mariner. Consequently, a low water plane is more satisfactory because the chart then will show the least depth to be expected regardless of cyclic variations in the water surface elevation. Hence, fathom or bathymetric curves express depth of water below the adopted datum.

14-4. Characteristics of Contours A knowledge of the characteristics of contours is essential if the topography is to be correctly delineated and the resulting map properly interpreted.

The principal characteristics are as follows:

1. Contours spaced closely together represent a steep slope. When spaced far apart, they indicate a flat slope (see Fig. 14-1).

2. If the terrain is rough and uneven, the contours will be irregular. If the ground surface is even, as on earthwork slopes, the contours will be uniformly spaced and parallel (see Fig. 14-21).

3. Contours that portray summits or depressions are closed lines. Usually the identification of adjacent contours or the presence of a pond or lake will indicate whether the feature is a summit or a depression. To dispel confusion a depression contour (see Fig. 14-18*b*) may be used. It can be concluded that all contours are closed lines, either within or outside the borders of a map.

4. Contours are typically positioned on the opposite sides of a ridge or drainage line, as shown in Fig. 14-4, *c-2*).

5. Contours do not cross each other nor do they merge.

6. Contours are perpendicular to the direction of the maximum slope.

7. Contours cross a watershed or ridge line at right angles. The concave side of the curve is toward the higher ground.

8. In valleys and ravines the contours run up the valley on one side, cross the stream at right angles, and run downward on the other side. The curving portion of the contour as it crosses the valley is convex toward the higher ground.

Figure 14-2 shows both a perspective view and the corresponding topographic map of the same terrain. The principal features include a river, lying between two hills and emptying into a bay of the sea.

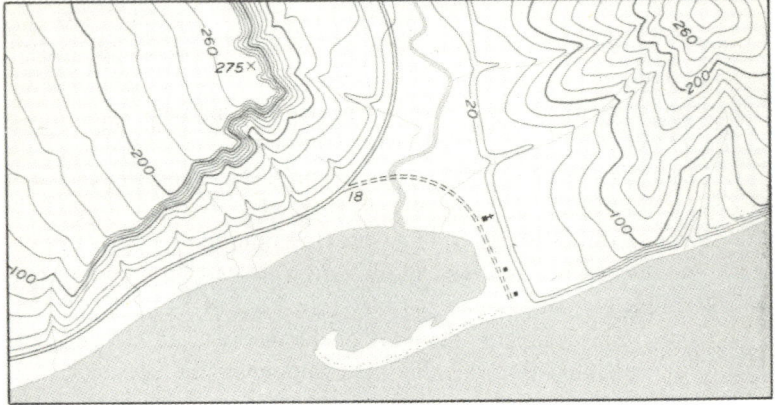

Fig. 14-2. Topographic Map. *U.S. Geological Survey*

Nearly all of the characteristics of contour lines listed above are shown in this map, which is drawn to a relatively small scale with a contour interval of 20 ft.

14-5. Other Relief Symbols Several other devices are used to indicate relief. They include hachures, shading, color tints, and form lines. The most dramatic expression of topography is provided by the terrain model shown in Fig. 14-3.

Hachures are short lines drawn in the direction of the ground slope (see Fig. 14-18b). They are not often used on modern topographic maps except in those cases where the scale of the map is too small to permit successful contouring of such features as gravel or borrow pits, mine dumps, and highway or railroad cuts and fills. If drawn properly, hachures convey a good conception of the terrain

Fig. 14-3. Terrain Model. *Tennessee Valley Authority*

but their value is largely pictorial. They are of little use if elevations are to be scaled from a map.

Shading is accomplished by the proper placement on the map of different shades of gray tints. The map can be regarded as a picture of a relief model illuminated by a light source directly overhead or from the northwest. If vertical illumination is assumed, less light will fall upon the slopes than on level land. Therefore, the effect is similar to hachuring because the steeper slopes will appear darker on the map. If oblique illumination is assumed, the illusion of solid, three-dimensional topography is especially striking. This is particularly so in mountainous country. Relief shading can be provided as an overprint on a conventional contour map in order to assist the layman with the interpretation of contours.

A *color tint system* is in common use for aerial navigation charts and on small-scale maps of the world. A scale of graded color tints or a system of different colors is chosen to show different zones of elevation. Each zone is bounded by contours which are usually shown on the map. Color tints, when used in conjunction with contour lines, give pictorial effect by accentuating the areas of different elevation.

When the land surface is too irregular or intricate to contour, as in the case of sand dunes, open strip mines, and lava beds, various *symbol patterns* are used. These are made up of dots, hachures, and form lines in a manner that is expressive of the typical appearance of the particular area. *Form lines* are similar to contour lines but are less accurate and have no definite interval.

14-6. Systems of Contour Points A contour can be drawn on a topographic map which is being prepared if the horizontal position and elevation of properly selected ground points are known. The manner of obtaining the required data provides a basis for defining four systems of contour points. They are as follows:

System A. Figure 14-4, (*a-1*), (*a-2*). System A consists in an established system of squares, usually marked by stakes in the ground. The elevations of the corners are then determined, thus forming a system of coordinate points from which the contour lines may be drawn.

System B. Figure 14-4, (*b-1*), (*b-2*). If a series of points having the same elevation are located on the ground and plotted on a map, the line joining these points will be a contour line. Thus, if the series of points having the elevation 914 ft are plotted, the 914 contour line is found by drawing a smooth line through these points.

System C. Figure 14-4, (*c-1*), (*c-2*). Although System B provides an accurate contour map, it requires the location of many points. Where such great accuracy is not necessary, a more expeditious method is that in which a few controlling points only are located and the contour lines are interpolated and sketched in, to represent the ground surface. Such points are summits, depressions, changes in slope, and especially points along ridge and valley lines.

System D. Figure 14-4, (*d-1*), (*d-2*). To establish System D a transit traverse is first run, with stakes set at 100-ft intervals, over which profile levels are taken. At these points, cross sections are taken to locate the contour points, valley lines, etc. From this system of points the contour lines may be drawn.

[14-6]   CARTOGRAPHIC SURVEYING   405

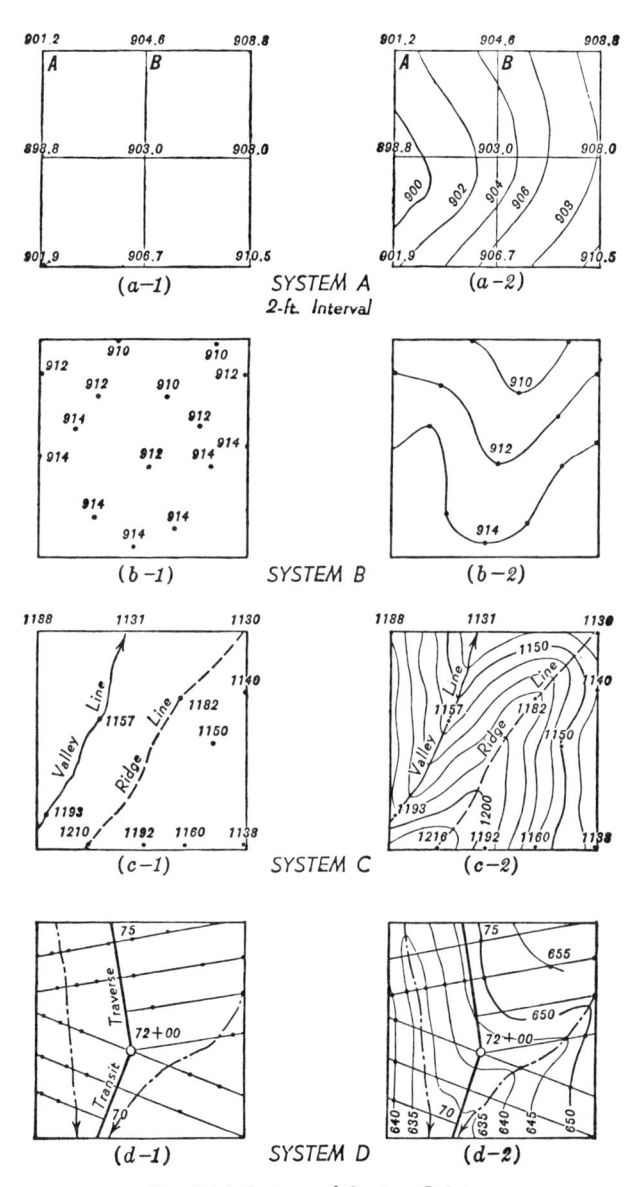

Fig. 14-4. Systems of Contour Points

14-7. Contour Interpolation For Systems A and C, it is necessary to interpolate between the plotted points to locate the positions of the contour lines. This interpolation may be done by estimation, by calculation, or by graphical means.

1. *Estimation.* The method of estimation is used where the highest accuracy is not desired, where the ground forms are quite regular, and where the scale of the map is intermediate or small. Thus, the contours of Fig. 14-4, (c-2), have been interpolated by this method. For example, on the valley line between the two points whose elevations are 1157 and 1193, it is at once noticed that the next higher contour above 1157 is 1160; also, that the additional contours, 1170, 1180, and 1190, fall on the stream line below 1193. Accordingly, the four points where these contours cross the stream are spaced by estimation and marked on the map. Possibly the first trial is seen to be erroneous, and a second spacing will be desirable. Similarly, the contour lines are spaced between the other controlling points, after which the contour lines may be sketched in to complete the map.

2. *Calculation.* The method of calculation is used where high accuracy is desired, and where the scale of the map is intermediate or large.

EXAMPLE 14-1: It is desired to space the contours between points A and B along the top line of Fig. 14-4, (a-1). It is noticed that the difference in elevation is 3.4 ft; also that the difference in elevation between point A and the next higher contour (902) is 0.8 ft. The distance (d) from A to the 902-ft contour line is given by the relation, $d = 0.8/3.4 \times AB = 0.24 AB$. If it is assumed that the distance $AB = 100$ ft, then $d = 24$ ft and the 902 contour line is plotted at that distance from A. Likewise, the 904 contour line is plotted at the distance $2.8/3.4 \times 100 = 82$ ft from A. These calculations are readily made on the slide rule.

3. *Graphical Means.* If many interpolations are to be made, and if a relatively high accuracy is desired, it will be more rapid and convenient to provide a proportional scale by which the contour points may be interpolated. Such a scale is drawn on tracing cloth or tracing paper, showing parallel lines (to any convenient scale) to represent the desired contour interval. Figure 14-5 illustrates such a scale drawn for plotting 2-ft interval contour lines.

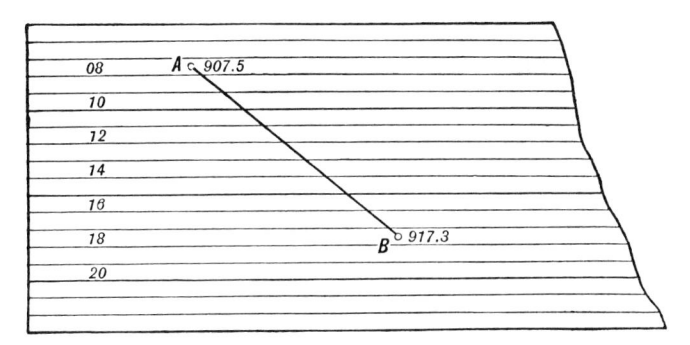

Fig. 14-5. Contour Interpolation Device.

The scale is used as follows: Suppose A and B are two points, 907.5 and 917.3 ft, respectively, and it is desired to interpolate the 2-ft contour lines between these points. The tracing-paper scale is shifted over the points A and B until the point A shows at elevation 07.5 on the scale, and point B shows at elevation 17.3. In this position, the 908, 910, 912 (etc.) contour lines are found at the intersections of the 08, 10, 12-ft (etc.) lines with the plotted line AB. These points are fixed by pricking through the tracing paper into the drawing.

Obviously it is feasible to plot only a few points on any given contour line and the line must be drawn freehand between such points. Contours are fairly regular curved lines except for the rough surfaces of outcropping rock strata; contour lines, therefore, are smooth curves which conform to one another more or less closely, depending on the regularity of the ground forms. Thus, the lines joining the contour points of Fig. 14-4, (*b-1*), (*b-2*), are not straight but are smooth curves.

It will usually aid in giving the proper character to the contour lines if every *fifth* contour line is carefully sketched in. These lines will then serve as guides when the intermediate lines are being drawn.

PLANE TABLE

14-8. Plane Table Plane table surveys are operations combining the stadia method (Art. 4-3) and field drafting on a portable drawing board mounted on a tripod. The plane table instrument (see Fig. 14-6), consisting of an alidade, board, and tripod, is essentially a mapping device.

Fig. 14-6. Plane Table Instrument. U.S. Geological Survey

The alidade consists of a telescope mounted on a column which is fixed to a steel straightedge about 2½ × 15 in. The telescope is similar in all respects to the transit telescope and is supported by a horizontal axis which rests on short standards above the supporting column. By this arrangement the telescope may be rotated in a plane perpendicular to the horizontal axis through a given arc. A vertical arc, vernier, clamp, and tangent screw are provided, as well as a striding, or an attached level, which permits the instrument to be used for direct leveling. A compass needle is usually provided, housed in a small metal box on the straightedge blade. A small bubble tube having a spherical surface, placed on the blade, is used to level the table. A bubble tube mounted on the vernier frame facilitates the measurement of vertical angles.

The drawing board is a plane board available in different sizes for different field conditions. The most common size is 24 × 30 in. A flat metal disc is let into the underside of the board by which it is screwed onto a threaded bolt of the tripod, thus securing the board to the tripod.

The tripod is provided with a special device by which the board is first leveled and then rotated, thus orienting the instrument for field use.

The topographer or plane table operator in Fig. 14-6 has interrupted his field work to examine beneath a lens stereoscope (see Art. 12-18) a pair of overlapping aerial photographs of the immediate vicinity. The rodman holds a special folding stadia rod that is 14 ft long.

Shown in Fig. 14-7 is an optical-reading, self-indexing plane table alidade. A pendulum device automatically corrects for slight tilts of the plane table board.

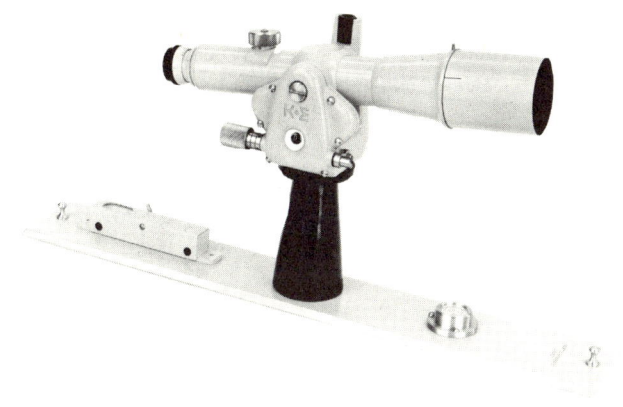

Fig. 14-7. Optical-reading Plane Table Alidade. *Keuffel & Esser Co.*

14-9. Setting Up and Orienting In plane table surveys, the locations of lines and points are plotted directly upon the drawing paper and, accordingly, a pencil dot on the paper will represent a small or a large dimension on the ground according to the scale to which the map is drawn. For example, a fine pencil-point dot may be 0.01 in. in diameter. If the scale of the drawing is 1 in. = 100 ft, evidently such a dot represents an area 1 ft in diameter on the ground. Thus it is evident that in centering the drawing board over a ground station, it is not necessary to use the care that is required in setting up a transit. Usually the table is centered over the ground point by estimation only. However, on careful work and if the scale of the map is large, e.g., 1 in. = 20 ft, a plumb bob may be suspended from the underside of the table, thus to center a plotted point over the corresponding ground point.

In setting up the table, it is first turned into its proper relation to the field being surveyed, and placed over the ground station as indicated above. The tripod legs are firmly planted in the ground and spread at such an angle as to bring the drawing board to a comfortable height. This height should be such that the observer can work easily on all parts of the drawing without leaning against the board. The table is then leveled by whatever devices are provided for that purpose.

As the transit must be oriented by means of a backsight, so the plane table must be properly oriented before any mapping is done. In small-scale mapping and rough surveys, this may be done by use of the compass needle, but on most work a backsight is used, as is the case with the transit.

If the compass needle is used, a line representing the magnetic meridian is drawn on the plane table sheet, at the initial station. The alidade straightedge is then placed along this reference meridian and the table is turned in azimuth until the needle indicates north. The table is then clamped in this position and mapping proceeds. At the next instrument station, after the table is set up and leveled, the board is again oriented as explained above.

If a backsight is used to orient the table, the procedure is as follows: Suppose the instrument is at station *B* and is to be oriented by a backsight to station *A*, Fig. 14-8. After the table is set up and

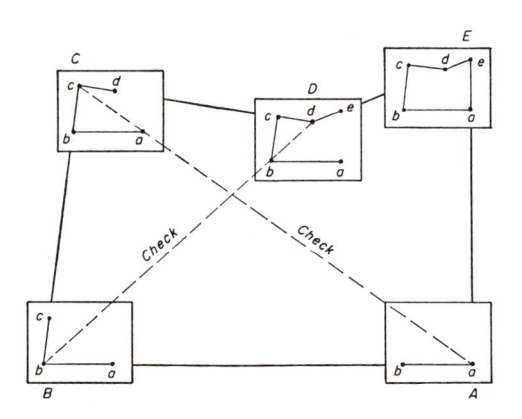

Fig. 14-8. Orienting and Traversing.

leveled, the alidade is placed along line *ba* and the table is turned in azimuth until the observer, looking through the telescope, sights station *A*. The board is then clamped and thus properly oriented.

14-10. Traversing It has been shown that in transit work the location of instrument stations is effected by the use of traversing or triangulation; likewise, in plane table surveys the instrument stations may be located by traversing or by *graphical triangulation*. Also, the plane table makes convenient the use of two other principles of locating points; namely, *intersection* and *resection*.

The procedure of traversing with the plane table is simple and is illustrated by Fig. 14-8. Having chosen a suitable scale for the map, a point (*a*) is chosen arbitrarily to represent the initial ground point, *A*, of the traverse. The table is then set up and oriented so that the sheet has a proper relation to the field. The straightedge of the alidade is then pivoted about the point, *a*, on the map and a sight is taken toward the forward station *B*, and a line is drawn along the straightedge in that direction. The distance *AB* is measured either by stadia or tape and the corresponding map distance is scaled to locate point *b*.

The table is then moved to station *B*, set up, and oriented by a backsight on *A* as explained in the previous article. A foresight is then taken on *C* by pivoting the alidade about *b*, sighting *C*, and drawing a line on the map in that direction. The distance *BC* is then measured, and point *c* plotted on the map. In this manner the traverse proceeds, and if the field is a closed one, the final position of *a* should fall on the initial position. The error of closure is, therefore, at once apparent as soon as the last observation is made.

14-11. Intersection The principle of intersection is a simple application of the condition that the position of a point is fixed by lines of direction drawn toward it from any two other known points. The method is especially convenient in plane table work. Thus, in Fig. 14-9 let *A, B, C,* represent instrument stations, located by the traverse *ABC*. At station *A* direction lines (*rays*) are drawn from *a* to two distant objects *E* (a fence corner) and *D* (a tree). At station *B* rays are again drawn from *b* toward *E* and *D*. These rays intersect those previously drawn from station *A*, and these intersections fix the positions *e* and *d,* being the plotted positions of *E* and *D*. A third station *C* is occupied and rays again drawn from *c* toward *E* and *D;* if these rays pass through the points previously located, the plotted positions *e* and *d* are proved or checked.

It will be noticed that these points are thus located without the necessity of any distance measurements or of a rod being held there. Obviously, any number of details like *E* or *D*, which are visible from two or more stations on the traverse *ABC,* can be located in this manner.

Fig. 14-9. Intersection.

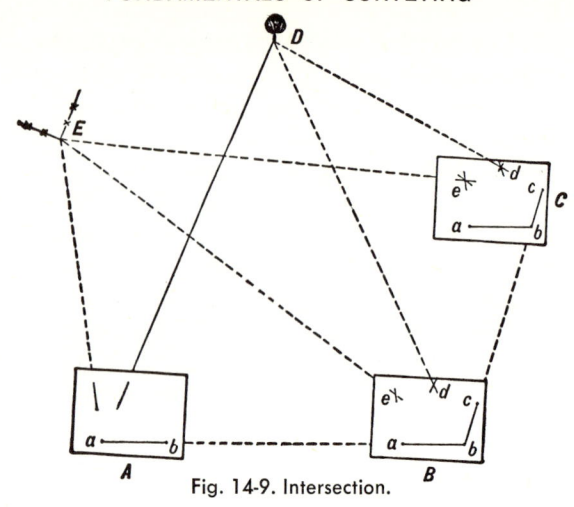

14-12. Resection The principle of intersection as described in the preceding article is useful in locating a distant point from known plane table stations. The principle of resection is opposite to that of intersection; i.e., it is used to locate the plane table station from known distant points.

In mapping details it is frequently desirable that the instrument should occupy an advantageous point that has not previously been located and, therefore, is not plotted on the map. The plane table offers a number of solutions for this problem, but it will be sufficient here to describe a simple one only.

Referring to Fig. 14-10 it may be supposed that it is desired to locate the position of C from sights on two points that have previously been located or plotted on the map. Also, it is supposed that

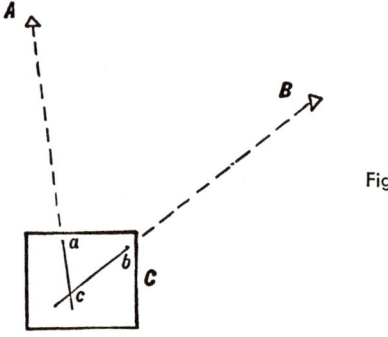

Fig. 14-10. Resection.

A and *B* are two adjacent stations or objects whose positions have been located and plotted at *a* and *b*, respectively. The procedure of locating point *c* on the map and orienting the board is as follows: The plane table is set up at station *B* and a ray is drawn toward station *C*, but no distance is measured. Next, the instrument is set up at *C* and oriented by a backsight along ray *cb*, toward station *B*. Then the alidade is pivoted about point *a* on the map until object *A* is sighted in the field, then ray *ac* is drawn to resect ray *bc* at point *c*. This fixes the position of *c* on the map as being the located position of station *C* over which the instrument is set up. This procedure is called location by *resection*.

14-13. Locating Details From what has been said about the plane table instrument and methods thus far, it will be a simple matter to describe its use in locating details. The instrument is set up at a given station and oriented as has been explained. The stadia method is universally used in this work, although on maps to a large scale, certain definite and important objects such as lot corners, water hydrants, sewer manholes, etc., may be located by taped measurements.

The rodman presents his rod at a given object to be mapped. The observer pivots the alidade about the plotted position of the station occupied, sights the rod, and draws a short portion of a ray toward it. He then shifts the alidade to a convenient position on the table, reads the stadia intercept, and observes either a vertical angle or a direct, level-rod reading to determine the elevation. He makes whatever reductions are necessary, and scales the distance to plot the point, writing the computed elevation beside it. Meanwhile, the rodman chooses another point and the work thus proceeds.

As the detail points are plotted, the observer draws the map, sketches the contours, and by suitable symbols represents the features of the area to be mapped. Thus the map is drawn, in all its essential details, in the field while the terrain is in the view of the observer.

TOPOGRAPHIC SURVEYING

14-14. Topographic Survey The major purpose of a topographic survey is the determination of the configuration of the surface of the earth and the location of natural and man-made features thereon.

The objective of this section is to outline the various methods for conducting the *topographic survey*. The scope of the subject is broad and embraces a wide variety of operations ranging from a simple survey of a building site to the national topographic mapping program. This treatment will be limited, however, to surveys of relatively small areas in connection with civil engineering construction projects.

The methods employed on topographic surveys of such limited scope are directly related to the scale of the resulting map. Three scales of maps are commonly recognized, viz., large, intermediate, and small. Large-scale maps include those drawn to scales between 1 in. = 10 ft and 1 in. = 100 ft; intermediate-scale maps are those drawn to scales between 1 in. = 100 ft and 1 in. = 1000 ft; small-scale maps include all others.

14-15. Horizontal and Vertical Control Topographic surveying requires initially the establishment of a network of strategically positioned stations over the area which is to be mapped. These points and their horizontal coordinates and elevations represent the horizontal and vertical control, respectively. They constitute the framework to which the location of all cultural, drainage, and topographic details is tied. Without such control the topographic survey would lack coordination and its separate parts would not fit together.

Horizontal control is provided by triangulation and traverse. Vertical control is established by the methods of differential leveling, although occasionally stadia leveling may provide entirely satisfactory results. This latter topic requires explanation.

In Art. 4-4 was presented the theory of determining the difference in elevation between two points from a stadia reading and a vertical angle. The application of this concept to a succession of instrumental stations results in the execution of a line of *stadia levels*. The procedure is now described.

The instrument is set up at a convenient distance from a benchmark and the stadia interval is read by placing the bottom cross-wire on a full footmark and noting where the upper cross-wire cuts the rod. The middle cross-wire is then sighted at a point on the rod whose height above the ground is the same as the height of the horizontal axis of the instrument above the ground. This distance is called the *height of instrument,* H.I., although this term does not have the same meaning as that used in differential leveling. The vertical angle is then read, and from these data the elevation of the transit station can be determined. The rodman then goes ahead to a turning point, and a similar series of readings is taken to determine its elevation.

Sometimes, because of obstructions, it is not possible to sight the middle cross-wire at the H.I. on the rod. If then some other point is sighted, a corresponding correction is made to the computed difference in elevation.

EXAMPLE 14-2: *Given:* Elevation of B.M. = 500.00 ft; H.I. = 4.2 ft; rod intercept = 5.13; vertical angle = $+4°14'$. Find the elevation of the instrument station.

$$\text{Difference in elevation} = -5.13 \times 7.36 = -37.8$$

Elevation of instrument station = 500.00 − 37.8 = 462.2

EXAMPLE 14-3: *Given:* Elevation of B.M. = 500.00 ft; H.I. = 4.2; $k = 102.0$; rod intercept = 6.34; vertical angle = $-5°20'$; reading taken on the 8.0-ft mark on the rod. Find the elevation of the instrument station.

$$\text{Difference in elevation} = +\left(\frac{102.0}{100} \times 6.34 \times 9.25\right)$$
$$+ 8.0 - 4.2 = 63.6 \text{ ft}$$

Elevation of instrument station = 500.00 + 63.6 = 563.6 ft

In case it is of no importance to know the elevation of the instrument station, it is immaterial what point (H.I.) on the rod is sighted in measuring the vertical angle, provided the same point (H.I.) is sighted both on a backsight and the corresponding foresight.

Of course, if the ground is level so that a direct level reading can be taken on the rod, no vertical angle is required, and the differences in elevation are computed as in ordinary differential leveling.

14-16. Scale and Contour Interval Many considerations may govern the scale of a map, but, in general, that will be a proper scale which is just sufficiently large to permit all desired features to be shown clearly and all dimensions to be scaled with the desired accuracy. Engineers, accustomed to the large-scale drawings of structures, are apt to choose a much larger scale for a map than is necessary. Possibly little harm results from such practice except that oftentimes dimensions are scaled from such maps with a precision which the accuracy of the map does not warrant.

The proper choice of the contour interval for a topographic survey depends on the slopes of the terrain to be represented, the scale of the map, and the purpose of the survey. In hilly regions, if the contour interval is too small in relation to the scale of the map, the contour lines become so crowded as to become illegible and to obscure other features. Also, the accuracy and number of the field measurements should bear a consistent relation to the contour interval. Consequently, the cost of a map to a small interval is much greater, for a given area, than one to a large interval.

No definite rule can be given that will meet all conditions, but for most purposes and field conditions, the contour intervals for large-scale maps (1 in. = 10 ft to 100 ft) will be taken as ½, 1, 2, or 5 ft for slopes which range from flat to hilly. For intermediate-scale maps (1 in. = 100 ft to 1000 ft) and for corresponding slopes the contour interval may be taken as 1, 2, 5, or 10 ft.

14-17. Selecting Contour Points The field work of finding the contour points for each of the four systems (Art. 14-6), except that for System C, is rather mechanical and offers little opportunity for the exercise of judgment on the part of the rodman. Where controlling points are being selected, however, the rodman is required continually to use his best judgment in selecting the points to be observed. In general, these will be at the marked changes in slope and along ridge and valley lines. However, small irregularities in the ground surface require the rodman to consider which points shall be observed and which shall be disregarded.

The location of details is conditioned by four factors, (a) the scale of map, (b) the instrument used, (c) the system of contour points, and (d) the character of the terrain. Obviously these factors vary so widely that it would be impossible to discuss all the varieties of field conditions that occur. Accordingly, the subject will be treated with respect to each of the systems of contour points described in Art. 14-6.

14-18. System A—Coordinates The system of points shown in Fig. 14-4, (*a-1*) and (*a-2*), is used when the available personnel are relatively unacquainted with mapping procedures and the tract is open and of limited extent. The survey of a building site would be a suitable example. The elevations of the coordinate points may be found with the engineer's level or transit. Sometimes, the hand level is employed. In irregular terrain the accuracy of the map will be increased if the contours are drawn in the field either on a sketch board or a plane table.

14-19. System *B*—Tracing Out Contours The delineation of relief by tracing out each individual contour is accurate but very time consuming. The method is used when the map scale is large and the project area is rather small and open.

A transit is set up on a control point and oriented along the line to another traverse station. The elevation of the line of sight is determined by a backsight on a benchmark. The proper foresight to locate a point on a given contour is found by subtracting the elevation of that contour from the H.I. Suppose, for example, the H.I. is 437.6 ft and the 430 contour is to be traced. This means that a foresight reading of 7.6 must be obtained if the rod is on that contour. As the rodman proceeds to move along the contour, regular determinations of his position are made by reading the stadia intercept and the horizontal circle of the transit.

The use of a plane table is much more efficient because the rod positions can be plotted directly on the map in the field.

14-20. System *C*—Controlling Points This system of points consists of such controlling features as summits, ridge and valley lines, and all important changes in slopes. The points are located in the field by the use of either the transit or the plane table. The procedure with each instrument will now be described.

The transit party usually consists of three men—an instrumentman, a recorder, and a rodman. Sometimes two rodmen and one or more axmen are employed. The organization of the party should be flexible to meet the field conditions. Thus, the three-man party may be organized as a transitman, recorder, and rodman; or, if the distances are great, it may be organized as a transitman and two rodmen.

For the usual field conditions, the procedure of locating details is somewhat as follows: the instrumentman sets up over a control station and orients the transit by a sight on an adjacent station; he then determines his H.I. above the ground. The rodman selects a point to be located and presents his stadia board for a reading. The transitman sights the board using the upper motion of the transit, sets the bottom cross-wire on a full foot mark, and reads the stadia interval by noting where the upper cross-wire cuts the rod. He then sets the middle cross-wire at a point on the rod that has the same height above the ground as the H.I. of the instrument, and motions the rodman to another point. While the rodman is finding another point, the instrumentman reads the vertical angle and the azimuth of the point observed. The recorder keeps the notes and, if time permits, computes the elevations and horizontal distances of all inclined sights. Sketches are often better than written descriptions of points and should be used freely. This procedure is varied somewhat if direct level readings are possible. Typical field notes are shown in Fig. 14-11. The instrument is assumed to be in position at station C,

	LOCATING DETAILS LAKE COMO RESERVOIR Inst. at C, Elev. 633.3 H.I.= 4.4					Gurley Transit No.7 July 12, 1968	G.B. Righter, ⚲ 43 H.R. Linder, Notes Dan Cohen, Rod
Object	Az	Rod	Vert. Angle	Hor. Dist.	Diff. in Elev.	Elev.	
B	5°35'						
1	208°40'	2.22	-2°21'	222	-9.1	624.2	
2	258°30'	1.10	(3.9)	110	+0.5	633.8	
3	333°40'	2.43	-1°14'	243	-5.2	628.1	
4	347°15'	4.98	-2°23'	497	-20.7	612.6	
5	8°25'	2.32	(on 8.0) -3°05'	431	-26.8	606.5	
6	14°15'	3.63	3°01' etc.	362	-19.1	614.2	
B	5°37'	check	↓				

Fig. 14-11. Field Notes for Stadia Surveying.

whose position has been located by traverse and whose elevation is 633.3. The transit is oriented by a backsight on station B, the azimuth being 5°35'. The H.I. is found to be 4.4 ft, and $k = 100.00$.

A reading is then taken on object 1, which is a point on the property line fence. The bottom cross-wire is set on a full footmark and the intercept is read at the upper cross-wire. The middle cross-wire is then set to read 4.4 on the rod, and the instrumentman then motions the rodman forward. Since the azimuths of side details are subsequently plotted with protractors commonly subdivided into 30′ spaces, such side azimuths are read to the nearest 5′ only and without the assistance of the vernier.

At point 2 a horizontal sight was taken. The foresight reading of 3.9 is enclosed in parentheses to indicate this. Since the H.I. is 4.4, the difference in elevation is +0.5 ft.

At point 5, it was impossible, for some reason, to sight the rod at 4.4, and a reading was taken at 8.0 ft. Accordingly, a correction of 3.6 ft is applied to the calculated difference in elevation.

Before leaving the station a check reading is taken on B and found to be 5°37′. This assures the transitman that the instrument has not been disturbed while his observations were being taken.

The reductions of the notes, shown in the last three columns, are usually made in the office or drafting room.

Fig. 14-12. Mapping with Plane Table. *U.S. Geological Survey*

In the case of surveys with the plane table (see Fig. 14-12), the contouring is performed in the field and few data are recorded.

On a topographic mapping project of any importance or magnitude that requires the use of a plane table, the field sheet will usually be a planimetric base map compiled from aerial photographs. For small jobs the contouring may be performed directly on the photographs.

An important principle relating to the correct portrayal of ground forms by means of contour lines is that the salient or controlling features of the topography should be sketched in before the contour lines are drawn. Thus, in Fig. 14-13 the ridge and valley lines were located and the contour crossings determined along these lines before any contours were drawn.

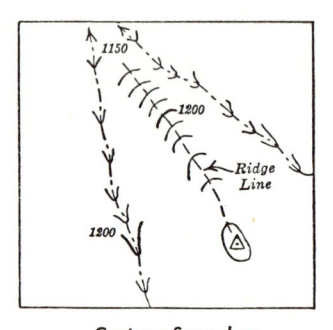

Contours Spaced on Ridge and Valley Lines   *Finished Sketch*

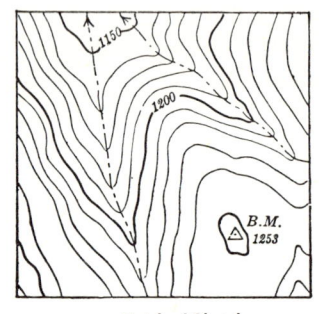

Fig. 14-13. Sketching Contours.

14-21. System *D*—Hand Level Method The method used for locating points according to System *D* makes use of the hand level instrument. The field party usually includes the topographer and two tapemen. The equipment consists of a hand level, a 100-ft steel tape, a rod (10 to 15 ft long) graduated in one-foot divisions, and a 5-ft staff. The record is kept in the regular transit notebook, or in a special topography book made up of cross-ruled pages.

This method is used principally in connection with route surveys and hence it is assumed that a located line has been established by means of a transit traverse, stakes being set at 100-ft intervals, and that profile levels have been run over the line.

In the field, cross sections are taken at the 100-ft stations or at irregular intervals, depending on the conditions. The procedure may be described by reference to Fig. 14-14. At station 40 the elevation (521.5) of the ground at the stake is known from the profile levels. Assuming that the contour points to the right of the centerline are to be located, the head tapeman takes the hand level and moves out on a line perpendicular to the centerline. The rear tapeman takes the rod and remains at the center stake. The ground slopes uphill and, by sighting on the rod held at the center stake, the head tapeman is

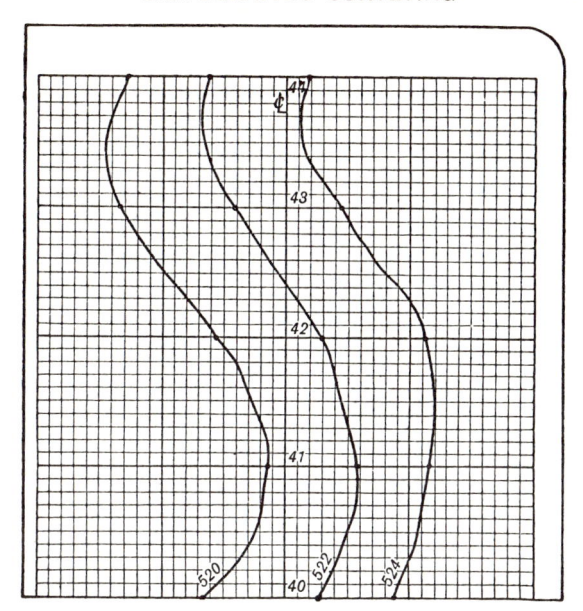

Fig. 14-14. Locating Contours with Hand Level.

directed to a point which is 0.5 ft higher than station 40, and thus locates a point on the 522 contour. The distance from the center stake is 27 ft, and this point is plotted by the topographer in his book. The rear tapeman then takes the position at this point and the head tapeman goes forward to find a point on the next higher (524-ft) contour. The distance between the two contours is 58 ft, or the latter point is 85 ft from the center stake. The method of taping may be that of measuring from one contour point to the next, or the topographer may serve as rear tapeman, thus permitting the rear tapeman to become a rodman, and all distances are measured continuously from the center stake.

Locating points to left of the center line, the slope is downhill, and either the tapemen reverse positions, or the head tapeman carries the rod and the rear tapeman the hand level.

Distinguishing marks are usually placed on the rod at each contour interval above and below the H.I. of the hand level. The H.I. of the hand level may be the natural height of the levelman's eyes above the ground, or a 5-ft staff may be used to support the hand level. The latter practice is better, because it permits the tapemen to reverse positions without confusion.

14-22. Aerial Mapping Stereoscopic compilation or stereocompilation refers to the process of preparing a map from aerial photographs with stereoplotting instruments like those mentioned in Art. 12-28.

Since the late 1930's practically all standard topographic maps produced by federal agencies, and special-purpose maps produced by engineering firms have utilized aerial photographs in one way or another. If the terrain is relatively flat and a small contour interval is specified, the photographs are used only for preparing planimetric base maps and the contours are added in the field with the plane table. When it is feasible and economical to do so, as in open and rolling country, a complete stereocompilation of the topographic map is performed.

MAP DRAFTING

14-23. Plats and Maps The distinction should be made between a plat and a map. A plat is a dimension drawing showing the data which pertain to a land survey or a subdivision of land. A map is a graphical representation, to scale, of the relative positions and character of the features of a given area. A plat is primarily a legal instrument being made a matter of public record and used in the description and conveyance of real estate, having all important dimensions recorded. Although a plat (see Fig. 14-15) is usually drawn to scale, little if any use is made of this condition. A map, however, shows no recorded dimensions, but a large part of its usefulness depends on the accuracy with which distances, areas, and elevations can be derived from it.

Fig. 14-15. Drawing a Plat.
Bureau of Land Management

Excessive ornateness is not desired on maps that are to serve practical purposes; nevertheless the draftsman should strive for such harmonious effects that the finished map will have a pleasing appearance. This is attained by the use of good taste in the matters of form, proportion, and color in the symbols and lettering used.

14-24. Plotting Plotting refers to the transfer of survey data to the map or plat. In the preparation of a topographic map the first plotting task is to define accurately the positions of the horizontal control points on the map. This is best accomplished by preparing a rectangular grid and plotting each traverse point by means of its computed coordinates.

The plotting of side details, including contour points, can be effected most easily with the use of a circular protractor. It is properly oriented at a control station and fastened to the map sheet. A measuring scale can be utilized in conjunction with the protractor to locate the points for which the field record is as shown in Fig. 14-11.

14-25. Titles A title for a map usually provides the following items of information: (1) the organization or company for whom the map is made, (2) the name of the tract or feature that has been mapped, (3) the name of the engineer in charge, (4) the name of the draftsman, (5) the place or office where the map was drawn, (6) the date, and (7) the scale, both numerical and graphical.

In executing a title, proper emphasis should be given to the different items listed above by varying the weight and the size of the letters used. A principal purpose which a title serves is to identify a given map in a file with other similar maps. Hence, the most important item in the title is the name of the tract or feature that the map represents, and this item should be given the most emphasis. For example, in Fig. 14-16*b* the feature shown on this map is that portion of the location between stations 425 and 563 on Route 47. Accordingly, this item is given the most prominence in the title.

A map title is placed anywhere on the drawing where it will balance the map as a whole. It is not boxed in at the lower right-hand corner as are titles on mechanical drawings. The size of the letters should be consistent with the size of the drawing. A general tendency with beginners is to make the title too bold, with letters too large or the weight of the lines too heavy. Another common fault is too

STATE OF CALIFORNIA
DEPARTMENT OF PUBLIC WORKS
DIVISION OF SAN FRANCISCO BAY TOLL CROSSINGS

SAN FRANCISCO - MARIN CROSSING

ANGEL ISLAND-TOLL PLAZA

PLAN AND ELEVATION

LINE - B

SCALE IN FEET

(a)

STATE OF ILLINOIS
DEPARTMENT OF HIGHWAYS
LOCATED LINE OF

ROUTE 47-STA. 425 TO 563

_____ District Engineer   Ottawa, Ill.
_____ Draftsman   October 5, 1968
Scale 1 in. = 400 ft

(b)

Fig. 14-16. Map Titles.

much space between the lines, which gives the title a loose and disjointed appearance.

Mechanical lettering is now used very widely for map drafting. This practice has important advantages when more than one person performs drafting tasks in the preparation of the same map.

A meridian should appear on every map. It is indicated by an

Fig. 14-17. Meridian Arrow.

arrow, somewhat as shown in Fig. 14-17, which, unless otherwise specified, indicates true north.

14-26. Symbols Symbols are used to portray various features on maps. Different draftsmen may use different symbols to represent the same feature; it is desirable, therefore, that the most common features be represented by symbols that are widely accepted and understood. In this matter it is natural to follow the practice of those governmental organizations engaged in mapping work, such as the U.S. Geological Survey, the U.S. Coast and Geodetic Survey, and others. The symbols shown in Fig. 14-18 *a* and *b* are those most commonly used.

Since map scales vary so widely, the character of the symbols will vary somewhat to suit the scale of the map. Thus, on large-scale maps, tree symbols (in plan) may be drawn to scale, whereas on small-scale maps no attempt is made to draw them to scale. Care must be taken in executing the symbols that they shall not obscure and render illegible other features on the map.

On topographic maps it is quite necessary to draw the various symbols in different colors. The standard practice is as follows: *black* for lettering and the works of man, such as roads, railways, houses, and other structures; *brown* for all land forms, i.e., contours and hachures; *blue* for water features, such as streams, lakes, marsh, and ponds; and *green* for vegetation, including trees and grass.

In order to secure greater uniformity of appearance of the finished map and to save time and money through the elimination of much hand lettering, increasing use is being made of printed names, symbols, meridian arrows, etc. These are prepared on a tough, transparent, plastic sheet (see Fig. 14-19). It is a simple matter to cut out the required name or symbol from this sheet, put it in the proper position on the map, and press it down.

14-27. Topographic Expression The portrayal of relief on a topographic map by means of contour lines is both an art and a science. It is a science in the sense that various vertical and horizontal field dimensions are measured which, when plotted on a map base, control the placement of the contour lines. It is an art with respect to the range of discretion and judgment that the topographer may exercise in determining the configuration of the contour lines

ROADS AND RAILROADS

Hard surface, medium duty, four or more lanes wide ....

Hard surface, medium duty, two or three lanes wide ....

Loose surface, graded, and drained or hard surface less than 16 feet in width ....

Improved dirt ....

Unimproved dirt ....

Trail ....

Single track ....

Multiple main line track. If more than 2 tracks, number is shown by labeling ....

BOUNDARY LINES

Political Boundaries, County, Township, etc. — Ash Twp., Jackson Co., Mich.

Government Section Lines

Street or other Property Line — Stone — Iron pin

Fence — (State kind)

Section Corner — 3 | 2 State kind of / 10 | 11 — Monument

MISCELLANEOUS

Buildings { Large scale / Small scale

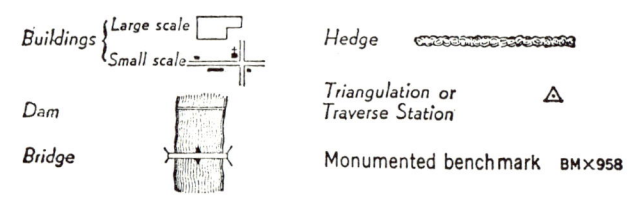

Dam

Bridge

Hedge

Triangulation or Traverse Station △

Monumented benchmark BM×958

Fig. 14-18a. Symbols.

between points of known horizontal position and elevation. Good topographic expression results when the contour lines convey to the map reader the typical characteristics of the ground surface. Expert detailed delineation of the topography so as to yield distinctive and correct topographic expression is an accomplishment that cannot be attained after only a few weeks of training and field experience.

TOPOGRAPHY

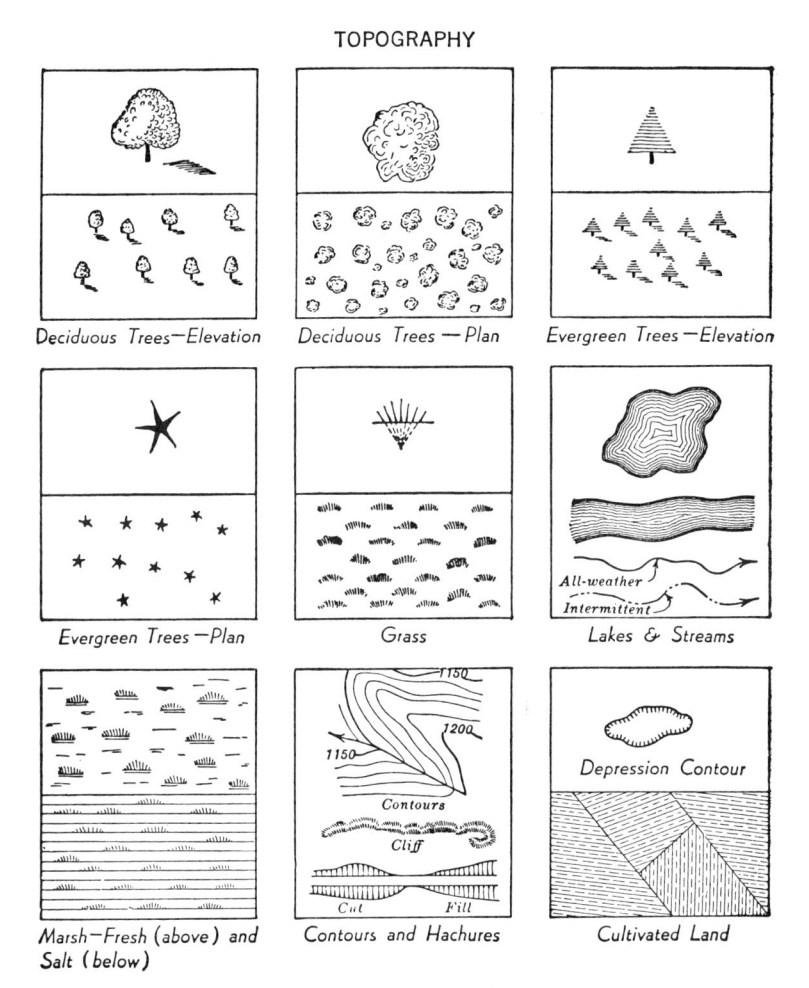

Deciduous Trees—Elevation

Deciduous Trees — Plan

Evergreen Trees—Elevation

Evergreen Trees—Plan

Grass

Lakes & Streams

Marsh—Fresh (above) and Salt (below)

Contours and Hachures

Cultivated Land

Fig. 14-18b. Symbols.

MAP USE AND PROCUREMENT

14-28. Map Studies Since a contour map is a representation of the earth's surface in its three dimensions, it provides the data for an endless variety of uses. For engineering purposes the principal uses of contour maps include representation of (a) reservoir areas and volumes, (b) drainage areas, (c) bridge and building sites, (d) earthwork structures, and (e) route projects.

Atlantic Ocean   Gulf of Mexico
94 Chaeudŏk   San'gwimi   Kŏmbal-li
ROAD   ROAD   ROAD   RD.   RD.
425425   425425   475475   475475
C a m d e n   B a y   R i v e r
METROPOLITAN AREAS
76^{2000m.}E. 76^{1000m.}E. 53₂₂ 53₂₁ 53₂₀

0°48'	0°15'	0°15'	1°06'
OR	OR	OR	OR
14 MILS	4 MILS	4 MILS	20 MILS
P'yŏngan-namdo	Tŏkch'ŏn-gun	P'yŏngan-namdo	

TÊTE DES CHÈTIVES   TÊTE DU GRAND PRÉ   Val Lèmina

Andrimäe Nônova Žiguri Grüšļi Čušli Buliņi

Fig. 14-19. Adhesive Map Type.

(a) *Reservoirs.* In the design of reservoirs for water supply, power, or irrigation projects, the studies are made on contour maps to locate the dam, to determine the volume of water to be impounded, to locate the boundary of the area to be inundated, and to find the drainage area. The necessary maps will be drawn to different scales suitable to the different studies that are made. Thus a large-scale map will be used to fix the dam site, an intermediate-scale map will be used to determine the area and the volume of the reservoir, and a small-scale map will be used to find the drainage area.

The method of finding the area and volume of the water to be stored may be indicated by reference to Fig. 14-20. As the water is impounded and rises by 10-ft stages to the elevations of 1030, 1040, and 1050, it will have as its shoreline the corresponding contour lines as shown. It may be noted that each contour line is a closed line within the reservoir area. If the areas of the 1030- and the 1040-ft contour lines are determined with a planimeter, averaged, and multiplied by the vertical distance between them (10 ft), the result is the volume of water included between these contours.

The total reservoir capacity, usually expressed in *acre-feet,* is the sum of the volumes between successive contours.

Obviously, the maximum flood line of the reservoir will be given by that contour having the elevation of the crest of the dam, increased by whatever head of water may exist above it.

(b) *Drainage Areas.* Any drainage area may be traced on a contour map by finding the ridge line around the watershed, as shown

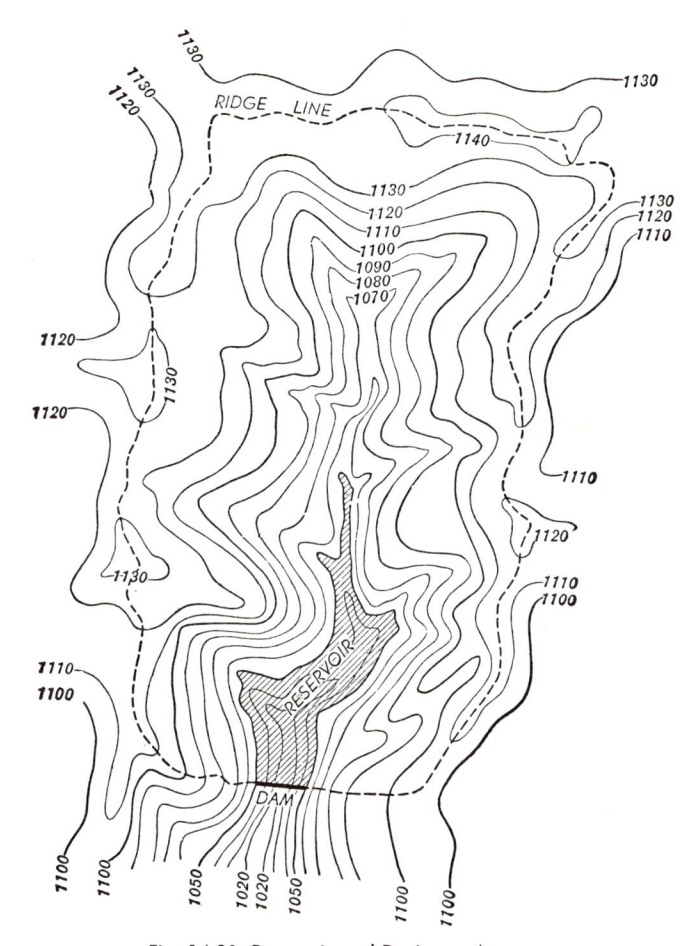

Fig. 14-20. Reservoir and Drainage Area.

in Fig. 14-20. This line may not always be evident at all places on the map, and some field measurements are sometimes necessary. The area is found by planimeter measurements.

(c) *Bridge and Building Sites.* In fixing the location of such important structures as dams, bridges, and buildings, use is frequently made of contour maps. Such maps are drawn to a large scale, from 1 in. = 10 ft to 1 in. = 100 ft and with contour intervals from ½ ft to 2 ft, by means of which the engineer is able to find the best location for his structure.

(*d*) *Earthwork Structures.* Earthwork estimates for route projects are usually made either from profiles or from cross-section notes taken in the field. But earthwork for other purposes is frequently estimated from contour maps. For example, Fig. 14-21 illustrates the

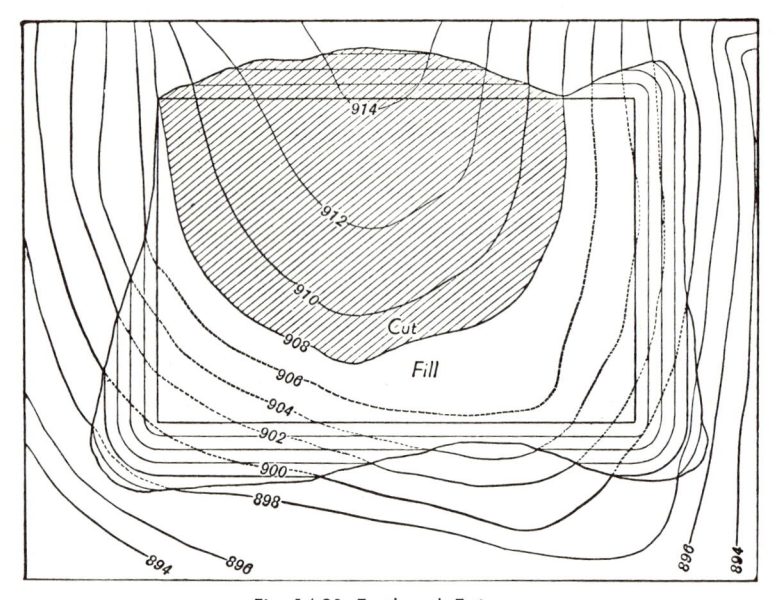

Fig. 14-21. Earthwork Estimate.

method of estimating the earthwork necessary to construct a parking space on a hillside.

The irregular lines are the contour lines of the original ground surface; the straight and circular lines are the contour lines of the proposed earthwork. The conditions are as follows: the area of the parking space is indicated by the heavy line rectangle; the elevation of the surface is 908 ft; the side slopes are 3 to 1.

From these conditions and from the principles of contours, the contour lines representing the proposed earthwork are drawn as shown. Since the elevation of the parking space is to be 908 ft, the contour lines above that elevation represent cut (shown crosshatched) and those below that elevation represent fill.

The volume of cut is found by determining with a planimeter the area of the closed 908-ft contour line in cut and that of the 910-ft contour line. The volume of earthwork between these two contours is then found as the average of the two areas multiplied by the contour interval, 2 ft. Similarly, the volume between the 910- and 912-ft contours is found, etc.

The volume of fill is found in a manner similar to that for the cut. It may be noted that, because of the ridge line through the parking space, the area of the 904-ft contour line is divided into two parts—one on the left and one on the right of the ridge. The same is true for the contour lines below 904 ft. The results may be recorded as shown below.

Contour Line	Area sq in.	Area sq ft	Volume, cu ft
CUT			
908	10.15	25,400	
			42,300
910	6.78	16,900	
			25,200
912	3.32	8,300	
			9,600
914	0.52	1,300	
			77,100 = 2,850 cu yd
FILL			
908	7.70	19,200	
			32,700
906	5.40	13,500	
			23,100
904	3.94	9,600	
			16,300
902	2.68	6,700	
			9,800
900	1.24	3,100	
			3,600
898	0.22	500	
			85,500 = 3,170 cu yd

(e) *Route Projects.* Studies for the location of such route projects as railways, highways, and canals are frequently made on contour maps. Many conditions that govern such locations need not be discussed here, but that one for which a contour map offers special aid is the establishing of a uniform or a maximum grade. The procedure may be described by reference to Fig. 14-22, where two existing highways are shown and it is desired to locate the centerline of

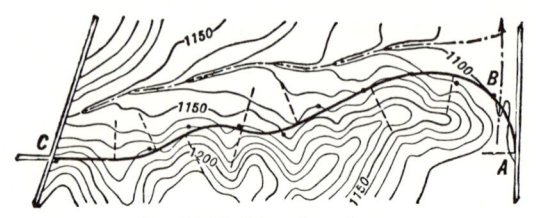

Fig. 14-22. Route Location.

a proposed connection between points A and C. The difference in elevation and the distance between these points make it desirable to establish a uniform grade of approximately 2.5%.

Since the contour interval is 10 ft, it is evident that a uniform grade of 2.5% will rise from one contour to the next in a horizontal distance of 400 ft. Accordingly, if the feet of a pair of dividers are set at a map distance of 400 ft apart and if one foot is placed on one contour and the other foot on the next adjacent contour, the line joining these two points will represent a grade of 2.5% on the ground. If the feet of the dividers are then turned about one foot, and the other placed on the next higher contour line, a third point on the uniform 2.5% grade will be found, and so on. By this means a series of points are located which, if joined by a smooth curved line, would represent a uniform grade of 2.5%. Such a line is sometimes called a *grade contour*. The location of the highway that will require the least earthwork will be the one that conforms most closely to the series of points thus located. Obviously the limitations as to curvature prevent the location from passing through all of the points, but it is made to conform to them as nearly as possible.

Thus, in the figure, beginning at point B, a series of points (round dots) have been located as described above, to fix the 2.5% grade contour; and the location, shown by the full line, has been made to follow the grade contour as closely as the conditions of alinement and ground forms would permit.

Obviously, this solution is not definite, and after a first location has been made another trial may show a better one. The process is not carried beyond two or three trials because the final adjustment is necessarily made in the field.

14-29. National Topographic Mapping Topographic maps can be classified into two broad categories, viz., special-purpose maps, and general-purpose maps. Special-purpose maps are prepared for specific projects and are compiled and published on various scales. Examples of special-purpose topographic maps are those prepared of reservoir areas and of metropolitan districts. General-purpose maps are designed to satisfy a wide range in public needs. The selection of the publication scale is based upon considerations of the economic character of the area, its cultural development, and the amount of map detail that is required for engineering, scientific, military, industrial, and commercial purposes.

The topographic maps published by the U.S. Geological Survey are general-purpose maps. Collectively these maps constitute the national topographic map series. They are of two general types differentiated chiefly by the publication scale as follows:

1. *Large-Scale Maps.* These maps cover areas of great public importance, such as metropolitan and industrial areas and other regions where detailed map information is needed. The publication scale is 1:24,000 (1 in. = 2000 ft) or 1:31,680 (1 in. = ½ mile), although the latter scale is used only infrequently for new mapping. Topographic maps published on these scales have dimensions of 7½′ in both latitude and longitude.

2. *Medium-Scale Maps.* These maps cover areas of average public importance. The publication scale is 1:62,500 (1 in. = nearly 1 mile). Geological Survey maps published on this scale measure 15′ in both latitude and longitude. This scale is generally considered quite adequate for all uses where detailed studies are not contemplated. Much of the past national mapping has been to this scale.

A knowledge of the accuracy specifications governing national topographic mapping is particularly important to engineers who may use the standard topographic maps as source data for the purpose of compiling special-purpose maps. If such specifications were better known, engineers would be much more cautious in the use of standard topographic maps, which have sometimes been enlarged to several times their intended usable scale in order to provide a base for detailed planning studies.

The accuracy specifications for special-purpose maps may cover a wide range in keeping with the requirements of the engineering firm that contracts to have such work done or performs it with its own forces. The specifications may be far above, equal to, or far below what are termed standard map accuracies.

The important features of accuracy specifications for currently executed federal topographic mapping in domestic United States areas are as follows:

1. *Horizontal Accuracy.* The horizontal position tolerance for 90% of all well-defined planimetric features is 40 ft of ground distance for both the 1:62,500 and 1:24,000 scale maps. This is equivalent to an error of 1/50 in. on the map at the publication scale of 1:24,000, and somewhat less than 1/100 in. at the 1:62,500 scale.

2. *Vertical Accuracy.* Ninety per cent of all elevations interpolated from the map shall be correct within one-half the contour interval. However, in checking elevations taken from the map, the apparent vertical error may be decreased by assuming a horizontal displacement within the permissible horizontal error for a map of that scale. Thus any contour may be shifted (imaginarily) either uphill or downhill by the map distance equivalent to a ground distance of 40 ft.

14-30. Procurement Engineers and others seeking map information not readily available from local sources are encouraged to direct their inquiries to the Map Information Office, U.S. Geological Survey, Washington, D.C. This facility can provide assistance in the procurement of maps and charts from the various federal surveying and mapping agencies. Available for free distribution are index maps showing the progress in each state of standard quadrangle topographic mapping. The prices of all maps are exceedingly nominal.

HYDROGRAPHIC SURVEYING

14-31. General A *hydrographic survey* is one whose principal purpose is to secure information concerning the physical features of water areas. Such information is essential for the preparation of modern nautical charts on which are shown available depths, improved channels, breakwaters, piers, aids to navigation, harbor facilities, shoals, menaces to navigation, magnetic declinations, sailing courses, and other details of concern to mariners. Also, a hydrographic survey may deal with various subaqueous investigations which are conducted to secure information needed for the construction, development, and improvement of port facilities; to obtain data necessary for the design of piers and other subaqueous structures; to determine the loss in capacity of lakes or reservoirs because of silting; and to ascertain the quantities of dredged material.

The fundamental principles of conducting the hydrographic survey of a harbor or an inland lake are substantially the same as those employed in executing a comprehensive survey of a large tidal estuary or in sounding a vast oceanic area, like the Gulf of Mexico, but there are marked differences in the sounding vessels, instrumental equipment, and the surveying techniques. This section deals mainly with the basic procedures for executing a hydrographic survey of limited scope and with the relationship of such surveys to the practice of civil engineering in such aspects as waterway improvement, dock and harbor construction, beach erosion control, and sewage disposal.

14-32. The Hydrographic Survey A hydrographic survey is distinguished by the measurements and observations that are made to determine and subsequently portray the submarine or underwater topography, as well as to locate various marine features of interest to the navigator. The chief elements of a hydrographic surveying project will now be briefly outlined. The detailed treatment of such operations as those of making and locating soundings will be presented in subsequent articles.

(a) *Reconnaissance.* Although the principal operation of the hydrographic survey is the obtaining of hydrography or making the soundings, this phase of the project cannot be performed until certain preliminary steps are undertaken. The first of these is a careful reconnaissance of the area to be surveyed in order to select the most expeditious manner of prosecuting the survey and to plan all operations so that the project mission is satisfactorily completed in accordance with the general instructions and specifications governing such work. The use of aerial photography can be of considerable assistance in this preliminary study.

(b) *Horizontal Control.* The next step is the establishment of horizontal control or the framework by which land and marine features are held in their true relationship to each other. Triangulation and, to a lesser extent, traverse are most commonly executed to provide horizontal control. It is not practicable to make a general statement regarding the accuracy of such control. In the case of original surveys over large bodies of water, second- or third-order triangulation may be required. For detached surveys of small and isolated reservoirs it may be entirely satisfactory to develop a control system by a combination of stadia and graphical triangulation procedures with the plane table.

Previously established control is a very important asset in any hydrographic survey. Every effort should be made to obtain and utilize data from earlier surveys in the area. Sometimes a sufficient number of former survey stations can be recovered to satisfy the requirements of a revision survey and no new horizontal control will be necessary.

(c) *Vertical Control.* Before sounding operations are begun, it is essential to execute the vertical control in order that the stage or elevation of the water surface can be known when the soundings are obtained. Vertical control data are also needed for the limited topography shown on all nautical charts. When surveys are conducted in tidal bodies of water whose low water level is not known, it is necessary to establish a tide station and begin observations of the tidal fluctuations so as to define a plane of reference for the soundings. This datum is then tied to one or more nearby benchmarks by leveling.

(d) *Topographic Survey.* Topographic surveys are conducted of the chart area back of the shoreline. Since the navigator's only interest in this area is in any prominent landmarks that it may contain, only a relatively narrow fringe of topography is shown on the chart.

(e) *Hydrography.* The measurement of water depths is the most important operation in nautical charting or in hydrographic surveys that are related to civil engineering.

(f) *Preparation of the Nautical Chart.* This (Fig. 14-23) usually represents the final product of the hydrographic survey. In the case

[14-32] CARTOGRAPHIC SURVEYING 437

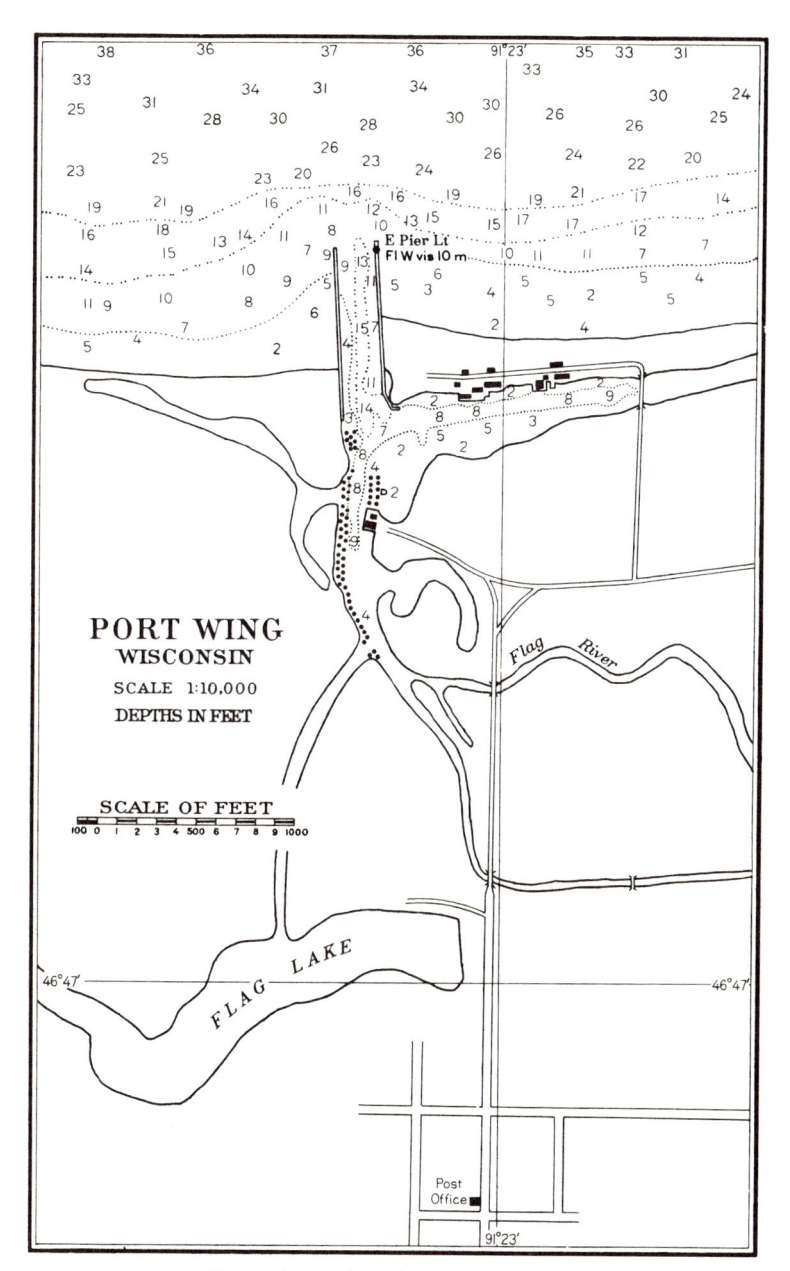

Fig. 14-23. Harbor Chart. U.S. Lake Survey

of subsurface surveys for engineering purposes, the end result may be the calculation of quantities of silt, dredged material, or the preparation of underwater profiles needed for subaqueous construction.

14-33. Sounding Datums In topographic surveying it is necessary to determine the position of various points on land and also their elevations above some datum, usually mean sea level. In hydrographic surveying the depth of water with respect to some particular *stage* or height of the water surface is required, as well as the horizontal position of the sounding. It should seem obvious, therefore, that observed soundings must be reduced to some plane of reference if any uniformity in the record of soundings is to be obtained.

In the case of nautical charting surveys it is common practice to reduce observed depths to the *plane of low water*. These reduced depths are then portrayed on the chart and will indicate to the navigator the least depth that will exist at any time. The *Low Water Datum* for Lake Michigan as used by the U.S. Lake Survey is defined as the surface of the lake when it is at elevation 578.5 ft above mean sea level. This corresponds quite closely with the lowest recorded water surface elevation for that lake. To illustrate the reduction of a measured depth to a particular plane of reference, consider soundings made in Chicago Harbor during a day on which the stage of the lake read 5.8 on the gage (Fig. 14-24) and it was known that the zero of the gage was at elevation 576.0. All measured depths must be reduced by (576.0 + 5.8) −578.5 or 3.3 ft in order to refer them to the Low Water Datum.

In tidal bodies of water it is the usual practice to refer soundings to *Mean Low Water* (MLW) or *Mean Lower Low Water* (MLLW), depending upon the character of the rise and fall of the sea.

In addition to the nautical applications of tidal data, there are some interesting and important engineering aspects. The harbor engineer must consider carefully tidal data when planning or designing water-front structures. During construction the tidal range will influence the freeboard requirements of cofferdams and caissons, the type of dock construction, and the cutoff elevations for timber piles. The predicted times of high and low water will affect the scheduling of certain construction operations. Soundings that are made to determine depths in tidewater reaches of rivers and in sea harbors for pipeline crossings, submarine cables, sewer outfalls, and other construction must be reduced to the appropriate tidal datum plane.

Fig. 14-24. Recording Lake Gage. *U.S. Lake Survey*

14-34. Topographic and Shoreline Surveys Although the depths in the water area are most important data on a nautical chart, the topographic features of a seacoast or the shore of a lake are indispensable in providing orientation for the mariner and improving the appearance of the sheet. Topographic surveys provide this information. In the past most of such surveys were made with the plane table and to some extent by the transit-stadia method. Today such ground survey methods are used only for hydrographic surveys of limited extent or for securing the information necessary to make periodic cultural revisions in the shore area of a harbor chart.

For new and extensive hydrographic survey projects the aerial photogrammetric method has very largely superseded ground mapping procedures because of the pronounced economies in time and cost which are made possible by aerial mapping. The extent of reefs and shoals is also made possible by the examination of aerial photographs by one skilled in their interpretation.

14-35. Equipment for Hydrography The modern survey ships of the U.S. Coast and Geodetic Survey and the U.S. Naval Oceanographic Office are complete, mobile, and self-sustaining surveying and chart producing plants. They carry the equipment and personnel needed for carrying out the entire hydrographic charting mission, including the procurement of air photography, the execution of horizontal and vertical control, the development of hydrography, and the lithographic reproduction of the finished chart. The remarks that follow are confined to brief descriptions of the kind of equipment that would be used in conducting sounding operations in a lake, reservoir, or harbor with a small boat or launch rather than from a survey ship operating miles off the seacoast.

(a) *Boats.* Various types of launches and small boats are used in hydrographic surveying. Most fishing or working-type launches are satisfactory, because they are seaworthy, have reliable motor performance at low speeds for sounding work, and can be adapted to the various operations associated with hydrographic surveying.

For limited surveys in protected areas small boats like the dinghy can be used. This boat has a rounded bottom and sufficient keel to aid in maintaining a steady course. It can be powered with an outboard motor.

(b) *Sounding Pole.* For measuring water depths up to 12 ft the soundings can be more easily obtained with a sounding pole. Such a pole can be made of a 15-ft length of 1½-in. rounded lumber, with painted gradations at foot or half-foot intervals, and have a metal shoe at each end to hasten sinking.

(c) *Leadline.* A leadline or sounding line consists of a suitable length of good quality sash cord, at the end of which a weight or sounding lead is attached. The leadline can be graduated by fathom or foot marks in various distinctive patterns so that no difficulty is experienced in reading the required water depth. In sounding operations the lead is lowered until it touches the bottom; and at the moment that the line is vertical and taut, the depth is determined from the markings on the leadline.

Even a well-seasoned leadline will change its length as the result of normal use. At regular intervals its length should be verified by comparison with a steel tape and, if necessary, suitable corrections should be applied to the observed depths. The following notes show the data from a comparison test made at the end of a day's sounding operations in Lake Benjamin at the University of Illinois Summer Surveying Camp in northern Minnesota:

TEST OF LEADLINE

July 27, 1954   By Hooper and Kinch

Leadline	Tape (ft)	Leadline	Tape (ft)	Leadline	Tape (ft)
5	4.9	50	47.5	95	91.3
10	9.7	55	52.5	100	96.2
15	14.5	60	57.4	105	100.9
20	19.3	65	62.3	110	105.9
25	23.9	70	67.1	115	110.6
30	28.6	75	71.9	120	115.4
35	33.4	80	76.7	125	120.1
40	38.2	85	81.5	130	124.9
45	42.8	90	86.4	135	129.8

From a plotting of the leadline readings against the computed leadline corrections, a table of corrections was prepared. This table indicates the applicable correction, to the nearest full foot, which is to be applied to soundings in various ranges as follows:

LEADLINE CORRECTIONS

Observed Depth (ft)	Correction (ft) (to be subtracted)
0–14	0
15–29	1
30–49	2
50–84	3
85–119	4
120–140	5

(d) *Fathometer.* On modern hydrographic surveys of any importance or extent, measurements of water depth are made with an echo sounding instrument called a *fathometer*. *Echo sounding* is a method for obtaining water depths by determining the time required for sound waves to travel from a point near the surface of the water to the bottom and back. A fathometer is designed to produce a signal, transmit it downward, receive and amplify the echo, measure the intervening time interval, and automatically convert this interval into feet or fathoms of depth. The fathometer may indicate the depth visually or record it graphically on a roll of specially prepared chart paper. Hence, every line of echo soundings provides a virtual profile of the lake or harbor bottom beneath the course of the survey launch even though the boat proceeds at full speed. Water depths can be easily scaled from the resulting *fathogram*. If the instrument (Fig. 14-25) is set to operate in the 0 to 70-ft range, the appropriate depth scale reads 47 ft at the bottom of the trench.

(e) *Sextant.* The sextant (see Fig. 14-26) is a portable instrument that is used for measuring horizontal and vertical angles from a ship. In celestial navigation the sextant is used to measure the altitude of

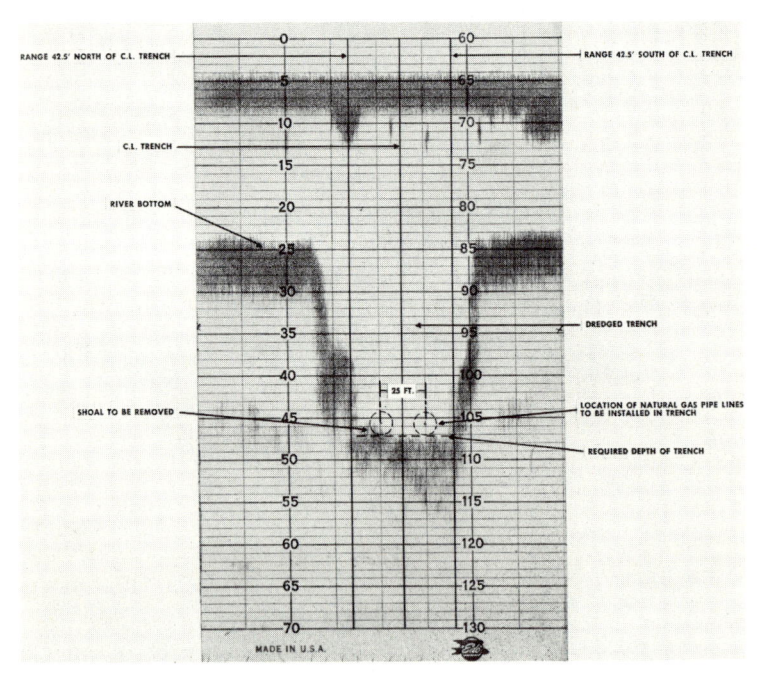

Fig. 14-25. Fathogram Showing Cross Section of Trench Carrying Pipeline across Hudson River. *Edo Corp.*

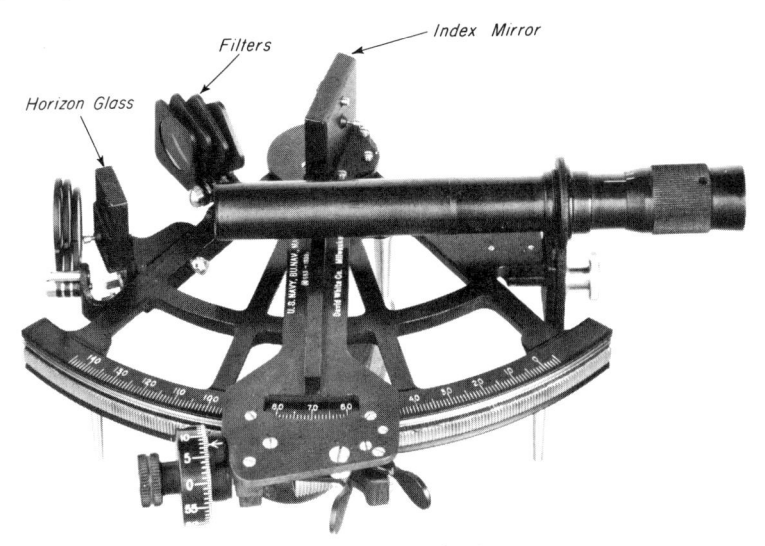

Fig. 14-26. Sextant. David White Co.

various prominent stars above the horizon. In hydrographic surveying the sextant is used to measure the horizontal angles between designated objects on land. This makes possible the subsequent solution of the three-point problem for determining the position of the sounding vessel at various selected times.

(f) *Three-Arm Protractor.* The three-arm protractor is used extensively in hydrographic surveying to plot boat positions during sounding operations. A transparent protractor constructed of celluloid is very satisfactory for all but the most accurate plottings. This protractor consists of a solid disc about 12 in. in diameter containing a circle graduated in degrees, and one fixed and two movable arms extending about 13 in. beyond the edge of the disc. Each of the movable arms has a vernier permitting angles to be set with an accuracy of 2'.

14-36. Sounding Operations Sounding operations constitute the basic element of the hydrographic survey. However, the determination of depth is useless unless the horizontal position of the point at which the depth was measured is simultaneously obtained.

Although a variety of methods can be employed to locate the soundings, only three principal ones will be mentioned.

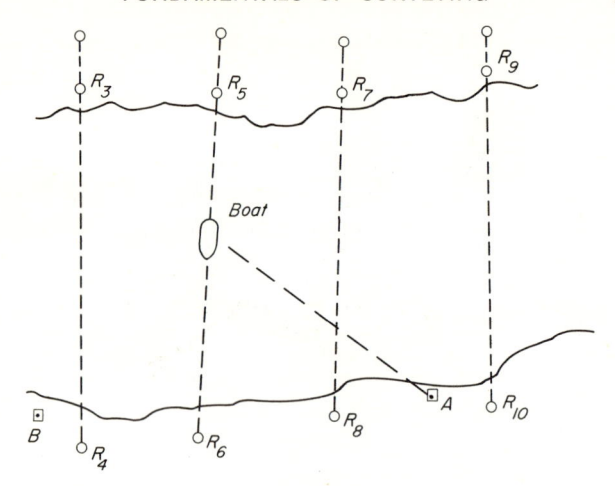

Fig. 14-27. Range Line and One Angle from Shore.

(a) *By Range Line and One Angle from Shore.* Figure 14-27 shows a common method for locating soundings on small lakes. The sounding craft is kept on a range line as defined by shore signals, and at various intervals a sounding is obtained at the same moment that the bow of the boat or any other appropriate part is "cut in" by a sight with a plane table at the shore station, A. It is necessary that the plane table operator and the boat party synchronize their watches before beginning operations and that they note and record the time for each sounding. Only in this way will it be possible to identify the position of each sounding when the depths are subsequently entered on the plane table sheet.

Although a transit can be utilized for locating soundings by this method, the plane table is much more suitable because it can be used to delineate the shoreline, obtain the topography, and secure other related information needed to complete the hydrographic survey.

In Fig. 14-28 a sounding rig working in a large reservoir is being kept on range by radio telephone from the transitman on shore while a plane table operator periodically measures the distance to the boat by stadia.

(b) *By Two Angles from Shore.* Where it is impracticable to establish ranges because of steep or heavily wooded shores, or where river currents make it difficult to keep the boat on range, the soundings may be located by angles read simultaneously from two

Fig. 14-28. Locating Soundings. *Tennessee Valley Authority*

transit positions on shore. At a given signal from the boat party both transitmen sight some definite object, such as the leadsman, on the sounding craft and read the horizontal angle. Figures 14-29 and 14-30 show the records of soundings and the transit observations at one shore station, respectively.

(c) *By Two Angles from the Boat.* An important and widely used method for locating a sounding is that known as the sextant three-point fix. This procedure involves the simultaneous measurement on board the sounding vessel of the two horizontal angles between three selected shore signals (Fig. 14-31) whose positions are known. The angles are measured with sextants at the same moment that the sounding is obtained. The position of the vessel is then immediately determined by using a three-arm protractor which effects a graphical solution of the three-point problem. The advantages of this method are that all the hydrographic surveying operations are performed on board the survey vessel, the frequency of soundings and their coverage of the area are conveniently discernible, and the boat may be directed to those parts of the lake, river, or harbor where additional development of hydrography seems necessary. Figure 14-32 shows the record of soundings and sextant angles when this method is employed.

The three preceding methods are equally applicable to leadline and echo-sounding operations. However, if a fathometer is used,

No.	Time h m	Obser. Depth (ft)	Lead-line Corr.	Datum Corr.	Total Corr.	Corr. Depth (ft)	SOUNDINGS
47	10 07	22	-1	-1	-2	20	Thuma-Chief
48	10 09	26	1	1	2	24	Talbot-Leadsman
49	10 11	30	2	1	3	27	Boyer-Notes
50	10 13	32	2	1	3	29	Clark-Oars
51	10 15	40	2	1	3	37	Leadline No.3
52	10 17	47	2	1	3	44	Staff gage:1.42ft
53	10 20	52	3	1	4	48	Elev. zero of staff gage=1347.79(msl)
54	10 22	62	3	1	4	58	Datum for soundings = Elev.1348.0
55	10 24	60	3	1	4	56	
56	10 27	58	3	1	4	54	
57	10 31	49	2	1	3	46	
58	10 33	40	2	1	3	37	
59	10 36	30	2	1	3	27	
60	10 38	22	1	1	2	20	
61	10 42	18	1	1	2	16	
62	10 45	15	1	1	2	13	
63	10 47	12	0	1	1	11	
64	10 49	7	0	1	1	6	
65	10 51	3	0	1	1	2	(Near shore by Sta. G)

Lake Benjamin
Windy
72°F
Beltrami, County
Minn.
July 29, 1955

Fig. 14-29. Sounding Notes (Two Angles from Shore Method).

the sextant three-point fix is particularly useful for locating soundings. At the precise moment the sextant angles are read, a button is pressed by the fathometer attendant and a mark is thereby registered on the fathogram. A pencil notation is then immediately made on the fathogram identifying the number of the position determination. This makes possible the subsequent scaling of intermediate depths from the fathogram between the consecutive sextant fixes.

14-37. Reduction of Soundings The observed soundings must be reduced or converted to values which would have been obtained if the water surface had been coincident with the selected datum or plane of reference. In tidal waters the stage or gage height of the sea above the datum is subtracted from the measured depth. The rapid fluctuation of the tide may make necessary the application of

LOCATION OF SOUNDINGS					
Station Occupied – ATHENS				Station Sighted – CASEY (0°-00')	
Location of Soundings			Lake Benjamin K & E No.5149 Cloudy 92°F		Beltrami County, Minn. Aug. 12, 1955
No.	Time h m	Azimuth ° '			
82	13 10	299 10	Hooper – transit		
83	13 12	302 15	Chamberlin – notes		
84	13 14	304 05			
85	13 17	308 55			
86	13 20	314 00			
87	13 22	316 50			
88	13 24	320 25			
89	13 27	323 15			
90	13 31	328 10			
91	13 33	330 25			
92	13 34	332 10			
93	13 37	336 30			
94	13 40	338 05			
95	13 43	340 55			
96	13 46	343 10			
97	13 50	348 15			
98	13 52	350 25			

Fig. 14-30. Transit Notes (Two Angles from Shore Method).

several different reduction factors in the course of one day's sounding operations.

In the case of waters not subjected to tidal influences, a similar but less variable correction must be applied. It is equal to the difference in elevation between the actual water surface and the selected datum and is usually negative. Figure 14-29 indicates the application of this correction to the soundings made in a small inland lake. It is

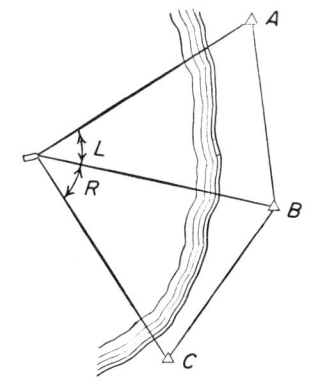

Fig. 14-31. Two Angles from Boat.

Angles			Signals			Obser. Depth (ft)	Total Corr.	Corr. Depth (ft)	Lake Benjamin Calm 88°F U. of Ill. Survey. Camp Beltrami County, Minn. June 22, 1955
Left °	'	Right °	'	Left	Center	Right			
29	16	73	20	A	B	F	6	−1	5	D. White Sextant No. 7120
29	36	74	08	A	B	F	11	1	10	Leadline No. 3
29	15	77	41	A	B	F	20	2	18	Eck-Chief
31	10	85	25	A	B	F	35	3	32	Easley-notes
107	50	33	27	G	I	J	55	4	51	Stokes-right angle
97	02	30	55	G	I	J	81	4	77	Hursh-left angle
90	43	29	00	G	I	J	96	5	91	Rees-leadsman
82	36	28	18	G	I	J	85	5	80	Schnoor-oars
74	00	27	03	G	I	J	66	4	62	Staff gage: 1.27 ft
65	48	27	03	G	I	J	30	3	27	Elev. zero of gage = 1347.79(msl)
62	16	26	38	G	I	J	20	2	18	Datum for soundings = 1348.0
59	00	26	44	G	I	J	6	1	5	Note: Total sounding correction
57	02	26	39	G	I	J	0	(shore)	0	equals leadline correction
58	02	22	10	G	I	J	12	1	11	minus one foot for datum
60	59	22	26	G	I	J	24	2	22	correction
65	01	22	43	G	I	J	36	3	33	
72	59	22	25	A	C	F	71	4	67	
82	58	18	54	A	C	F	128	6	122	
84	21	17	19	A	C	F	135	6	129	

Fig. 14-32. Record of Soundings and Sextant Angles (Three-Point Method).

to be noted that this correction is combined with the leadline correction of Art. 14-35 and the total correction is subtracted from the observed depths. Since depths were measured to the nearest foot only, all corrections were likewise expressed to the nearest foot.

14-38. Plotting Soundings The plotting of boat or sounding positions requires no explanation except in the case of a sextant three-point fix (two angles from the boat).

It is necessary to call attention to the use of the boat sheet without which a sextant fix could not be made on board the survey vessel as the hydrographic survey is conducted. The *boat sheet* is the work sheet used by the hydrographer to plot the soundings as soon as they are made. Its use enables the hydrographer to cover an area with properly spaced lines of soundings, to judge the completeness of the survey, and to determine where additional soundings or investigations are needed.

To permit the fixing of the successive positions of the sounding craft the boat sheet must contain the plotted positions of all shore signals. When the left and right sextant angles are simultaneously measured between the objects A, B, and C (Fig. 14-31), these angles are recorded and immediately set off on a three-arm protractor. The protractor is then moved over the boat sheet until the three arms pass through the plotted positions of the appropriate signals. When this is effected, a pencil mark is made at the center of the protractor. This indicates the position of the boat or the sounding.

The hydrographer plots the successive positions of the sounding craft as each new fix is obtained so as to determine whether the area is being systematically covered. Also, each day the depths from the fathogram are entered on the boat sheet.

The depth curves commonly drawn on nautical charts are depicted in Fig. 14-23. There they are the one-, two-, and three-fathom curves. Such a chart is usually multicolor.

PROBLEM

Below and on pp. 450 and 451 are field data from the topographic survey of a rural tract by transit-stadia. Make the proper reduction computations and prepare a map having a scale of 1 in. = 100 ft and a contour interval of 2 ft. Design a suitable title. Map should be preferably finished in ink of appropriate colors.

TRAVERSE

Course	Distance	Azimuth
AB	449.4	305°02′
BC	397.2	55°49′
CD	507.8	107°38′
DA	551.2	233°28′

DETAILS

Inst. at Station *A*   H.I. = 4.2   Elev. 842.9
$k = 100$, $(F + c)$ neglect

Sta.	Az.	Rod	Angle	Remarks
B	305°02′			
1	248°30′	0.25	(4.6)	East line of 20-ft drive, on line with N. highway fence.
2	267°35′	2.16	−1°16′	Fence corner
3	306°10′	1.20		S.W. corner of house, 40 × 30 ft
4	320°10′	0.90	+1°28′	S.E. corner of house
5	10°35′	1.38	−1°25′	S.W. corner of barn, 30 × 60 ft
6	30°30′	1.60	−1°28′	S.E. corner of barn
7	94°25′	1.34	(4.8)	Fence corner
8	100°35′	0.55	(4.0)	On north fence line of a road 60 ft wide.
B	305°01′	check		

The area within the enclosure about the house and barn is in grass; the area north and west of the enclosure is cultivated; north and east is pasture, except that north of the stream is a wood lot.

Inst. at Station *B*   H.I. = 4.2   Elev. 844.7

Sta.	Az.	Rod	Angle	Remarks
C	55°49′			
9	111°35′	3.02	−0°37′	N.W. corner of shed 20 × 20 ft
10	120°05′	3.16	(4.3)	N.W. corner of house
11	150°00′	3.07	−1°15′	Fence corner
12	201°25′	3.18		Center of road, 60 ft wide
13	203°50′	2.90	−2°28′	Fence corner
14	318°20′	1.72	(0.7)	Point on fence line
15	342°35′	3.82	(3.7)	Fence corner
C	55°50′	check		

Inst. at Station *C*   H.I. = 4.0   Elev. 831.1

Sta.	Az.	Rod	Angle	Remarks
B	235°49′			
16	5°35′	1.42	(8.5)	Fence corner
17	61°00′	2.91	−2°20′	On north fence in creek
18	149°00′	3.38	+0°50′	Fence corner
19	177°15′	2.90	+1°34′	On fence at end of 20 ft drive
20	211°25′	3.40	+1°45′	Fence corner
21	287°20′	4.65	+1°46′	Fence corner
B	235°47′	check		Pts. *16* and *19* are connected by a fence.

CARTOGRAPHIC SURVEYING

Inst. at Station *D*   H.I. = 4.2   Elev. 822.8

Sta.	Az.	Rod	Angle	Remarks
A	233°28′			
22	28°35′	3.36	+1°36′	Fence corner
23	33°05′	3.52		Centerline of N. and S. road
24	105°50′	1.68	(5.8)	On east fence in creek
25	152°30′	4.15		Center of cross roads
26	154°30′	3.74	+0°50′	Fence corner
27	215°00′	0.44	(6.2)	Bend in creek
28	246°05′	3.38	+2°12′	Fence corner
A	233°30′	check		

15 Engineering Surveying

15-1. Introduction Of the four major categories of professional surveying and mapping officially recognized (see Art. 1-17) by the American Society of Civil Engineers only one, *engineering surveying,* remains to be treated. Land surveying and cartographic surveying were the subjects of Chapters 13 and 14, respectively, and some insight into the field practices of geodetic surveying has been provided by Chapter 9.

Engineering surveying, according to ASCE, essentially includes the study and selection of sites for engineering construction, the procurement of design data, and the layout of engineering works. Hence, the operations of engineering surveying are closely allied to the various stages of project development. These stages are briefly described as follows:

1. *Investigation and Planning.* This phase of activity is concerned with the assessment of the relative merit of a suggested project and its construction suitability. An exploratory study, termed the reconnaissance, is made to provide a basis for a preliminary evaluation and to determine if detailed studies are warranted. Surveys in sufficient detail are executed to supplement information obtained from published maps and previous studies. The report covering this stage of activity presents a project plan and indicates whether it is physically feasible and economically justified.

2. *Design.* Surveying operations associated with design involve the full range of measuring capabilities needed to obtain dimensional data in adequate detail and with appropriate accuracy. Sometimes these surveys must comply with certain prescribed specifications.

3. *Construction Layout.* This activity consists of setting stakes and reference marks to denote the position of the various parts of the project (see Fig. 15-1).

Fig. 15-1. Tunnel Survey. *Chicago Metropolitan Sanitary District*

4. *Postconstruction.* Periodic surveys are made during the course of construction to determine the extent of finished work or the partial payment quantities. Upon the completion of all construction, final pay quantities are calculated from survey data. Occasionally, postconstruction surveys are made at regular intervals to measure structural behavior, such as the settlement of a building.

Engineering surveying accounts for a very small fraction of the total construction cost of most projects. Attempting to reduce survey costs by using inadequate methods, inexperienced personnel, or engaging a minimum-cost surveying consultant of uncertain capabilities is poor economy, because higher construction costs inevitably result. In general, well-planned and carefully executed engineering surveys save both time and money.

The scope of this chapter includes primarily a brief treatment of the principles of route surveying and the procedures for staking out construction projects. Certain special measurement operations, such as industrial or shop surveying, will be capsuled.

ROUTE SURVEYING

15-2. General Route surveying deals with all the field work and office studies performed in connection with the investigation of any route of transportation and the detailed layout of it. Route projects, such as those for railroads and highways, are so designed as to satisfy specific geometric criteria with respect to horizontal and vertical alinement.

The horizontal alinement of such projects consists of straight lines, termed *tangents,* connected by curves. The curves are usually arcs of circles or of *spirals.* The use of spiraled or *easement curves,* which provide a gradual transition between the tangents and the circular arcs, is not treated in this book.

The vertical alinement consists of straight sections of grade line connected by *vertical curves.* These curves are always parabolic in form because certain characteristics of the parabola facilitate the calculation and layout of the curve. The transverse section of highway pavement is likewise built to a parabolic form.

Although the treatment of route surveying presented here is directed primarily to highways and railroads, the principles are also generally applicable to such projects as pipelines, waterways, and electric transmission lines.

Highway and railroad surveys are executed to secure the essential data needed for the design of new construction or the improvement of existing facilities. Even though such surveys are not always carried out in the three classic steps traditionally associated with the location and construction of any route of transportation or communication, it will be instructive to describe them briefly here.

1. *Reconnaissance* is the study of the general feasibility of one or more possible corridors connecting specific termini. A favorable reconnaissance report recommends the detailed investigation of the most promising route.

2. The *preliminary survey* is a comprehensive study of the most feasible route. Its result is usually a *paper location* that defines on a map the position of the centerline for the subsequent location survey.

3. The *location survey* includes the staking of the centerline and

the procurement of all field data needed for design of the facility and the acquisition of right-of-way.

In the succeeding articles the elements of horizontal and vertical alinement will be first presented. Later, some of the related operations in highway surveying will be mentioned.

15-3. Horizontal Curves The various definitions and curve elements applicable to *circular curves* are as follows:

Degree of Curve, D. As used in this text, degree of curve (see Fig. 15-2) is defined as the central angle subtended by a 100-ft chord.

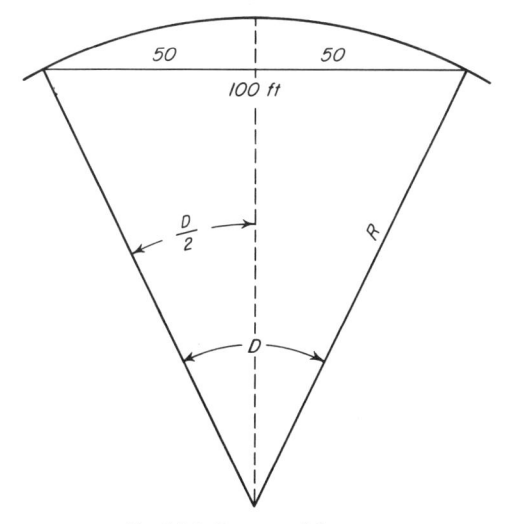

Fig. 15-2. Degree of Curve.

This is termed the *railroad,* or *chord, definition* of degree of curve. When D is $1°$, the curve is called a one-degree curve; when D is $2°$, it is called a two-degree curve, etc.

Point of Curve, P.C., is the point at which the curve departs from the tangent as one proceeds around the curve in the direction in which the stationing increases (see Fig. 15-3).

Point of Tangent, P.T., opposite the *P.C.,* marks the end of the curve and the beginning of the tangent.

456   FUNDAMENTALS OF SURVEYING   [15-3]

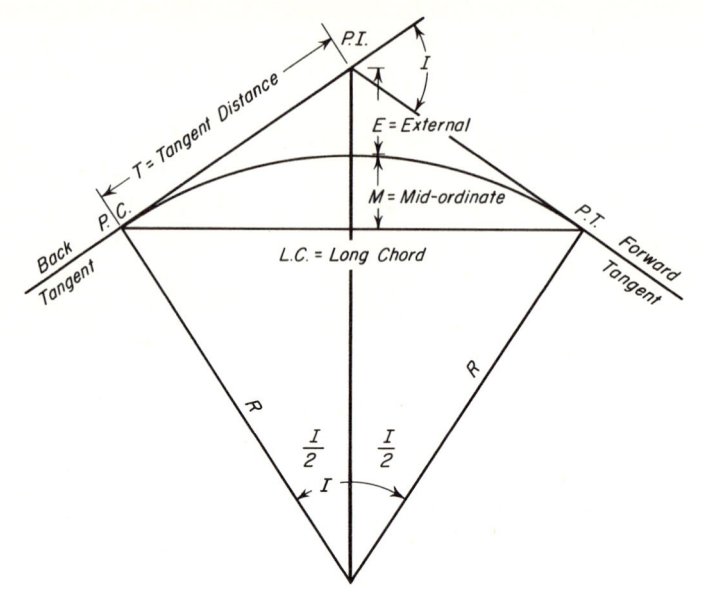

Fig. 15-3. Elements of a Circular Curve.

Point of Intersection, P.I. The tangents to a curve, produced, meet at a point called the *Point of Intersection, P.I.*

The *Intersection Angle, I,* is the angle formed by the intersection of the two tangents at the *P.I.*

The *Tangent Distance, T,* is the distance along the tangent from the *P.C.* or *P.T.* to the *P.I.* From Fig. 15-3 it is evident that

$$T = R \tan \frac{I}{2} \qquad (15\text{-}1)$$

The External, E, is the distance from the mid-point of the curve to the *P.I.* Evidently,

$$\frac{E + R}{R} = \sec \frac{I}{2}, \quad \text{or} \quad E = R \sec \frac{I}{2} - R$$

Then

$$E = R \left(\sec \frac{I}{2} - 1 \right) \qquad (15\text{-}2)$$

The Mid-Ordinate, M, is the perpendicular distance from the midpoint of the curve to the long chord.

Then $\dfrac{R - M}{R} = \cos \dfrac{I}{2}$ or $M = R - R \cos \dfrac{I}{2}$

Hence, $$M = R\left(1 - \cos \dfrac{I}{2}\right) \tag{15-3}$$

The Long Chord, L.C., is the chord joining the *P.C.* and *P.T.*

$$\dfrac{L.C.}{2} = R \sin \dfrac{I}{2} \quad \text{or} \quad L.C. = 2R \sin \dfrac{I}{2} \tag{15-4}$$

From Fig. 15-2 it is evident that

$$R = \dfrac{50}{\sin D/2} \tag{15-5}$$

If D is $1°$, R becomes 5729.65 ft. Also, since R varies inversely as $\sin D/2$, R will vary inversely (and almost exactly so) as D for the small values of D associated with the relatively flat curves of modern route engineering practice.

In general, for *approximate* calculations

$$R = \dfrac{5730}{D} \tag{15-6}$$

or

$$D = \dfrac{5730}{R} \tag{15-7}$$

Equation 15-7 permits the degree of curve to be readily obtained if the curve is defined by its radius. It is to be noted that a sharp curve has a short radius and a flat curve a long radius. The degree of curve on modern, high-speed highways is usually less than $4°$.

It is seldom necessary to calculate R from Eq. 15-5. When D is given and the curve elements involving R are to be calculated, the value of R should be taken from Table VIII.

The Length of Curve, L, is the sum of the chord distances around the curve from the *P.C.* to the *P.T.* This will invariably involve subchord lengths (see Fig. 15-4) adjacent to the *P.C.* and *P.T.* as well as the 100-ft chord lengths between full stations on the curve.

From the definition of D it follows that

$$L = \dfrac{I}{D} \times 100 \tag{15-8}$$

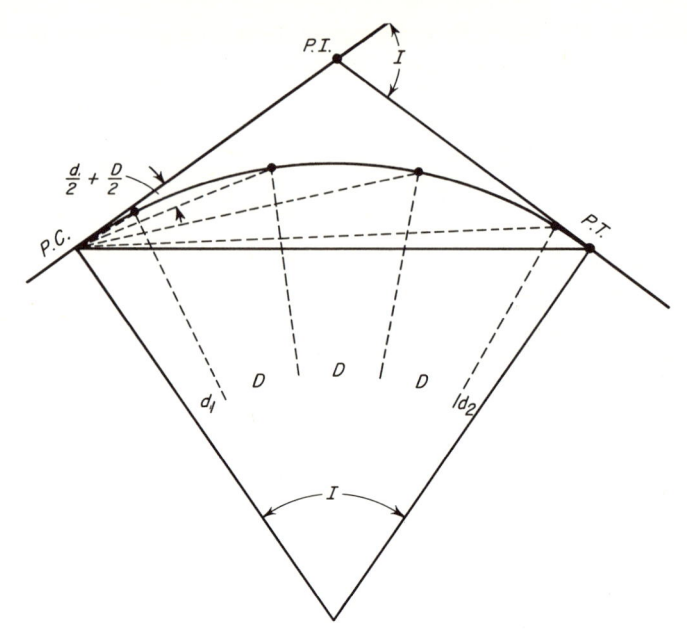

Fig. 15-4. Principle of Deflection Angles.

15-4. Tables of T and E It is to be noticed from an examination of Eqs. 15-1 and 15-2 that T and E, respectively, vary directly as R. This means they vary inversely (and almost exactly so) as D. Accordingly, values of T and E for any degree of curve, for a particular value of I, can be easily obtained by dividing the values of T and E for a 1° curve, having the same value of I, by the value of D. Hence,

$$T = \frac{T_{1°}}{D} \tag{15-9}$$

and

$$E = \frac{E_{1°}}{D} \tag{15-10}$$

Values of $T_{1°}$ and $E_{1°}$ are tabulated in Table IX for various values of I. For flat curves the agreement in the values of T calculated from Eqs. 15-1 and 15-9 should be very close. The value obtained from Eq. 15-1 is exact.

15-5. Principle of Deflection Angles The deflection-angle method is employed almost exclusively in laying out circular curves. It is based on the geometric principle that the angle between a tangent and a chord at a point on a circle is equal to one-half the angle subtended by the chord. Figure 15-4 depicts the essential relationships between deflection angles at the *P.C.* and the corresponding central angles. The first subchord is denoted by c and the first deflection angle, $(d/2)$, is calculated by

$$\left(\frac{d}{2}\right) = \frac{c}{100} \times \frac{D}{2} \qquad (15\text{-}11)$$

The increment of deflection for a full 100-ft chord is $D/2$. The last increment of deflection (for the final subchord) is likewise calculated from Eq. 15-11.

15-6. Calculation and Layout Before a curve can be staked out, it is necessary to extend the two established tangents to an intersection at the *P.I.*, measure *I*, and select the value of *D*. Sometimes the value of *D* is fixed by topographic or other considerations. It is important to observe that the station of the *P.I.* is determined by continuing the stationing along the *back tangent* to that point, and that the station of the *P.T.* (along the *forward tangent*) is ascertained by adding the calculated length of curve to the stationing of the *P.C.* From these data the necessary computations are made, after which the field work is executed. The successive steps in this procedure are as follows:

1. The various functions of the curve are computed by the use of formulas of Art. 15-3 or by the use of suitable tables (see Tables VIII and IX).

2. The deflection angles are computed and properly arranged in the field notebook.

3. The distance *T* is measured from the *P.I.* along each of the tangents to set the *P.T.* and the *P.C.*

4. The transit is set up at the *P.C.* and properly oriented.

5. The deflection angles are turned off with the transit and corresponding chords are measured, thus to establish the successive points along the curve.

The above procedure will be illustrated by the use of an example.

EXAMPLE 15-1: *Given* the following data for a circular curve: $I = 26°40'$; $D = 4°00'$; $P.I. = $ station $45 + 59.5$.

Required: the data and description of the field procedure to stake out this curve.

1. The various functions are found by the use of the formulas of Art. 15-3 and T and E are checked with Table IX.

$$R = 1432.7 \text{ ft} \quad \text{(Table VIII)}$$

$$T = R \tan \frac{I}{2} = 339.5 \text{ ft}$$

also

$$T = \frac{1358.0}{4} = 339.5 \text{ ft} \quad \text{(Table IX)}$$

$$E = R \sec \frac{I}{2} - R = 39.7 \text{ ft}$$

also

$$E = \frac{158.7}{4} = 39.7 \text{ ft} \quad \text{(Table IX)}$$

$$M = R\left(1 - \cos \frac{I}{2}\right) = 38.6 \text{ ft}$$

$$L.C. = 2R \sin \frac{I}{2} = 660.7 \text{ ft}$$

$$L = \frac{I}{D} \times 100 = 666.7 \text{ ft}$$

2. It is customary to arrange the notes for transit route surveys from the bottom of the page upward. This arrangement permits field sketches to be entered on the right-hand page of the notebook in a natural relation to the forward direction of the centerline of the survey. Thus a feature which appears on the right of the survey centerline may be sketched on the right of the notebook centerline. Following this procedure and the principles stated above, the deflection angles are calculated and arranged as shown in Fig. 15-5.

There it will be seen that the station number of the P.C. is found by subtracting the tangent distance from the P.I., after which the forward stationing proceeds along the curve to the P.T. Thus:

$$\begin{array}{r} 45 + 59.5 = P.I. \\ -3 + 39.5 = T \\ \hline 42 + 20.0 = P.C. \\ +6 + 66.7 = L \\ \hline 48 + 86.7 = P.T. \end{array}$$

Since the station of the P.C. is $42 + 20.0$, the distance from the P.C. to the first station on the curve (i.e., station $43 + 00$) is 80.0 ft.

[15-6] ENGINEERING SURVEYING 461

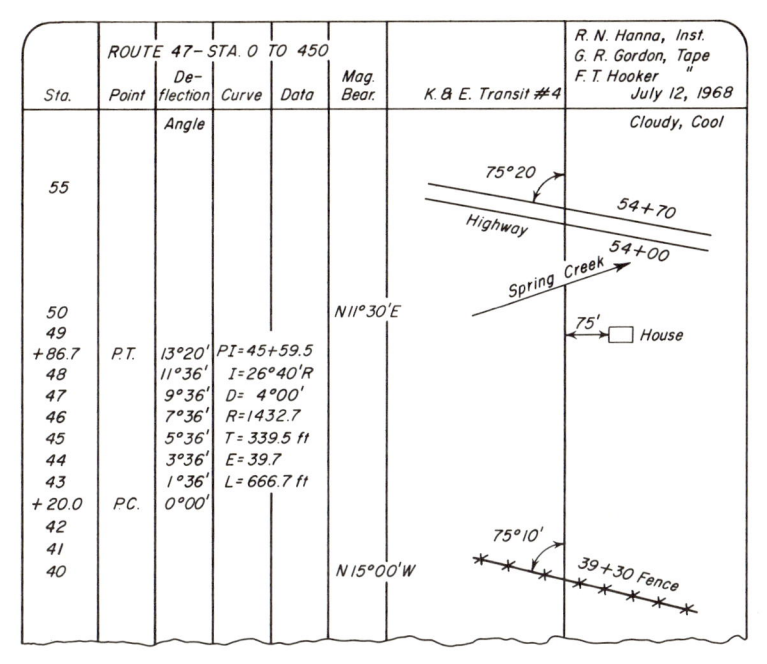

Sta.	Point	Deflection Angle	Curve Data	Mag. Bear.	K.&E. Transit #4	R. N. Hanna, Inst. G. R. Gordon, Tape F. T. Hooker " July 12, 1968
						Cloudy, Cool
55					75°20' Highway	54+70 54+00
50				N11°30'E	Spring Creek	
49						75' ☐ House
+86.7	P.T.	13°20'	PI=45+59.5			
48		11°36'	I=26°40'R			
47		9°36'	D= 4°00'			
46		7°36'	R=1432.7			
45		5°36'	T= 339.5 ft			
44		3°36'	E= 39.7			
43		1°36'	L= 666.7 ft			
+20.0	P.C.	0°00'				
42					75°10'	
41						39+30 Fence
40				N15°00'W		

Fig. 15-5. Notes for a Curve.

Accordingly, the deflection angle at the *P.C.* for a point on the curve 80 ft distant will be $0.8D/2 = 1°36'$. The next deflection angle for station 44 will be $1.8D/2 = 1°36' + 2°00' = 3°36'$; etc. The deflection angle for the *P.T.* will equal the deflection angle for station 48 plus that for a distance of 86.7 ft, or

$$11°36' + \frac{0.867D}{2} = 11°36' + 1°44' = 13°20'$$

Since this value is seen to be equal to *I/2*, a check is thus provided on the computation of all deflection angles.

3. The tapemen measure the distance *T* along the forward tangent from the *P.I.*, and set the *P.T.*, marked with its proper station number 48 + 86.7. Then they measure the same distance *T* along the initial tangent from the *P.I.* and set the *P.C.*, being station 42 + 20.0.

4. The transitman then sets his instrument up at the *P.C.*, sets the vernier *A* to read zero, and sights the *P.I.* with the telescope normal.

5. The transitman now turns the first deflection angle, 1°36', on

the *A* vernier and the tapemen measure a distance of 80.0 ft from the *P.C.*, thus to locate the first station, 43 + 00, on the curve. Next the transitman turns off the deflection angle for station 44 + 00 as shown in the notes, 3°36′, and the tapemen measure a full 100-ft station from the stake previously set to locate station 44 + 00. In like manner the successive stations are located around the curve. When the *P.T.* is reached and located, as were the previous stations, a check is provided by its proximity to the point previously set as described under step 3 above.

15-7. Intermediate Setup on Curve Occasionally, because of obstructions or the great length of the curve, it may be necessary to make intermediate setups. The procedures for making two intermediate setups on the curve of Art. 15-6 are described as follows:

(a) *At station 44 + 00 with backsight on P.C.* With telescope inverted and *A* vernier set at 0°00′, backsight on *P.C.* using lower motion. Release upper motion clamp, bring telescope to direct position, and set off 5°36′ in order to locate station 45 + 00.

The theory supporting the procedure is explained with the use of Fig. 15-6. The angle at 44 + 00 between the auxiliary tangent to the curve at that point and the chord to the *P.C.* is the same as the angle, 3°36′, that was used to set 44 + 00. Following the backsight on the *P.C.*, the telescope can be considered to be rotated in azi-

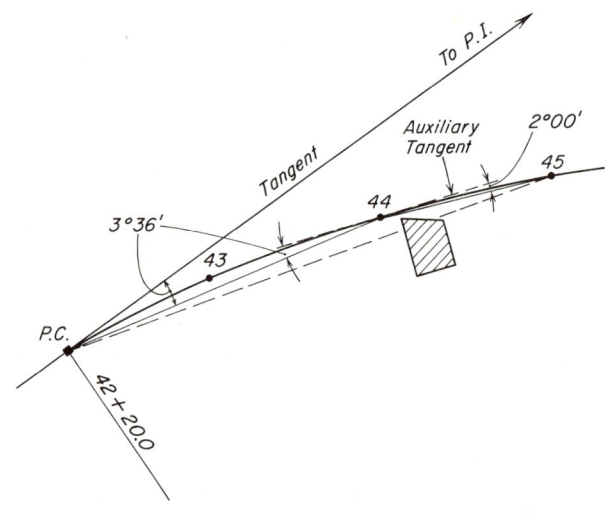

Fig. 15-6. Intermediate Setup.

muth until it reaches the auxiliary tangent. It is then brought to the normal position and rotated in azimuth by the amount, $D/2$, so that the reading used to set station $45 + 00$ is $5°36'$ or the same angular quantity opposite that point in the original notes.

(b) *At station $45 + 00$ with backsight on $44 + 00$.* With telescope inverted and A vernier set at $3°36'$ (with zero to inside of curve), backsight on $44 + 00$ using lower motion. Release upper motion clamp, bring telescope to direct position, and set off $7°36'$ in order to locate station $46 + 00$.

The procedure described above can be summarized by the following general rule: *With the transit in position at any station on a curve, the backsight is taken with the telescope inverted and the A vernier set to read the deflection angle of the point sighted. The telescope is then brought to normal position and the following stations on the curve are located by using the deflection angles previously computed and recorded in the notebook.*

15-8. Parabola as a Vertical Curve Three mathematical properties of the parabola render it especially convenient to use as a *vertical*

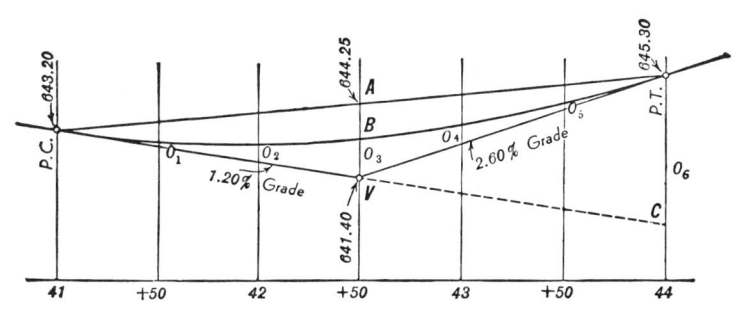

Fig. 15-7. Vertical Curve.

curve to connect two intersecting grades. These properties are illustrated in Fig. 15-7 and may be stated as follows:

1. That portion of the axis shown as AV is bisected by the curve at B.

2. Offsets from a tangent to the curve vary as the square of the distance from the point of tangency.

3. For a parabola used as a vertical curve, the second differences of the elevations of points spaced at equal horizontal intervals along the curve are equal.

The applications of these principles may be indicated as follows:
(a) The distance $AB = BV$.
(b) By the method given in this article, the offset, O_3, at the vertex is found directly from the given data. Then if the distances from the point of tangency, P.C., are expressed in stations, we have the relations

$$\frac{O_2}{O_3} = \frac{1.0^2}{1.5^2} \quad \text{or} \quad O_2 = \frac{4}{9} O_3; \quad \text{also} \quad \frac{O_1}{O_3} = \frac{0.5^2}{1.5^2} \quad \text{or} \quad O_1 = \frac{1}{9} O_3;$$

also

$$\frac{O_6}{O_3} = \frac{3.0^2}{1.5^2} \quad \text{or} \quad O_6 = 4 \times O_3$$

The offsets O_4 and O_5 are equal to O_2 and O_1, respectively.

(c) The second differences between the curve elevations are given in the table of Art. 15-9 and are seen to be constant except for the small variations due to dropping thousandths in the computed elevations.

15-9. Calculation of Vertical Curve The known data for the computation of a vertical curve include the station and elevation of the vertex, and the gradients of the two intersecting grade lines. Figure 5-28b shows a plotted profile which usually provides the basis for determining the gradients. The engineer selects the length of curve as some whole number of 100-ft stations. The length of the vertical curve is the horizontal distance from P.C. to P.T. Sometimes the length of curve is fixed by certain design criteria, such as the minimum sight distance over a summit vertical curve.

It is worthwhile emphasizing here that the term *grade* has two different meanings in engineering practice. It can indicate the slope, or gradient, of a line. Hence, a 1% grade is one that rises or falls 1 ft per 100 ft of horizontal distance. The term grade also indicates the final or finish elevation of some part of an engineering project.

Different methods are available for computing and checking the elevations for a vertical curve, but the one given here is simple and convenient. An example will illustrate the method (see Fig. 15-7).

EXAMPLE 15-2: *Given:* a -1.20% grade intersects a $+2.60\%$ grade at station $42 + 50$; elevation 641.40.

Required: Connect these grades with a vertical curve 300 ft long, using 50-ft stations.

First, the elevations of the 50-ft stations along the original grade lines are calculated.

Next, the value of the offset from the vertex to the curve is found to be one-half the distance from the vertex to the chord connecting the P.C. and P.T.

$$\text{Elev. of } A = \frac{\text{elev. of } P.C. + \text{elev. of } P.T.}{2} = 644.25. \text{ Then } AV =$$

$644.25 - 641.40 = 2.85$ ft; and $VB = 1.425$. The value, VB, may be taken as the offset from the tangent at V.

From Fig. 15-7 it is evident that the distances along the tangent from the P.C. to stations $41 + 50$ and 42 are $\frac{1}{3}$ and $\frac{2}{3}$, respectively, of the distance from P.C. to V. Since the offset O_3 at V is 1.425 ft, then, according to one of the principles stated above,

$$\text{the offset } O_1 = \left(\frac{1}{3}\right)^2 \text{ of } 1.425 = 0.158 \text{ ft}$$

$$\text{and the offset } O_2 = \left(\frac{2}{3}\right)^2 \text{ of } 1.425 = 0.632 \text{ ft}$$

As stated previously, offsets O_5 and O_4 are equal to O_1 and O_2, respectively.

Having computed the offsets, the elevations of the corresponding points on the curve are readily found, and the results are conveniently arranged as shown.

Sta.	Grade Elev.	Offsets	Curve Elev.	1st Diff.	2nd Diff.
41 P.C.	643.20	0.00	643.20		
				+0.44	
+50	642.60	.158	642.76		+0.31
				+0.13	
42	642.00	.632	642.63		+0.32
				−0.19	
+50	641.40	1.425	642.82		+0.32
				−0.51	
43	642.70	.632	643.33		+0.32
				−0.83	
+50	644.00	.158	644.16		+0.31
				−1.14	
44 P.T.	645.30	0.00	645.30		

According to another principle of Art. 15-8, the second differences of the curve elevations should be equal, and this condition is found to be true. This check is valid only when the points are at equal distances along the curve.

The parabola is used in the design of crowned pavements. Knowing the total rise or *crown* at the center of a pavement and the width, it is a simple matter to find the elevation of any intermediate point along a cross section of the roadway.

Thus in Fig. 15-8 the crown is represented by O_2, the rise, or

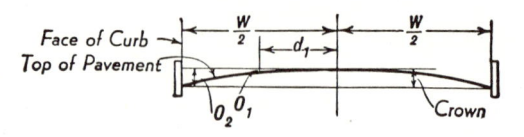

Fig. 15-8. Pavement Crown.

height of the center of the pavement above the gutter elevation. The offset O_1, at any other distance d_1, is given by the relation

$$\frac{O_1}{O_2} = \frac{d_1^2}{(W/2)^2}$$

For example, a pavement has the following dimensions: $W = 28$ ft, $d_1 = 7$ ft, and $O_2 = 4$ in. Then $O_1 = \frac{1}{4} \times O_2 = 1$ in.

15-10. Other Operations Once the project centerline has been established by the location survey it is necessary to secure additional data needed for final design and construction. The following summary has particular significance for surveys for major highways.

1. *Horizontal Control.* The beginning and end of the project should be connected with the national horizontal control network. This will permit the calculation on a state plane coordinate system of the highway centerline and provide adequate assurance that no blunders have been made in measuring distances and angles.

2. *Vertical Control.* Project elevations should be referenced to the 1929 mean sea-level datum and suitable ties should be made to governmental benchmarks at both ends of the project and, if feasible, at intermediate points also. It is customary to execute a line of "bench levels" along the route setting permanent benchmarks at intervals not to exceed one mile and in the vicinity of each structure. Supplementary or temporary benchmarks (**TBM**) are frequently set

at intervals of one-fourth to one-third mile. Leveling should be of third-order quality.

3. *Profile Leveling.* Profile levels (see Art. 5-21) originate from and close upon permanent and temporary benchmarks. In addition to determining ground elevations, to the nearest 0.1 ft, at all full stations, rod readings should be obtained at all significant changes in the surface profile and at intersections of the centerline with railroads and other highways.

4. *Cross-Section Leveling.* Cross sections (see Art. 5-22) are usually taken at all full stations and at such intermediate points as are required to ensure adequate coverage for the subsequent calculation of earthwork. Cross sections extend at least to the right-of-way line. Various expedients are utilized when cross sectioning through brush and rugged terrain. For example, a fiber-glass telescoping rod up to 25 ft in length may be useful. *Right-angle mirrors* (see Fig. 15-9) can be particularly helpful on wide sections.

Fig. 15-9. Right-Angle Mirror. *Keuffel & Esser Co.*

5. *Planimetric Detail.* The position is determined by aerial or ground (see Art. 6-13) methods of all dwellings, buildings, fences, streams, roads, large ornamental trees, and any other pertinent drainage and cultural features within the limits of the proposed right-of-way. This information is sometimes misleadingly termed "topography."

6. *Slope Staking.* If earthwork is to conform to a given alinement and side slope, it is necessary to set *slope stakes* to guide the contractor in his work. Thus, at a given cross section for a roadway, slope stakes are set where the proposed side slopes meet the ground surface, as illustrated in Fig. 15-10.

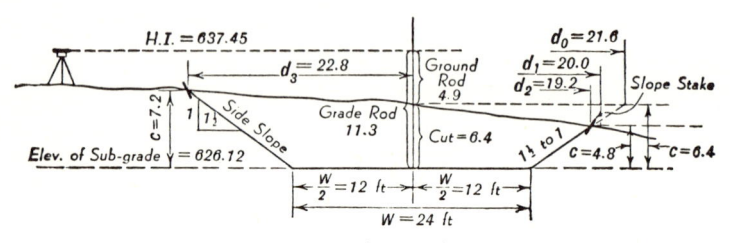

Fig. 15-10. Slope Staking.

If, following the execution of cross-section levels (see Art. 5-22), all cross sections have been plotted and the design cross section of the proposed roadway superimposed (see Fig. 5-30), it is necessary merely to scale the horizontal distance from the centerline to the intersection of the side slopes with the original ground surface and subsequently to lay off this distance in the field with a tape. However, slope stakes can also be located without such plotted cross sections by a trial-and-error procedure which will now be explained.

Figure 15-10 shows the conditions for a roadway in excavation, 24 ft wide, with side slopes to the ratio of 1½ (horizontal) to 1 (vertical). The height of instrument has been found by the usual process of differential leveling as indicated in the notes of Fig. 5-16. The elevation of *subgrade* is determined by the position of the grade line of the roadway as fixed from the profile study by the engineer.

The field party consists of four men: the engineer who supervises the work generally and who may keep the notes; a levelman; a rodman who, in addition to his rod, carries a 50-ft metallic tape; and a stakeman who aids the rodman in measuring distances and marks and drives stakes as directed.

It is more convenient to find the difference in elevation between the ground surface and the subgrade by means of rod readings than by computed elevations. This procedure makes use of a quantity called the *grade rod* which is found by subtracting the elevation of subgrade from the H.I., all rod readings and measured distances being taken to the nearest one-tenth of a foot only. Thus, the grade rod for the station shown is $637.4 - 626.1 = 11.3$. In other words, the grade rod is the reading that would be found on a rod if it could be held on the subgrade.

The *ground rod* at the center stake is 4.9 and obviously the cut is 6.4.

Having thus found the amount of cut at the center stake the party proceeds to locate a slope stake either to the right or left. On the right it is obvious that if the ground were level, as indicated by the dashed line, the distance to the slope stake would be $W/2 + 3/2 \times 6.4 = 12 + 9.6 = 21.6$ ft. But, as the rodman goes out from the center stake he notices that the ground slopes downward and that the position of the slope stake will be found at a distance less than 21.6 ft. Hence, he estimates a distance out d_1, say 20.0 ft, and holds his rod for a ground-rod reading. Suppose this reading is 6.5; then the indicated cut (grade rod–ground rod) is 4.8, for which the calculated distance out from the center, d_2, is $12 + 3/2 \times 4.8 = 19.2$ ft. However, the rodman estimated the distance and held the rod at a measured distance of 20.0 ft. There is a discrepancy, therefore, of 0.8 ft between the measured distance and the calculated distance from the centerline. Accordingly, the rod must be moved inward and, if the ground is assumed to be level for that small distance, the cut will be, as before, 4.8 ft, and the correct location of the slope stake will be 19.2 ft.

Sometimes two or three trials are necessary before the correct location of the slope stake is found, but it is always fixed at that point where the *measured* distance is equal to the *calculated* distance from the centerline.

In a similar manner the slope stake on the left side is found at a point where the cut (i.e., the distance above subgrade) is 7.2 ft and $d_3 = 22.8$ ft.

The amount of the cut at the center is marked on the back of the center stake. Each slope stake has the cut (or fill) marked on the side facing the centerline and the station number on the back.

The procedure is similar in the case of a fill, except that the ground rod will be greater than the grade rod, and it is this condition that enables the party to determine whether any doubtful point marks a cut or a fill.

To avoid confusion it is always customary to subtract the ground rod from the grade rod. If the result is positive, it represents a cut; if negative, it represents a fill.

7. *Land Ties.* The location of all section and property corners is determined with respect to the highway centerline. At the intersection with a section line, for example, the stationing is determined, the angle between the section line and the centerline is measured, and the distances along the section line to the nearest flanking land corners are ascertained.

8. *Interchange Sites.* It is desirable to prepare a contour map of all interchange sites so as to facilitate the study of drainage and all design features.

9. *Drainage Surveys.* Information concerning the stream profile, tributary area, dimensions of culverts under nearby highways and railroads upstream and downstream from the project centerline is secured.

10. *Utility Surveys.* Such surveys consist of locating power, telephone and telegraph lines, transmission lines, sewer, water, and oil pipelines wherever they cross the project centerline or run diagonally or parallel to the centerline within the limits of the right-of-way. Elevations are obtained for the inverts of sewers and for the tops of the manholes.

11. *Miscellaneous.* In order to preserve the accepted project centerline during the interval between the initial stakeout and the beginning of construction, all P.I.'s must be durably monumented. Adequate reference ties to ensure the accurate recovery of these markers should be secured and documented. On long tangents additional monuments termed points on tangent, P.O.T.'s, are usually set and suitably referenced.

Changes in alinement affecting the stationing generally require the use of a *station equation* in order to avoid changing existing stationing. Hence, the station equation at a particular centerline point might be *237 + 16.42 Back = 237 + 81.05 Ahead.* This means that station 237 + 00 is 16.42 ft back of this point and station 238 + 00 is 18.95 ft ahead.

CONSTRUCTION SURVEYING

15-11. General *Construction surveying* is primarily concerned with the establishment of certain lines and grades that guide and control construction operations. In short, construction surveying deals with the transfer of design dimensions from the engineering drawings to the ground so that the project is built in the correct position and with the proper relationship between its component parts.

There are no principles involved in the practice of construction surveying that have not been treated in prior chapters of this book. However, the project engineer will occasionally find that the problems of layout will tax his ingenuity and he will be unable to discover a precedent for a successful solution either in a textbook or in his past experience. It will remain for him to assess the field problem as intelligently as possible and devise a solution that will be sufficiently accurate and economical (see Fig. 15-11).

Fig. 15-11. Surveying for Hydroelectric Plant. *Harry R. Feldman, Inc.*

It is emphasized that before any layout work is begun the engineer should examine the major stakes or monuments that will control construction operations. For example, it is of critical importance that benchmarks be checked to reveal possible disturbance. The reference ties (see Art. 6-4) to key markers should be remeasured. Furthermore, it is essential to make certain that the correct elevation is used for any benchmark. An undetected mistake in the primary horizontal or vertical control could have serious consequences.

In addition to the preceding precautionary measures, all tapes and leveling rods should be checked and levels and transits tested and, if necessary, adjusted.

Layout surveys take many forms and require various grades of accuracy in their execution. Included are staking operations for horizontal and vertical curves, land subdivisions, buildings, sewers, earthwork, and bridges. Only the essential features of a few representative types will be treated in this chapter.

15-12. Construction Grid System On major construction projects, such as at an industrial plant site, it is customary to establish a *construction grid system* to facilitate layout surveys and the recovery of important reference points which may become lost or disturbed. The grid axes are straight lines exactly at right angles to each other and should have their ends securely monumented with heavy poured-in-place concrete posts with metal tablets embedded in their tops. Along both axes are set stout $2'' \times 2''$ tacked stakes at intervals of exactly 100 ft. The station values assigned to the west and south ends of the axes, which are preferably alined with the cardinal directions, are sufficiently large to preclude the possibility of negative station values being developed if the survey is extended to the south or west. In order to determine the relationship between the construction grid system coordinates and those of the state-wide plane coordinate system (see Art. 11-23), a line of connecting transit-traverse is run to the nearest government traverse or triangulation stations.

It is highly desirable that the elevation, preferably on the 1929 Mean Sea Level Datum, be determined for several strategically situated points in the construction area. This is particularly important when the industrial plant will have a complex of underground utilities that must be set correctly in elevation.

15-13. Grade Stakes In constructing any project to a given grade it is necessary that *grade stakes* be set to guide the contractor in his work. A grade stake is one driven until its top has the same elevation as the grade of the finished work, or until it has a known relation to that grade. In many cases, also, it fixes the alinement of the project. Examples are grade stakes for street pavements, sidewalks, sewers, railways, and highways.

1. *Street Pavements.* Grade stakes for street pavements are usually set outside (i.e, away from the centerline) of the curb about 2 or 3 ft, and are driven to fix the elevation of the top of the curb.

The stakes are set at 50-ft intervals when the grade is uniform, and at 25-ft intervals on vertical curves, and are carefully set to fix the alinement of the back of the curb.

The level party then sets up the level in a convenient location, its H.I. being determined by differential levels from a nearby benchmark. The difference between the H.I. and the grade elevation of any given grade stake is the rod reading, or grade rod, for that stake. The stake is then driven down until, after repeated trials, the top of the stake has the desired elevation. Finally, the alinement is fixed on the stakes by tacks carefully lined in with a transit.

The tops of grade stakes are usually colored with red or blue keel to distinguish them from other stakes and to assure the contractor that they are at grade. If the location of the stakes is on a high bank such that much excavation would be necessary to set the stakes to grade, they may be set at a height of, say, 2 ft above grade. The contractor then measures down this amount to fix the elevation of the forms for the pavement.

2. *Sewers.* The grade line for a sewer is commonly established by fixing a line of sight, or by stretching a string, a known distance above the grade. At regular intervals of perhaps 50 ft along the centerline of the sewer, two stout stakes are driven, one on either side of the centerline. Having determined the grade rod for a given station, as indicated above, the rod is slid up or down along the stakes and a mark is made to establish the desired elevation. Then a cross piece or "batter board" (see Fig. 15-12) is nailed to the two stakes, its top level and the elevation indicated by the mark. Thus, a series of batter boards are established and, if a string is stretched taut over the tops of these boards, it fixes a grade line at a known distance above the grade of the sewer. The workmen then measure this distance down from any point along the string to establish a point on the grade of the sewer.

3. *Railways.* Railway track is brought to its final grade by means of grade stakes. These are usually driven to the elevation of the top of rail and the track is then raised by tamping ballast underneath until the top of rail is level with the grade stakes.

15-14. Building Layout Before the lines of a building are established, it is frequently necessary to execute a property survey to

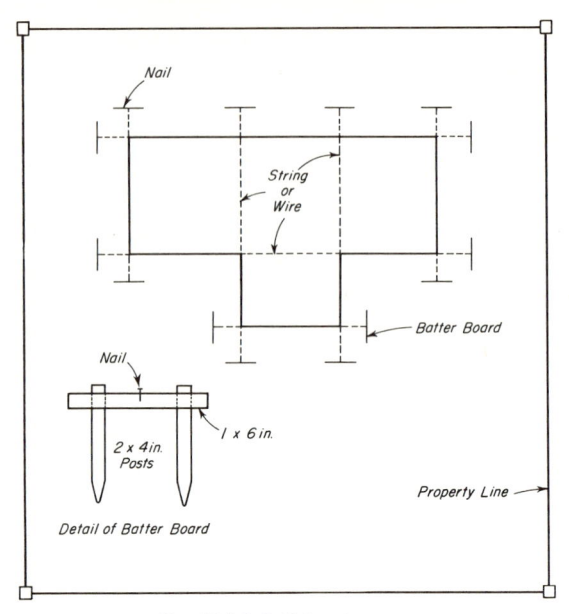

Fig. 15-12. Building Layout.

locate the boundaries of the tract. Then the project engineer will proceed with the detailed location of the building with respect to the property lines.

Since stakes placed at the corners of the building would be continuously disturbed during excavation and construction operations, batter boards are erected. The upper edges of these boards are sometimes set at some special elevation, such as that of the top of the foundation wall. Figure 15-12 shows that strings or wires connecting these boards serve to define certain building lines, which are of great importance to the contractor. Nails driven into the top of the boards mark the exact position of the lines.

For building layouts of limited extent and for excavation measurements an inexpensive instrument combining the major characteristics of level and transit is frequently used by contractors. A *builders transit-level* is shown in Fig. 15-13.

15-15. Bridges and Tunnels The treatment of bridge and tunnel surveys and those for very large structures like dams lies beyond the scope of this book. Nevertheless, some of the basic field oper-

Fig. 15-13. Builder's Transit-Level. *Keuffel & Esser* Co.

ations associated with defining line and grade are illustrated in the earlier chapters where the student was introduced to the use of surveying equipment. In Fig. 15-14 is depicted an unusual instrumental setup during pier construction.

15-16. Laser Instrument A *laser instrument* developed for surveying applications is shown in Fig. 15-15. It consists of a low-power laser mounted between the trunnions of an engineer's transit. The telescope is beneath the laser and is parallel with it in the vertical plane.

The laser emits a continuous red light beam which remains essentially parallel. The Transit-Lite is used to project a straight reference line or plane to guide dredging and tunneling operations. Equipment operators can easily perform alinement checks by either looking at the beam or referencing it to a target.

Fig. 15-14. Unsual Instrumental Setup. Harry R. Feldman, Inc.

Fig. 15-15. Transit-Lite Laser Instrument. *Spectra Physics*

INDUSTRIAL SURVEYING

15-17. Optical Tooling Occasionally the engineer may be faced with the problem of providing very precise dimensional control in the erection and alinement of turbines, jigs, and other machine elements. The term, *optical tooling,* refers to surveying techniques that have been introduced into the aircraft and other industries to make accurate dimensional layouts possible. *Industrial surveying,* as it is also sometimes called, originally utilized the conventional transit and level to define lines and planes of reference in the shop. Modern industrial layouts and shop practices frequently permit dimensional tolerances of only a few thousandths of an inch. The stringency of these requirements led to the development of new instruments for conducting such work. Some are wholly new in design, whereas others, like the *jig transit,* represent modifications of conventional surveying equipment.

The jig transit (see Fig. 15-16) is designed especially for optical

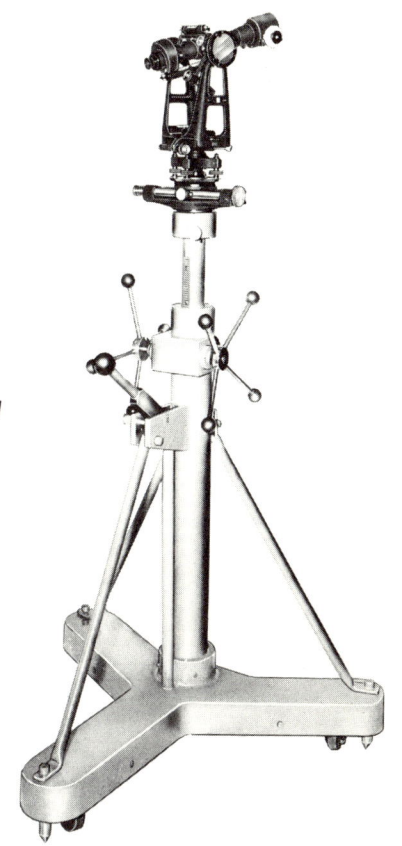

Fig. 15-16. Jig Transit. *Keuffel & Esser* Co.

tooling. It is used principally to establish with precision vertical planes in industrial layout work. It differs from an ordinary transit in that it has no horizontal or vertical circles, no compass, and only one horizontal motion. It is commonly mounted on a heavy metal stand. Since many shop sights are very short, the ability of the telescope to focus on a point as close as 3 ft from the instrument center is most essential.

15-18. Electron Accelerator Surveys New techniques have been developed in recent years in connection with layout surveys for

Fig. 15-17. Electron Accelerator Layout. *Harry R. Feldman, Inc.*

electron accelerators. The accuracy requirements have been extremely exacting. Depicted in Fig. 15-17 are typical operations needed to set control monuments on a circle 244 ft in diameter with an uncertainty of not more than 0.0005 ft.

PROBLEMS

15-1. Given: $I = 32°10'R$, $D = 4°00'$, P.I. $= 62 + 05.2$. Prepare a page of notes needed for the layout of this curve from the P.C. with stakes at the full stations.

15.2. Given: $I = 20°22'L$, $D = 2°10'$, P.I. $= 37 + 18.9$. Prepare a page of notes needed for the layout of this curve from the P.C. with stakes at the full and half-stations.

15-3. For the preceding problem, calculate the station number of the mid-point of the curve, the deflection angle needed to locate it, and the value of the external distance.

15-4. Given: $I = 16°23'L$, $D = 2°15'$, P.I. $= 81 + 14.5$.

(a) Prepare a page of notes needed for the layout of this curve from the P.C. with stakes at the full stations. (b) Assuming obstructions are encountered, explain completely how to make an intermediate setup at

(1) station $79 + 00$ with backsight on P.C.
(2) station $82 + 00$ with backsight on $80 + 00$

15-5. A +2.40% grade meets a −1.70% grade at station 32 + 50, elevation 522.14. Calculate the elevations of all full stations on a vertical curve 600 ft long.

15-6. A −1.20% grade meets a +2.50% grade at station 51 + 00, elevation 127.60. Calculate the elevations of all full and half-stations on a vertical curve 450 ft long.

15-7. The elevation of the top of pavement on the centerline at station 20 + 00 is 645.30. The grade is −0.80%, the pavement is 32 ft wide face to face of curb, and the crown is 4 in. What will be the grade elevation of a point at station 20 + 40 and 12 ft distant, at right angles, from the centerline?

REFERENCES

1. Hickerson, Thomas F., *Route Location and Design,* 5th ed., McGraw-Hill Book Co., New York, 1967.
2. Meyer, Carl F., *Route Surveying,* 3rd ed., International Textbook Co., Scranton, Pa., 1962.
3. Surveying and Mapping Division, *Report on Highway and Bridge Surveys,* Manual No. 44, American Society of Civil Engineers, New York, 1962.
4. Surveying and Mapping Division, *Report on Pipeline Location,* Manual No. 46, American Society of Civil Engineers, New York, 1965.

Tables

TABLE I GREENWICH HOUR ANGLE OF POLARIS
for 0ʰ Greenwich Civil Time
1965

Day	Jan. ° ′	Feb. ° ′	March ° ′	April ° ′	May ° ′	June ° ′
1	72 07.6	102 50.5	130 34.4	161 12.7	190 46.3	221 13.8
2	73 07.0	103 50.0	131 33.8	162 12.0	191 45.4	222 12.7
3	74 06.4	104 49.4	132 33.2	163 11.2	192 44.4	223 11.5
4	75 05.9	105 48.9	133 32.5	164 10.4	193 43.4	224 10.4
5	76 05.3	106 48.3	134 31.9	165 09.5	194 42.4	225 09.3
6	77 04.7	107 47.8	135 31.3	166 08.7	195 41.4	226 08.1
7	78 04.2	108 47.2	136 30.6	167 07.9	196 40.4	227 07.0
8	79 03.6	109 46.7	137 30.0	168 07.1	197 39.4	228 05.8
9	80 03.0	110 46.1	138 29.3	169 06.2	198 38.4	229 04.7
10	81 02.5	111 45.5	139 28.6	170 05.4	199 37.4	230 03.5
11	82 01.9	112 45.0	140 28.0	171 04.6	200 36.4	231 02.4
12	83 01.4	113 44.4	141 27.3	172 03.7	201 35.3	232 01.2
13	84 00.8	114 43.9	142 26.6	173 02.9	202 34.3	233 00.1
14	85 00.3	115 43.3	143 26.0	174 02.0	203 33.3	233 58.9
15	85 59.7	116 42.7	144 25.3	175 01.1	204 32.3	234 57.7
16	86 59.2	117 42.2	145 24.6	176 00.2	205 31.2	235 56.6
17	87 58.6	118 41.6	146 23.9	176 59.3	206 30.1	236 55.4
18	88 58.1	119 41.0	147 23.2	177 58.5	207 29.0	237 54.2
19	89 57.5	120 40.4	148 22.5	178 57.6	208 28.0	238 53.1
20	90 57.0	121 39.8	149 21.8	179 65.7	209 26.9	239 51.9
21	91 56.5	122 39.2	150 21.0	180 55.8	210 25.8	240 50.7
22	92 55.9	123 38.6	151 20.3	181 54.8	211 24.8	241 49.5
23	93 55.4	124 38.0	152 19.6	182 53.9	212 23.7	242 48.3
24	94 54.9	125 37.5	153 18.8	183 53.0	213 22.6	243 47.1
25	95 54.3	126 36.9	154 18.1	184 52.1	214 21.5	244 46.0
26	96 53.8	127 36.3	155 17.4	185 51.1	215 20.4	245 44.8
27	97 53.2	128 35.7	156 16.6	186 50.2	216 19.3	246 43.6
28	98 52.7	129 35.0	157 15.8	187 49.2	217 18.2	247 42.4
29	99 52.1		158 15.1	188 48.3	218 17.1	248 41.2
30	100 51.6		159 14.3	189 47.3	219 16.0	249 40.0
31	101 51.1		160 13.5		220 14.9	

TABLE II INCREASE IN GHA FOR ELAPSED TIME SINCE 0ʰ GCT

Hrs.	Corr. ° ′	Min.	Corr. ° ′	Min.	Corr. ° ′	Sec.	Corr. ′	Sec.	Corr. ′
1	15 02.5	1	0 15.0	31	7 46.3	1	0.3	31	7.8
2	30 04.9	2	0 30.1	32	8 01.3	2	0.5	32	8.0
3	45 07.4	3	0 45.1	33	8 16.4	3	0.8	33	8.3
4	60 09.9	4	1 00.2	34	8 31.4	4	1.0	34	8.5
5	75 12.3	5	1 15.2	35	8 46.4	5	1.3	35	8.8
6	90 14.8	6	1 30.2	36	9 01.5	6	1.5	36	9.0
8	105 17.2	7	1 45.3	37	9 16.5	7	1.8	37	9.3
7	120 19.7	8	2 00.3	38	9 31.6	8	2.0	38	9.5
9	135 22.2	9	2 15.4	39	9 46.6	9	2.3	39	9.8
10	150 24.6	10	2 30.4	40	10 01.6	10	2.5	40	10.0
11	165 27.1	11	2 45.5	41	10 16.7	11	2.8	41	10.3
12	180 29.6	12	3 00.5	42	10 31.7	12	3.0	42	10.5
13	195 32.0	13	3 15.5	43	10 46.8	13	3.3	43	10.8
14	210 34.5	14	3 30.6	44	11 01.8	14	3.5	44	11.0
15	225 37.0	15	3 45.6	45	11 16.8	15	3.8	45	11.3
16	240 39.4	16	4 00.7	46	11 31.9	16	4.0	46	11.5
17	255 41.9	17	4 15.7	47	11 46.9	17	4.3	47	11.8
18	270 44.4	18	4 30.7	48	12 02.0	18	4.5	48	12.0
19	285 46.8	19	4 45.8	49	12 12.0	19	4.8	49	12.3
20	300 49.3	20	5 00.8	50	12 32.1	20	5.0	50	12.5
21	315 51.7	21	5 15.9	51	12 47.1	21	5.3	51	12.8
22	330 54.2	22	5 30.9	52	13 02.1	22	5.5	52	13.0
23	345 56.7	23	5 45.9	53	13 17.2	23	5.8	53	13.3
24	360 59.1	24	6 01.0	54	13 32.2	24	6.0	54	13.5
		25	6 16.0	55	13 47.3	25	6.3	55	13.8
		26	6 31.1	56	14 02.3	26	6.5	56	14.0
		27	6 46.1	57	14 17.3	27	6.8	57	14.3
		28	7 01.1	58	14 32.4	28	7.0	58	14.5
		29	7 16.2	59	14 47.4	29	7.3	59	14.8
		30	7 31.2	60	15 02.5	30	7.5	60	15.0

TABLE III BEARING OF POLARIS AT ALL LOCAL HOUR ANGLES
1965

For local hour angles 0° to 180° the star is west of north, and from 180° to 360° it is east of north.

Lat. LHA	30°	32°	34°	36°	38°	40°	42°	44°	46°	48°	Lat. LHA
0	0 0.0	0 0.0	0 0.0	0 0.0	0 0.0	0 0.0	0 0.0	0 0.0	0 0.0	0 0.0	360
5	0 5.8	0 5.9	0 6.0	0 6.2	0 6.4	0 6.5	0 6.8	0 7.0	0 7.2	0 7.5	355
10	0 11.5	0 11.7	0 12.0	0 12.3	0 12.7	0 13.0	0 13.5	0 13.9	0 14.4	0 15.0	350
15	0 17.1	0 17.5	0 17.9	0 18.4	0 18.9	0 19.4	0 20.1	0 20.7	0 21.5	0 22.3	345
20	0 22.6	0 23.1	0 23.7	0 24.3	0 24.9	0 25.7	0 26.5	0 27.4	0 28.4	0 29.5	340
25	0 27.9	0 28.5	0 29.2	0 30.0	0 30.8	0 31.7	0 32.7	0 33.8	0 35.1	0 36.4	335
30	0 33.0	0 33.8	0 34.6	0 35.4	0 36.4	0 37.5	0 38.7	0 40.0	0 41.5	0 43.1	330
35	0 37.9	0 38.7	0 39.6	0 40.6	0 41.7	0 43.0	0 44.3	0 45.8	0 47.5	0 49.4	325
40	0 42.4	0 43.4	0 44.4	0 45.5	0 46.8	0 48.1	0 49.7	0 51.3	0 53.2	0 55.3	320
45	0 46.7	0 47.7	0 48.8	0 50.0	0 51.4	0 52.9	0 54.6	0 56.4	0 58.5	1 0.8	315
50	0 50.5	0 51.6	0 52.8	0 54.2	0 55.6	0 57.3	0 59.1	1 1.1	1 3.3	1 5.7	310
55	0 54.0	0 55.1	0 56.4	0 57.9	0 59.4	1 1.2	1 3.1	1 5.2	1 7.6	1 10.2	305
60	0 57.0	0 58.3	0 59.6	1 1.1	1 2.8	1 4.6	1 6.6	1 8.9	1 11.4	1 14.1	300
65	0 59.6	0 0.9	1 2.3	1 3.9	1 5.6	1 7.5	1 9.6	1 12.0	1 14.6	1 17.5	295
70	1 1.8	1 3.1	1 4.6	1 6.2	1 8.0	1 9.9	1 12.1	1 14.5	1 17.2	1 20.2	290
75	1 3.5	1 4.8	1 6.3	1 8.0	1 9.8	1 11.8	1 14.0	1 16.5	1 19.3	1 22.3	285
80	1 4.6	1 6.0	1 7.5	1 9.2	1 11.1	1 13.1	1 15.4	1 17.9	1 20.7	1 23.8	280
85	1 5.3	1 6.7	1 8.3	1 9.9	1 11.8	1 13.9	1 16.2	1 18.7	1 21.5	1 24.6	275

TABLE III BEARING OF POLARIS AT ALL LOCAL HOUR ANGLES—(Continued)
1965

For local hour angles 0° to 180° the star is west of north, and from 180° to 360° it is east of north.

Lat. LHA	30°	32°	34°	36°	38°	40°	42°	44°	46°	48°	Lat. LHA
90	1 5.5	1 6.9	1 8.5	1 10.1	1 12.0	1 14.1	1 16.4	1 18.9	1 21.7	1 24.8	270
95	1 5.2	1 6.6	1 8.1	1 9.8	1 11.7	1 13.7	1 16.0	1 18.5	1 21.3	1 24.3	265
100	1 4.4	1 5.8	1 7.3	1 8.9	1 10.8	1 12.8	1 15.0	1 17.5	1 20.2	1 23.2	260
105	1 3.1	1 4.5	1 5.9	1 7.5	1 9.3	1 11.3	1 13.5	1 15.9	1 18.6	1 21.5	255
110	1 1.4	1 2.7	1 4.1	1 5.6	1 7.4	1 9.3	1 11.4	1 13.7	1 16.3	1 19.2	250
115	0 59.2	1 0.4	1 1.7	1 3.3	1 4.9	1 6.7	1 8.8	1 11.0	1 13.5	1 16.3	245
120	0 56.5	0 57.7	0 59.0	1 0.4	1 2.0	1 3.7	1 5.6	1 7.8	1 10.1	1 12.8	240
125	0 53.4	0 54.5	0 55.7	0 57.1	0 58.6	1 0.2	1 2.0	1 4.0	1 6.3	1 8.7	235
130	0 49.9	0 50.9	0 52.1	0 53.3	0 54.7	0 56.3	0 57.9	0 59.8	1 1.9	1 4.2	230
135	0 46.0	0 47.0	0 48.0	0 49.2	0 50.5	0 51.9	0 53.4	0 55.2	0 57.1	0 59.2	225
140	0 41.8	0 42.7	0 43.6	0 44.7	0 45.8	0 47.1	0 48.5	0 50.1	0 51.8	0 53.8	220
145	0 37.3	0 38.1	0 38.9	0 39.8	0 40.9	0 42.0	0 43.3	0 44.7	0 46.2	0 47.9	215
150	0 32.5	0 33.2	0 33.9	0 34.7	0 35.6	0 36.6	0 37.7	0 38.9	0 40.3	0 41.7	210
155	0 27.5	0 28.0	0 28.6	0 29.3	0 30.1	0 30.9	0 31.8	0 32.9	0 34.0	0 35.3	205
160	0 22.2	0 22.7	0 23.2	0 23.7	0 24.3	0 25.0	0 25.8	0 26.6	0 27.5	0 28.5	200
165	0 16.8	0 17.2	0 17.5	0 17.9	0 18.4	0 18.9	0 19.5	0 20.1	0 20.8	0 21.6	195
170	0 11.3	0 11.5	0 11.8	0 12.0	0 12.4	0 12.7	0 13.1	0 13.5	0 14.0	0 14.5	190
175	0 5.7	0 5.8	0 5.9	0 6.0	0 6.2	0 6.4	0 6.6	0 6.8	0 7.0	0 7.3	185
180	0 0.0	0 0.0	0 0.0	0 0.0	0 0.0	0 0.0	0 0.0	0 0.0	0 0.0	0 0.0	180

This table has been computed for a polar distance of 0°56.8'. For other polar distances the correction from Table V should be applied.

TABLE IV POLAR DISTANCE OF POLARIS
1965

Date	Polar Distance	Date	Polar Distance
	° ′		° ′
January 1	0 56.7	July 1	0 57.2
February 1	0 56.6	August 1	0 57.1
March 1	0 56.7	September 1	0 57.0
April 1	0 56.8	October 1	0 56.9
May 1	0 57.0	November 1	0 56.7
June 1	0 57.1	December 1	0 56.5

TABLE V CORRECTIONS TO PRELIMINARY BEARINGS OF POLARIS AS OBTAINED FROM TABLE III

Polar Distance	Bearing					
	0′	20′	40′	1°	1° 20′	1° 40′
° ′	′	′	′	′	′	′
0 57.2	0.0	+0.1	+0.3	+0.4	+0.6	+0.7
0 57.0	0.0	+0.1	+0.1	+0.2	+0.3	+0.4
0 56.8	0.0	0.0	0.0	0.0	0.0	0.0
0 56.6	0.0	−0.1	−0.1	−0.2	−0.3	−0.4
0 56.4	0.0	−0.1	−0.3	−0.4	−0.6	−0.7

TABLE VI CORRECTIONS TO BE APPLIED TO ALTITUDE OF POLARIS TO OBTAIN LATITUDE
1965

t	Cor.	t	Cor.	t	Cor.	t	Cor.
°	′	°	′	°	′	°	′
0	−56.7	45	−39.8	90	+ 0.5	135	+40.3
1	56.7	46	39.1	91	1.5	136	41.0
2	56.6	47	38.4	92	2.4	137	41.7
3	56.6	48	37.7	93	3.4	138	42.3
4	56.5	49	36.9	94	4.4	139	43.0
5	−56.4	50	−36.1	95	+ 5.4	140	+43.6
6	56.3	51	35.4	96	6.4	141	44.2
7	56.2	52	34.6	97	7.4	142	44.8
8	56.1	53	33.8	98	8.3	143	45.4
9	55.9	54	33.0	99	9.3	144	46.0
10	−55.8	55	−32.2	100	+10.3	145	+46.6
11	55.6	56	31.4	101	11.3	146	47.1
12	55.4	57	30.5	102	12.2	147	47.7
13	55.3	58	29.7	103	13.2	148	48.2
14	55.0	59	28.8	104	14.1	149	48.7
15	−54.7	60	−28.0	105	+15.1	150	+49.2
16	54.4	61	27.1	106	16.0	151	49.7
17	54.2	62	26.2	107	17.0	152	50.1
18	53.8	63	25.4	108	17.9	153	50.6
19	53.5	64	24.5	109	18.9	154	51.0
20	−53.2	65	−23.6	110	+19.8	155	+51.4
21	52.9	66	22.7	111	20.7	156	51.8
22	52.5	67	21.7	112	21.6	157	52.2
23	52.1	68	20.8	113	22.5	158	52.6
24	51.7	69	19.9	114	23.4	159	53.0
25	−51.3	70	−19.0	115	+24.3	160	+53.3
26	50.8	71	18.0	116	25.2	161	53.6
27	50.4	72	17.1	117	26.1	162	53.9
28	49.9	73	16.1	118	27.0	163	54.2
29	49.4	74	15.2	119	27.8	164	54.5
30	−49.0	75	−14.2	120	+28.7	165	+54.8
31	48.4	76	13.3	121	29.5	166	55.0
32	47.9	77	12.3	122	30.4	167	55.2
33	47.4	78	11.3	123	31.2	168	55.4
34	46.8	79	10.4	124	32.0	169	55.6
35	−46.3	80	− 9.4	125	+32.8	170	+55.8
36	45.7	81	8.4	126	33.6	171	56.0
37	45.1	82	7.4	127	34.4	172	56.1
38	44.5	83	6.4	128	35.2	173	56.3
39	43.8	84	5.5	129	35.9	174	56.4
40	−43.2	85	− 4.5	130	+36.7	175	+56.5
41	42.6	86	3.5	131	37.5	176	56.5
42	41.9	87	2.5	132	38.2	177	56.6
43	41.2	88	1.5	133	38.9	178	56.6
44	40.5	89	− 0.5	134	39.6	179	56.7
45	39.8	90	+ 0.5	135	40.3	180	56.7

TABLE VII STADIA TABLE

Minutes	0° Hor. Dist.	0° Diff. Elev.	1° Hor. Dist.	1° Diff. Elev.	2° Hor. Dist.	2° Diff. Elev.	3° Hor. Dist.	3° Diff. Elev.
0	100.00	.00	99.97	1.74	99.88	3.49	99.73	5.23
2	100.00	.06	99.97	1.80	99.87	3.55	99.72	5.28
4	100.00	.12	99.97	1.86	99.87	3.60	99.71	5.34
6	100.00	.17	99.96	1.92	99.87	3.66	99.71	5.40
8	100.00	.23	99.96	1.98	99.86	3.72	99.70	5.46
10	100.00	.29	99.96	2.04	99.86	3.78	99.69	5.52
12	100.00	.35	99.96	2.09	99.85	3.84	99.69	5.57
14	100.00	.41	99.95	2.15	99.85	3.89	99.68	5.63
16	100.00	.47	99.95	2.21	99.84	3.95	99.68	5.69
18	100.00	.52	99.95	2.27	99.84	4.01	99.67	5.75
20	100.00	.58	99.95	2.33	99.83	4.07	99.66	5.80
22	100.00	.64	99.94	2.38	99.83	4.13	99.66	5.86
24	100.00	.70	99.94	2.44	99.82	4.18	99.65	5.92
26	99.99	.76	99.94	2.50	99.82	4.24	99.64	5.98
28	99.99	.81	99.93	2.56	99.81	4.30	99.63	6.04
30	99.99	.87	99.93	2.62	99.81	4.36	99.63	6.09
32	99.99	.93	99.93	2.67	99.80	4.42	99.62	6.15
34	99.99	.99	99.93	2.73	99.80	4.47	99.61	6.21
36	99.99	1.05	99.92	2.79	99.79	4.53	99.61	6.27
38	99.99	1.11	99.92	2.85	99.79	4.59	99.60	6.32
40	99.99	1.16	99.92	2.91	99.78	4.65	99.59	6.38
42	99.99	1.22	99.91	2.97	99.78	4.71	99.58	6.44
44	99.98	1.28	99.91	3.02	99.77	4.76	99.58	6.50
46	99.98	1.34	99.90	3.08	99.77	4.82	99.57	6.56
48	99.98	1.40	99.90	3.14	99.76	4.88	99.56	6.61
50	99.98	1.45	99.90	3.20	99.76	4.94	99.55	6.67
52	99.98	1.51	99.89	3.26	99.75	4.99	99.55	6.73
54	99.98	1.57	99.89	3.31	99.74	5.05	99.54	6.79
56	99.97	1.63	99.89	3.37	99.74	5.11	99.53	6.84
58	99.97	1.69	99.88	3.43	99.73	5.17	99.52	6.90
60	99.97	1.74	99.88	3.49	99.73	5.23	99.51	6.96

TABLE VII STADIA TABLE (Continued)

Minutes	4° Hor. Dist.	4° Diff. Elev.	5° Hor. Dist.	5° Diff. Elev.	6° Hor. Dist.	6° Diff. Elev.	7° Hor. Dist.	7° Diff. Elev.
0	99.51	6.96	99.24	8.68	98.91	10.40	98.51	12.10
2	99.51	7.02	99.23	8.74	98.90	10.45	98.50	12.15
4	99.50	7.07	99.22	8.80	98.88	10.51	98.49	12.21
6	99.49	7.13	99.21	8.85	98.87	10.57	98.47	12.27
8	99.48	7.19	99.20	8.91	98.86	10.62	98.46	12.32
10	99.47	7.25	99.19	8.97	98.85	10.68	98.44	12.38
12	99.46	7.30	99.18	9.03	98.83	10.74	98.43	12.43
14	99.46	7.36	99.17	9.08	98.82	10.79	98.41	12.49
16	99.45	7.42	99.16	9.14	98.81	10.85	98.40	12.55
18	99.44	7.48	99.15	9.20	98.80	10.91	98.39	12.60
20	99.43	7.53	99.14	9.25	98.78	10.96	98.37	12.66
22	99.42	7.59	99.13	9.31	98.77	11.02	98.36	12.72
24	99.41	7.65	99.11	9.37	98.76	11.08	98.34	12.77
26	99.40	7.71	99.10	9.43	98.74	11.13	98.33	12.83
28	99.39	7.76	99.09	9.48	98.73	11.19	98.31	12.88
30	99.38	7.82	99.08	9.54	98.72	11.25	98.30	12.94
32	99.38	7.88	99.07	9.60	98.71	11.30	98.28	13.00
34	99.37	7.94	99.06	9.65	98.69	11.36	98.27	13.05
36	99.36	7.99	99.05	9.71	98.68	11.42	98.25	13.11
38	99.35	8.05	99.04	9.77	98.67	11.47	98.24	13.17
40	99.34	8.11	99.03	9.83	98.65	11.53	98.22	13.22
42	99.33	8.17	99.01	9.88	98.64	11.59	98.20	13.28
44	99.32	8.22	99.00	9.94	98.63	11.64	98.19	13.33
46	99.31	8.28	98.99	10.00	98.61	11.70	98.17	13.39
48	99.30	8.34	98.98	10.05	98.60	11.76	98.16	13.45
50	99.29	8.40	98.97	10.11	98.58	11.81	98.14	13.50
52	99.28	8.45	98.96	10.17	98.57	11.87	98.13	13.56
54	99.27	8.51	98.94	10.22	98.56	11.93	98.11	13.61
56	99.26	8.57	98.93	10.28	98.54	11.98	98.10	13.67
58	99.25	8.63	98.92	10.34	98.53	12.04	98.08	13.73
60	99.24	8.68	98.91	10.40	98.51	12.10	98.06	13.78

TABLE VII STADIA TABLE (Continued)

Minutes	8° Hor. Dist.	8° Diff. Elev.	9° Hor. Dist.	9° Diff. Elev.	10° Hor. Dist.	10° Diff. Elev.	11° Hor. Dist.	11° Diff. Elev.
0	98.06	13.78	97.55	15.45	96.98	17.10	96.36	18.73
2	98.05	13.84	97.53	15.51	96.96	17.16	96.34	18.78
4	98.03	13.89	97.52	15.56	96.94	17.21	96.32	18.84
6	98.01	13.95	97.50	15.62	96.92	17.26	96.29	18.89
8	98.00	14.01	97.48	15.67	96.90	17.32	96.27	18.95
10	97.98	14.06	97.46	15.73	96.88	17.37	96.25	19.00
12	97.97	14.12	97.44	15.78	96.86	17.43	96.23	19.05
14	97.95	14.17	97.43	15.84	96.84	17.48	96.21	19.11
16	97.93	14.23	97.41	15.89	96.82	17.54	96.18	19.16
18	97.92	14.28	97.39	15.95	96.80	17.59	96.16	19.21
20	97.90	14.34	97.37	16.00	96.78	17.65	96.14	19.27
22	97.88	14.40	97.35	16.06	96.76	17.70	96.12	19.32
24	97.87	14.45	97.33	16.11	96.74	17.76	96.09	19.38
26	97.85	14.51	97.31	16.17	96.72	17.81	96.07	19.43
28	97.83	14.56	97.29	16.22	96.70	17.86	96.05	19.48
30	97.82	14.62	97.28	16.28	96.68	17.92	96.03	19.54
32	97.80	14.67	97.26	16.33	96.66	17.97	96.00	19.59
34	97.78	14.73	97.24	16.39	96.64	18.03	95.98	19.64
36	97.76	14.79	97.22	16.44	96.62	18.08	95.96	19.70
38	97.75	14.84	97.20	16.50	96.60	18.14	95.93	19.75
40	97.73	14.90	97.18	16.55	96.57	18.19	95.91	19.80
42	97.71	14.95	97.16	16.61	96.55	18.24	95.89	19.86
44	97.69	15.01	97.14	16.66	96.53	18.30	95.86	19.91
46	97.68	15.06	97.12	16.72	96.51	18.35	95.84	19.96
48	97.66	15.12	97.10	16.77	96.49	18.41	95.82	20.02
50	97.64	15.17	97.08	16.83	96.47	18.46	95.79	20.07
52	97.62	15.23	97.06	16.88	96.45	18.51	95.77	20.12
54	97.61	15.28	97.04	16.94	96.42	18.57	95.75	20.18
56	97.59	15.34	97.02	16.99	96.40	18.62	95.72	20.23
58	97.57	15.40	97.00	17.05	96.38	18.68	95.70	20.28
60	97.55	15.45	96.98	17.10	96.36	18.73	95.68	20.34

TABLE VII STADIA TABLE (Continued)

Minutes	12° Hor. Dist.	12° Diff. Elev.	13° Hor. Dist.	13° Diff. Elev.	14° Hor. Dist.	14° Diff. Elev.	15° Hor. Dist.	15° Diff. Elev.
0	95.68	20.34	94.94	21.92	94.15	23.47	93.30	25.00
2	95.65	20.39	94.91	21.97	94.12	23.52	93.27	25.05
4	95.63	20.44	94.89	22.02	94.09	23.58	93.24	25.10
6	95.61	20.50	94.86	22.08	94.07	23.63	93.21	25.15
8	95.58	20.55	94.84	22.13	94.04	23.68	93.18	25.20
10	95.56	20.60	94.81	22.18	94.01	23.73	93.16	25.25
12	95.53	20.66	94.79	22.23	93.98	23.78	93.13	25.30
14	95.51	20.71	94.76	22.28	93.95	23.83	93.10	25.35
16	95.49	20.76	94.73	22.34	93.93	23.88	93.07	25.40
18	95.46	20.81	94.71	22.39	93.90	23.93	93.04	25.45
20	95.44	20.87	94.68	22.44	93.87	23.99	93.01	25.50
22	95.41	20.92	94.66	22.49	93.84	24.04	92.98	25.55
24	95.39	20.97	94.63	22.54	93.82	24.09	92.95	25.60
26	95.36	21.03	94.60	22.60	93.79	24.14	92.92	25.65
28	95.34	21.08	94.58	22.65	93.76	24.19	92.89	25.70
30	95.32	21.13	94.55	22.70	93.73	24.24	92.86	25.75
32	95.29	21.18	94.52	22.75	93.70	24.29	92.83	25.80
34	95.27	21.24	94.50	22.80	93.67	24.34	92.80	25.85
36	95.24	21.29	94.47	22.85	93.65	24.39	92.77	25.90
38	95.22	21.34	94.44	22.91	93.62	24.44	92.74	25.95
40	95.19	21.39	94.42	22.96	93.59	24.49	92.71	26.00
42	95.17	21.45	94.39	23.01	93.56	24.55	92.68	26.05
44	95.14	21.50	94.36	23.06	93.53	24.60	92.65	26.10
46	95.12	21.55	94.34	23.11	93.50	24.65	92.62	26.15
48	95.09	21.60	94.31	23.16	93.47	24.70	92.59	26.20
50	95.07	21.66	94.28	23.22	93.45	24.75	92.56	26.25
52	95.04	21.71	94.26	23.27	93.42	24.80	92.53	26.30
54	95.02	21.76	94.23	23.32	93.39	24.85	92.49	26.35
56	94.99	21.81	94.20	23.37	93.36	24.90	92.46	26.40
58	94.97	21.87	94.17	23.42	93.33	24.95	92.43	26.45
60	94.94	21.92	94.15	23.47	93.30	25.00	92.40	26.50

TABLE VIII RADII FOR CIRCULAR CURVES—Chord Definition

Deg D	Radius R	Log R	Deg D	Radius R	Log R	Deg D	Radius R	Log R
° ′			° ′			° ′		
0 0	∞	∞	1 0	5729.65	3.758128	2 0	2864.93	3.457115
1	343774.68	5.536274	1	5635.72	.750950	1	2841.26	.453511
2	171887.34	.235244	2	5544.83	.743888	2	2817.97	.449937
3	114591.56	.059153	3	5456.82	.736939	3	2795.06	.446392
4	85943.67	4.934214	4	5371.56	.730100	4	2772.53	.442876
5	68754.94	4.837304	5	5288.92	3.723367	5	2750.35	3.439388
6	57295.79	.758123	6	5208.79	.716737	6	2728.52	.435928
7	49110.68	.691176	7	5131.05	.710206	7	2707.04	.432495
8	42971.84	.633184	8	5055.59	.703772	8	2685.89	.429089
9	38197.20	.582031	9	4982.33	.697432	9	2665.0$	.425710
10	34377.48	4.536274	10	4911.15	3.691183	10	2644.58	3.422356
11	31252.26	.494881	11	4841.98	.685023	11	2624.39	.419029
12	28647.90	.457093	12	4774.74	.678949	12	2604.51	.415727
13	26444.22	.422331	13	4709.33	.672959	13	2584.93	.412449
14	24555.35	.390146	14	4645.69	.667051	14	2565.65	.409197
15	22918.33	4.360183	15	4583.75	3.661221	15	2546.64	3.405968
16	21485.94	.332154	16	4523.44	.655469	16	2527.92	.402763
17	20222.06	.305825	17	4464.70	.649792	17	2509.47	.399582
18	19098.61	.281002	18	4407.46	.644189	18	2491.29	.396424
19	18093.43	.257521	19	4351.67	.638656	19	2473.37	.393289
20	17188.76	4.235244	20	4297.28	3.633194	20	2455.70	3.390176
21	16370.25	.214055	21	4244.23	.627799	21	2438.29	.387085
22	15626.15	.193852	22	4192.47	.622470	22	2421.12	.384016
23	14946.75	.174547	23	4141.96	.617206	23	2404.19	.380969
24	14323.97	.156064	24	4092.66	.612005	24	2387.50	.377943
25	13751.02	4.138335	25	4044.51	3.606866	25	2371.04	3.374938
26	13222.13	.121302	26	3997.48	.601787	26	2354.80	.371954
27	12732.43	.104911	27	3951.54	.596766	27	2338.78	.368990
28	12277.70	.089117	28	3906.64	.591803	28	2322.98	.366046
29	11854.33	.073877	29	3862.74	.586896	29	2307.39	.363122
30	11459.19	4.059154	30	3819.83	3.582044	30	2292.01	3.360217
31	11089.54	.044914	31	3777.85	.577245	31	2276.84	.357332
32	10743.00	.031125	32	3736.79	.572499	32	2261.86	.354466
33	10417.45	.017762	33	3696.61	.567804	33	2247.08	.351618
34	10111.06	.004797	34	3657.29	.563160	34	2232.49	.348789
35	9822.18	3.992208	35	3618.80	3.558564	35	2218.09	3.345979
36	9549.34	.979973	36	3581.10	.554017	36	2203.87	.343187
37	9291.25	.968074	37	3544.19	.549517	37	2189.84	.340412
38	9046.75	.956492	38	3508.02	.545063	38	2175.98	.337655
39	8814.78	.945212	39	3472.59	.540654	39	2162.30	.334915
40	8594.42	3.934216	40	3437.87	3.536289	40	2148.79	3.332193
41	8384.80	.923493	41	3403.83	.531968	41	2135.44	.329488
42	8185.16	.913027	42	3370.46	.527690	42	2122.26	.326799
43	7994.81	.902808	43	3337.74	.523453	43	2109.24	.324127
44	7813.11	.892824	44	3305.65	.519257	44	2096.39	.321471
45	7639.49	3.883064	45	3274.17	3.515101	45	2083.68	3.318832
46	7473.42	.873519	46	3243.29	.510985	46	2071.13	.316208
47	7314.41	.864179	47	3212.98	.506908	47	2058.73	.313600
48	7162.03	.855036	48	3183.23	.502868	48	2046.48	.311008
49	7015.87	.846081	49	3154.03	.498866	49	2034.37	.308431
50	6875.55	3.837308	50	3125.36	3.494900	50	2022.41	3.305869
51	6740.74	.828708	51	3097.20	.490970	51	2010.59	.303323
52	6611.12	.820275	52	3069.55	.487075	52	1998.90	.300791
53	6486.38	.812002	53	3042.39	.483215	53	1987.35	.298274
54	6366.26	.803885	54	3015.71	.479389	54	1975.93	.295771
55	6250.51	3.795916	55	2989.48	3.475596	55	1964.64	3.293283
56	6138.90	.788091	56	2963.72	.471836	56	1953.48	.290809
57	6031.20	.780404	57	2938.39	.468109	57	1942.44	.288349
58	5927.22	.772851	58	2913.49	.464413	58	1931.53	.285902
59	5826.76	.765427	59	2889.01	.460749	59	1920.75	.283470
60	5729.65	3.758128	60	2864.93	3.457115	60	1910.08	3.281051

TABLE VIII RADII FOR CIRCULAR CURVES—Chord Definition
(Continued)

Deg D	Radius R	Log R	Deg D	Radius R	Log R	Deg D	Radius R	Log R
° ′			° ′			° ′		
3 0	1910.08	3.281051	4 0	1432.69	3.156151	5 0	1146.28	3.059290
1	1899.53	.278645	1	1426.74	.154346	1	1142.47	.057846
2	1889.09	.276253	2	1420.85	.152548	2	1138.69	.056407
3	1878.77	.273874	3	1415.01	.150758	3	1134.94	.054972
4	1868.56	.271508	4	1409.21	.148975	4	1131.21	.053542
5	1858.47	3.269155	5	1403.46	3.147200	5	1127.50	3.052116
6	1848.48	.266814	6	1397.76	.145431	6	1123. 2	.050696
7	1838.59	.264486	7	1392.10	.143670	7	1120.16	.049280
8	1828.82	.262170	8	1386.49	.141916	8	1116.52	.047868
9	1819.14	.259867	9	1380.92	.140169	9	1112.91	.046462
10	1809.57	3.257576	10	1375.40	3.138430	10	1109.33	3.045059
11	1800.10	.255296	11	1369.92	.136697	11	1105.76	.043662
12	1790.73	.253029	12	1364.49	.134971	12	1102.22	.042268
13	1781.45	.250774	13	1359.10	.133251	13	1098.70	.040880
14	1772.27	.248530	14	1353.75	.131539	14	1095.20	.039495
15	1763.18	3.246297	15	1348.45	3.129833	15	1091.73	3.038115
16	1754.19	.244077	16	1343.18	.128134	16	1088.28	.036740
17	1745.29	.241867	17	1337.96	.126442	17	1084.85	.035368
18	1736.48	.239669	18	1332.77	.124756	18	1081.44	.034002
19	1727.75	.237481	19	1327.63	.123077	19	1078.05	.032639
20	1719.12	3.235305	20	1322.53	3.121404	20	1074.68	3.031281
21	1710.57	.233140	21	1317.46	.119738	21	1071.34	.029927
22	1702.10	.230985	22	1312.43	.118078	22	1068.01	.028577
23	1693.72	.228841	23	1307.45	.116424	23	1064.71	.027231
24	1685.42	.226707	24	1302.50	.114777	24	1061.43	.025890
25	1677.20	3.224584	25	1297.58	3.113136	25	1058.16	3.024552
26	1669.06	.222472	26	1292.71	.111501	26	1054.92	.023219
27	1661.00	.220369	27	1287.87	.109872	27	1051.70	.021890
28	1653.02	.218277	28	1283.07	.108249	28	1048.49	.020565
29	1645.11	.216194	29	1278.30	.106632	29	1045.31	.019244
30	1637.28	3.214122	30	1273.57	3.105022	30	1042.14	3.017927
31	1629.52	.212060	31	1268.87	.103417	31	1039.00	.016614
32	1621.84	.210007	32	1264.21	.101818	32	1035.87	.015305
33	1614.22	.207964	33	1259.58	.100225	33	1032.76	.013999
34	1606.68	.205930	34	1254.98	.098638	34	1029.67	.012698
35	1599.21	3.203906	35	1250.42	3.097057	35	1026.60	3.011401
36	1591.81	.201892	36	1245.89	.095481	36	1023.55	.010107
37	1584.48	.199886	37	1241.40	.093912	37	1020.51	.008818
38	1577.21	.197890	38	1236.94	.092347	38	1017.49	.007532
39	1570.01	.195903	39	1232.51	.090789	39	1014.50	.006250
40	1562.88	3.193925	40	1228.11	3.089236	40	1011.51	3.004972
41	1555.81	.191956	41	1223.74	.087688	41	1008.55	.003698
42	1548.80	.189996	42	1219.40	.086147	42	1005.60	.002427
43	1541.86	.188045	43	1215.09	.084610	43	1002.67	.001160
44	1534.98	.186103	44	1210.82	.083079	44	999.76	2.999897
45	1528.16	3.184169	45	1206.57	3.081553	45	996.87	2.998637
46	1521.40	.182244	46	1202.36	.080033	46	993.99	.997381
47	1514.70	.180327	47	1198.17	.078518	47	991.13	.996129
48	1508.06	.178419	48	1194.01	.077008	48	988.28	.994880
49	1501.48	.176519	49	1189.88	.075504	49	985.45	.993635
50	1494.95	3.174627	50	1185.78	3.074005	50	982.64	2.992393
51	1488.48	.172744	51	1181.71	.072511	51	979.84	.991155
52	1482.07	.170868	52	1177.66	.071022	52	977.06	.989921
53	1475.71	.169001	53	1173.65	.069538	53	974.29	.988690
54	1469.41	.167142	54	1169.66	.068059	54	971.54	.987463
55	1463.16	3.165291	55	1165.70	3.066585	55	968.81	2.986239
56	1456.96	.163447	56	1161.76	.065116	56	966.09	.985018
57	1450.81	.161612	57	1157.85	.063653	57	963.39	.983801
58	1444.72	.159784	58	1153.97	.062194	58	960.70	.982587
59	1438.68	.157963	59	1150.11	.060740	59	958.02	.981377
60	1432.69	3.156151	60	1146.28	3.059290	60	955.37	2.980170

TABLE IX TANGENTS AND EXTERNALS TO A 1° CURVE*
—Chord Definition

Angle	Tangent	External	Angle	Tangent	External	Angle	Tangent	External
1° 00′	50.00	.22	11° 00′	551.70	26.50	21° 00′	1061.9	97.57
10	58.34	.30	10	560.11	27.31	10	1070.6	99.16
20	66.67	.39	20	568.53	28.14	20	1079.2	100.75
30	75.01	.49	30	576.95	28.97	30	1087.8	102.35
40	83.34	.61	40	585.36	29.82	40	1096.4	103.97
50	91.68	.73	50	593.79	30.68	50	1105.1	105.60
2° 00′	100.01	.87	12° 00′	602.21	31.56	22° 00′	1113.7	107.24
10	108.35	1.02	10	610.64	32.45	10	1122.4	108.90
20	116.68	1.19	20	619.07	33.35	20	1131.0	110.57
30	125.02	1.36	30	627.50	34.26	30	1139.7	112.25
40	133.36	1.55	40	635.93	35.18	40	1148.4	113.95
50	141.70	1.75	50	644.37	36.12	50	1157.0	115.66
3° 00′	150.04	1.96	13° 00′	652.81	37.07	23° 00′	1165.7	117.38
10	158.38	2.19	10	661.25	38.03	10	1174.4	119.12
20	166.72	2.43	20	669.70	39.01	20	1183.1	120.87
30	175.06	2.67	30	678.15	39.99	30	1191.8	122.63
40	183.40	2.93	40	686.60	40.99	40	1200.5	124.41
50	191.74	3.21	50	695.06	42.00	50	1209.2	126.20
4° 00′	200.08	3.49	14° 00′	703.51	43.03	24° 00′	1217.9	128.00
10	208.43	3.79	10	711.97	44.07	10	1226.6	129.82
20	216.77	4.10	20	720.44	45.12	20	1235.3	131.65
30	225.12	4.42	30	728.90	46.18	30	1244.0	133.50
40	233.47	4.76	40	737.37	47.25	40	1252.8	135.35
50	241.81	5.10	50	745.85	48.34	50	1261.5	137.23
5° 00′	250.16	5.46	15° 00′	754.32	49.44	25° 00′	1270.2	139.11
10	258.51	5.83	10	762.80	50.55	10	1279.0	141.01
20	266.86	6.21	20	771.29	51.68	20	1287.7	142.93
30	275.21	6.61	30	779.77	52.89	30	1296.5	144.85
40	283.57	7.01	40	788.26	53.97	40	1305.3	146.79
50	291.92	7.43	50	796.75	55.13	50	1314.0	148.75
6° 00′	300.28	7.86	16° 00′	805.25	56.31	26° 00′	1322.8	150.71
10	308.64	8.31	10	813.75	57.50	10	1331.6	152.69
20	316.99	8.76	20	822.25	58.70	20	1340.4	154.69
30	325.35	9.23	30	830.76	59.91	30	1349.2	156.70
40	333.71	9.71	40	839.27	61.14	40	1358.0	158.72
50	342.08	10.20	50	847.78	62.38	50	1366.8	160.76
7° 00′	350.44	10.71	17° 00′	856.30	63.63	27° 00′	1375.6	162.81
10	358.81	11.22	10	864.82	64.90	10	1384.4	164.86
20	367.17	11.75	20	873.35	66.18	20	1393.2	166.95
30	375.54	12.29	30	881.88	67.47	30	1402.0	169.04
40	383.91	12.85	40	890.41	68.77	40	1410.9	171.15
50	392.28	13.41	50	898.95	70.09	50	1419.7	173.27
8° 00′	400.66	13.99	18° 00′	907.49	71.42	28° 00′	1428.6	175.41
10	409.03	14.58	10	916.03	72.76	10	1437.4	177.55
20	417.41	15.18	20	924.58	74.12	20	1446.3	179.72
30	425.79	15.80	30	933.13	75.49	30	1455.1	181.89
40	434.17	16.43	40	941.69	76.86	40	1464.0	184.08
50	442.55	17.07	50	950.25	78.26	50	1472.9	186.29
9° 00′	450.93	17.72	19° 00′	958.81	79.67	29° 00′	1481.8	188.51
10	459.32	18.38	10	967.38	81.09	10	1490.7	190.74
20	467.71	19.06	20	975.96	82.53	20	1499.6	192.99
30	476.10	19.75	30	984.53	83.97	30	1508.5	195.25
40	484.49	20.45	40	993.12	85.43	40	1517.4	197.53
50	492.88	21.16	50	1001.7	86.90	50	1526.3	199.82
10° 00′	501.28	21.89	20° 00′	1010.3	88.39	30° 00′	1535.3	202.12
10	509.68	22.62	10	1018.9	89.89	10	1544.2	204.44
20	518.08	23.38	20	1027.5	91.40	20	1553.1	206.77
30	526.48	24.14	30	1036.1	92.92	30	1562.1	209.12
40	534.89	24.91	40	1044.7	94.46	40	1571.0	211.48
50	543.29	25.70	50	1053.3	96.01	50	1580.0	213.86

TABLE IX TANGENTS AND EXTERNALS TO A 1° CURVE*
—Chord Definition (Continued)

Angle	Tangent	External	Angle	Tangent	External	Angle	Tangent	External
31° 00′	1589.0	216.3	41° 00′	2142.2	387.4	51° 00′	2732.9	618.4
10	1598.0	218.7	10	2151.7	390.7	10	2743.1	622.8
20	1606.9	221.1	20	2161.2	394.1	20	2753.4	627.2
30	1615.9	223.5	30	2170.8	397.4	30	2763.7	631.7
40	1624.9	226.0	40	2180.3	400.8	40	2773.9	636.2
50	1633.9	228.4	50	2189.9	404.2	50	2784.2	640.7
32° 00′	1643.0	230.9	42° 00′	2199.4	407.6	52° 00′	2794.5	645.2
10	1652.0	233.4	10	2209.0	411.1	10	2804.9	649.7
20	1661.0	235.9	20	2218.6	414.5	20	2815.2	654.3
30	1670.0	238.4	30	2228.1	418.0	30	2825.6	658.8
40	1679.1	241.0	40	2237.7	421.4	40	2835.9	663.4
50	1688.1	243.5	50	2247.3	425.0	50	2846.3	668.0
33° 00′	1697.2	246.1	43° 00′	2257.0	428.5	53° 00′	2856.7	672.7
10	1706.3	248.7	10	2266.6	432.0	10	2867.1	677.3
20	1715.3	251.3	20	2276.2	435.6	20	2877.5	682.0
30	1724.4	253.9	30	2285.9	439.2	30	2888.0	686.7
40	1733.5	256.5	40	2295.6	442.8	40	2898.4	691.4
50	1742.6	259.1	50	2305.2	446.4	50	2908.9	696.1
34° 00′	1751.7	261.8	44° 00′	2314.9	450.0	54° 00′	2929.4	700.9
10	1760.8	264.5	10	2324.6	453.6	10	2929.9	707.7
20	1770.0	267.2	20	2334.3	457.3	20	2940.4	710.5
30	1779.1	269.9	30	2344.1	461.0	30	2951.0	715.3
40	1788.2	272.6	40	2353.8	464.6	40	2961.5	720.1
50	1797.4	275.3	50	2363.5	468.4	50	2972.1	725.0
35° 00′	1806.6	278.1	45° 00	2373.3	472.1	55° 00	2982.7	729.9
10	1815.7	280.8	10	2383.1	475.8	10	2993.3	734.8
20	1824.9	283.6	20	2392.8	479.6	20	3003.9	739.7
30	1834.1	286.4	30	2402.6	483.4	30	3014.5	744.6
40	1843.3	289.2	40	2412.4	487.2	40	3025.2	749.6
50	1852.5	292.0	50	2422.3	491.0	50	3035.8	754.6
36° 00	1861.7	294.9	46° 00	2432.1	494.8	56° 00	3046.5	759.6
10	1870.9	297.7	10	2441.9	498.7	10	3057.2	764.6
20	1880.1	300.6	20	2451.8	502.5	20	3067.9	769.7
30	1889.4	303.5	30	2461.7	506.4	30	3078.7	774.7
40	1898.6	306.4	40	2471.5	510.3	40	3089.4	779.8
50	1907.9	309.3	50	2481.4	514.3	50	3100.2	784.9
37° 00	1917.1	312.2	47° 00	2491.3	518.2	57° 00	3110.9	790.1
10	1926.4	315.2	10	2501.2	522.2	10	3121.7	795.2
20	1935.7	318.1	20	2511.2	526.1	20	3132.6	800.4
30	1945.0	321.1	30	2521.1	530.1	30	3143.4	805.6
40	1954.3	324.1	40	2531.1	434.2	40	3154.2	810.9
50	1963.6	327.1	50	2541.0	538.2	50	3165.1	816.1
38° 00	1972.9	330.2	48° 00	2551.0	542.2	58° 00	3176.0	821.4
10	1982.2	333.2	10	2561.0	546.3	10	3186.9	826.7
20	1991.5	336.3	20	2571.0	550.4	20	3197.8	832.0
30	2000.9	339.3	30	2581.0	554.5	30	3208.8	837.3
40	2010.2	342.4	40	2591.0	558.6	40	3219.7	842.7
50	2019.6	345.5	50	2601.1	562.8	50	3230.7	848.1
39° 00	2029.0	348.6	49° 00	2611.2	566.9	59° 00	3241.7	853.5
10	2038.4	351.8	10	2621.2	571.1	10	3252.7	858.9
20	2047.8	354.9	20	2631.3	575.3	20	3263.7	864.3
30	2057.2	358.1	30	2641.4	579.5	30	3274.8	869.8
40	2066.6	361.3	40	2651.5	583.8	40	3285.8	875.3
50	2076.0	364.5	50	2661.6	588.0	50	3296.9	880.8
40° 00	2085.4	367.7	50° 00′	2671.8	592.3	60° 00′	3308.0	886.4
10	2094.9	371.0	10	2681.9	596.6	10	3319.1	892.0
20	2104.3	374.2	20	2692.1	600.9	20	3330.3	897.5
30	2113.8	377.5	30	2702.3	605.3	30	3341.4	903.2
40	2123.3	380.8	40	2712.5	609.6	40	3352.6	908.8
50	2132.7	384.1	50	2722.7	614.0	50	3363.8	914.5

* This table is published as a part of Keuffel and Esser Company's Engineer Field Books and is printed here by permission of Keuffel and Esser Company.

TABLE X INCLINATION CORRECTION FOR 100-FT TAPE

Difference of elevation, v, in feet and tenths of a foot

Correction in feet

v	0.0	0.1	0.2	0.3	0.4	0.5	0.6	0.7	0.8	0.9
0	0.000	0.000	0.000	0.000	0.001	0.001	0.002	0.002	0.003	0.004
1	0.005	0.006	0.007	0.008	0.010	0.011	0.013	0.014	0.016	0.018
2	0.020	0.022	0.024	0.026	0.029	0.031	0.034	0.036	0.039	0.042
3	0.045	0.048	0.051	0.054	0.058	0.061	0.065	0.068	0.072	0.076
4	0.080	0.084	0.088	0.092	0.097	0.101	0.106	0.110	0.115	0.120
5	0.125	0.130	0.135	0.141	0.146	0.151	0.157	0.163	0.168	0.174
6	0.180	0.186	0.192	0.199	0.205	0.211	0.218	0.225	0.231	0.238
7	0.245	0.252	0.260	0.267	0.274	0.282	0.289	0.297	0.305	0.313
8	0.321	0.329	0.337	0.345	0.353	0.362	0.370	0.379	0.388	0.397
9	0.406	0.415	0.424	0.433	0.443	0.452	0.462	0.472	0.481	0.491
10	0.501	0.511	0.522	0.532	0.542	0.553	0.563	0.574	0.585	0.596

TABLE XI ELEVATION FACTORS

Elevation (Feet)	Elevation Factor	Elevation (Feet)	Elevation Factor
Sea level	1.0000000	3000	0.9998565
500	0.9999761	3500	0.9998326
1000	0.9999522	4000	0.9998087
1500	0.9999283	4500	0.9997848
2000	0.9999043	5000	0.9997609
2500	0.9998804	5500	0.9997370

TABLE XII ELEVATION CORRECTIONS, IN FEET, PER 1000 FT

Elevation (Feet)	Correction Factor	Elevation (Feet)	Correction Factor
Sea level	0.00	3000	0.1435
500	0.0239	3500	0.1674
1000	0.0478	4000	0.1913
1500	0.0717	4500	0.2152
2000	0.0957	5000	0.2391
2500	0.1196	5500	0.2630

TABLES

TABLE XIII NATURAL SINES AND COSINES

′	0° Sin	0° Cosin	1° Sin	1° Cosin	2° Sin	2° Cosin	3° Sin	3° Cosin	4° Sin	4° Cosin	′
0	.00000	One.	.01745	.99985	.03490	.99939	.05234	.99863	.06976	.99756	60
1	.00029	One.	.01774	.99984	.03519	.99938	.05263	.99861	.07005	.99754	59
2	.00058	One.	.01803	.99984	.03548	.99937	.05292	.99860	.07034	.99752	58
3	.00087	One.	.01832	.99983	.03577	.99936	.05321	.99858	.07063	.99750	57
4	.00116	One.	.01862	.99983	.03606	.99935	.05350	.99857	.07092	.99748	56
5	.00145	One.	.01891	.99982	.03635	.99934	.05379	.99855	.07121	.99746	55
6	.00175	One.	.01920	.99982	.03664	.99933	.05408	.99854	.07150	.99744	54
7	.00204	One.	.01949	.99981	.03693	.99932	.05437	.99852	.07179	.99742	53
8	.00233	One.	.01978	.99980	.03723	.99931	.05466	.99851	.07208	.99740	52
9	.00262	One.	.02007	.99980	.03752	.99930	.05495	.99849	.07237	.99738	51
10	.00291	One.	.02036	.99979	.03781	.99929	.05524	.99847	.07266	.99736	50
11	.00320	.99999	.02065	.99979	.03810	.99927	.05553	.99846	.07295	.99734	49
12	.00349	.99999	.02094	.99978	.03839	.99926	.05582	.99844	.07324	.99731	48
13	.00378	.99999	.02123	.99977	.03868	.99925	.05611	.99842	.07353	.99729	47
14	.00407	.99999	.02152	.99977	.03897	.99924	.05640	.99841	.07382	.99727	46
15	.00436	.99999	.02181	.99976	.03926	.99923	.05669	.99839	.07411	.99725	45
16	.00465	.99999	.02211	.99976	.03955	.99922	.05698	.99838	.07440	.99723	44
17	.00495	.99999	.02240	.99975	.03984	.99921	.05727	.99836	.07469	.99721	43
18	.00524	.99999	.02269	.99974	.04013	.99919	.05756	.99834	.07498	.99719	42
19	.00553	.99998	.02298	.99974	.04042	.99918	.05785	.99833	.07527	.99716	41
20	.00582	.99998	.02327	.99973	.04071	.99917	.05814	.99831	.07556	.99714	40
21	.00611	.99998	.02356	.99972	.04100	.99916	.05844	.99829	.07585	.99712	39
22	.00640	.99998	.02385	.99972	.04129	.99915	.05873	.99827	.07614	.99710	38
23	.00669	.99998	.02414	.99971	.04159	.99913	.05902	.99826	.07643	.99708	37
24	.00698	.99998	.02443	.99970	.04188	.99912	.05931	.99824	.07672	.99705	36
25	.00727	.99997	.02472	.99969	.04217	.99911	.05960	.99822	.07701	.99703	35
26	.00756	.99997	.02501	.99969	.04246	.99910	.05989	.99821	.07730	.99701	34
27	.00785	.99997	.02530	.99968	.04275	.99909	.06018	.99819	.07759	.99699	33
28	.00814	.99997	.02560	.99967	.04304	.99907	.06047	.99817	.07788	.99696	32
29	.00844	.99996	.02589	.99966	.04333	.99906	.06076	.99815	.07817	.99694	31
30	.00873	.99996	.02618	.99966	.04362	.99905	.06105	.99813	.07846	.99692	30
31	.00902	.99996	.02647	.99965	.04391	.99904	.06134	.99812	.07875	.99689	29
32	.00931	.99996	.02676	.99964	.04420	.99902	.06163	.99810	.07904	.99687	28
33	.00960	.99995	.02705	.99963	.04449	.99901	.06192	.99808	.07933	.99685	27
34	.00989	.99995	.02734	.99963	.04478	.99900	.06221	.99806	.07962	.99683	26
35	.01018	.99995	.02763	.99962	.04507	.99898	.06250	.99804	.07991	.99680	25
36	.01047	.99995	.02792	.99961	.04536	.99897	.06279	.99803	.08020	.99678	24
37	.01076	.99994	.02821	.99960	.04565	.99896	.06308	.99801	.08049	.99676	23
38	.01105	.99994	.02850	.99959	.04594	.99894	.06337	.99799	.08078	.99673	22
39	.01134	.99994	.02879	.99959	.04623	.99893	.06366	.99797	.08107	.99671	21
40	.01164	.99993	.02908	.99958	.04653	.99892	.06395	.99795	.08136	.99668	20
41	.01193	.99993	.02938	.99957	.04682	.99890	.06424	.99793	.08165	.99666	19
42	.01222	.99993	.02967	.99956	.04711	.99889	.06453	.99792	.08194	.99664	18
43	.01251	.99992	.02996	.99955	.04740	.99888	.06482	.99790	.08223	.99661	17
44	.01280	.99992	.03025	.99954	.04769	.99886	.06511	.99788	.08252	.99659	16
45	.01309	.99991	.03054	.99953	.04798	.99885	.06540	.99786	.08281	.99657	15
46	.01338	.99991	.03083	.99952	.04827	.99883	.06569	.99784	.08310	.99654	14
47	.01367	.99991	.03112	.99952	.04856	.99882	.06598	.99782	.08339	.99652	13
48	.01396	.99990	.03141	.99951	.04885	.99881	.06627	.99780	.08368	.99649	12
49	.01425	.99990	.03170	.99950	.04914	.99879	.06656	.99778	.08397	.99647	11
50	.01454	.99989	.03199	.99949	.04943	.99878	.06685	.99776	.08426	.99644	10
51	.01483	.99989	.03228	.99948	.04972	.99876	.06714	.99774	.08455	.99642	9
52	.01513	.99989	.03257	.99947	.05001	.99875	.06743	.99772	.08484	.99639	8
53	.01542	.99988	.03286	.99946	.05030	.99873	.06773	.99770	.08513	.99637	7
54	.01571	.99988	.03316	.99945	.05059	.99872	.06802	.99768	.08542	.99635	6
55	.01600	.99987	.03345	.99944	.05088	.99870	.06831	.99766	.08571	.99632	5
56	.01629	.99987	.03374	.99943	.05117	.99869	.06860	.99764	.08600	.99630	4
57	.01658	.99986	.03403	.99942	.05146	.99867	.06889	.99762	.08629	.99627	3
58	.01687	.99986	.03432	.99941	.05175	.99866	.06918	.99760	.08658	.99625	2
59	.01716	.99985	.03461	.99940	.05205	.99864	.06947	.99758	.08687	.99622	1
60	.01745	.99985	.03490	.99939	.05234	.99863	.06976	.99756	.08716	.99619	0
′	Cosin	Sin	Cosin	Sin	Cosin	Sin	Cosin	Sin	Cosin	Sin	′
	89°		88°		87°		86°		85°		

TABLE XIII NATURAL SINES AND COSINES (Continued)

′	5° Sin	5° Cosin	6° Sin	6° Cosin	7° Sin	7° Cosin	8° Sin	8° Cosin	9° Sin	9° Cosin	′
0	.08716	.99619	.10453	.99452	.12187	.99255	.13917	.99027	.15643	.98769	60
1	.08745	.99617	.10482	.99449	.12216	.99251	.13946	.99023	.15672	.98764	59
2	.08774	.99614	.10511	.99446	.12245	.99248	.13975	.99019	.15701	.98760	58
3	.08803	.99612	.10540	.99443	.12274	.99244	.14004	.99015	.15730	.98755	57
4	.08831	.99609	.10569	.99440	.12302	.99240	.14033	.99011	.15758	.98751	56
5	.08860	.99607	.10597	.99437	.12331	.99237	.14061	.99006	.15787	.98746	55
6	.08889	.99604	.10626	.99434	.12360	.99233	.14090	.99002	.15816	.98741	54
7	.08918	.99602	.10655	.99431	.12389	.99230	.14119	.98998	.15845	.98737	53
8	.08947	.99599	.10684	.99428	.12418	.99226	.14148	.98994	.15873	.98732	52
9	.08976	.99596	.10713	.99424	.12447	.99222	.14177	.98990	.15902	.98728	51
10	.09005	.99594	.10742	.99421	.12476	.99219	.14205	.98986	.15931	.98723	50
11	.09034	.99591	.10771	.99418	.12504	.99215	.14234	.98982	.15959	.98718	49
12	.09063	.99588	.10800	.99415	.12533	.99211	.14263	.98978	.15988	.98714	48
13	.09092	.99586	.10829	.99412	.12562	.99208	.14292	.98973	.16017	.98709	47
14	.09121	.99583	.10858	.99409	.12591	.99204	.14320	.98969	.16046	.98704	46
15	.09150	.99580	.10887	.99406	.12620	.99200	.14349	.98965	.16074	.98700	45
16	.09179	.99578	.10916	.99402	.12649	.99197	.14378	.98961	.16103	.98695	44
17	.09208	.99575	.10945	.99399	.12678	.99193	.14407	.98957	.16132	.98690	43
18	.09237	.99572	.10973	.99396	.12706	.99189	.14436	.98953	.16160	.98686	42
19	.09266	.99570	.11002	.99393	.12735	.99186	.14464	.98948	.16189	.98681	41
20	.09295	.99567	.11031	.99390	.12764	.99182	.14493	.98944	.16218	.98676	40
21	.09324	.99564	.11060	.99386	.12793	.99178	.14522	.98940	.16246	.98671	39
22	.09353	.99562	.11089	.99383	.12822	.99175	.14551	.98936	.16275	.98667	38
23	.09382	.99559	.11118	.99380	.12851	.99171	.14580	.98931	.16304	.98662	37
24	.09411	.99556	.11147	.99377	.12880	.99167	.14608	.98927	.16333	.98657	36
25	.09440	.99553	.11176	.99374	.12908	.99163	.14637	.98923	.16361	.98652	35
26	.09469	.99551	.11205	.99370	.12937	.99160	.14666	.98919	.16390	.98648	34
27	.09498	.99548	.11234	.99367	.12966	.99156	.14695	.98914	.16419	.98643	33
28	.09527	.99545	.11263	.99364	.12995	.99152	.14723	.98910	.16447	.98638	32
29	.09556	.99542	.11291	.99360	.13024	.99148	.14752	.98906	.16476	.98633	31
30	.09585	.99540	.11320	.99357	.13053	.99144	.14781	.98902	.16505	.98629	30
31	.09614	.99537	.11349	.99354	.13081	.99141	.14810	.98897	.16533	.98624	29
32	.09642	.99534	.11378	.99351	.13110	.99137	.14838	.98893	.16562	.98619	28
33	.09671	.99531	.11407	.99347	.13139	.99133	.14867	.98889	.16591	.98614	27
34	.09700	.99528	.11436	.99344	.13168	.99129	.14896	.98884	.16620	.98609	26
35	.09729	.99526	.11465	.99341	.13197	.99125	.14925	.98880	.16648	.98604	25
36	.09758	.99523	.11494	.99337	.13226	.99122	.14954	.98876	.16677	.98600	24
37	.09787	.99520	.11523	.99334	.13254	.99118	.14982	.98871	.16706	.98595	23
38	.09816	.99517	.11552	.99331	.13283	.99114	.15011	.98867	.16734	.98590	22
39	.09845	.99514	.11580	.99327	.13312	.99110	.15040	.98863	.16763	.98585	21
40	.09874	.99511	.11609	.99324	.13341	.99106	.15069	.98858	.16792	.98580	20
41	.09903	.99508	.11638	.99320	.13370	.99102	.15097	.98854	.16820	.98575	19
42	.09932	.99506	.11667	.99317	.13399	.99098	.15126	.98849	.16849	.98570	18
43	.09961	.99503	.11696	.99314	.13427	.99094	.15155	.98845	.16878	.98565	17
44	.09990	.99500	.11725	.99310	.13456	.99091	.15184	.98841	.16906	.98561	16
45	.10019	.99497	.11754	.99307	.13485	.99087	.15212	.98836	.16935	.98556	15
46	.10048	.99494	.11783	.99303	.13514	.99083	.15241	.98832	.16964	.98551	14
47	.10077	.99491	.11812	.99300	.13543	.99079	.15270	.98827	.16992	.98546	13
48	.10106	.99488	.11840	.99297	.13572	.99075	.15299	.98823	.17021	.98541	12
49	.10135	.99485	.11869	.99293	.13600	.99071	.15327	.98818	.17050	.98536	11
50	.10164	.99482	.11898	.99290	.13629	.99067	.15356	.98814	.17078	.98531	10
51	.10192	.99479	.11927	.99286	.13658	.99063	.15385	.98809	.17107	.98526	9
52	.10221	.99476	.11956	.99283	.13687	.99059	.15414	.98805	.17136	.98521	8
53	.10250	.99473	.11985	.99279	.13716	.99055	.15442	.98800	.17164	.98516	7
54	.10279	.99470	.12014	.99276	.13744	.99051	.15471	.98796	.17193	.98511	6
55	.10308	.99467	.12043	.99272	.13773	.99047	.15500	.98791	.17222	.98506	5
56	.10337	.99464	.12071	.99269	.13802	.99043	.15529	.98787	.17250	.98501	4
57	.10366	.99461	.12100	.99265	.13831	.99039	.15557	.98782	.17279	.98496	3
58	.10395	.99458	.12129	.99262	.13860	.99035	.15586	.98778	.17308	.98491	2
59	.10424	.99455	.12158	.99258	.13889	.99031	.15615	.98773	.17336	.98486	1
60	.10453	.99452	.12187	.99255	.13917	.99027	.15643	.98769	.17365	.98481	0
′	Cosin	Sin	Cosin	Sin	Cosin	Sin	Cosin	Sin	Cosin	Sin	′
	84°		83°		82°		81°		80°		

TABLE XIII NATURAL SINES AND COSINES (Continued)

′	10° Sin	10° Cosin	11° Sin	11° Cosin	12° Sin	12° Cosin	13° Sin	13° Cosin	14° Sin	14° Cosin	′
0	.17365	.98481	.19081	.98163	.20791	.97815	.22495	.97437	.24192	.97030	60
1	.17393	.98476	.19109	.98157	.20820	.97809	.22523	.97430	.24220	.97023	59
2	.17422	.98471	.19138	.98152	.20848	.97803	.22552	.97424	.24249	.97015	58
3	.17451	.98466	.19167	.98146	.20877	.97797	.22580	.97417	.24277	.97008	57
4	.17479	.98461	.19195	.98140	.20905	.97791	.22608	.97411	.24305	.97001	56
5	.17508	.98455	.19224	.98135	.20933	.97784	.22637	.97404	.24333	.96994	55
6	.17537	.98450	.19252	.98129	.20962	.97778	.22665	.97398	.24362	.96987	54
7	.17565	.98445	.19281	.98124	.20990	.97772	.22693	.97391	.24390	.96980	53
8	.17594	.98440	.19309	.98118	.21019	.97766	.22722	.97384	.24418	.96973	52
9	.17623	.98435	.19338	.98112	.21047	.97760	.22750	.97378	.24446	.96966	51
10	.17651	.98430	.19366	.98107	.21076	.97754	.22778	.97371	.24474	.96959	50
11	.17680	.98425	.19395	.98101	.21104	.97748	.22807	.97365	.24503	.96952	49
12	.17708	.98420	.19423	.98096	.21132	.97742	.22835	.97358	.24531	.96945	48
13	.17737	.98414	.19452	.98090	.21161	.97735	.22863	.97351	.24559	.96937	47
14	.17766	.98409	.19481	.98084	.21189	.97729	.22892	.97345	.24587	.96930	46
15	.17794	.98404	.19509	.98079	.21218	.97723	.22920	.97338	.24615	.96923	45
16	.17823	.98399	.19538	.98073	.21246	.97717	.22948	.97331	.24644	.96916	44
17	.17852	.98394	.19566	.98067	.21275	.97711	.22977	.97325	.24672	.96909	43
18	.17880	.98389	.19595	.98061	.21303	.97705	.23005	.97318	.24700	.96902	42
19	.17909	.98383	.19623	.98056	.21331	.97698	.23033	.97311	.24728	.96894	41
20	.17937	.98378	.19652	.98050	.21360	.97692	.23062	.97304	.24756	.96887	40
21	.17966	.98373	.19680	.98044	.21388	.97686	.23090	.97298	.24784	.96880	39
22	.17995	.98368	.19709	.98039	.21417	.97680	.23118	.97291	.24813	.96873	38
23	.18023	.98362	.19737	.98033	.21445	.97673	.23146	.97284	.24841	.96866	37
24	.18052	.98357	.19766	.98027	.21474	.97667	.23175	.97278	.24869	.96858	36
25	.18081	.98352	.19794	.98021	.21502	.97661	.23203	.97271	.24897	.96851	35
26	.18109	.98347	.19823	.98016	.21530	.97655	.23231	.97264	.24925	.96844	34
27	.18138	.98341	.19851	.98010	.21559	.97648	.23260	.97257	.24954	.96837	33
28	.18166	.98336	.19880	.98004	.21587	.97642	.23288	.97251	.24982	.96829	32
29	.18195	.98331	.19908	.97998	.21616	.97636	.23316	.97244	.25010	.96822	31
30	.18224	.98325	.19937	.97992	.21644	.97630	.23345	.97237	.25038	.96815	30
31	.18252	.98320	.19965	.97987	.21672	.97623	.23373	.97230	.25066	.96807	29
32	.18281	.98315	.19994	.97981	.21701	.97617	.23401	.97223	.25094	.96800	28
33	.18309	.98310	.20022	.97975	.21729	.97611	.23429	.97217	.25122	.96793	27
34	.18338	.98304	.20051	.97969	.21758	.97604	.23458	.97210	.25151	.96786	26
35	.18367	.98299	.20079	.97963	.21786	.97598	.23486	.97203	.25179	.96778	25
36	.18395	.98294	.20108	.97958	.21814	.97592	.23514	.97196	.25207	.96771	24
37	.18424	.98288	.20136	.97952	.21843	.97585	.23542	.97189	.25235	.96764	23
38	.18452	.98283	.20165	.97946	.21871	.97579	.23571	.97182	.25263	.96756	22
39	.18481	.98277	.20193	.97940	.21899	.97573	.23599	.97176	.25291	.96749	21
40	.18509	.98272	.20222	.97934	.21928	.97566	.23627	.97169	.25320	.96742	20
41	.18538	.98267	.20250	.97928	.21956	.97560	.23656	.97162	.25348	.96734	19
42	.18567	.98261	.20279	.97922	.21985	.97553	.23684	.97155	.25376	.96727	18
43	.18595	.98256	.20307	.97916	.22013	.97547	.23712	.97148	.25404	.96719	17
44	.18624	.98250	.20336	.97910	.22041	.97541	.23740	.97141	.25432	.96712	16
45	.18652	.98245	.20364	.97905	.22070	.97534	.23769	.97134	.25460	.96705	15
46	.18681	.98240	.20393	.97899	.22098	.97528	.23797	.97127	.25488	.96697	14
47	.18710	.98234	.20421	.97893	.22126	.97521	.23825	.97120	.25516	.96690	13
48	.18738	.98229	.20450	.97887	.22155	.97515	.23853	.97113	.25545	.96682	12
49	.18767	.98223	.20478	.97881	.22183	.97508	.23882	.97106	.25573	.96675	11
50	.18795	.98218	.20507	.97875	.22212	.97502	.23910	.97100	.25601	.96667	10
51	.18824	.98212	.20535	.97869	.22240	.97496	.23938	.97093	.25629	.96660	9
52	.18852	.98207	.20563	.97863	.22268	.97489	.23966	.97086	.25657	.96653	8
53	.18881	.98201	.20592	.97857	.22297	.97483	.23995	.97079	.25685	.96645	7
54	.18910	.98196	.20620	.97851	.22325	.97476	.24023	.97072	.25713	.96638	6
55	.18938	.98190	.20649	.97845	.22353	.97470	.24051	.97065	.25741	.96630	5
56	.18967	.98185	.20677	.97839	.22382	.97463	.24079	.97058	.25769	.96623	4
57	.18995	.98179	.20706	.97833	.22410	.97457	.24108	.97051	.25798	.96615	3
58	.19024	.98174	.20734	.97827	.22438	.97450	.24136	.97044	.25826	.96608	2
59	.19052	.98168	.20763	.97821	.22467	.97444	.24164	.97037	.25854	.96600	1
60	.19081	.98163	.20791	.97815	.22495	.97437	.24192	.97030	.25882	.96593	0
′	Cosin	Sin	Cosin	Sin	Cosin	Sin	Cosin	Sin	Cosin	Sin	′
	79°		78°		77°		76°		75°		

TABLE XIII NATURAL SINES AND COSINES (Continued)

′	15° Sin	15° Cosin	16° Sin	16° Cosin	17° Sin	17° Cosin	18° Sin	18° Cosin	19° Sin	19° Cosin	′
0	.25882	.96593	.27564	.96126	.29237	.95630	.30902	.95106	.32557	.94552	60
1	.25910	.96585	.27592	.96118	.29265	.95622	.30929	.95097	.32584	.94542	59
2	.25938	.96578	.27620	.96110	.29293	.95613	.30957	.95088	.32612	.94533	58
3	.25966	.96570	.27648	.96102	.29321	.95605	.30985	.95079	.32639	.94523	57
4	.25994	.96562	.27676	.96094	.29348	.95596	.31012	.95070	.32667	.94514	56
5	.26022	.96555	.27704	.96086	.29376	.95588	.31040	.95061	.32694	.94504	55
6	.26050	.96547	.27731	.96078	.29404	.95579	.31068	.95052	.32722	.94495	54
7	.26079	.96540	.27759	.96070	.29432	.95571	.31095	.95043	.32749	.94485	53
8	.26107	.96532	.27787	.96062	.29460	.95562	.31123	.95033	.32777	.94476	52
9	.26135	.96524	.27815	.96054	.29487	.95554	.31151	.95024	.32804	.94466	51
10	.26163	.96517	.27843	.96046	.29515	.95545	.31178	.95015	.32832	.94457	50
11	.26191	.96509	.27871	.96037	.29543	.95536	.31206	.95006	.32859	.94447	49
12	.26219	.96502	.27899	.96029	.29571	.95528	.31233	.94997	.32887	.94438	48
13	.26247	.96494	.27927	.96021	.29599	.95519	.31261	.94988	.32914	.94428	47
14	.26275	.96486	.27955	.96013	.29626	.95511	.31289	.94979	.32942	.94418	46
15	.26303	.96479	.27983	.96005	.29654	.95502	.31316	.94970	.32969	.94409	45
16	.26331	.96471	.28011	.95997	.29682	.95493	.31344	.94961	.32997	.94399	44
17	.26359	.96463	.28039	.95989	.29710	.95485	.31372	.94952	.33024	.94390	43
18	.26387	.96456	.28067	.95981	.29737	.95476	.31399	.94943	.33051	.94380	42
19	.26415	.96448	.28095	.95972	.29765	.95467	.31427	.94933	.33079	.94370	41
20	.26443	.96440	.28123	.95964	.29793	.95459	.31454	.94924	.33106	.94361	40
21	.26471	.96433	.28150	.95956	.29821	.95450	.31482	.94915	.33134	.94351	39
22	.26500	.96425	.28178	.95948	.29849	.95441	.31510	.94906	.33161	.94342	38
23	.26528	.96417	.28206	.95940	.29876	.95433	.31537	.94897	.33189	.94332	37
24	.26556	.96410	.28234	.95931	.29904	.95424	.31565	.94888	.33216	.94322	36
25	.26584	.96402	.28262	.95923	.29932	.95415	.31593	.94878	.33244	.94313	35
26	.26612	.96394	.28290	.95915	.29960	.95407	.31620	.94869	.33271	.94303	34
27	.26640	.96386	.28318	.95907	.29987	.95398	.31648	.94860	.33298	.94293	33
28	.26668	.96379	.28346	.95898	.30015	.95389	.31675	.94851	.33326	.94284	32
29	.26696	.96371	.28374	.95890	.30043	.95380	.31703	.94842	.33353	.94274	31
30	.26724	.96363	.28402	.95882	.30071	.95372	.31730	.94832	.33381	.94264	30
31	.26752	.96355	.28429	.95874	.30098	.95363	.31758	.94823	.33408	.94254	29
32	.26780	.96347	.28457	.95865	.30126	.95354	.31786	.94814	.33436	.94245	28
33	.26808	.96340	.28485	.95857	.30154	.95345	.31813	.94805	.33463	.94235	27
34	.26836	.96332	.28513	.95849	.30182	.95337	.31841	.94795	.33490	.94225	26
35	.26864	.96324	.28541	.95841	.30209	.95328	.31868	.94786	.33518	.94215	25
36	.26892	.96316	.28569	.95832	.30237	.95319	.31896	.94777	.33545	.94206	24
37	.26920	.96308	.28597	.95824	.30265	.95310	.31923	.94768	.33573	.94196	23
38	.26948	.96301	.28625	.95816	.30292	.95301	.31951	.94758	.33600	.94186	22
39	.26976	.96293	.28652	.95807	.30320	.95293	.31979	.94749	.33627	.94176	21
40	.27004	.96285	.28680	.95799	.30348	.95284	.32006	.94740	.33655	.94167	20
41	.27032	.96277	.28708	.95791	.30376	.95275	.32034	.94730	.33682	.94157	19
42	.27060	.96269	.28736	.95782	.30403	.95266	.32061	.94721	.33710	.94147	18
43	.27088	.96261	.28764	.95774	.30431	.95257	.32089	.94712	.33737	.94137	17
44	.27116	.96253	.28792	.95766	.30459	.95248	.32116	.94702	.33764	.94127	16
45	.27144	.96246	.28820	.95757	.30486	.95240	.32144	.94693	.33792	.94118	15
46	.27172	.96238	.28847	.95749	.30514	.95231	.32171	.94684	.33819	.94108	14
47	.27200	.96230	.28875	.95740	.30542	.95222	.32199	.94674	.33846	.94098	13
48	.27228	.96222	.28903	.95732	.30570	.95213	.32227	.94665	.33874	.94088	12
49	.27256	.96214	.28931	.95724	.30597	.95204	.32254	.94656	.33901	.94078	11
50	.27284	.96206	.28959	.95715	.30625	.95195	.32282	.94646	.33929	.94068	10
51	.27312	.96198	.28987	.95707	.30653	.95186	.32309	.94637	.33956	.94058	9
52	.27340	.96190	.29015	.95698	.30680	.95177	.32337	.94627	.33983	.94049	8
53	.27368	.96182	.29042	.95690	.30708	.95168	.32364	.94618	.34011	.94039	7
54	.27396	.96174	.29070	.95681	.30736	.95159	.32392	.94609	.34038	.94029	6
55	.27424	.96166	.29098	.95673	.30763	.95150	.32419	.94599	.34065	.94019	5
56	.27452	.96158	.29126	.95664	.30791	.95142	.32447	.94590	.34093	.94009	4
57	.27480	.96150	.29154	.95656	.30819	.95133	.32474	.94580	.34120	.93999	3
58	.27508	.96142	.29182	.95647	.30846	.95124	.32502	.94571	.34147	.93989	2
59	.27536	.96134	.29209	.95639	.30874	.95115	.32529	.94561	.34175	.93979	1
60	.27564	.96126	.29237	.95630	.30902	.95106	.32557	.94552	.34202	.93969	0
′	Cosin	Sin	Cosin	Sin	Cosin	Sin	Cosin	Sin	Cosin	Sin	′
	74°		73°		72°		71°		70°		

TABLE XIII NATURAL SINES AND COSINES (Continued)

′	20° Sin	20° Cosin	21° Sin	21° Cosin	22° Sin	22° Cosin	23° Sin	23° Cosin	24° Sin	24° Cosin	′
0	.34202	.93969	.35837	.93358	.37461	.92718	.39073	.92050	.40674	.91355	60
1	.34229	.93959	.35864	.93348	.37488	.92707	.39100	.92039	.40700	.91343	59
2	.34257	.93949	.35891	.93337	.37515	.92697	.39127	.92028	.40727	.91331	58
3	.34284	.93939	.35918	.93327	.37542	.92686	.39153	.92016	.40753	.91319	57
4	.34311	.93929	.35945	.93316	.37569	.92675	.39180	.92005	.40780	.91307	56
5	.34339	.93919	.35973	.93306	.37595	.92664	.39207	.91994	.40806	.91295	55
6	.34366	.93909	.36000	.93295	.37622	.92653	.39234	.91982	.40833	.91283	54
7	.34393	.93899	.36027	.93285	.37649	.92642	.39260	.91971	.40860	.91272	53
8	.34421	.93889	.36054	.93274	.37676	.92631	.39287	.91959	.40886	.91260	52
9	.34448	.93879	.36081	.93264	.37703	.92620	.39314	.91948	.40913	.91248	51
10	.34475	.93869	.36108	.93253	.37730	.92609	.39341	.91936	.40939	.91236	50
11	.34503	.93859	.36135	.93243	.37757	.92598	.39367	.91925	.40966	.91224	49
12	.34530	.93849	.36162	.93232	.37784	.92587	.39394	.91914	.40992	.91212	48
13	.34557	.93839	.36190	.93222	.37811	.92576	.39421	.91902	.41019	.91200	47
14	.34584	.93829	.36217	.93211	.37838	.92565	.39448	.91891	.41045	.91188	46
15	.34612	.93819	.36244	.93201	.37865	.92554	.39474	.91879	.41072	.91176	45
16	.34639	.93809	.36271	.93190	.37892	.92543	.39501	.91868	.41098	.91164	44
17	.34666	.93799	.36298	.93180	.37919	.92532	.39528	.91856	.41125	.91152	43
18	.34694	.93789	.36325	.93169	.37946	.92521	.39555	.91845	.41151	.91140	42
19	.34721	.93779	.36352	.93159	.37973	.92510	.39581	.91833	.41178	.91128	41
20	.34748	.93769	.36379	.93148	.37999	.92499	.39608	.91822	.41204	.91116	40
21	.34775	.93750	.36406	.93137	.38026	.92488	.39635	.91810	.41231	.91104	39
22	.34803	.93748	.36434	.93127	.38053	.92477	.39661	.91799	.41257	.91092	38
23	.34830	.93738	.36461	.93116	.38080	.92466	.39688	.91787	.41284	.91080	37
24	.34857	.93728	.36488	.93106	.38107	.92455	.39715	.91775	.41310	.91068	36
25	.34884	.93718	.36515	.93095	.38134	.92444	.39741	.91764	.41337	.91056	35
26	.34912	.93708	.36542	.93084	.38161	.92432	.39768	.91752	.41363	.91044	34
27	.34939	.93698	.36569	.93074	.38188	.92421	.39795	.91741	.41390	.91032	33
28	.34966	.93688	.36596	.93063	.38215	.92410	.39822	.91729	.41416	.91020	32
29	.34993	.93677	.36623	.93052	.38241	.92399	.39848	.91718	.41443	.91008	31
30	.35021	.93667	.36650	.93042	.38268	.92388	.39875	.91706	.41469	.90996	30
31	.35048	.93657	.36677	.93031	.38295	.92377	.39902	.91694	.41496	.90984	29
32	.35075	.93647	.36704	.93020	.38322	.92366	.39928	.91683	.41522	.90972	28
33	.35102	.93637	.36731	.93010	.38349	.92355	.39955	.91671	.41549	.90960	27
34	.35130	.93626	.36758	.92999	.38376	.92343	.39982	.91660	.41575	.90948	26
35	.35157	.93616	.36785	.92988	.38403	.92332	.40008	.91648	.41602	.90936	25
36	.35184	.93606	.36812	.92978	.38430	.92321	.40035	.91636	.41628	.90924	24
37	.35211	.93596	.36839	.92967	.38456	.92310	.40062	.91625	.41655	.90911	23
38	.35239	.93585	.36867	.92956	.38483	.92299	.40088	.91613	.41681	.90899	22
39	.35266	.93575	.36894	.92945	.38510	.92287	.40115	.91601	.41707	.90887	21
40	.35293	.93565	.36921	.92935	.38537	.92276	.40141	.91590	.41734	.90875	20
41	.35320	.93555	.36948	.92924	.38564	.92265	.40168	.91578	.41760	.90863	19
42	.35347	.93544	.36975	.92913	.38591	.92254	.40195	.91566	.41787	.90851	18
43	.35375	.93534	.37002	.92902	.38617	.92243	.40221	.91555	.41813	.90839	17
44	.35402	.93524	.37029	.92892	.38644	.92231	.40248	.91543	.41840	.90826	16
45	.35429	.93514	.37056	.92881	.38671	.92220	.40275	.91531	.41866	.90814	15
46	.35456	.93503	.37083	.92870	.38698	.92209	.40301	.91519	.41892	.90802	14
47	.35484	.93493	.37110	.92859	.38725	.92198	.40328	.91508	.41919	.90790	13
48	.35511	.93483	.37137	.92849	.38752	.92186	.40355	.91496	.41945	.90778	12
49	.35538	.93472	.37164	.92838	.38778	.92175	.40381	.91484	.41972	.90766	11
50	.35565	.93462	.37191	.92827	.38805	.92164	.40408	.91472	.41998	.90753	10
51	.35592	.93452	.37218	.92816	.38832	.92152	.40434	.91461	.42024	.90741	9
52	.35619	.93441	.37245	.92805	.38859	.92141	.40461	.91449	.42051	.90729	8
53	.35647	.93431	.37272	.92794	.38886	.92130	.40488	.91437	.42077	.90717	7
54	.35674	.93420	.37299	.92784	.38912	.92119	.40514	.91425	.42104	.90704	6
55	.35701	.93410	.37326	.92773	.38939	.92107	.40541	.91414	.42130	.90692	5
56	.35728	.93400	.37353	.92762	.38966	.92096	.40567	.91402	.42156	.90680	4
57	.35755	.93389	.37380	.92751	.38993	.92085	.40594	.91390	.42183	.90668	3
58	.35782	.93379	.37407	.92740	.39020	.92073	.40621	.91378	.42209	.90655	2
59	.35810	.93368	.37434	.92729	.39046	.92062	.40647	.91366	.42235	.90643	1
60	.35837	.93358	.37461	.92718	.39073	.92050	.40674	.91355	.42262	.90631	0
′	Cosin	Sin	Cosin	Sin	Cosin	Sin	Cosin	Sin	Cosin	Sin	′
	69°		68°		67°		66°		65°		

TABLE XIII NATURAL SINES AND COSINES (Continued)

′	25° Sin	25° Cosin	26° Sin	26° Cosin	27° Sin	27° Cosin	28° Sin	28° Cosin	29° Sin	29° Cosin	′
0	.42262	.90631	.43837	.89879	.45399	.89101	.46947	.88295	.48481	.87462	60
1	.42288	.90618	.43863	.89867	.45425	.89087	.46973	.88281	.48506	.87448	59
2	.42315	.90606	.43889	.89854	.45451	.89074	.46999	.88267	.48532	.87434	58
3	.42341	.90594	.43916	.89841	.45477	.89061	.47024	.88254	.48557	.87420	57
4	.42367	.90582	.43942	.89828	.45503	.89048	.47050	.88240	.48583	.87406	56
5	.42394	.90569	.43968	.89816	.45529	.89035	.47076	.88226	.48608	.87391	55
6	.42420	.90557	.43994	.89803	.45554	.89021	.47101	.88213	.48634	.87377	54
7	.42446	.90545	.44020	.89790	.45580	.89008	.47127	.88199	.48659	.87363	53
8	.42473	.90532	.44046	.89777	.45606	.88995	.47153	.88185	.48684	.87349	52
9	.42499	.90520	.44072	.89764	.45632	.88981	.47178	.88172	.48710	.87335	51
10	.42525	.90507	.44098	.89752	.45658	.88968	.47204	.88158	.48735	.87321	50
11	.42552	.90495	.44124	.89739	.45684	.88955	.47229	.88144	.48761	.87306	49
12	.42578	.90483	.44151	.89726	.45710	.88942	.47255	.88130	.48786	.87292	48
13	.42604	.90470	.44177	.89713	.45736	.88928	.47281	.88117	.48811	.87278	47
14	.42631	.90458	.44203	.89700	.45762	.88915	.47306	.88103	.48837	.87264	46
15	.42657	.90446	.44229	.89687	.45787	.88902	.47332	.88089	.48862	.87250	45
16	.42683	.90433	.44255	.89674	.45813	.88888	.47358	.88075	.48888	.87235	44
17	.42709	.90421	.44281	.89662	.45839	.88875	.47383	.88062	.48913	.87221	43
18	.42736	.90408	.44307	.89649	.45865	.88862	.47409	.88048	.48938	.87207	42
19	.42762	.90396	.44333	.89636	.45891	.88848	.47434	.88034	.48964	.87193	41
20	.42788	.90383	.44359	.89623	.45917	.88835	.47460	.88020	.48989	.87178	40
21	.42815	.90371	.44385	.89610	.45942	.88822	.47486	.88006	.49014	.87164	39
22	.42841	.90358	.44411	.89597	.45968	.88808	.47511	.87993	.49040	.87150	38
23	.42867	.90346	.44437	.89584	.45994	.88795	.47537	.87979	.49065	.87136	37
24	.42894	.90334	.44464	.89571	.46020	.88782	.47562	.87965	.49090	.87121	36
25	.42920	.90321	.44490	.89558	.46046	.88768	.47588	.87951	.49116	.87107	35
26	.42946	.90309	.44516	.89545	.46072	.88755	.47614	.87937	.49141	.87093	34
27	.42972	.90296	.44542	.89532	.46097	.88741	.47639	.87923	.49166	.87079	33
28	.42999	.90284	.44568	.89519	.46123	.88728	.47665	.87909	.49192	.87064	32
29	.43025	.90271	.44594	.89506	.46149	.88715	.47690	.87896	.49217	.87050	31
30	.43051	.90259	.44620	.89493	.46175	.88701	.47716	.87882	.49242	.87036	30
31	.43077	.90246	.44646	.89480	.46201	.88688	.47741	.87868	.49268	.87021	29
32	.43104	.90233	.44672	.89467	.46226	.88674	.47767	.87854	.49293	.87007	28
33	.43130	.90221	.44698	.89454	.46252	.88661	.47793	.87840	.49318	.86993	27
34	.43156	.90208	.44724	.89441	.46278	.88647	.47818	.87826	.49344	.86978	26
35	.43182	.90196	.44750	.89428	.46304	.88634	.47844	.87812	.49369	.86964	25
36	.43209	.90183	.44776	.89415	.46330	.88620	.47869	.87798	.49394	.86949	24
37	.43235	.90171	.44802	.89402	.46355	.88607	.47895	.87784	.49419	.86935	23
38	.43261	.90158	.44828	.89389	.46381	.88593	.47920	.87770	.49445	.86921	22
39	.43287	.90146	.44854	.89376	.46407	.88580	.47946	.87756	.49470	.86906	21
40	.43313	.90133	.44880	.89363	.46433	.88566	.47971	.87743	.49495	.86892	20
41	.43340	.90120	.44906	.89350	.46458	.88553	.47997	.87729	.49521	.86878	19
42	.43366	.90108	.44932	.89337	.46484	.88539	.48022	.87715	.49546	.86863	18
43	.43392	.90095	.44958	.89324	.46510	.88526	.48048	.87701	.49571	.86849	17
44	.43418	.90082	.44984	.89311	.46536	.88512	.48073	.87687	.49596	.86834	16
45	.43445	.90070	.45010	.89298	.46561	.88499	.48099	.87673	.49622	.86820	15
46	.43471	.90057	.45036	.89285	.46587	.88485	.48124	.87659	.49647	.86805	14
47	.43497	.90045	.45062	.89272	.46613	.88472	.48150	.87645	.49672	.86791	13
48	.43523	.90032	.45088	.89259	.46639	.88458	.48175	.87631	.49697	.86777	12
49	.43549	.90019	.45114	.89245	.46664	.88445	.48201	.87617	.49723	.86762	11
50	.43575	.90007	.45140	.89232	.46690	.88431	.48226	.87603	.49748	.86748	10
51	.43602	.89994	.45166	.89219	.46716	.88417	.48252	.87589	.49773	.86733	9
52	.43628	.89981	.45192	.89206	.46742	.88404	.48277	.87575	.49798	.86719	8
53	.43654	.89968	.45218	.89193	.46767	.88390	.48303	.87561	.49824	.86704	7
54	.43680	.89956	.45243	.89180	.46793	.88377	.48328	.87546	.49849	.86690	6
55	.43706	.89943	.45269	.89167	.46819	.88363	.48354	.87532	.49874	.86675	5
56	.43733	.89930	.45295	.89153	.46844	.88349	.48379	.87518	.49899	.86661	4
57	.43759	.89918	.45321	.89140	.46870	.88336	.48405	.87504	.49924	.86646	3
58	.43785	.89905	.45347	.89127	.46896	.88322	.48430	.87490	.49950	.86632	2
59	.43811	.89892	.45373	.89114	.46921	.88308	.48456	.87476	.49975	.86617	1
60	.43837	.89879	.45399	.89101	.46947	.88295	.48481	.87462	.50000	.86603	0
′	Cosin	Sin	Cosin	Sin	Cosin	Sin	Cosin	Sin	Cosin	Sin	′
	64°		63°		62°		61°		60°		

TABLE XIII NATURAL SINES AND COSINES (Continued)

′	30° Sin	30° Cosin	31° Sin	31° Cosin	32° Sin	32° Cosin	33° Sin	33° Cosin	34° Sin	34° Cosin	′
0	.50000	.86603	.51504	.85717	.52992	.84805	.54464	.83867	.55919	.82904	60
1	.50025	.86588	.51529	.85702	.53017	.84789	.54488	.83851	.55943	.82887	59
2	.50050	.86573	.51554	.85687	.53041	.84774	.54513	.83835	.55968	.82871	58
3	.50076	.86559	.51579	.85672	.53066	.84759	.54537	.83819	.55992	.82855	57
4	.50101	.86544	.51604	.85657	.53091	.84743	.54561	.83804	.56016	.82839	56
5	.50126	.86530	.51628	.85642	.53115	.84728	.54586	.83788	.56040	.82822	55
6	.50151	.86515	.51653	.85627	.53140	.84712	.54610	.83772	.56064	.82806	54
7	.50176	.86501	.51678	.85612	.53164	.84697	.54635	.83756	.56088	.82790	53
8	.50201	.86486	.51703	.85597	.53189	.84681	.54659	.83740	.56112	.82773	52
9	.50227	.86471	.51728	.85582	.53214	.84666	.54683	.83724	.56136	.82757	51
10	.50252	.86457	.51753	.85567	.53238	.84650	.54708	.83708	.56160	.82741	50
11	.50277	.86442	.51778	.85551	.53263	.84635	.54732	.83692	.56184	.82724	49
12	.50302	.86427	.51803	.85536	.53288	.84619	.54756	.83676	.56208	.82708	48
13	.50327	.86413	.51828	.85521	.53312	.84604	.54781	.83660	.56232	.82692	47
14	.50352	.86398	.51852	.85506	.53337	.84588	.54805	.83645	.56256	.82675	46
15	.50377	.86384	.51877	.85491	.53361	.84573	.54829	.83629	.56280	.82659	45
16	.50403	.86369	.51902	.85476	.53386	.84557	.54854	.83613	.56305	.82643	44
17	.50428	.86354	.51927	.85461	.53411	.84542	.54878	.83597	.56329	.82626	43
18	.50453	.86340	.51952	.85446	.53435	.84526	.54902	.83581	.56353	.82610	42
19	.50478	.86325	.51977	.85431	.53460	.84511	.54927	.83565	.56377	.82593	41
20	.50503	.86310	.52002	.85416	.53484	.84495	.54951	.83549	.56401	.82577	40
21	.50528	.86295	.52026	.85401	.53509	.84480	.54975	.83533	.56425	.82561	39
22	.50553	.86281	.52051	.85385	.53534	.84464	.54999	.83517	.56449	.82544	38
23	.50578	.86266	.52076	.85370	.53558	.84448	.55024	.83501	.56473	.82528	37
24	.50603	.86251	.52101	.85355	.53583	.84433	.55048	.83485	.56497	.82511	36
25	.50628	.86237	.52126	.85340	.53607	.84417	.55072	.83469	.56521	.82495	35
26	.50654	.86222	.52151	.85325	.53632	.84402	.55097	.83453	.56545	.82478	34
27	.50679	.86207	.52175	.85310	.53656	.84386	.55121	.83437	.56569	.82462	33
28	.50704	.86192	.52200	.85294	.53681	.84370	.55145	.83421	.56593	.82446	32
29	.50729	.86178	.52225	.85279	.53705	.84355	.55169	.83405	.56617	.82429	31
30	.50754	.86163	.52250	.85264	.53730	.84339	.55194	.83389	.56641	.82413	30
31	.50779	.86148	.52275	.85249	.53754	.84324	.55218	.83373	.56665	.82396	29
32	.50804	.86133	.52299	.85234	.53779	.84308	.55242	.83356	.56689	.82380	28
33	.50829	.86119	.52324	.85218	.53804	.84292	.55266	.83340	.56713	.82363	27
34	.50854	.86104	.52349	.85203	.53828	.84277	.55291	.83324	.56736	.82347	26
35	.50879	.86089	.52374	.85188	.53853	.84261	.55315	.83308	.56760	.82330	25
36	.50904	.86074	.52399	.85173	.53877	.84245	.55339	.83292	.56784	.82314	24
37	.50929	.86059	.52423	.85157	.53902	.84230	.55363	.83276	.56808	.82297	23
38	.50954	.86045	.52448	.85142	.53926	.84214	.55388	.83260	.56832	.82281	22
39	.50979	.86030	.52473	.85127	.53951	.84198	.55412	.83244	.56856	.82264	21
40	.51004	.86015	.52498	.85112	.53975	.84182	.55436	.83228	.56880	.82248	20
41	.51029	.86000	.52522	.85096	.54000	.84167	.55460	.83212	.56904	.82231	19
42	.51054	.85985	.52547	.85081	.54024	.84151	.55484	.83195	.56928	.82214	18
43	.51079	.85970	.52572	.85066	.54049	.84135	.55509	.83179	.56952	.82198	17
44	.51104	.85956	.52597	.85051	.54073	.84120	.55533	.83163	.56976	.82181	16
45	.51129	.85941	.52621	.85035	.54097	.84104	.55557	.83147	.57000	.82165	15
46	.51154	.85926	.52646	.85020	.54122	.84088	.55581	.83131	.57024	.82148	14
47	.51179	.85911	.52671	.85005	.54146	.84072	.55605	.83115	.57047	.82132	13
48	.51204	.85896	.52696	.84989	.54171	.84057	.55630	.83098	.57071	.82115	12
49	.51229	.85881	.52720	.84974	.54195	.84041	.55654	.83082	.57095	.82098	11
50	.51254	.85866	.52745	.84959	.54220	.84025	.55678	.83066	.57119	.82082	10
51	.51279	.85851	.52770	.84943	.54244	.84009	.55702	.83050	.57143	.82065	9
52	.51304	.85836	.52794	.84928	.54269	.83994	.55726	.83034	.57167	.82048	8
53	.51329	.85821	.52819	.84913	.54293	.83978	.55750	.83017	.57191	.82032	7
54	.51354	.85806	.52844	.84897	.54317	.83962	.55775	.83001	.57215	.82015	6
55	.51379	.85792	.52869	.84882	.54342	.83946	.55799	.82985	.57238	.81999	5
56	.51404	.85777	.52893	.84866	.54366	.83930	.55823	.82969	.57262	.81982	4
57	.51429	.85762	.52918	.84851	.54391	.83915	.55847	.82953	.57286	.81965	3
58	.51454	.85747	.52943	.84836	.54415	.83899	.55871	.82936	.57310	.81949	2
59	.51479	.85732	.52967	.84820	.54440	.83883	.55895	.82920	.57334	.81932	1
60	.51504	.85717	.52992	.84805	.54464	.83867	.55919	.82904	.57358	.81915	0
′	Cosin	Sin	Cosin	Sin	Cosin	Sin	Cosin	Sin	Cosin	Sin	′
	59°		58°		57°		56°		55°		

TABLE XIII NATURAL SINES AND COSINES (Continued)

′	35° Sin	35° Cosin	36° Sin	36° Cosin	37° Sin	37° Cosin	38° Sin	38° Cosin	39° Sin	39° Cosin	′
0	.57358	.81915	.58779	.80902	.60182	.79864	.61566	.78801	.62932	.77715	60
1	.57381	.81899	.58802	.80885	.60205	.79846	.61589	.78783	.62955	.77696	59
2	.57405	.81882	.58826	.80867	.60228	.79829	.61612	.78765	.62977	.77678	58
3	.57429	.81865	.58849	.80850	.60251	.79811	.61635	.78747	.63000	.77660	57
4	.57453	.81848	.58873	.80833	.60274	.79793	.61658	.78729	.63022	.77641	56
5	.57477	.81832	.58896	.80816	.60298	.79776	.61681	.78711	.63045	.77623	55
6	.57501	.81815	.58920	.80799	.60321	.79758	.61704	.78694	.63068	.77605	54
7	.57524	.81798	.58943	.80782	.60344	.79741	.61726	.78676	.63090	.77586	53
8	.57548	.81782	.58967	.80765	.60367	.79723	.61749	.78658	.63113	.77568	52
9	.57572	.81765	.58990	.80748	.60390	.79706	.61772	.78640	.63135	.77550	51
10	.57596	.81748	.59014	.80730	.60414	.79688	.61795	.78622	.63158	.77531	50
11	.57619	.81731	.59037	.80713	.60437	.79671	.61818	.78604	.63180	.77513	49
12	.57643	.81714	.59061	.80696	.60460	.79653	.61841	.78586	.63203	.77494	48
13	.57667	.81698	.59084	.80679	.60483	.79635	.61864	.78568	.63225	.77476	47
14	.57691	.81681	.59108	.80662	.60506	.79618	.61887	.78550	.63248	.77458	46
15	.57715	.81664	.59131	.80644	.60529	.79600	.61909	.78532	.63271	.77439	45
16	.57738	.81647	.59154	.80627	.60553	.79583	.61932	.78514	.63293	.77421	44
17	.57762	.81631	.59178	.80610	.60576	.79565	.61955	.78496	.63316	.77402	43
18	.57786	.81614	.59201	.80593	.60599	.79547	.61978	.78478	.63338	.77384	42
19	.57810	.81597	.59225	.80576	.60622	.79530	.62001	.78460	.63361	.77366	41
20	.57833	.81580	.59248	.80558	.60645	.79512	.62024	.78442	.63383	.77347	40
21	.57857	.81563	.59272	.80541	.60668	.79494	.62046	.78424	.63406	.77329	39
22	.57881	.81546	.59295	.80524	.60691	.79477	.62069	.78405	.63428	.77310	38
23	.57904	.81530	.59318	.80507	.60714	.79459	.62092	.78387	.63451	.77292	37
24	.57928	.81513	.59342	.80489	.60738	.79441	.62115	.78369	.63473	.77273	36
25	.57952	.81496	.59365	.80472	.60761	.79424	.62138	.78351	.63496	.77255	35
26	.57976	.81479	.59389	.80455	.60784	.79406	.62160	.78333	.63518	.77236	34
27	.57999	.81462	.59412	.80438	.60807	.79388	.62183	.78315	.63540	.77218	33
28	.58023	.81445	.59436	.80420	.60830	.79371	.62206	.78297	.63563	.77199	32
29	.58047	.81428	.59459	.80403	.60853	.79353	.62229	.78279	.63585	.77181	31
30	.58070	.81412	.59482	.80386	.60876	.79335	.62251	.78261	.63608	.77162	30
31	.58094	.81395	.59506	.80368	.60899	.79318	.62274	.78243	.63630	.77144	29
32	.58118	.81378	.59529	.80351	.60922	.79300	.62297	.78225	.63653	.77125	28
33	.58141	.81361	.59552	.80334	.60945	.79282	.62320	.78206	.63675	.77107	27
34	.58165	.81344	.59576	.80316	.60968	.79264	.62342	.78188	.63698	.77088	26
35	.58189	.81327	.59599	.80299	.60991	.79247	.62365	.78170	.63720	.77070	25
36	.58212	.81310	.59622	.80282	.61015	.79229	.62388	.78152	.63742	.77051	24
37	.58236	.81293	.59646	.80264	.61038	.79211	.62411	.78134	.63765	.77033	23
38	.58260	.81276	.59669	.80247	.61061	.79193	.62433	.78116	.63787	.77014	22
39	.58283	.81259	.59693	.80230	.61084	.79176	.62456	.78098	.63810	.76996	21
40	.58307	.81242	.59716	.80212	.61107	.79158	.62479	.78079	.63832	.76977	20
41	.58330	.81225	.59739	.80195	.61130	.79140	.62502	.78061	.63854	.76959	19
42	.58354	.81208	.59763	.80178	.61153	.79122	.62524	.78043	.63877	.76940	18
43	.58378	.81191	.59786	.80160	.61176	.79105	.62547	.78025	.63899	.76921	17
44	.58401	.81174	.59809	.80143	.61199	.79087	.62570	.78007	.63922	.76903	16
45	.58425	.81157	.59832	.80125	.61222	.79069	.62592	.77988	.63944	.76884	15
46	.58449	.81140	.59856	.80108	.61245	.79051	.62615	.77970	.63966	.76866	14
47	.58472	.81123	.59879	.80091	.61268	.79033	.62638	.77952	.63989	.76847	13
48	.58496	.81106	.59902	.80073	.61291	.79016	.62660	.77934	.64011	.76828	12
49	.58519	.81089	.59926	.80056	.61314	.78998	.62683	.77916	.64033	.76810	11
50	.58543	.81072	.59949	.80038	.61337	.78980	.62706	.77897	.64056	.76791	10
51	.58567	.81055	.59972	.80021	.61360	.78962	.62728	.77879	.64078	.76772	9
52	.58590	.81038	.59995	.80003	.61383	.78944	.62751	.77861	.64100	.76754	8
53	.58614	.81021	.60019	.79986	.61406	.78926	.62774	.77843	.64123	.76735	7
54	.58637	.81004	.60042	.79968	.61429	.78908	.62796	.77824	.64145	.76717	6
55	.58661	.80987	.60065	.79951	.61451	.78891	.62819	.77806	.64167	.76698	5
56	.58684	.80970	.60089	.79934	.61474	.78873	.62842	.77788	.64190	.76679	4
57	.58708	.80953	.60112	.79916	.61497	.78855	.62864	.77769	.64212	.76661	3
58	.58731	.80936	.60135	.79899	.61520	.78837	.62887	.77751	.64234	.76642	2
59	.58755	.80919	.60158	.79881	.61543	.78819	.62909	.77733	.64256	.76623	1
60	.58779	.80902	.60182	.79864	.61566	.78801	.62932	.77715	.64279	.76604	0
′	Cosin	Sin	Cosin	Sin	Cosin	Sin	Cosin	Sin	Cosin	Sin	′
	54°		53°		52°		51°		50°		

TABLE XIII NATURAL SINES AND COSINES (Continued)

′	40° Sin	40° Cosin	41° Sin	41° Cosin	42° Sin	42° Cosin	43° Sin	43° Cosin	44° Sin	44° Cosin	′
0	.64279	.76604	.65606	.75471	.66913	.74314	.68200	.73135	.69466	.71934	60
1	.64301	.76586	.65628	.75452	.66935	.74295	.68221	.73116	.69487	.71914	59
2	.64323	.76567	.65650	.75433	.66956	.74276	.68242	.73096	.69508	.71894	58
3	.64346	.76548	.65672	.75414	.66978	.74256	.68264	.73076	.69529	.71873	57
4	.64368	.76530	.65694	.75395	.66999	.74237	.68285	.73056	.69549	.71853	56
5	.64390	.76511	.65716	.75375	.67021	.74217	.68306	.73036	.69570	.71833	55
6	.64412	.76492	.65738	.75356	.67043	.74198	.68327	.73016	.69591	.71813	54
7	.64435	.76473	.65759	.75337	.67064	.74178	.68349	.72996	.69612	.71792	53
8	.64457	.76455	.65781	.75318	.67086	.74159	.68370	.72976	.69633	.71772	52
9	.64479	.76436	.65803	.75299	.67107	.74139	.68391	.72957	.69654	.71752	51
10	.64501	.76417	.65825	.75280	.67129	.74120	.68412	.72937	.69675	.71732	50
11	.64524	.76398	.65847	.75261	.67151	.74100	.68434	.72917	.69696	.71711	49
12	.64546	.76380	.65869	.75241	.67172	.74080	.68455	.72897	.69717	.71691	48
13	.64568	.76361	.65891	.75222	.67194	.74061	.68476	.72877	.69737	.71671	47
14	.64590	.76342	.65913	.75203	.67215	.74041	.68497	.72857	.69758	.71650	46
15	.64612	.76323	.65935	.75184	.67237	.74022	.68518	.72837	.69779	.71630	45
16	.64635	.76304	.65956	.75165	.67258	.74002	.68539	.72817	.69800	.71610	44
17	.64657	.76286	.65978	.75146	.67280	.73983	.68561	.72797	.69821	.71590	43
18	.64679	.76267	.66000	.75126	.67301	.73963	.68582	.72777	.69842	.71569	42
19	.64701	.76248	.66022	.75107	.67323	.73944	.68603	.72757	.69862	.71549	41
20	.64723	.76229	.66044	.75088	.67344	.73924	.68624	.72737	.69883	.71529	40
21	.64746	.76210	.66066	.75069	.67366	.73904	.68645	.72717	.69904	.71508	39
22	.64768	.76192	.66088	.75050	.67387	.73885	.68666	.72697	.69925	.71488	38
23	.64790	.76173	.66109	.75030	.67409	.73865	.68688	.72677	.69946	.71468	37
24	.64812	.76154	.66131	.75011	.67430	.73846	.68709	.72657	.69966	.71447	36
25	.64834	.76135	.66153	.74992	.67452	.73826	.68730	.72637	.69987	.71427	35
26	.64856	.76116	.66175	.74973	.67473	.73806	.68751	.72617	.70008	.71407	34
27	.64878	.76097	.66197	.74953	.67495	.73787	.68772	.72597	.70029	.71386	33
28	.64901	.76078	.66218	.74934	.67516	.73767	.68793	.72577	.70049	.71366	32
29	.64923	.76059	.66240	.74915	.67538	.73747	.68814	.72557	.70070	.71345	31
30	.64945	.76041	.66262	.74896	.67559	.73728	.68835	.72537	.70091	.71325	30
31	.64967	.76022	.66284	.74876	.67580	.73708	.68857	.72517	.70112	.71305	29
32	.64989	.76003	.66306	.74857	.67602	.73688	.68878	.72497	.70132	.71284	28
33	.65011	.75984	.66327	.74838	.67623	.73669	.68899	.72477	.70153	.71264	27
34	.65033	.75965	.66349	.74818	.67645	.73649	.68920	.72457	.70174	.71243	26
35	.65055	.75946	.66371	.74799	.67666	.73629	.68941	.72437	.70195	.71223	25
36	.65077	.75927	.66393	.74780	.67688	.73610	.68962	.72417	.70215	.71203	24
37	.65100	.75908	.66414	.74760	.67709	.73590	.68983	.72397	.70236	.71182	23
38	.65122	.75889	.66436	.74741	.67730	.73570	.69004	.72377	.70257	.71162	22
39	.65144	.75870	.66458	.74722	.67752	.73551	.69025	.72357	.70277	.71141	21
40	.65166	.75851	.66480	.74703	.67773	.73531	.69046	.72337	.70298	.71121	20
41	.65188	.75832	.66501	.74683	.67795	.73511	.69067	.72317	.70319	.71100	19
42	.65210	.75813	.66523	.74664	.67816	.73491	.69088	.72297	.70339	.71080	18
43	.65232	.75794	.66545	.74644	.67837	.73472	.69109	.72277	.70360	.71059	17
44	.65254	.75775	.66566	.74625	.67859	.73452	.69130	.72257	.70381	.71039	16
45	.65276	.75756	.66588	.74606	.67880	.73432	.69151	.72236	.70401	.71019	15
46	.65298	.75738	.66610	.74586	.67901	.73413	.69172	.72216	.70422	.70998	14
47	.65320	.75719	.66632	.74567	.67923	.73393	.69193	.72196	.70443	.70978	13
48	.65342	.75700	.66653	.74548	.67944	.73373	.69214	.72176	.70463	.70957	12
49	.65364	.75680	.66675	.74528	.67965	.73353	.69235	.72156	.70484	.70937	11
50	.65386	.75661	.66697	.74509	.67987	.73333	.69256	.72136	.70505	.70916	10
51	.65408	.75642	.66718	.74489	.68008	.73314	.69277	.72116	.70525	.70896	9
52	.65430	.75623	.66740	.74470	.68029	.73294	.69298	.72095	.70546	.70875	8
53	.65452	.75604	.66762	.74451	.68051	.73274	.69319	.72075	.70567	.70855	7
54	.65474	.75585	.66783	.74431	.68072	.73254	.69340	.72055	.70587	.70834	6
55	.65496	.75566	.66805	.74412	.68093	.73234	.69361	.72035	.70608	.70813	5
56	.65518	.75547	.66827	.74392	.68115	.73215	.69382	.72015	.70628	.70793	4
57	.65540	.75528	.66848	.74373	.68136	.73195	.69403	.71995	.70649	.70772	3
58	.65562	.75509	.66870	.74353	.68157	.73175	.69424	.71974	.70670	.70752	2
59	.65584	.75490	.66891	.74334	.68179	.73155	.69445	.71954	.70690	.70731	1
60	.65606	.75471	.66913	.74314	.68200	.73135	.69466	.71934	.70711	.70711	0
′	Cosin	Sin	Cosin	Sin	Cosin	Sin	Cosin	Sin	Cosin	Sin	′
	49°		48°		47°		46°		45°		

TABLE XIV NATURAL TANGENTS AND COTANGENTS

	0° Tan	0° Cotan	1° Tan	1° Cotan	2° Tan	2° Cotan	3° Tan	3° Cotan	
0	.00000	Infinite.	.01746	57.2900	.03492	28.6363	.05241	19.0811	60
1	.00029	3437.75	.01775	56.3506	.03521	28.3994	.05270	18.9755	59
2	.00058	1718.87	.01804	55.4415	.03550	28.1664	.05299	18.8711	58
3	.00087	1145.92	.01833	54.5613	.03579	27.9372	.05328	18.7678	57
4	.00116	859.436	.01862	53.7086	.03609	27.7117	.05357	18.6656	56
5	.00145	687.549	.01891	52.8821	.03638	27.4899	.05387	18.5645	55
6	.00175	572.957	.01920	52.0807	.03667	27.2715	.05416	18.4645	54
7	.00204	491.106	.01949	51.3032	.03696	27.0566	.05445	18.3655	53
8	.00233	429.718	.01978	50.5485	.03725	26.8450	.05474	18.2677	52
9	.00262	381.971	.02007	49.8157	.03754	26.6367	.05503	18.1708	51
10	.00291	343.774	.02036	49.1039	.03783	26.4316	.05533	18.0750	50
11	.00320	312.521	.02066	48.4121	.03812	26.2296	.05562	17.9802	49
12	.00349	286.478	.02095	47.7395	.03842	26.0307	.05591	17.8863	48
13	.00378	264.441	.02124	47.0853	.03871	25.8348	.05620	17.7934	47
14	.00407	245.552	.02153	46.4489	.03900	25.6418	.05649	17.7015	46
15	.00436	229.182	.02182	45.8294	.03929	25.4517	.05678	17.6106	45
16	.00465	214.858	.02211	45.2261	.03958	25.2644	.05708	17.5205	44
17	.00495	202.219	.02240	44.6386	.03987	25.0798	.05737	17.4314	43
18	.00524	190.984	.02269	44.0661	.04016	24.8978	.05766	17.3432	42
19	.00553	180.932	.02298	43.5081	.04046	24.7185	.05795	17.2558	41
20	.00582	171.885	.02328	42.9641	.04075	24.5418	.05824	17.1693	40
21	.00611	163.700	.02357	42.4335	.04104	24.3675	.05854	17.0837	39
22	.00640	156.259	.02386	41.9158	.04133	24.1957	.05883	16.9990	38
23	.00669	149.465	.02415	41.4106	.04162	24.0263	.05912	16.9150	37
24	.00698	143.237	.02444	40.9174	.04191	23.8593	.05941	16.8319	36
25	.00727	137.507	.02473	40.4358	.04220	23.6945	.05970	16.7496	35
26	.00756	132.219	.02502	39.9655	.04250	23.5321	.05999	16.6681	34
27	.00785	127.321	.02531	39.5059	.04279	23.3718	.06029	16.5874	33
28	.00815	122.774	.02560	39.0568	.04308	23.2137	.06058	16.5075	32
29	.00844	118.540	.02589	38.6177	.04337	23.0577	.06087	16.4283	31
30	.00873	114.589	.02619	38.1885	.04366	22.9038	.06116	16.3499	30
31	.00902	110.892	.02648	37.7686	.04395	22.7519	.06145	16.2722	29
32	.00931	107.426	.02677	37.3579	.04424	22.6020	.06175	16.1952	28
33	.00960	104.171	.02706	36.9560	.04454	22.4541	.06204	16.1190	27
34	.00989	101.107	.02735	36.5627	.04483	22.3081	.06233	16.0435	26
35	.01018	98.2179	.02764	36.1776	.04512	22.1640	.06262	15.9687	25
36	.01047	95.4895	.02793	35.8006	.04541	22.0217	.06291	15.8945	24
37	.01076	92.9085	.02822	35.4313	.04570	21.8813	.06321	15.8211	23
38	.01105	90.4633	.02851	35.0695	.04599	21.7426	.06350	15.7483	22
39	.01135	88.1436	.02881	34.7151	.04628	21.6056	.06379	15.6762	21
40	.01164	85.9398	.02910	34.3678	.04658	21.4704	.06408	15.6048	20
41	.01193	83.8435	.02939	34.0273	.04687	21.3369	.06437	15.5340	19
42	.01222	81.8470	.02968	33.6935	.04716	21.2049	.06467	15.4638	18
43	.01251	79.9434	.02997	33.3662	.04745	21.0747	.06496	15.3943	17
44	.01280	78.1263	.03026	33.0452	.04774	20.9460	.06525	15.3254	16
45	.01309	76.3900	.03055	32.7303	.04803	20.8188	.06554	15.2571	15
46	.01338	74.7292	.03084	32.4213	.04833	20.6932	.06584	15.1893	14
47	.01367	73.1390	.03114	32.1181	.04862	20.5691	.06613	15.1222	13
48	.01396	71.6151	.03143	31.8205	.04891	20.4465	.06642	15.0557	12
49	.01425	70.1533	.03172	31.5284	.04920	20.3253	.06671	14.9898	11
50	.01455	68.7501	.03201	31.2416	.04949	20.2056	.06700	14.9244	10
51	.01484	67.4019	.03230	30.9599	.04978	20.0872	.06730	14.8596	9
52	.01513	66.1055	.03259	30.6833	.05007	19.9702	.06759	14.7954	8
53	.01542	64.8580	.03288	30.4116	.05037	19.8546	.06788	14.7317	7
54	.01571	63.6567	.03317	30.1446	.05066	19.7403	.06817	14.6685	6
55	.01600	62.4992	.03346	29.8823	.05095	19.6273	.06847	14.6059	5
56	.01629	61.3829	.03376	29.6245	.05124	19.5156	.06876	14.5438	4
57	.01658	60.3058	.03405	29.3711	.05153	19.4051	.06905	14.4823	3
58	.01687	59.2659	.03434	29.1220	.05182	19.2959	.06934	14.4212	2
59	.01716	58.2612	.03463	28.8771	.05212	19.1879	.06963	14.3607	1
60	.01746	57.2900	.03492	28.6363	.05241	19.0811	.06993	14.3007	0
	Cotan	Tan	Cotan	Tan	Cotan	Tan	Cotan	Tan	
	89°		88°		87°		86°		

TABLE XIV   NATURAL TANGENTS AND COTANGENTS (Continued)

′	4° Tan	4° Cotan	5° Tan	5° Cotan	6° Tan	6° Cotan	7° Tan	7° Cotan	′
0	.06993	14.3007	.08749	11.4301	.10510	9.51436	.12278	8.14435	60
1	.07022	14.2411	.08778	11.3919	.10540	9.48781	.12308	8.12481	59
2	.07051	14.1821	.08807	11.3540	.10569	9.46141	.12338	8.10536	58
3	.07080	14.1235	.08837	11.3163	.10599	9.43515	.12367	8.08600	57
4	.07110	14.0655	.08866	11.2789	.10628	9.40904	.12397	8.06674	56
5	.07139	14.0079	.08895	11.2417	.10657	9.38307	.12426	8.04756	55
6	.07168	13.9507	.08925	11.2048	.10687	9.35724	.12456	8.02848	54
7	.07197	13.8940	.08954	11.1681	.10716	9.33155	.12485	8.00948	53
8	.07227	13.8378	.08983	11.1316	.10746	9.30599	.12515	7.99058	52
9	.07256	13.7821	.09013	11.0954	.10775	9.28058	.12544	7.97176	51
10	.07285	13.7267	.09042	11.0594	.10805	9.25530	.12574	7.95302	50
11	.07314	13.6719	.09071	11.0237	.10834	9.23016	.12603	7.93438	49
12	.07344	13.6174	.09101	10.9882	.10863	9.20516	.12633	7.91582	48
13	.07373	13.5634	.09130	10.9529	.10893	9.18028	.12662	7.89734	47
14	.07402	13.5098	.09159	10.9178	.10922	9.15554	.12692	7.87895	46
15	.07431	13.4566	.09189	10.8829	.10952	9.13093	.12722	7.86064	45
16	.07461	13.4039	.09218	10.8483	.10981	9.10646	.12751	7.84242	44
17	.07490	13.3515	.09247	10.8139	.11011	9.08211	.12781	7.82428	43
18	.07519	13.2996	.09277	10.7797	.11040	9.05789	.12810	7.80622	42
19	.07548	13.2480	.09306	10.7457	.11070	9.03379	.12840	7.78825	41
20	.07578	13.1969	.09335	10.7119	.11099	9.00983	.12869	7.77035	40
21	.07607	13.1461	.09365	10.6783	.11128	8.98598	.12899	7.75254	39
22	.07636	13.0958	.09394	10.6450	.11158	8.96227	.12929	7.73480	38
23	.07665	13.0458	.09423	10.6118	.11187	8.93867	.12958	7.71715	37
24	.07695	12.9962	.09453	10.5789	.11217	8.91520	.12988	7.69957	36
25	.07724	12.9469	.09482	10.5462	.11246	8.89185	.13017	7.68208	35
26	.07753	12.8981	.09511	10.5136	.11276	8.86862	.13047	7.66466	34
27	.07782	12.8496	.09541	10.4813	.11305	8.84551	.13076	7.64732	33
28	.07812	12.8014	.09570	10.4491	.11335	8.82252	.13106	7.63005	32
29	.07841	12.7536	.09600	10.4172	.11364	8.79964	.13136	7.61287	31
30	.07870	12.7062	.09629	10.3854	.11394	8.77689	.13165	7.59575	30
31	.07899	12.6591	.09658	10.3538	.11423	8.75425	.13195	7.57872	29
32	.07929	12.6124	.09688	10.3224	.11452	8.73172	.13224	7.56176	28
33	.07958	12.5660	.09717	10.2913	.11482	8.70931	.13254	7.54487	27
34	.07987	12.5199	.09746	10.2602	.11511	8.68701	.13284	7.52806	26
35	.08017	12.4742	.09776	10.2294	.11541	8.66482	.13313	7.51132	25
36	.08046	12.4288	.09805	10.1988	.11570	8.64275	.13343	7.49465	24
37	.08075	12.3838	.09834	10.1683	.11600	8.62078	.13372	7.47806	23
38	.08104	12.3390	.09864	10.1381	.11629	8.59893	.13402	7.46154	22
39	.08134	12.2946	.09893	10.1080	.11659	8.57718	.13432	7.44509	21
40	.08163	12.2505	.09923	10.0780	.11688	8.55555	.13461	7.42871	20
41	.08192	12.2067	.09952	10.0483	.11718	8.53402	.13491	7.41240	19
42	.08221	12.1632	.09981	10.0187	.11747	8.51259	.13521	7.39616	18
43	.08251	12.1201	.10011	9.98931	.11777	8.49128	.13550	7.37999	17
44	.08280	12.0772	.10040	9.96007	.11806	8.47007	.13580	7.36389	16
45	.08309	12.0346	.10069	9.93101	.11836	8.44896	.13609	7.34786	15
46	.08339	11.9923	.10099	9.90211	.11865	8.42795	.13639	7.33190	14
47	.08368	11.9504	.10128	9.87338	.11895	8.40705	.13669	7.31600	13
48	.08397	11.9087	.10158	9.84482	.11924	8.38625	.13698	7.30018	12
49	.08427	11.8673	.10187	9.81641	.11954	8.36555	.13728	7.28442	11
50	.08456	11.8262	.10216	9.78817	.11983	8.34496	.13758	7.26873	10
51	.08485	11.7853	.10246	9.76009	.12013	8.32446	.13787	7.25310	9
52	.08514	11.7448	.10275	9.73217	.12042	8.30406	.13817	7.23754	8
53	.08544	11.7045	.10305	9.70441	.12072	8.28376	.13846	7.22204	7
54	.08573	11.6645	.10334	9.67680	.12101	8.26355	.13876	7.20661	6
55	.08602	11.6248	.10363	9.64935	.12131	8.24345	.13906	7.19125	5
56	.08632	11.5853	.10393	9.62205	.12160	8.22344	.13935	7.17594	4
57	.08661	11.5461	.10422	9.59490	.12190	8.20352	.13965	7.16071	3
58	.08690	11.5072	.10452	9.56791	.12219	8.18370	.13995	7.14553	2
59	.08720	11.4685	.10481	9.54106	.12249	8.16398	.14024	7.13042	1
60	.08749	11.4301	.10510	9.51436	.12278	8.14435	.14054	7.11537	0
′	Cotan	Tan	Cotan	Tan	Cotan	Tan	Cotan	Tan	′
	85°		84°		83°		82°		

TABLE XIV NATURAL TANGENTS AND COTANGENTS (Continued)

′	8° Tan	8° Cotan	9° Tan	9° Cotan	10° Tan	10° Cotan	11° Tan	11° Cotan	′
0	.14054	7.11537	.15838	6.31375	.17633	5.67128	.19438	5.14455	60
1	.14084	7.10038	.15868	6.30189	.17663	5.66165	.19468	5.13658	59
2	.14113	7.08546	.15898	6.29007	.17693	5.65205	.19498	5.12862	58
3	.14143	7.07059	.15928	6.27829	.17723	5.64248	.19529	5.12069	57
4	.14173	7.05579	.15958	6.26655	.17753	5.63295	.19559	5.11279	56
5	.14202	7.04105	.15988	6.25486	.17783	5.62344	.19589	5.10490	55
6	.14232	7.02637	.16017	6.24321	.17813	5.61397	.19619	5.09704	54
7	.14262	7.01174	.16047	6.23160	.17843	5.60452	.19649	5.08921	53
8	.14291	6.99718	.16077	6.22003	.17873	5.59511	.19680	5.08139	52
9	.14321	6.98268	.16107	6.20851	.17903	5.58573	.19710	5.07360	51
10	.14351	6.96823	.16137	6.19703	.17933	5.57638	.19740	5.06584	50
11	.14381	6.95385	.16167	6.18559	.17963	5.56706	.19770	5.05809	49
12	.14410	6.93952	.16196	6.17419	.17993	5.55777	.19801	5.05037	48
13	.14440	6.92525	.16226	6.16283	.18023	5.54851	.19831	5.04267	47
14	.14470	6.91104	.16256	6.15151	.18053	5.53927	.19861	5.03499	46
15	.14499	6.89688	.16286	6.14023	.18083	5.53007	.19891	5.02734	45
16	.14529	6.88278	.16316	6.12899	.18113	5.52090	.19921	5.01971	44
17	.14559	6.86874	.16346	6.11779	.18143	5.51176	.19952	5.01210	43
18	.14588	6.85475	.16376	6.10664	.18173	5.50264	.19982	5.00451	42
19	.14618	6.84082	.16405	6.09552	.18203	5.49356	.20012	4.99695	41
20	.14648	6.82694	.16435	6.08444	.18233	5.48451	.20042	4.98940	40
21	.14678	6.81312	.16465	6.07340	.18263	5.47548	.20073	4.98188	39
22	.14707	6.79936	.16495	6.06240	.18293	5.46648	.20103	4.97438	38
23	.14737	6.78564	.16525	6.05143	.18323	5.45751	.20133	4.96690	37
24	.14767	6.77199	.16555	6.04051	.18353	5.44857	.20164	4.95945	36
25	.14796	6.75838	.16585	6.02962	.18384	5.43966	.20194	4.95201	35
26	.14826	6.74483	.16615	6.01878	.18414	5.43077	.20224	4.94460	34
27	.14856	6.73133	.16645	6.00797	.18444	5.42192	.20254	4.93721	33
28	.14886	6.71789	.16674	5.99720	.18474	5.41309	.20285	4.92984	32
29	.14915	6.70450	.16704	5.98646	.18504	5.40429	.20315	4.92249	31
30	.14945	6.69116	.16734	5.97576	.18534	5.39552	.20345	4.91516	30
31	.14975	6.67787	.16764	5.96510	.18564	5.38677	.20376	4.90785	29
32	.15005	6.66463	.16794	5.95448	.18594	5.37805	.20406	4.90056	28
33	.15034	6.65144	.16824	5.94390	.18624	5.36936	.20436	4.89330	27
34	.15064	6.63831	.16854	5.93335	.18654	5.36070	.20466	4.88605	26
35	.15094	6.62523	.16884	5.92283	.18684	5.35206	.20497	4.87882	25
36	.15124	6.61219	.16914	5.91236	.18714	5.34345	.20527	4.87162	24
37	.15153	6.59921	.16944	5.90191	.18745	5.33487	.20557	4.86444	23
38	.15183	6.58627	.16974	5.89151	.18775	5.32631	.20588	4.85727	22
39	.15213	6.57339	.17004	5.88114	.18805	5.31778	.20618	4.85013	21
40	.15243	6.56055	.17033	5.87080	.18835	5.30928	.20648	4.84300	20
41	.15272	6.54777	.17063	5.86050	.18865	5.30080	.20679	4.83590	19
42	.15302	6.53503	.17093	5.85024	.18895	5.29235	.20709	4.82882	18
43	.15332	6.52234	.17123	5.84001	.18925	5.28393	.20739	4.82175	17
44	.15362	6.50970	.17153	5.82982	.18955	5.27553	.20770	4.81471	16
45	.15391	6.49710	.17183	5.81966	.18986	5.26715	.20800	4.80769	15
46	.15421	6.48456	.17213	5.80953	.19016	5.25880	.20830	4.80068	14
47	.15451	6.47206	.17243	5.79944	.19046	5.25048	.20861	4.79370	13
48	.15481	6.45961	.17273	5.78938	.19076	5.24218	.20891	4.78673	12
49	.15511	6.44720	.17303	5.77936	.19106	5.23391	.20921	4.77978	11
50	.15540	6.43484	.17333	5.76937	.19136	5.22566	.20952	4.77286	10
51	.15570	6.42253	.17363	5.75941	.19166	5.21744	.20982	4.76595	9
52	.15600	6.41026	.17393	5.74949	.19197	5.20925	.21013	4.75906	8
53	.15630	6.39804	.17423	5.73960	.19227	5.20107	.21043	4.75219	7
54	.15660	6.38587	.17453	5.72974	.19257	5.19293	.21073	4.74534	6
55	.15689	6.37374	.17483	5.71992	.19287	5.18480	.21104	4.73851	5
56	.15719	6.36165	.17513	5.71013	.19317	5.17671	.21134	4.73170	4
57	.15749	6.34961	.17543	5.70037	.19347	5.16863	.21164	4.72490	3
58	.15779	6.33761	.17573	5.69064	.19378	5.16058	.21195	4.71813	2
59	.15809	6.32566	.17603	5.68094	.19408	5.15256	.21225	4.71137	1
60	.15838	6.31375	.17633	5.67128	.19438	5.14455	.21256	4.70463	0
′	Cotan	Tan	Cotan	Tan	Cotan	Tan	Cotan	Tan	′
	81°		80°		79°		78°		

TABLE XIV NATURAL TANGENTS AND COTANGENTS (Continued)

′	12° Tan	12° Cotan	13° Tan	13° Cotan	14° Tan	14° Cotan	15° Tan	15° Cotan	′
0	.21256	4.70463	.23087	4.33148	.24933	4.01078	.26795	3.73205	60
1	.21286	4.69791	.23117	4.32573	.24964	4.00582	.26826	3.72771	59
2	.21316	4.69121	.23148	4.32001	.24995	4.00086	.26857	3.72338	58
3	.21347	4.68452	.23179	4.31430	.25026	3.99592	.26888	3.71907	57
4	.21377	4.67786	.23209	4.30860	.25056	3.99099	.26920	3.71476	56
5	.21408	4.67121	.23240	4.30291	.25087	3.98607	.26951	3.71046	55
6	.21438	4.66458	.23271	4.29724	.25118	3.98117	.26982	3.70616	54
7	.21469	4.65797	.23301	4.29159	.25149	3.97627	.27013	3.70188	53
8	.21499	4.65138	.23332	4.28595	.25180	3.97139	.27044	3.69761	52
9	.21529	4.64480	.23363	4.28032	.25211	3.96651	.27076	3.69335	51
10	.21560	4.63825	.23393	4.27471	.25242	3.96165	.27107	3.68909	50
11	.21590	4.63171	.23424	4.26911	.25273	3.95680	.27138	3.68485	49
12	.21621	4.62518	.23455	4.26352	.25304	3.95196	.27169	3.68061	48
13	.21651	4.61868	.23485	4.25795	.25335	3.94713	.27201	3.67638	47
14	.21682	4.61219	.23516	4.25239	.25366	3.94232	.27232	3.67217	46
15	.21712	4.60572	.23547	4.24685	.25397	3.93751	.27263	3.66796	45
16	.21743	4.59927	.23578	4.24132	.25428	3.93271	.27294	3.66376	44
17	.21773	4.59283	.23608	4.23580	.25459	3.92793	.27326	3.65957	43
18	.21804	4.58641	.23639	4.23030	.25490	3.92316	.27357	3.65538	42
19	.21834	4.58001	.23670	4.22481	.25521	3.91839	.27388	3.65121	41
20	.21864	4.57363	.23700	4.21933	.25552	3.91364	.27419	3.64705	40
21	.21895	4.56726	.23731	4.21387	.25583	3.90890	.27451	3.64289	39
22	.21925	4.56091	.23762	4.20842	.25614	3.90417	.27482	3.63874	38
23	.21956	4.55458	.23793	4.20298	.25645	3.89945	.27513	3.63461	37
24	.21986	4.54826	.23823	4.19756	.25676	3.89474	.27545	3.63048	36
25	.22017	4.54196	.23854	4.19215	.25707	3.89004	.27576	3.62636	35
26	.22047	4.53568	.23885	4.18675	.25738	3.88536	.27607	3.62224	34
27	.22078	4.52941	.23916	4.18137	.25769	3.88068	.27638	3.61814	33
28	.22108	4.52316	.23946	4.17600	.25800	3.87601	.27670	3.61405	32
29	.22139	4.51693	.23977	4.17064	.25831	3.87136	.27701	3.60996	31
30	.22169	4.51071	.24008	4.16530	.25862	3.86671	.27732	3.60588	30
31	.22200	4.50451	.24039	4.15997	.25893	3.86208	.27764	3.60181	29
32	.22231	4.49832	.24069	4.15465	.25924	3.85745	.27795	3.59775	28
33	.22261	4.49215	.24100	4.14934	.25955	3.85284	.27826	3.59370	27
34	.22292	4.48600	.24131	4.14405	.25986	3.84824	.27858	3.58966	26
35	.22322	4.47986	.24162	4.13877	.26017	3.84364	.27889	3.58562	25
36	.22353	4.47374	.24193	4.13350	.26048	3.83906	.27921	3.58160	24
37	.22383	4.46764	.24223	4.12825	.26079	3.83449	.27952	3.57758	23
38	.22414	4.46155	.24254	4.12301	.26110	3.82992	.27983	3.57357	22
39	.22444	4.45548	.24285	4.11778	.26141	3.82537	.28015	3.56957	21
40	.22475	4.44942	.24316	4.11256	.26172	3.82083	.28046	3.56557	20
41	.22505	4.44338	.24347	4.10736	.26203	3.81630	.28077	3.56159	19
42	.22536	4.43735	.24377	4.10216	.26235	3.81177	.28109	3.55761	18
43	.22567	4.43134	.24408	4.09699	.26266	3.80726	.28140	3.55364	17
44	.22597	4.42534	.24439	4.09182	.26297	3.80276	.28172	3.54968	16
45	.22628	4.41936	.24470	4.08666	.26328	3.79827	.28203	3.54573	15
46	.22658	4.41340	.24501	4.08152	.26359	3.79378	.28234	3.54179	14
47	.22689	4.40745	.24532	4.07639	.26390	3.78931	.28266	3.53785	13
48	.22719	4.40152	.24562	4.07127	.26421	3.78485	.28297	3.53393	12
49	.22750	4.39560	.24593	4.06616	.26452	3.78040	.28329	3.53001	11
50	.22781	4.38969	.24624	4.06107	.26483	3.77595	.28360	3.52609	10
51	.22811	4.38381	.24655	4.05599	.26515	3.77152	.28391	3.52219	9
52	.22842	4.37793	.24686	4.05092	.26546	3.76709	.28423	3.51829	8
53	.22872	4.37207	.24717	4.04586	.26577	3.76268	.28454	3.51441	7
54	.22903	4.36623	.24747	4.04081	.26608	3.75828	.28486	3.51053	6
55	.22934	4.36040	.24778	4.03578	.26639	3.75388	.28517	3.50666	5
56	.22964	4.35459	.24809	4.03076	.26670	3.74950	.28549	3.50279	4
57	.22995	4.34879	.24840	4.02574	.26701	3.74512	.28580	3.49894	3
58	.23026	4.34300	.24871	4.02074	.26733	3.74075	.28612	3.49509	2
59	.23056	4.33723	.24902	4.01576	.26764	3.73640	.28643	3.49125	1
60	.23087	4.33148	.24933	4.01078	.26795	3.73205	.28675	3.48741	0
	Cotan	Tan	Cotan	Tan	Cotan	Tan	Cotan	Tan	
′	77°		76°		75°		74°		′

TABLE XIV NATURAL TANGENTS AND COTANGENTS (Continued)

′	16° Tan	16° Cotan	17° Tan	17° Cotan	18° Tan	18° Cotan	19° Tan	19° Cotan	′
0	.28675	3.48741	.30573	3.27085	.32492	3.07768	.34433	2.90421	60
1	.28706	3.48359	.30605	3.26745	.32524	3.07464	.34465	2.90147	59
2	.28738	3.47977	.30637	3.26406	.32556	3.07160	.34498	2.89873	58
3	.28769	3.47596	.30669	3.26067	.32588	3.06857	.34530	2.89600	57
4	.28800	3.47216	.30700	3.25729	.32621	3.06554	.34563	2.89327	56
5	.28832	3.46837	.30732	3.25392	.32653	3.06252	.34596	2.89055	55
6	.28864	3.46458	.30764	3.25055	.32685	3.05950	.34628	2.88783	54
7	.28895	3.46080	.30796	3.24719	.32717	3.05649	.34661	2.88511	53
8	.28927	3.45703	.30828	3.24383	.32749	3.05349	.34693	2.88240	52
9	.28958	3.45327	.30860	3.24049	.32782	3.05049	.34726	2.87970	51
10	.28990	3.44951	.30891	3.23714	.32814	3.04749	.34758	2.87700	50
11	.29021	3.44576	.30923	3.23381	.32846	3.04450	.34791	2.87430	49
12	.29053	3.44202	.30955	3.23048	.32878	3.04152	.34824	2.87161	48
13	.29084	3.43829	.30987	3.22715	.32911	3.03854	.34856	2.86892	47
14	.29116	3.43456	.31019	3.22384	.32943	3.03556	.34889	2.86624	46
15	.29147	3.43084	.31051	3.22053	.32975	3.03260	.34922	2.86356	45
16	.29179	3.42713	.31083	3.21722	.33007	3.02963	.34954	2.86089	44
17	.29210	3.42343	.31115	3.21392	.33040	3.02667	.34987	2.85822	43
18	.29242	3.41973	.31147	3.21063	.33072	3.02372	.35020	2.85555	42
19	.29274	3.41604	.31178	3.20734	.33104	3.02077	.35052	2.85289	41
20	.29305	3.41236	.31210	3.20406	.33136	3.01783	.35085	2.85023	40
21	.29337	3.40869	.31242	3.20079	.33169	3.01489	.35118	2.84758	39
22	.29368	3.40502	.31274	3.19752	.33201	3.01196	.35150	2.84494	38
23	.29400	3.40136	.31306	3.19426	.33233	3.00903	.35183	2.84229	37
24	.29432	3.39771	.31338	3.19100	.33266	3.00611	.35216	2.83965	36
25	.29463	3.39406	.31370	3.18775	.33298	3.00319	.35248	2.83702	35
26	.29495	3.39042	.31402	3.18451	.33330	3.00028	.35281	2.83439	34
27	.29526	3.38679	.31434	3.18127	.33363	2.99738	.35314	2.83176	33
28	.29558	3.38317	.31466	3.17804	.33395	2.99447	.35346	2.82914	32
29	.29590	3.37955	.31498	3.17481	.33427	2.99158	.35379	2.82653	31
30	.29621	3.37594	.31530	3.17159	.33460	2.98868	.35412	2.82391	30
31	.29653	3.37234	.31562	3.16838	.33492	2.98580	.35445	2.82130	29
32	.29685	3.36875	.31594	3.16517	.33524	2.98292	.35477	2.81870	28
33	.29716	3.36516	.31626	3.16197	.33557	2.98004	.35510	2.81610	27
34	.29748	3.36158	.31658	3.15877	.33589	2.97717	.35543	2.81350	26
35	.29780	3.35800	.31690	3.15558	.33621	2.97430	.35576	2.81091	25
36	.29811	3.35443	.31722	3.15240	.33654	2.97144	.35608	2.80833	24
37	.29843	3.35087	.31754	3.14922	.33686	2.96858	.35641	2.80574	23
38	.29875	3.34732	.31786	3.14605	.33718	2.96573	.35674	2.80316	22
39	.29906	3.34377	.31818	3.14288	.33751	2.96288	.35707	2.80059	21
40	.29938	3.34023	.31850	3.13972	.33783	2.96004	.35740	2.79802	20
41	.29970	3.33670	.31882	3.13656	.33816	2.95721	.35772	2.79545	19
42	.30001	3.33317	.31914	3.13341	.33848	2.95437	.35805	2.79289	18
43	.30033	3.32965	.31946	3.13027	.33881	2.95155	.35838	2.79033	17
44	.30065	3.32614	.31978	3.12713	.33913	2.94872	.35871	2.78778	16
45	.30097	3.32264	.32010	3.12400	.33945	2.94591	.35904	2.78523	15
46	.30128	3.31914	.32042	3.12087	.33978	2.94309	.35937	2.78269	14
47	.30160	3.31565	.32074	3.11775	.34010	2.94028	.35969	2.78014	13
48	.30192	3.31216	.32106	3.11464	.34043	2.93748	.36002	2.77761	12
49	.30224	3.30868	.32139	3.11153	.34075	2.93468	.36035	2.77507	11
50	.30255	3.30521	.32171	3.10842	.34108	2.93189	.36068	2.77254	10
51	.30287	3.30174	.32203	3.10532	.34140	2.92910	.36101	2.77002	9
52	.30319	3.29829	.32235	3.10223	.34173	2.92632	.36134	2.76750	8
53	.30351	3.29483	.32267	3.09914	.34205	2.92354	.36167	2.76498	7
54	.30382	3.29139	.32299	3.09606	.34238	2.92076	.36199	2.76247	6
55	.30414	3.28795	.32331	3.09298	.34270	2.91799	.36232	2.75996	5
56	.30446	3.28452	.32363	3.08991	.34303	2.91523	.36265	2.75746	4
57	.30478	3.28109	.32396	3.08685	.34335	2.91246	.36298	2.75496	3
58	.30509	3.27767	.32428	3.08379	.34368	2.90971	.36331	2.75246	2
59	.30541	3.27426	.32460	3.08073	.34400	2.90696	.36364	2.74997	1
60	.30573	3.27085	.32492	3.07768	.34433	2.90421	.36397	2.74748	0
′	Cotan	Tan	Cotan	Tan	Cotan	Tan	Cotan	Tan	′
	73°		72°		71°		70°		

TABLES

TABLE XIV NATURAL TANGENTS AND COTANGENTS (Continued)

′	20° Tan	20° Cotan	21° Tan	21° Cotan	22° Tan	22° Cotan	23° Tan	23° Cotan	′
0	.36397	2.74748	.38386	2.60509	.40403	2.47509	.42447	2.35585	60
1	.36430	2.74499	.38420	2.60283	.40436	2.47302	.42482	2.35395	59
2	.36463	2.74251	.38453	2.60057	.40470	2.47095	.42516	2.35205	58
3	.36496	2.74004	.38487	2.59831	.40504	2.46888	.42551	2.35015	57
4	.36529	2.73756	.38520	2.59606	.40538	2.46682	.42585	2.34825	56
5	.36562	2.73509	.38553	2.59381	.40572	2.46476	.42619	2.34636	55
6	.36595	2.73263	.38587	2.59156	.40606	2.46270	.42654	2.34447	54
7	.36628	2.73017	.38620	2.58932	.40640	2.46065	.42688	2.34258	53
8	.36661	2.72771	.38654	2.58708	.40674	2.45860	.42722	2.34069	52
9	.36694	2.72526	.38687	2.58484	.40707	2.45655	.42757	2.33881	51
10	.36727	2.72281	.38721	2.58261	.40741	2.45451	.42791	2.33693	50
11	.36760	2.72036	.38754	2.58038	.40775	2.45246	.42826	2.33505	49
12	.36793	2.71792	.38787	2.57815	.40809	2.45043	.42860	2.33317	48
13	.36826	2.71548	.38821	2.57593	.40843	2.44839	.42894	2.33130	47
14	.36859	2.71305	.38854	2.57371	.40877	2.44636	.42929	2.32943	46
15	.36892	2.71062	.38888	2.57150	.40911	2.44433	.42963	2.32756	45
16	.36925	2.70819	.38921	2.56928	.40945	2.44230	.42998	2.32570	44
17	.36958	2.70577	.38955	2.56707	.40979	2.44027	.43032	2.32383	43
18	.36991	2.70335	.38988	2.56487	.41013	2.43825	.43067	2.32197	42
19	.37024	2.70094	.39022	2.56266	.41047	2.43623	.43101	2.32012	41
20	.37057	2.69853	.39055	2.56046	.41081	2.43422	.43136	2.31826	40
21	.37090	2.69612	.39089	2.55827	.41115	2.43220	.43170	2.31641	39
22	.37123	2.69371	.39122	2.55608	.41149	2.43019	.43205	2.31456	38
23	.37157	2.69131	.39156	2.55389	.41183	2.42819	.43239	2.31271	37
24	.37190	2.68892	.39190	2.55170	.41217	2.42618	.43274	2.31086	36
25	.37223	2.68653	.39223	2.54952	.41251	2.42418	.43308	2.30902	35
26	.37256	2.68414	.39257	2.54734	.41285	2.42218	.43343	2.30718	34
27	.37289	2.68175	.39290	2.54516	.41319	2.42019	.43378	2.30534	33
28	.37322	2.67937	.39324	2.54299	.41353	2.41819	.43412	2.30351	32
29	.37355	2.67700	.39357	2.54082	.41387	2.41620	.43447	2.30167	31
30	.37388	2.67462	.39391	2.53865	.41421	2.41421	.43481	2.29984	30
31	.37422	2.67225	.39425	2.53648	.41455	2.41223	.43516	2.29801	29
32	.37455	2.66989	.39458	2.53432	.41490	2.41025	.43550	2.29619	28
33	.37488	2.66752	.39492	2.53217	.41524	2.40827	.43585	2.29437	27
34	.37521	2.66516	.39526	2.53001	.41558	2.40629	.43620	2.29254	26
35	.37554	2.66281	.39559	2.52786	.41592	2.40432	.43654	2.29073	25
36	.37588	2.66046	.39593	2.52571	.41626	2.40235	.43689	2.28891	24
37	.37621	2.65811	.39626	2.52357	.41660	2.40038	.43724	2.28710	23
38	.37654	2.65576	.39660	2.52142	.41694	2.39841	.43758	2.28528	22
39	.37687	2.65342	.39694	2.51929	.41728	2.39645	.43793	2.28348	21
40	.37720	2.65109	.39727	2.51715	.41763	2.39449	.43828	2.28167	20
41	.37754	2.64875	.39761	2.51502	.41797	2.39253	.43862	2.27987	19
42	.37787	2.64642	.39795	2.51289	.41831	2.39058	.43897	2.27806	18
43	.37820	2.64410	.39829	2.51076	.41865	2.38863	.43932	2.27626	17
44	.37853	2.64177	.39862	2.50864	.41899	2.38668	.43966	2.27447	16
45	.37887	2.63945	.39896	2.50652	.41933	2.38473	.44001	2.27267	15
46	.37920	2.63714	.39930	2.50440	.41968	2.38279	.44036	2.27088	14
47	.37953	2.63483	.39963	2.50229	.42002	2.38084	.44071	2.26909	13
48	.37986	2.63252	.39997	2.50018	.42036	2.37891	.44105	2.26730	12
49	.38020	2.63021	.40031	2.49807	.42070	2.37697	.44140	2.26552	11
50	.38053	2.62791	.40065	2.49597	.42105	2.37504	.44175	2.26374	10
51	.38086	2.62561	.40098	2.49386	.42139	2.37311	.44210	2.26196	9
52	.38120	2.62332	.40132	2.49177	.42173	2.37118	.44244	2.26018	8
53	.38153	2.62103	.40166	2.48967	.42207	2.36925	.44279	2.25840	7
54	.38186	2.61874	.40200	2.48758	.42242	2.36733	.44314	2.25663	6
55	.38220	2.61646	.40234	2.48549	.42276	2.36541	.44349	2.25486	5
56	.38253	2.61418	.40267	2.48340	.42310	2.36349	.44384	2.25309	4
57	.38286	2.61190	.40301	2.48132	.42345	2.36158	.44418	2.25132	3
58	.38320	2.60963	.40335	2.47924	.42379	2.35967	.44453	2.24956	2
59	.38353	2.60736	.40369	2.47716	.42413	2.35776	.44488	2.24780	1
60	.38386	2.60509	.40403	2.47509	.42447	2.35585	.44523	2.24604	0
′	Cotan	Tan	Cotan	Tan	Cotan	Tan	Cotan	Tan	′
	69°		68°		67°		66°		

TABLE XIV NATURAL TANGENTS AND COTANGENTS (Continued)

	24°		25°		26°		27°		
′	Tan	Cotan	Tan	Cotan	Tan	Cotan	Tan	Cotan	′
0	.44523	2.24604	.46631	2.14451	.48773	2.05030	.50953	1.96261	60
1	.44558	2.24428	.46666	2.14288	.48809	2.04879	.50989	1.96120	59
2	.44593	2.24252	.46702	2.14125	.48845	2.04728	.51026	1.95979	58
3	.44627	2.24077	.46737	2.13963	.48881	2.04577	.51063	1.95838	57
4	.44662	2.23902	.46772	2.13801	.48917	2.04426	.51099	1.95698	56
5	.44697	2.23727	.46808	2.13639	.48953	2.04276	.51136	1.95557	55
6	.44732	2.23553	.46843	2.13477	.48989	2.04125	.51173	1.95417	54
7	.44767	2.23378	.46879	2.13316	.49026	2.03975	.51209	1.95277	53
8	.44802	2.23204	.46914	2.13154	.49062	2.03825	.51246	1.95137	52
9	.44837	2.23030	.46950	2.12993	.49098	2.03675	.51283	1.94997	51
10	.44872	2.22857	.46985	2.12832	.49134	2.03526	.51319	1.94858	50
11	.44907	2.22683	.47021	2.12671	.49170	2.03376	.51356	1.94718	49
12	.44942	2.22510	.47056	2.12511	.49206	2.03227	.51393	1.94579	48
13	.44977	2.22337	.47092	2.12350	.49242	2.03078	.51430	1.94440	47
14	.45012	2.22164	.47128	2.12190	.49278	2.02929	.51467	1.94301	46
15	.45047	2.21992	.47163	2.12030	.49315	2.02780	.51503	1.94162	45
16	.45082	2.21819	.47199	2.11871	.49351	2.02631	.51540	1.94023	44
17	.45117	2.21647	.47234	2.11711	.49387	2.02483	.51577	1.93885	43
18	.45152	2.21475	.47270	2.11552	.49423	2.02335	.51614	1.93746	42
19	.45187	2.21304	.47305	2.11392	.49459	2.02187	.51651	1.93608	41
20	.45222	2.21132	.47341	2.11233	.49495	2.02039	.51688	1.93470	40
21	.45257	2.20961	.47377	2.11075	.49532	2.01891	.51724	1.93332	39
22	.45292	2.20790	.47412	2.10916	.49568	2.01743	.51761	1.93195	38
23	.45327	2.20619	.47448	2.10758	.49604	2.01596	.51798	1.93057	37
24	.45362	2.20449	.47483	2.10600	.49640	2.01449	.51835	1.92920	36
25	.45397	2.20278	.47519	2.10442	.49677	2.01302	.51872	1.92782	35
26	.45432	2.20108	.47555	2.10284	.49713	2.01155	.51909	1.92645	34
27	.45467	2.19938	.47590	2.10126	.49749	2.01008	.51946	1.92508	33
28	.45502	2.19769	.47626	2.09969	.49786	2.00862	.51983	1.92371	32
29	.45538	2.19599	.47662	2.09811	.49822	2.00715	.52020	1.92235	31
30	.45573	2.19430	.47698	2.09654	.49858	2.00569	.52057	1.92098	30
31	.45608	2.19261	.47733	2.09498	.49894	2.00423	.52094	1.91962	29
32	.45643	2.19092	.47769	2.09341	.49931	2.00277	.52131	1.91826	28
33	.45678	2.18923	.47805	2.09184	.49967	2.00131	.52168	1.91690	27
34	.45713	2.18755	.47840	2.09028	.50004	1.99986	.52205	1.91554	26
35	.45748	2.18587	.47876	2.08872	.50040	1.99841	.52242	1.91418	25
36	.45784	2.18419	.47912	2.08716	.50076	1.99695	.52279	1.91282	24
37	.45819	2.18251	.47948	2.08560	.50113	1.99550	.52316	1.91147	23
38	.45854	2.18084	.47984	2.08405	.50149	1.99406	.52353	1.91012	22
39	.45889	2.17916	.48019	2.08250	.50185	1.99261	.52390	1.90876	21
40	.45924	2.17749	.48055	2.08094	.50222	1.99116	.52427	1.90741	20
41	.45960	2.17582	.48091	2.07939	.50258	1.98972	.52464	1.90607	19
42	.45995	2.17416	.48127	2.07785	.50295	1.98828	.52501	1.90472	18
43	.46030	2.17249	.48163	2.07630	.50331	1.98684	.52538	1.90337	17
44	.46065	2.17083	.48198	2.07476	.50368	1.98540	.52575	1.90203	16
45	.46101	2.16917	.48234	2.07321	.50404	1.98396	.52613	1.90069	15
46	.46136	2.16751	.48270	2.07167	.50441	1.98253	.52650	1.89935	14
47	.46171	2.16585	.48306	2.07014	.50477	1.98110	.52687	1.89801	13
48	.46206	2.16420	.48342	2.06860	.50514	1.97966	.52724	1.89667	12
49	.46242	2.16255	.48378	2.06706	.50550	1.97823	.52761	1.89533	11
50	.46277	2.16090	.48414	2.06553	.50587	1.97681	.52798	1.89400	10
51	.46312	2.15925	.48450	2.06400	.50623	1.97538	.52836	1.89266	9
52	.46348	2.15760	.48486	2.06247	.50660	1.97395	.52873	1.89133	8
53	.46383	2.15596	.48521	2.06094	.50696	1.97253	.52910	1.89000	7
54	.46418	2.15432	.48557	2.05942	.50733	1.97111	.52947	1.88867	6
55	.46454	2.15268	.48593	2.05790	.50769	1.96969	.52985	1.88734	5
56	.46489	2.15104	.48629	2.05637	.50806	1.96827	.53022	1.88602	4
57	.46525	2.14940	.48665	2.05485	.50843	1.96685	.53059	1.88469	3
58	.46560	2.14777	.48701	2.05333	.50879	1.96544	.53096	1.88337	2
59	.46595	2.14614	.48737	2.05182	.50916	1.96402	.53134	1.88205	1
60	.46631	2.14451	.48773	2.05030	.50953	1.96261	.53171	1.88073	0
	Cotan	Tan	Cotan	Tan	Cotan	Tan	Cotan	Tan	
′	65°		64°		63°		62°		′

TABLE XIV NATURAL TANGENTS AND COTANGENTS (Continued)

′	28° Tan	28° Cotan	29° Tan	29° Cotan	30° Tan	30° Cotan	31° Tan	31° Cotan	′
0	.53171	1.88073	.55431	1.80405	.57735	1.73205	.60086	1.66428	60
1	.53208	1.87941	.55469	1.80281	.57774	1.73089	.60126	1.66318	59
2	.53246	1.87809	.55507	1.80158	.57813	1.72973	.60165	1.66209	58
3	.53283	1.87677	.55545	1.80034	.57851	1.72857	.60205	1.66099	57
4	.53320	1.87546	.55583	1.79911	.57890	1.72741	.60245	1.65990	56
5	.53358	1.87415	.55621	1.79788	.57929	1.72625	.60284	1.65881	55
6	.53395	1.87283	.55659	1.79665	.57968	1.72509	.60324	1.65772	54
7	.53432	1.87152	.55697	1.79542	.58007	1.72393	.60364	1.65663	53
8	.53470	1.87021	.55736	1.79419	.58046	1.72278	.60403	1.65554	52
9	.53507	1.86891	.55774	1.79296	.58085	1.72163	.60443	1.65445	51
10	.53545	1.86760	.55812	1.79174	.58124	1.72047	.60483	1.65337	50
11	.53582	1.86630	.55850	1.79051	.58162	1.71932	.60522	1.65228	49
12	.53620	1.86499	.55888	1.78929	.58201	1.71817	.60562	1.65120	48
13	.53657	1.86369	.55926	1.78807	.58240	1.71702	.60602	1.65011	47
14	.53694	1.86239	.55964	1.78685	.58279	1.71588	.60642	1.64903	46
15	.53732	1.86109	.56003	1.78563	.58318	1.71473	.60681	1.64795	45
16	.53769	1.85979	.56041	1.78441	.58357	1.71358	.60721	1.64687	44
17	.53807	1.85850	.56079	1.78319	.58396	1.71244	.60761	1.64579	43
18	.53844	1.85720	.56117	1.78198	.58435	1.71129	.60801	1.64471	42
19	.53882	1.85591	.56156	1.78077	.58474	1.71015	.60841	1.64363	41
20	.53920	1.85462	.56194	1.77955	.58513	1.70901	.60881	1.64256	40
21	.53957	1.85333	.56232	1.77834	.58552	1.70787	.60921	1.64148	39
22	.53995	1.85204	.56270	1.77713	.58591	1.70673	.60960	1.64041	38
23	.54032	1.85075	.56309	1.77592	.58631	1.70560	.61000	1.63934	37
24	.54070	1.84946	.56347	1.77471	.58670	1.70446	.61040	1.63826	36
25	.54107	1.84818	.56385	1.77351	.58709	1.70332	.61080	1.63719	35
26	.54145	1.84689	.56424	1.77230	.58748	1.70219	.61120	1.63612	34
27	.54183	1.84561	.56462	1.77110	.58787	1.70106	.61160	1.63505	33
28	.54220	1.84433	.56501	1.76990	.58826	1.69992	.61200	1.63398	32
29	.54258	1.84305	.56539	1.76869	.58865	1.69879	.61240	1.63292	31
30	.54296	1.84177	.56577	1.76749	.58905	1.69766	.61280	1.63185	30
31	.54333	1.84049	.56616	1.76629	.58944	1.69653	.61320	1.63079	29
32	.54371	1.83922	.56654	1.76510	.58983	1.69541	.61360	1.62972	28
33	.54409	1.83794	.56693	1.76390	.59022	1.69428	.61400	1.62866	27
34	.54446	1.83667	.56731	1.76271	.59061	1.69316	.61440	1.62760	26
35	.54484	1.83540	.56769	1.76151	.59101	1.69203	.61480	1.62654	25
36	.54522	1.83413	.56808	1.76032	.59140	1.69091	.61520	1.62548	24
37	.54560	1.83286	.56846	1.75913	.59179	1.68979	.61561	1.62442	23
38	.54597	1.83159	.56885	1.75794	.59218	1.68866	.61601	1.62336	22
39	.54635	1.83033	.56923	1.75675	.59258	1.68754	.61641	1.62230	21
40	.54673	1.82906	.56962	1.75556	.59297	1.68643	.61681	1.62125	20
41	.54711	1.82780	.57000	1.75437	.59336	1.68531	.61721	1.62019	19
42	.54748	1.82654	.57039	1.75319	.59376	1.68419	.61761	1.61914	18
43	.54786	1.82528	.57078	1.75200	.59415	1.68308	.61801	1.61808	17
44	.54824	1.82402	.57116	1.75082	.59454	1.68196	.61842	1.61703	16
45	.54862	1.82276	.57155	1.74964	.59494	1.68085	.61882	1.61598	15
46	.54900	1.82150	.57193	1.74846	.59533	1.67974	.61922	1.61493	14
47	.54938	1.82025	.57232	1.74728	.59573	1.67863	.61962	1.61388	13
48	.54975	1.81899	.57271	1.74610	.59612	1.67752	.62003	1.61283	12
49	.55013	1.81774	.57309	1.74492	.59651	1.67641	.62043	1.61179	11
50	.55051	1.81649	.57348	1.74375	.59691	1.67530	.62083	1.61074	10
51	.55089	1.81524	.57386	1.74257	.59730	1.67419	.62124	1.60970	9
52	.55127	1.81399	.57425	1.74140	.59770	1.67309	.62164	1.60865	8
53	.55165	1.81274	.57464	1.74022	.59809	1.67198	.62204	1.60761	7
54	.55203	1.81150	.57503	1.73905	.59849	1.67088	.62245	1.60657	6
55	.55241	1.81025	.57541	1.73788	.59888	1.66978	.62285	1.60553	5
56	.55279	1.80901	.57580	1.73671	.59928	1.66867	.62325	1.60449	4
57	.55317	1.80777	.57619	1.73555	.59967	1.66757	.62366	1.60345	3
58	.55355	1.80653	.57657	1.73438	.60007	1.66647	.62406	1.60241	2
59	.55393	1.80529	.57696	1.73321	.60046	1.66538	.62446	1.60137	1
60	.55431	1.80405	.57735	1.73205	.60086	1.66428	.62487	1.60033	0
′	Cotan	Tan	Cotan	Tan	Cotan	Tan	Cotan	Tan	
	61°		60°		59°		58°		

TABLE XIV NATURAL TANGENTS AND COTANGENTS (Continued)

′	32° Tan	32° Cotan	33° Tan	33° Cotan	34° Tan	34° Cotan	35° Tan	35° Cotan	′
0	.62487	1.60033	.64941	1.53986	.67451	1.48256	.70021	1.42815	60
1	.62527	1.59930	.64982	1.53888	.67493	1.48163	.70064	1.42726	59
2	.62568	1.59826	.65024	1.53791	.67536	1.48070	.70107	1.42638	58
3	.62608	1.59723	.65065	1.53693	.67578	1.47977	.70151	1.42550	57
4	.62649	1.59620	.65106	1.53595	.67620	1.47885	.70194	1.42462	56
5	.62689	1.59517	.65148	1.53497	.67663	1.47792	.70238	1.42374	55
6	.62730	1.59414	.65189	1.53400	.67705	1.47699	.70281	1.42286	54
7	.62770	1.59311	.65231	1.53302	.67748	1.47607	.70325	1.42198	53
8	.62811	1.59208	.65272	1.53205	.67790	1.47514	.70368	1.42110	52
9	.62852	1.59105	.65314	1.53107	.67832	1.47422	.70412	1.42022	51
10	.62892	1.59002	.65355	1.53010	.67875	1.47330	.70455	1.41934	50
11	.62933	1.58900	.65397	1.52913	.67917	1.47238	.70499	1.41847	49
12	.62973	1.58797	.65438	1.52816	.67960	1.47146	.70542	1.41759	48
13	.63014	1.58695	.65480	1.52719	.68002	1.47053	.70586	1.41672	47
14	.63055	1.58593	.65521	1.52622	.68045	1.46962	.70629	1.41584	46
15	.63095	1.58490	.65563	1.52525	.68088	1.46870	.70673	1.41497	45
16	.63136	1.58388	.65604	1.52429	.68130	1.46778	.70717	1.41409	44
17	.63177	1.58286	.65646	1.52332	.68173	1.46686	.70760	1.41322	43
18	.63217	1.58184	.65688	1.52235	.68215	1.46595	.70804	1.41235	42
19	.63258	1.58083	.65729	1.52139	.68258	1.46503	.70848	1.41148	41
20	.63299	1.57981	.65771	1.52043	.68301	1.46411	.70891	1.41061	40
21	.63340	1.57879	.65813	1.51946	.68343	1.46320	.70935	1.40974	39
22	.63380	1.57778	.65854	1.51850	.68386	1.46229	.70979	1.40887	38
23	.63421	1.57676	.65896	1.51754	.68429	1.46137	.71023	1.40800	37
24	.63462	1.57575	.65938	1.51658	.68471	1.46046	.71066	1.40714	36
25	.63503	1.57474	.65980	1.51562	.68514	1.45955	.71110	1.40627	35
26	.63544	1.57372	.66021	1.51466	.68557	1.45864	.71154	1.40540	34
27	.63584	1.57271	.66063	1.51370	.68600	1.45773	.71198	1.40454	33
28	.63625	1.57170	.66105	1.51275	.68642	1.45682	.71242	1.40367	32
29	.63666	1.57069	.66147	1.51179	.68685	1.45592	.71285	1.40281	31
30	.63707	1.56969	.66189	1.51084	.68728	1.45501	.71329	1.40195	30
31	.63748	1.56868	.66230	1.50988	.68771	1.45410	.71373	1.40109	29
32	.63789	1.56767	.66272	1.50893	.68814	1.45320	.71417	1.40022	28
33	.63830	1.56667	.66314	1.50797	.68857	1.45229	.71461	1.39936	27
34	.63871	1.56566	.66356	1.50702	.68900	1.45139	.71505	1.39850	26
35	.63912	1.56466	.66398	1.50607	.68942	1.45049	.71549	1.39764	25
36	.63953	1.56366	.66440	1.50512	.68985	1.44958	.71593	1.39679	24
37	.63994	1.56265	.66482	1.50417	.69028	1.44868	.71637	1.39593	23
38	.64035	1.56165	.66524	1.50322	.69071	1.44778	.71681	1.39507	22
39	.64076	1.56065	.66566	1.50228	.69114	1.44688	.71725	1.39421	21
40	.64117	1.55966	.66608	1.50133	.69157	1.44598	.71769	1.39336	20
41	.64158	1.55866	.66650	1.50038	.69200	1.44508	.71813	1.39250	19
42	.64199	1.55766	.66692	1.49944	.69243	1.44418	.71857	1.39165	18
43	.64240	1.55666	.66734	1.49849	.69286	1.44329	.71901	1.39079	17
44	.64281	1.55567	.66776	1.49755	.69329	1.44239	.71946	1.38994	16
45	.64322	1.55467	.66818	1.49661	.69372	1.44149	.71990	1.38909	15
46	.64363	1.55368	.66860	1.49566	.69416	1.44060	.72034	1.38824	14
47	.64404	1.55269	.66902	1.49472	.69459	1.43970	.72078	1.38738	13
48	.64446	1.55170	.66944	1.49378	.69502	1.43881	.72122	1.38653	12
49	.64487	1.55071	.66986	1.49284	.69545	1.43792	.72167	1.38568	11
50	.64528	1.54972	.67028	1.49190	.69588	1.43703	.72211	1.38484	10
51	.64569	1.54873	.67071	1.49097	.69631	1.43614	.72255	1.38399	9
52	.64610	1.54774	.67113	1.49003	.69675	1.43525	.72299	1.38314	8
53	.64652	1.54675	.67155	1.48909	.69718	1.43436	.72344	1.38229	7
54	.64693	1.54576	.67197	1.48816	.69761	1.43347	.72388	1.38145	6
55	.64734	1.54478	.67239	1.48722	.69804	1.43258	.72432	1.38060	5
56	.64775	1.54379	.67282	1.48629	.69847	1.43169	.72477	1.37976	4
57	.64817	1.54281	.67324	1.48536	.69891	1.43080	.72521	1.37891	3
58	.64858	1.54183	.67366	1.48442	.69934	1.42992	.72565	1.37807	2
59	.64899	1.54085	.67409	1.48349	.69977	1.42903	.72610	1.37722	1
60	.64941	1.53986	.67451	1.48256	.70021	1.42815	.72654	1.37638	0
′	Cotan	Tan	Cotan	Tan	Cotan	Tan	Cotan	Tan	′
	57°		56°		55°		54°		

TABLE XIV NATURAL TANGENTS AND COTANGENTS (Continued)

′	36° Tan	36° Cotan	37° Tan	37° Cotan	38° Tan	38° Cotan	39° Tan	39° Cotan	′
0	.72654	1.37638	.75355	1.32704	.78129	1.27994	.80978	1.23490	60
1	.72699	1.37554	.75401	1.32624	.78175	1.27917	.81027	1.23416	59
2	.72743	1.37470	.75447	1.32544	.78222	1.27841	.81075	1.23343	58
3	.72788	1.37386	.75492	1.32464	.78269	1.27764	.81123	1.23270	57
4	.72832	1.37302	.75538	1.32384	.78316	1.27688	.81171	1.23196	56
5	.72877	1.37218	.75584	1.32304	.78363	1.27611	.81220	1.23123	55
6	.72921	1.37134	.75629	1.32224	.78410	1.27535	.81268	1.23050	54
7	.72966	1.37050	.75675	1.32144	.78457	1.27458	.81316	1.22977	53
8	.73010	1.36967	.75721	1.32064	.78504	1.27382	.81364	1.22904	52
9	.73055	1.36883	.75767	1.31984	.78551	1.27306	.81413	1.22831	51
10	.73100	1.36800	.75812	1.31904	.78598	1.27230	.81461	1.22758	50
11	.73144	1.36716	.75858	1.31825	.78645	1.27153	.81510	1.22685	49
12	.73189	1.36633	.75904	1.31745	.78692	1.27077	.81558	1.22612	48
13	.73234	1.36549	.75950	1.31666	.78739	1.27001	.81606	1.22539	47
14	.73278	1.36466	.75996	1.31586	.78786	1.26925	.81655	1.22467	46
15	.73323	1.36383	.76042	1.31507	.78834	1.26849	.81703	1.22394	45
16	.73368	1.36300	.76088	1.31427	.78881	1.26774	.81752	1.22321	44
17	.73413	1.36217	.76134	1.31348	.78928	1.26698	.81800	1.22249	43
18	.73457	1.36134	.76180	1.31269	.78975	1.26622	.81849	1.22176	42
19	.73502	1.36051	.76226	1.31190	.79022	1.26546	.81898	1.22104	41
20	.73547	1.35968	.76272	1.31110	.79070	1.26471	.81946	1.22031	40
21	.73592	1.35885	.76318	1.31031	.79117	1.26395	.81995	1.21959	39
22	.73637	1.35802	.76364	1.30952	.79164	1.26319	.82044	1.21886	38
23	.73681	1.35719	.76410	1.30873	.79212	1.26244	.82092	1.21814	37
24	.73726	1.35637	.76456	1.30795	.79259	1.26169	.82141	1.21742	36
25	.73771	1.35554	.76502	1.30716	.79306	1.26093	.82190	1.21670	35
26	.73816	1.35472	.76548	1.30637	.79354	1.26018	.82238	1.21598	34
27	.73861	1.35389	.76594	1.30558	.79401	1.25943	.82287	1.21526	33
28	.73906	1.35307	.76640	1.30480	.79449	1.25867	.82336	1.21454	32
29	.73951	1.35224	.76686	1.30401	.79496	1.25792	.82385	1.21382	31
30	.73996	1.35142	.76733	1.30323	.79544	1.25717	.82434	1.21310	30
31	.74041	1.35060	.76779	1.30244	.79591	1.25642	.82483	1.21238	29
32	.74086	1.34978	.76825	1.30166	.79639	1.25567	.82531	1.21166	28
33	.74131	1.34896	.76871	1.30087	.79686	1.25492	.82580	1.21094	27
34	.74176	1.34814	.76918	1.30009	.79734	1.25417	.82629	1.21023	26
35	.74221	1.34732	.76964	1.29931	.79781	1.25343	.82678	1.20951	25
36	.74267	1.34650	.77010	1.29853	.79829	1.25268	.82727	1.20879	24
37	.74312	1.34568	.77057	1.29775	.79877	1.25193	.82776	1.20808	23
38	.74357	1.34487	.77103	1.29696	.79924	1.25118	.82825	1.20736	22
39	.74402	1.34405	.77149	1.29618	.79972	1.25044	.82874	1.20665	21
40	.74447	1.34323	.77196	1.29541	.80020	1.24969	.82923	1.20593	20
41	.74492	1.34242	.77242	1.29463	.80067	1.24895	.82972	1.20522	19
42	.74538	1.34160	.77289	1.29385	.80115	1.24820	.83022	1.20451	18
43	.74583	1.34079	.77335	1.29307	.80163	1.24746	.83071	1.20379	17
44	.74628	1.33998	.77382	1.29229	.80211	1.24672	.83120	1.20308	16
45	.74674	1.33916	.77428	1.29152	.80258	1.24597	.83169	1.20237	15
46	.74719	1.33835	.77475	1.29074	.80306	1.24523	.83218	1.20166	14
47	.74764	1.33754	.77521	1.28997	.80354	1.24449	.83268	1.20095	13
48	.74810	1.33673	.77568	1.28919	.80402	1.24375	.83317	1.20024	12
49	.74855	1.33592	.77615	1.28842	.80450	1.24301	.83366	1.19953	11
50	.74900	1.33511	.77661	1.28764	.80498	1.24227	.83415	1.19882	10
51	.74946	1.33430	.77708	1.28687	.80546	1.24153	.83465	1.19811	9
52	.74991	1.33349	.77754	1.28610	.80594	1.24079	.83514	1.19740	8
53	.75037	1.33268	.77801	1.28533	.80642	1.24005	.83564	1.19669	7
54	.75082	1.33187	.77848	1.28456	.80690	1.23931	.83613	1.19599	6
55	.75128	1.33107	.77895	1.28379	.80738	1.23858	.83662	1.19528	5
56	.75173	1.33026	.77941	1.28302	.80786	1.23784	.83712	1.19457	4
57	.75219	1.32946	.77988	1.28225	.80834	1.23710	.83761	1.19387	3
58	.75264	1.32865	.78035	1.28148	.80882	1.23637	.83811	1.19316	2
59	.75310	1.32785	.78082	1.28071	.80930	1.23563	.83860	1.19246	1
60	.75355	1.32704	.78129	1.27994	.80978	1.23490	.83910	1.19175	0
′	Cotan	Tan	Cotan	Tan	Cotan	Tan	Cotan	Tan	′
	53°		52°		51°		50°		

TABLE XIV NATURAL TANGENTS AND COTANGENTS (Continued)

′	40° Tan	40° Cotan	41° Tan	41° Cotan	42° Tan	42° Cotan	43° Tan	43° Cotan	′
0	.83910	1.19175	.86929	1.15037	.90040	1.11061	.93252	1.07237	60
1	.83960	1.19105	.86980	1.14969	.90093	1.10996	.93306	1.07174	59
2	.84009	1.19035	.87031	1.14902	.90146	1.10931	.93360	1.07112	58
3	.84059	1.18964	.87082	1.14834	.90199	1.10867	.93415	1.07049	57
4	.84108	1.18894	.87133	1.14767	.90251	1.10802	.93469	1.06987	56
5	.85158	1.18824	.87184	1.14699	.90304	1.10737	.93524	1.06925	55
6	.84208	1.18754	.87236	1.14632	.90357	1.10672	.93578	1.06862	54
7	.84258	1.18684	.87287	1.14565	.90410	1.10607	.93633	1.06800	53
8	.84307	1.18614	.87338	1.14498	.90463	1.10543	.93688	1.06738	52
9	.84357	1.18544	.87389	1.14430	.90516	1.10478	.93742	1.06676	51
10	.84407	1.18474	.87441	1.14363	.90569	1.10414	.93797	1.06613	50
11	.84457	1.18404	.87492	1.14296	.90621	1.10349	.93852	1.06551	49
12	.84507	1.18334	.87543	1.14229	.90674	1.10285	.93906	1.06489	48
13	.84556	1.18264	.87595	1.14162	.90727	1.10220	.93961	1.06427	47
14	.84606	1.18194	.87646	1.14095	.90781	1.10156	.94016	1.06365	46
15	.84656	1.18125	.87698	1.14028	.90834	1.10091	.94071	1.06303	45
16	.84706	1.18055	.87749	1.13961	.90887	1.10027	.94125	1.06241	44
17	.84756	1.17986	.87801	1.13894	.90940	1.09963	.94180	1.06179	43
18	.84806	1.17916	.87852	1.13828	.90993	1.09899	.94235	1.06117	42
19	.84856	1.17846	.87904	1.13761	.91046	1.09834	.94290	1.06056	41
20	.84906	1.17777	.87955	1.13694	.91099	1.09770	.94345	1.05994	40
21	.84956	1.17708	.88007	1.13627	.91153	1.09706	.94400	1.05932	39
22	.85006	1.17638	.88059	1.13561	.91206	1.09642	.94455	1.05870	38
23	.85057	1.17569	.88110	1.13494	.91259	1.09578	.94510	1.05809	37
24	.85107	1.17500	.88162	1.13428	.91313	1.09514	.94565	1.05747	36
25	.85157	1.17430	.88214	1.13361	.91366	1.09450	.94620	1.05685	35
26	.85207	1.17361	.88265	1.13295	.91419	1.09386	.94676	1.05624	34
27	.85257	1.17292	.88317	1.13228	.91473	1.09322	.94731	1.05562	33
28	.85308	1.17223	.88369	1.13162	.91526	1.09258	.94786	1.05501	32
29	.85358	1.17154	.88421	1.13096	.91580	1.09195	.94841	1.05439	31
30	.85408	1.17085	.88473	1.13029	.91633	1.09131	.94896	1.05378	30
31	.85458	1.17016	.88524	1.12963	.91687	1.09067	.94952	1.05317	29
32	.85509	1.16947	.88576	1.12897	.91740	1.09003	.95007	1.05255	28
33	.85559	1.16879	.88628	1.12831	.91794	1.08940	.95062	1.05194	27
34	.85609	1.16809	.88680	1.12765	.91847	1.08876	.95118	1.05133	26
35	.85660	1.16741	.88732	1.12699	.91901	1.08813	.95173	1.05072	25
36	.85710	1.16672	.88784	1.12633	.91955	1.08749	.95229	1.05010	24
37	.85761	1.16603	.88836	1.12567	.92008	1.08686	.95284	1.04949	23
38	.85811	1.16535	.88888	1.12501	.92062	1.08622	.95340	1.04888	22
39	.85862	1.16466	.88940	1.12435	.92116	1.08559	.95395	1.04827	21
40	.85912	1.16398	.88992	1.12369	.92170	1.08496	.95451	1.04766	20
41	.85963	1.16329	.89045	1.12303	.92224	1.08432	.95506	1.04705	19
42	.86014	1.16261	.89097	1.12238	.92277	1.08369	.95562	1.04644	18
43	.86064	1.16192	.89149	1.12172	.92331	1.08306	.95618	1.04583	17
44	.86115	1.16124	.89201	1.12106	.92385	1.08243	.95673	1.04522	16
45	.86166	1.16056	.89253	1.12041	.92439	1.08179	.95729	1.04461	15
46	.86216	1.15987	.89306	1.11975	.92493	1.08116	.95785	1.04401	14
47	.86267	1.15919	.89358	1.11909	.92547	1.08053	.95841	1.04340	13
48	.86318	1.15851	.89410	1.11844	.92601	1.07990	.95897	1.04279	12
49	.86368	1.15783	.89463	1.11778	.92655	1.07927	.95952	1.04218	11
50	.86419	1.15715	.89515	1.11713	.92709	1.07864	.96008	1.04158	10
51	.86470	1.15647	.89567	1.11648	.92763	1.07801	.96064	1.04097	9
52	.86521	1.15579	.89620	1.11582	.92817	1.07738	.96120	1.04036	8
53	.86572	1.15511	.89672	1.11517	.92872	1.07676	.96176	1.03976	7
54	.86623	1.15443	.89725	1.11452	.92926	1.07613	.96232	1.03915	6
55	.86674	1.15375	.89777	1.11387	.92980	1.07550	.96288	1.03855	5
56	.86725	1.15308	.89830	1.11321	.93034	1.07487	.96344	1.03794	4
57	.86776	1.15240	.89883	1.11256	.93088	1.07425	.96400	1.03734	3
58	.86827	1.15172	.89935	1.11191	.93143	1.07362	.96457	1.03674	2
59	.86878	1.15104	.89988	1.11126	.93197	1.07299	.96513	1.03613	1
60	.86929	1.15037	.90040	1.11061	.93252	1.07237	.96569	1.03553	0
′	Cotan	Tan	Cotan	Tan	Cotan	Tan	Cotan	Tan	′
	49°		48°		47°		46°		

TABLE XIV NATURAL TANGENTS AND COTANGENTS (Continued)

′	44° Tan	44° Cotan	′	′	44° Tan	44° Cotan	′	′	44° Tan	44° Cotan	′
0	.96569	1.03553	60	20	.97700	1.02355	40	40	.98843	1.01170	20
1	.96625	1.03493	59	21	.97756	1.02295	39	41	.98901	1.01112	19
2	.96681	1.03433	58	22	.97813	1.02236	38	42	.98958	1.01053	18
3	.96738	1.03372	57	23	.97870	1.02176	37	43	.99016	1.00994	17
4	.96794	1.03312	56	24	.97927	1.02117	36	44	.99073	1.00935	16
5	.96850	1.03252	55	25	.97984	1.02057	35	45	.99131	1.00876	15
6	.96907	1.03192	54	26	.98041	1.01998	34	46	.99189	1.00818	14
7	.96963	1.03132	53	27	.98098	1.01939	33	47	.99247	1.00759	13
8	.97020	1.03072	52	28	.98155	1.01879	32	48	.99304	1.00701	12
9	.97076	1.03012	51	29	.98213	1.01820	31	49	.99362	1.00642	11
10	.97133	1.02952	50	30	.98270	1.01761	30	50	.99420	1.00583	10
11	.97189	1.02892	49	31	.98327	1.01702	29	51	.99478	1.00525	9
12	.97246	1.02832	48	32	.98384	1.01642	28	52	.99536	1.00467	8
13	.97302	1.02772	47	33	.98441	1.01583	27	53	.99594	1.00408	7
14	.97359	1.02713	46	34	.98499	1.01524	26	54	.99652	1.00350	6
15	.97416	1.02653	45	35	.98556	1.01465	25	55	.99710	1.00291	5
16	.97472	1.02593	44	36	.98613	1.01406	24	56	.99768	1.00233	4
17	.97529	1.02533	43	37	.98671	1.01347	23	57	.99826	1.00175	3
18	.97586	1.02474	42	38	.98728	1.01288	22	58	.99884	1.00116	2
19	.97643	1.02414	41	39	.98786	1.01229	21	59	.99942	1.00058	1
20	.97700	1.02355	40	40	.98843	1.01170	20	60	1.00000	1.00000	0
′	Cotan	Tan	′	′	Cotan	Tan	′	′	Cotan	Tan	′
	45°				45°				45°		

TABLE XV GREEK LETTERS

Form		English Equivalent	Name
A	α	a	Alpha
B	β	b	Beta
Γ	γ	g	Gamma
Δ	δ	d	Delta
E	ϵ	e (short)	Epsilon
Z	ζ	z	Zeta
H	η	e (long)	Eta
Θ	θ	th	Theta
I	ι	i	Iota
K	κ	k, c	Kappa
Λ	λ	l	Lambda
M	μ	m	Mu
N	ν	n	Nu
Ξ	ξ	x	Xi
O	o	o	Omicron
Π	π	p	Pi
P	ρ	r	Rho
Σ	σ	s	Sigma
T	τ	t	Tau
Υ	υ	u	Upsilon
Φ	ϕ	ph	Phi
X	χ	ch	Chi
Ψ	ψ	ps	Psi
Ω	ω	o	Omega

TABLE XVI VALUES OF R, y', AND SCALE FACTORS— MINNESOTA NORTH ZONE

Lambert Projection for Minnesota—North Zone

Lat.		R (ft)	y' y Value on Central Meridian (ft)	Tabular Difference for 1" of Lat. (ft)	Scale in Units of 7th Place of Logs	Scale Expressed as a Ratio
47°	31'	19,100,580.81	370,817.94	101.31550	−355.2	0.9999182
	32	19,094,501.88	376,896.87	101.31550	−362.0	0.9999166
	33	19,088,422.95	382,975.80	101.31567	−368.4	0.9999152
	34	19,082,344.01	389,054.74	101.31583	−374.5	0.9999138
	35	19,076,265.06	395,133.69	101.31600	−380.2	0.9999125
47°	36'	19,070,186.10	401,212.65	101.31617	−385.6	0.9999112
	37	19,064,107.13	407,291.62	101.31633	−390.6	0.9999101
	38	19,058,028.15	413,370.60	101.31650	−395.2	0.9999090
	39	19,051,949.16	419,449.59	101.31683	−399.4	0.9999080
	40	19,045,870.15	425,528.60	101.31700	−403.3	0.9999071
47°	41'	19,039,791.13	431,607.62	101.31717	−406.8	0.9999063
	42	19,033,712.10	437,686.65	101.31750	−410.0	0.9999056
	43	19,027,633.05	443,765.70	101.31767	−412.8	0.9999050
	44	19,021,553.99	449,844.76	101.31783	−415.2	0.9999044
	45	19,015,474.92	455,923.83	101.31817	−417.3	0.9999039
47°	46'	19,009,395.83	462,002.92	101.31850	−419.0	0.9999035
	47	19,003,316.72	468,082.03	101.31867	−420.3	0.9999032
	48	18,997,237.60	474,161.15	101.31900	−421.2	0.9999030
	49	18,991,158.46	480,240.29	101.31933	−421.8	0.9999029
	50	18,985,079.30	486,319.45	101.31950	−422.1	0.9999028

TABLE XVII   VALUES OF θ—MINNESOTA NORTH ZONE

Lambert Projection for Minnesota—North Zone
1″ of long. = 0″.7412196637 of θ

Long.		θ			Long.		θ		
94°	21′	−0	55	35.4885	94°	41′	−1	10	24.9521
	22	−0	56	19.9617		42	−1	11	09.4253
	23	−0	57	04.4348		43	−1	11	53.8984
	24	−0	57	48.9080		44	−1	12	38.3716
	25	−0	58	33.3812		45	−1	13	22.8448
94°	26′	−0	59	17.8543	94°	46′	−1	14	07.3180
	27	−1	00	02.3276		47	−1	14	51.7912
	28	−1	00	46.8007		48	−1	15	36.2643
	29	−1	01	31.2739		49	−1	16	20.7375
	30	−1	02	15.7471		50	−1	17	05.2107
94°	31′	−1	03	00.2202	94°	51′	−1	17	49.6839
	32	−1	03	44.6935		52	−1	18	34.1571
	33	−1	04	29.1666		53	−1	19	18.6302
	34	−1	05	13.6398		54	−1	20	03.1034
	35	−1	05	58.1130		55	−1	20	47.5766
94°	36′	−1	06	42.5862	94°	56′	−1	21	32.0498
	37	−1	07	27.0594		57	−1	22	16.5230
	38	−1	08	11.5325		58	−1	23	00.9961
	39	−1	08	56.0057		59	−1	23	45.4693
	40	−1	09	40.4789	95°	00′	−1	24	29.9425

TABLE XVIII VALUES OF H AND V—ILLINOIS EAST ZONE

Transverse Mercator Projection
Illinois
East Zone

Lat.	Y_0 (feet)	ΔY_0 per second	H	ΔH per second	V	ΔV per second	a
40° 35'	1,426,385.98	101.19683	77.158010	319.28	1.216989	1.85	−0.509
40 36	1,432,457.79	101.19700	77.138853	319.42	1.217100	1.85	−0.507
40 37	1,438,529.61	101.19733	77.119688	319.52	1.217211	1.83	−0.505
40 38	1,444,601.45	101.19767	77.100517	319.62	1.217321	1.83	−0.503
40 39	1,450,673.31	101.19800	77.081340	319.73	1.217431	1.82	−0.501
40 40	1,456,745.19	101.19817	77.062156	319.85	1.217540	1.82	−0.499
40 41	1,462,817.08	101.19850	77.042965	319.95	1.217649	1.80	−0.497
40 42	1,468,888.99	101.19883	77.023768	320.05	1.217757	1.80	−0.495
40 43	1,474,960.92	101.19917	77.004565	320.18	1.217865	1.80	−0.493
40 44	1,481,032.87	101.19950	76.985354	320.27	1.217973	1.78	−0.491
40 45	1,487,104.84	101.19967	76.966138	320.40	1.218080	1.78	−0.489
40 46	1,493,176.82	101.20000	76.946914	320.48	1.218187	1.77	−0.487
40 47	1,499,248.82	101.20033	76.927685	320.62	1.218293	1.77	−0.485
40 48	1,505,320.84	101.20067	76.908448	320.72	1.218399	1.77	−0.483
40 49	1,511,392.88	101.20083	76.889205	320.82	1.218505	1.75	−0.481
40 50	1,517,464.93	101.20133	76.869956	320.93	1.218610	1.75	−0.479
40 51	1,523,537.01	101.20150	76.850700	321.05	1.218715	1.73	−0.477
40 52	1,529,609.10	101.20167	76.831437	321.15	1.218819	1.73	−0.475
40 53	1,535,681.20	101.20217	76.812168	321.25	1.218923	1.73	−0.473
40 54	1,541,753.33	101.20233	76.792893	321.38	1.219027	1.72	−0.471

TABLE XIX VALUES OF b AND c—ILLINOIS ZONES

Transverse Mercator Projection
Illinois
Both Zones

Δλ″	b	Δb	c
0	0.000	+0.212	0.000
100	+0.212	+0.212	0.000
200	+0.424	+0.210	−0.001
300	+0.634	+0.208	−0.002
400	+0.842	+0.207	−0.003
500	+1.049	+0.203	−0.005
600	+1.252	+0.201	−0.007
700	+1.453	+0.196	−0.010
800	+1.649	+0.192	−0.014
900	+1.841	+0.187	−0.018
1000	+2.028	+0.181	−0.022
1100	+2.209	+0.175	−0.027
1200	+2.384	+0.169	−0.032
1300	+2.553	+0.162	−0.038
1400	+2.715	+0.153	−0.043
1500	+2.868	+0.146	−0.049
1600	+3.014	+0.137	−0.055
1700	+3.151	+0.128	−0.061
1800	+3.279	+0.118	−0.067
1900	+3.397	+0.107	−0.073
2000	+3.504	+0.097	−0.079

TABLE XX VALUES OF g—ILLINOIS ZONES

Transverse Mercator Projection
Values of g

Latitude	\\(\Delta\lambda''\\)						
	0″	1000″	2000″	3000″	4000″	5000″	6000″
36°	0.00	0.00	0.02	0.08	0.19	0.38	0.65
37	0	0	0.02	0.08	0.19	0.38	0.65
38	0	0	0.02	0.08	0.19	0.38	0.65
39	0	0	0.02	0.08	0.19	0.37	0.64
40	0	0	0.02	0.08	0.19	0.37	0.64
41	0	0	0.02	0.08	0.19	0.37	0.63
42	0	0	0.02	0.08	0.18	0.36	0.63
43	0	0	0.02	0.08	0.18	0.36	0.62
44	0	0	0.02	0.08	0.18	0.35	0.61
45	0	0	0.02	0.08	0.18	0.35	0.60

TABLE XXI VALUES OF SCALE FACTORS— ILLINOIS EAST ZONE

Transverse Mercator Projection
Illinois
East Zone

x' (feet)	Scale in Units of 7th Place of Logs	Scale Expressed as a Ratio	x' (feet)	Scale in Units of 7th Place of Logs	Scale Expressed as a Ratio
0	−108.6	0.9999750	75,000	−80.7	0.9999814
5,000	−108.5	0.9999750	80,000	−76.8	0.9999823
10,000	−108.1	0.9999751	85,000	−72.7	0.9999833
15,000	−107.5	0.9999752	90,000	−68.4	0.9999843
20,000	−106.6	0.9999755	95,000	−63.8	0.9999853
25,000	−105.5	0.9999757	100,000	−58.9	0.9999864
30,000	−104.1	0.9999760	105,000	−53.9	0.9999876
35,000	−102.5	0.9999764	110,000	−48.5	0.9999888
40,000	−100.7	0.9999768	115,000	−42.9	0.9999901
45,000	−98.5	0.9999773	120,000	−37.1	0.9999915
50,000	−96.2	0.9999778	125,000	−31.0	0.9999929
55,000	−93.6	0.9999784	130,000	−24.7	0.9999943
60,000	−90.7	0.9999791	135,000	−18.1	0.9999958
65,000	−87.6	0.9999798	140,000	−11.3	0.9999974
70,000	−84.3	0.9999806	145,000	−4.2	0.9999990

TABLE XXII TRIGONOMETRIC FORMULAS

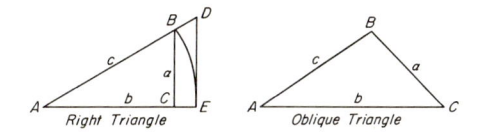

Right Triangle

Oblique Triangle

RIGHT TRIANGLES

$$\sin A = \frac{a}{c} = \cos B \qquad \sec A = \frac{c}{b} = \operatorname{cosec} B$$

$$\cos A = \frac{b}{c} = \sin B \qquad \operatorname{cosec} A = \frac{c}{a} = \sec B$$

$$\tan A = \frac{a}{b} = \cot B \qquad \cot A = \frac{b}{a} = \tan B$$

$$a = c \sin A = c \cos B = b \tan A = b \cot B = \sqrt{c^2 - b^2}$$

$$b = c \cos A = c \sin B = a \cot A = a \tan B = \sqrt{c^2 - a^2}$$

$$c = \frac{a}{\sin A} = \frac{a}{\cos B} = \frac{b}{\sin B} = \frac{b}{\cos A}$$

OBLIQUE TRIANGLES

Given	Sought	Formulas
A, B, a	b, c	$b = \dfrac{a}{\sin A} \cdot \sin B \qquad c = \dfrac{a}{\sin A} \cdot \sin(A+B)$
A, a, b	B, c	$\sin B = \dfrac{\sin A}{a} \cdot b \qquad c = \dfrac{a}{\sin A} \cdot \sin C$
C, a, b	$½ (A+B)$	$½ (A+B) = 90° - ½ C$
	$½ (A-B)$	$\operatorname{Tan} ½ (A-B) = \dfrac{a-b}{a+b} \cdot \tan ½ (A+B)$
a, b, c	A	Given $s = ½ (a+b+c)$, then $\sin ½ A = \sqrt{\dfrac{(s-b)(s-c)}{bc}}$ $\cos ½ A = \sqrt{\dfrac{s(s-a)}{bc}}, \quad \tan ½ A = \sqrt{\dfrac{(s-b)(s-c)}{s(s-a)}}$ $\sin A = 2\dfrac{\sqrt{s(s-a)(s-b)(s-c)}}{bc}$
	Area	Area $= \sqrt{s(s-a)(s-b)(s-c)}$
C, a, b	Area	Area $= ½\, ab \sin C$

Index

Accidental error, 219
Accretion, 393
Accuracy, 14, 220
Acre, 21
Acre-foot, 428
Adjustment, 228
  of angles in triangulation, 251
  of level, 122
  of transit, 66
  of traverse, 200
Adverse possession, 389
Aerial surveying, 5
Agonic line, 264
Alinement in taping, 37
Alluvium, 393
Altimeter, 152
Altitude, 109
Angle, adjustment of, 251
  deflection, 171
  of depression, 50
  of elevation, 50
  errors in, 75
  horizontal, 48
  interior, 171
  measurement, 47, 245
  to right, 171
  vertical, 50
Area computations, 201
Astronomical observations, 266
Astronomical triangle, 274
Atmospheric refraction, 112
Automatic level, 118
Auxiliary traverse, 212
Average-end-area, volume by, 208
Avulsion, 394
Axis of bubble tube, 59
Azimuth, 50
  astronomic, 263
  geodetic, 303
  grid, 301
  quality of determination of, 281
  true, 263
Azimuth mark, 244
Azimuth problem, 276
Azimuth specifications, 177

Backsight, in leveling, 130
  in transit operations, 53
Balancing a survey, 200
Barometric leveling, 152
Baseline, 241
  measurement of, 247
  in public lands system, 356
Baseline reduction calculations, 249
Baseline tape, 248
Batter board, 473
Bearing, 48
  grid, 50
  magnetic, 50
  true, 50
Bearing tree, 368
Benchmark, 130
  description of, 150
Blunder, 219
Boat sheet, 448
Borrow pit, 157, 210
Boundary surveying, 351
Breaking tape, 32
Bubble tube, 59
Builder's transit-level, 474

Cadastral surveying, 4, 351
Calculation (*see* Computation)
Camera, 315, 317
Cartographic surveying, 398
Cartography, 398
Celestial sphere, 266
Centesimal system, 47
Central-point figure, 237
Chain, 21, 363

INDEX

Chain of triangles, 236
Chaining, 24
Chaining pin, 25
Checks, 14
  in computations, 195
  in level notes, 134
  in taping, 40
Circular bubble vial, 60
City surveying, 355
Clamping handle, 25
Clarke spheroid, 3
Closed traverse, 164
Closure, error of, 198
Collimation level, 82
Color tint system, 403
Compass, 263
Compass rule, 200
Compensating error, 219
Computation, 189
  of area, 201
  checking of, 195
  of coordinates, 195
  of cutoff line, 211
  by desk calculator, 191
  by electronic calculator, 212
  of forward problem, 196
  of inverse problem, 196
  of latitudes and departures, 197
  of linear error of closure, 198
  of relative error of closure, 198
  rounding off in, 194
  significant figures in, 192
  of traverse, 197
  of triangulation, 253
  of volume, 208
Connecting traverse, 164
Constant error, 219
Construction surveying, 470
Contour, 400
  characteristics of, 401
  grade, 432
  interpolation of, 406
  sketching of, 420
Contour interval, 400
Contour point systems, 416
Contour points, 404
Convergency of meridians, 176
Coordinate systems, 195
  local, 288
  tangent plane, 290
  state plane, 294
Corner, 366
  closing township, 359
  interior quarter, 360

lost, 379
marking, 367
meander, 367
obliterated, 379
perpetuation of, 368
quarter-section, 360
restoration of, 383
section, 360
standard township, 359
witness, 367
Corner accessories, 368
Correction line, 357
Cross-section leveling, 155, 367
Cross-wires, 56
Culmination, 275
Cumulative error, 219
Curvature and refraction, 111
Curve computation, 459
Cutoff line, 211

Datum, 111
  horizontal, 285
  sounding, 438
Declination, of needle, 50, 264
  of star, 271
  variations of, 266
Deed, 353
Deflection angle, 459
Degree of curve, 455
Departure, 197
Depth curve, 449
Desk calculator, 191
Deviation, 221
Differential leveling, 113
Dip of needle, 264
Direct leveling, 112
Direction instrument, 79
Discrepancy, 220
Distance angles, 242
Double centering, 71
Dumpy level, 114

Earth's shape, 2, 112
Earthwork, 208
Eastings, 197
Echo sounding, 441
Electromagnetic measurements, 98
Electronic calculator, 212
Electronic surveying, 99
Electrotape, 103
Elevation, 109
Elevation factor, 304
Elongation of Polaris, 275
Engineering surveying, 1, 452

INDEX

Ephemerides, 277
Error, accidental, 219
　in angles, 75
　compensating, 219
　constant, 219
　cumulative, 219
　instrumental, 75, 220
　kinds of, 13
　in leveling, 137
　mean square, 223
　natural, 77, 220
　personal, 76, 220
　probable, 223
　propagation of, 226
　random, 219
　sources of, 12
　standard, 223
　systematic, 219
　in taping, 35
　in traverse, 200
Excess and deficiency, 388
External distance, 456
Eyepiece, 55

Factor, elevation, 304
　grid, 306
　scale, 304
Fathogram, 442
Fathom curve, 449
Fathometer, 441
Fiducial marks, 317
Field communications, 185
Field notes, 11
　for azimuth observation, 278
　for cross-section leveling, 157
　for curve layout, 461
　for differential leveling, 131
　for hydrographic surveying, 447, 448
　importance of, 179
　for precise leveling, 146
　for profile leveling, 156
　for taping, 42
　for stadia surveying, 418
Field of view, of telescope, 58
Field work, 1
Figures, significant, 192
Floating mark, 345
Focal length, 55
Foot, 20, 23
Foot plate, 63
Foresight, in leveling, 130
　in transit operations, 53

Geodetic control diagram, 178
Geodetic surveying, 3
Geodetic triangulation, 234
Geodesy, 2
Geodimeter, 104
Grade, 109
Grade contour, 432
Grade correction, 34
Grade rod, 468
Grade stake, 472
Gradient, 33
Grantee, 353
Grantor, 353
Graphical triangulation, 411
Great circle, 266
Greenwich hour angle, 271
Grid azimuth, 301
Grid bearing, 50
Grid coordinates, 307
Grid distance, 304
Grid factor, 306
Grid meridian, 48
Grid systems, 294, 299
Ground control, 325
Ground rod, 155, 469
Ground swing, 103
Guard stake, 166
Guide meridian, 357
Gunther's chain, 21
Gyro-theodolite, 262

Hachure, 402
Hand level, 119
Heat waves, 141
Height, 109
Height of instrument, 130, 415
Hertz, 100
Horizon, 268
Horizontal angle, measurement of, 63
Horizontal axis, of transit, 54
Horizontal control, 162, 235
Horizontal curve, 455
Horizontal line, 111
Horizontal refraction, 78
Hour angle, 273
Hour circle, 268
Hub, 166
Hydrographic surveying, 5, 398, 434
Hydrography, 436

Illumination of lens, 58
Image displacement, 322
Index error, 83
Index of refraction, 100

INDEX

Index reading, 66
Industrial surveying, 4, 476
Informational surveys, 2
Initial point, 356
Instrumental error, 75, 220
Intermediate setup, 462
International meter, 23
Intersection, of lines, 73
    with plane table, 411
    point of, 456
Interval, contour, 400
Intervisibility of stations, 240
Invar tape, 24, 248
Inverse problem, 196
Isogonic chart, 264

Jig transit, 477

Kelsh plotter, 347

Lambert grid, 294
Land descriptions, 373
Land surveying, 1, 351
    for deed, 386
    scope of, 378
Land surveys, kinds of, 353
    public, 355
    rural, 354, 382
    urban, 354, 386
Land ties, 470
Land transfers, 352
Lap of photograph, 315
Laser instrument, 475
Latitude, 271
Latitudes and departures, 197
Law of compensation, 15
Laying out an angle, 73
Layout of building, 473
Leadline, 440
Legal authority of surveyor, 396
Length, units of, 21
Length of curve, 457
Lens, focal length, 317
    objective, 55
Lettering, 424
Level, adjustments of, 122
    automatic, 118
    dumpy, 114
    hand, 119
    rod, 121
    self-leveling, 118
    tilting, 115
    wye, 115
Level circuit, 144

Level rod, 119, 145
Level surface, 111
Leveling, 111
    barometric, 113, 152
    for cross-sections, 155
    differential, 113, 128
    errors in, 137
    field notes for, 134
    field procedure for, 131
    grades of, 135
    indirect, 113
    line closure, 135, 229
    mistakes in, 141
    precise, 143
    profile, 154
    reciprocal, 142
    spirit, 113
    stadia, 414
    three-wire, 146
    trigonometric, 151
Leveling head, 52
Leveling rod, 119, 145
Liability of surveyor, 396
Light-wave instrument, 104
Line of sight, 54
Linear error of closure, 198
Linear measurement, 21
Local hour angle, 273
Location of details, 178, 413
Location survey, 454
Long chord, 457
Longitude, 271
Loop traverse, 164
Lost corner, 379
Lovar tape, 24
Lower plate, 51
Low-water datum, 438

Magnetic bearing, 50
Magnetic compass, 263
Magnetic declination, 264
Magnetic meridian, 48
Magnification of telescope, 57
Map, 398
    contour, 400
    control for, 414
    drawing of, 422
    procurement of, 434
    scale of, 433
    specifications for, 434
    studies of, 427
    symbols for, 425
    titles for, 423
Mean low water, 438

INDEX

Mean sea level, 111
Mean square error, 223
Meander line, 365
Measurement, of distance, 21
  of elevation, 109
  of horizontal angles, 63
  of vertical angles, 66
Megacycle, 100
Meridian, 48, 262, 268
  convergency of, 176
Meridian angle, 273
Meridian arrow, 424
Meter, international, 21
Metes and bounds, 373
Metrical photography, 316
Metrology, 20
Micron, 22
Microwave instrument, 101
Mid-ordinate, 457
Mil, 47
Millimicron, 22
Mine surveying, 5
Mistake, 219
  in leveling, 141
  in taping, 39
  in transit use, 78
Monumentation, 147
Mosaic, 341
Most probable value, 221
Municipal surveying, 355

National Bureau of Standards, 28
National Topographic Mapping, 433
Natural error, 77, 220
Nautical chart, 398
Network, of leveling, 148
  of triangulation, 255
Nomogram, 190
North American Datum of 1927, 256
Northings, 197
Notekeeping, 179

Objective lens, 55
Oblique photograph, 315
Obliterated corner, 379
Observations for meridian, 262
Open traverse, 164
Optical metrology, 4
Optical plummet, 64
Optical tooling, 476
Optical transit, 79
Orientation, 53, 262
Original surveys, 353

Parabolic curve, 463
Parallax, 56
  angle of, 332
  in stereoscopic views, 335
Parallax bar, 340, 345
Peg test, 126
Personal error, 76, 220
Philadelphia rod, 119
Photogrammetry, 315
Photograph, oblique, 315
  scale of, 319
  vertical, 315
Photographic mission, 343
Photo interpretation, 316
Plane of low water, 438
Plane surveying, 3
Plane table, 407
Plane triangulation, 234
Planimeter, 206
Planimetric detail, 467
Plat, 422
Plotting, 423
Plumb bob, 64
Point of intersection, 456
Polar distance, 272
Polaris, 275
Precision, 14, 220
  measures of, 223
Preliminary survey, 454
Principal meridian, 356
Prismoidal formula, 209
Probable error, 223
Profile leveling, 154, 467
Projections, 290, 293
Prolonging a straight line, 71
Proportionate measurement, 383
Public lands survey, 356
PZS triangle, 274

Quadrilateral, 237

Radial-line method, 325
Radiation, 179
Random error, 219
Random line, 72
Range, 358
Range pole, 25, 173
Real property, 352
Reciprocal leveling, 142
Reconnaissance, 240, 454
Rectangular system, 360
Reduction to sea level, 250
Reference mark, 166, 244
Refraction, atmospheric, 112
  horizontal, 78

INDEX

Relative error of closure, 198
Reliction, 393
Relief symbols, 402
Repeating an angle, 65
Repeating instrument, 79
Representative fraction, 321
Resection, 412
Residual, 221
Restoration of corner, 383
Resurvey, 353
Reticule, 81
Reversing point, 126
Right-angle mirror, 467
Riparian boundaries, 392
Rod, grade, 468
   leveling, 119
   stadia, 89
   waving, 139
Rod level, 121
Rounding off, 194
Route surveying, 454
Rule, compass, 200
   trapezoidal, 203
   Simpson's, 204

Safety, 182
Sag, 37
Scale factor, 304
Scale of map, 433
Sea level datum, 111
Section, 359
   subdivision of, 365
Sensitivity of bubble tube, 59, 127
Sexagesimal system, 47
Sextant, 442
Shading, 403
Shiner, 166
Shoreline survey, 439
Signals, 242
Significant figures, 192
Simpson's rule, 204
Sketching contours, 420
Slope correction, 33
Slope staking, 468
Solar ephemeris, 277
Sounding datums, 438
Sounding operations, 443
Space-coordinate equations, 336
Spirit leveling, 113
Split bubble, 83
Stadia, 89
Stadia levels, 414
Stake, 166, 472
Standard deviation, 223

Standard error, 223
Standardization, 27
Standard of length, 22
Standard parallel, 357
Standard time, 269
State coordinate systems, 286, 309
Station description, 244
Station equation, 470
Station marker, 243
Stereocomparagraph, 340, 346
Stereoplotter, 347
Stereoscope, 335
Stereoscopic vision, 332
Strength of figure, 241
Subdivisional survey, 354
Subgrade, 468
Subtense method, 95
Surveying, 1
   aerial, 5
   boundary, 351
   cadastral, 4, 351
   cartographic, 398
   city, 355
   construction, 470
   engineering, 452
   geodetic, 3
   hydrographic, 5, 398
   industrial, 476
   kinds of, 4
   land, 351
   mine, 5
   municipal, 355
   origins of, 5
   plane, 3
   professional practice of, 15
   route, 454
   topographic, 398, 413
Surveying agencies, 17
Surveying societies, 18
Surveying technician, 17
Surveyor, legal authority of, 396
   liability of, 396
   registration of, 397
Systematic error, 219

Tacheometry, 88
Tangent distance, 456
Tangent screw, 64
Tape, 24
Taping, 28
   accuracy of, 40
   errors in, 35
   mistakes in, 39
   specifications for, 41

INDEX

Taping arrow, 25
Taping corrections, 34
Taping tripod (buck), 248
Telescope, 54, 89
Tellurometer, 101
Templet, 330
Tension handle, 25
Tension in tape, 38
Theodolite, 78
Thermal expansion, 36
Three-arm protractor, 443
Three-point problem, 258
Three-wire leveling, 146
Tier, 358
Tilting level, 115
Tilt of camera, 315
Time, 269
Toise, 20
Topographic expression, 425
Topographic map, 399, 415
Topographic surveying, 398, 413
Topography, 467
Tower, 242
Township, 357
Transit, 51
  jig, 477
  optical, 79
Transit traverse, 163
Transverse mercator grid, 299
Trapezoidal rule, 203
Traverse, 163
  adjustment of, 200
  auxiliary, 212
  checking of, 175
  computation of, 195
  configuration of, 164
  grades of, 176
Triangulation, 234
  angle adjustment in, 251
  classification of, 237
  computation of, 253
  geodetic, 234
  graphical, 411
  network of, 254

plane, 234
signals in, 242
specifications for, 239
systems of, 235
uses of, 257
Tribrach, 80
Trigonometric leveling, 151
Trilateration, 258
Tripod, 51
True bearing, 50
True meridian, 48
Turning point, 121, 130

Uncertainty, 220
U.S. Coast and Geodetic Survey, 147, 235
U.S. public lands surveys, 355
Units of measurement, 21, 47
Upper plate, 51
Uses of transit, 71
Utility surveys, 470

Vara, 22
Vernier, 53, 60
Vertical angle, 50, 66
Vertical circle, of transit, 66
  of celestial sphere, 268
Vertical control, 146
Vertical curve, 463
View, field of, 58
Vision, stereoscopic, 332
Volume computations, 208

Waving rod, 139
Weighted measurements, 227
Wiggling-in, 72
Witness corner, 367
Woven tape, 24
Wye level, 115

Yard, 23

Zenith, 267
Zenith distance, 50